Comprehensive Organometallic Chemistry II

A Review of the Literature 1982–1994

Comprehensive Organometallic Chemistry II

A Review of the Literature 1982–1994

Editors-in-Chief

Edward W. Abel
University of Exeter, UK

F. Gordon A. Stone
Baylor University, Waco, TX, USA

Geoffrey Wilkinson
Imperial College of Science, Technology and Medicine, London, UK

Volume 3
COPPER AND ZINC GROUPS

Volume Editor

James L. Wardell
University of Aberdeen, UK

PERGAMON

UK   Elsevier Science Ltd., The Boulevard, Langford Lane, Kidlington, Oxford
                OX5 1GB, UK

USA   Elsevier Science Inc., 660 White Plains Road, Tarrytown, New York
                10591-5153, USA

JAPAN   Elsevier Science Japan, Tsunashima Building Annex, 3-20-12 Yushima,
                Bunkyo-ku, Tokyo 113, Japan

Copyright © 1995 Elsevier Science Ltd.

First edition 1995

Library of Congress Cataloging in Publication Data
Comprehensive organometallic chemistry II : a review of the literature
1982–1994 / editors-in-chief, Edward W. Abel, F. Gordon A. Stone,
Geoffrey Wilkinson
     p. cm.
   Includes indexes.
   1. Organometallic chemistry.   I. Abel, Edward W.   II. Stone, F. Gordon
A.   III.   Wilkinson, Geoffrey.
QD411.C652   1995
547'.05—dc20   95–7030

British Library Cataloguing in Publication Data
A catalogue record for this book is available from the British Library.

ISBN 0–08–040608–4 (set : alk. paper)
ISBN 0–08–042310–8 (Volume 3)

Important note
For safety reasons, readers should always consult the list of abbreviations
on p. xi before making use of the experimental details provided.

∞™   The paper used in this publication meets the minimum requirements of the American National Standard for Information Sciences—Permanence of Paper for Printed Library Materials, ANSI Z39.48–1984.

Chemical structures drawn by Synopsys Scientific Systems Ltd., Leeds, UK.

Printed and bound in Great Britain by BPC Wheatons Ltd., Exeter, UK.

Contents

Preface   vii

Preface to 'Comprehensive Organometallic Chemistry'   vii

Contributors to Volume 3   ix

Abbreviations   xi

Contents of All Volumes   xv

1   Gold   1
A. GROHMANN and H. SCHMIDBAUR, *Technische Universität München, Garching, Germany*

2   Copper and Silver   57
G. VAN KOTEN, S. L. JAMES and J. T. B. H. JASTRZEBSKI, *University of Utrecht, The Netherlands*

3   Mercury   135
A. G. DAVIES, *University College London, UK,* and J. L. WARDELL, *University of Aberdeen, UK*

4   Cadmium and Zinc   175
P. O'BRIEN, *Imperial College of Science, Technology and Medicine, London, UK*

Author Index   207

Subject Index   221

Preface

'Comprehensive Organometallic Chemistry', published in 1982, was well received and remains very highly cited in the primary journal literature. Since its publication, studies on the chemistry of molecules with carbon–metal bonds have continued to expand rapidly. This is due to many factors, ranging from the sheer intellectual challenge and excitement provided by the continuing production of novel results, which demand new ideas, through to the successful application of organometallic species in organic syntheses, the generation of living catalysts for polymerization, and the synthesis of precursors for materials employed in the electronic and ceramic industries. For many reasons, therefore, we judged it timely to update 'Comprehensive Organometallic Chemistry' with a new work.

Due to the scope and depth of this area of chemistry, to have merely revised each of the original nine volumes did not seem the most user-friendly or cost-effective procedure to follow. As a consequence of the sheer bulk of the literature of the subject, a revised edition would necessarily require either the elimination of much chemistry of archival value but which is still important, or the production of a set of volumes significantly larger in number than the original nine. Accordingly, we decided it would be best to use the original work as a basis for new volumes focusing on organometallic chemistry reported since 1982, with reference back to the original work when necessary. For ease of use the new volumes maintain the same general structure as employed previously but reflect the changes in substance and direction the field has undergone in the last ten years. Thus it is not surprising that the largest volume in the new work concerns the role of the transition elements in metal-mediated organic syntheses.

The expansion of organometallic chemistry since the early 1980s also led us to decide that an updating of 'Comprehensive Organometallic Chemistry' would be more effectively accomplished if each volume had one or two editors who would be responsible both for recruiting experts for the Herculean task of writing the many chapters of each volume and for overseeing the content. We are deeply indebted to the volume editors and their authors for the time and effort they have given to the project.

As with the original 'Comprehensive Organometallic Chemistry', published some thirteen years ago, we hope this new version will serve as a pivotal reference point for new work and will function to generate new ideas and perceptions for the continued advance of what will surely continue as a vibrant area of chemistry.

Edward W. Abel
Exeter, UK

F. Gordon A. Stone
Waco, Texas, USA

Geoffrey Wilkinson
London, UK

Preface to 'Comprehensive Organometallic Chemistry'

Although the discovery of the platinum complex that we now know to be the first π-alkene complex, $K[PtCl_3(C_2H_4)]$, by Zeise in 1827 preceded Frankland's discovery (1849) of diethylzinc, it was the latter that initiated the rapidly developing interest during the latter half of the nineteenth century in compounds with organic groups bound to the elements. This era may be considered to have reached its apex in the discovery by Grignard of the magnesium reagents which occupy a special place because of their ease of synthesis and reactivity. With the exception of trimethylplatinum chloride discovered by Pope, Peachy and Gibson in 1907 by use of the Grignard reagent, attempts to make stable transition metal alkyls and aryls corresponding to those of main group elements met with little success, although it is worth recalling that even in 1919 Hein and his co-workers were describing the 'polyphenyl-chromium' compounds now known to be arene complexes.

The other major area of organometallic compounds, namely metal compounds of carbon monoxide, originated in the work starting in 1868 of Schützenberger and later of Mond and his co-workers and was subsequently developed especially by Hieber and his students. During the first half of this century, aided by the use of magnesium and, later, lithium reagents the development of main group organo chemistry was quite rapid, while from about 1920 metal carbonyl chemistry and catalytic reactions of carbon monoxide began to assume importance.

In 1937 Krause and von Grosse published their classic book 'Die Chemie der Metallorganischen Verbindungen'. Almost 1000 pages in length, it listed scores of compounds, mostly involving metals of the main groups of the periodic table. Compounds of the transition elements could be dismissed in 40 pages. Indeed, even in 1956 the stimulating 197-page monograph 'Organometallic Compounds' by Coates adequately reviewed organo transition metal complexes within 27 pages.

Although exceedingly important industrial processes in which transition metals were used for catalysis of organic reactions were developed in the 1930s, mainly in Germany by Reppe, Koch, Roelen, Fischer and Tropsch and others, the most dramatic growth in our knowledge of organometallic chemistry, particularly of transition metals, has stemmed from discoveries made in the middle years of this century. The introduction in the same period of physical methods of structure determination (infrared, nuclear magnetic resonance, and especially single-crystal X-ray diffraction) as routine techniques to be used by preparative chemists allowed increasingly sophisticated exploitation of discoveries. Following the recognition of the structure of ferrocene, other major advances quickly followed, including the isolation of a host of related π-complexes, the synthesis of a plethora of organometallic compounds containing metal–metal bonds, the characterization of low-valent metal species in which hydrocarbons are the only ligands, and the recognition from dynamic NMR spectra that ligand site exchange and tautomerism were common features in organometallic and metal carbonyl chemistry. The discovery of alkene polymerization using aluminium alkyl–titanium chloride systems by Ziegler and Natta and of the Wacker palladium–copper catalysed ethylene oxidation led to enormous developments in these areas.

In the last two decades, organometallic chemistry has grown more rapidly in scope than have the classical divisions of chemistry, leading to publications in journals of all national chemical societies, the appearance of primary journals specifically concerned with the topic, and the growth of annual review volumes designed to assist researchers to keep abreast of accelerating developments.

Organometallic chemistry has become a mature area of science which will obviously continue to grow. We believe that this is an appropriate time to produce a comprehensive review of the subject, treating organo derivatives in the widest sense of both main group and transition elements. Although advances in transition metal chemistry have appeared to dominate progress in recent years, spectacular progress has, nevertheless, also been made in our knowledge of organo compounds of main group elements such as aluminium, boron, lithium and silicon.

In these Volumes we have assembled a compendium of knowledge covering contemporary organometallic and carbon monoxide chemistry. In addition to reviewing the chemistry of the elements individually, two Volumes survey the use of organometallic species in organic synthesis and in catalysis, especially of industrial utility. Within the other Volumes are sections devoted to such diverse topics as the nature of carbon–metal bonds, the dynamic behaviour of organometallic compounds in solution, heteronuclear metal–metal bonded compounds, and the impact of organometallic compounds on the environment. The Volumes provide a unique record, especially of the intensive studies conducted during the past 25 years. The last Volume of indexes of various kinds will assist readers seeking information on the properties and synthesis of compounds and on earlier reviews.

As Editors, we are deeply indebted to all those who have given their time and effort to this project. Our Contributors are among the most active research workers in those areas of the subject that they have reviewed and they have well justified international reputations for their scholarship. We thank them sincerely for their cooperation.

Finally, we believe that 'Comprehensive Organometallic Chemistry', as well as providing a lasting source of information, will provide the stimulus for many new discoveries since we do not believe it possible to read any of the articles without generating ideas for further research.

E. W. ABEL   F. G. A. STONE
Exeter   *Bristol*

G. WILKINSON
London

Contributors to Volume 3

Professor A. G. Davies
Department of Chemistry, University College London, 20 Gordon Street, London, WC1H 0AJ, UK

Dr. A. Grohmann
Technische Universität München, Lichtenbergstrasse 4, D-85747 Garching, Germany

Dr. S. L. James
Debye Institute, Department of Metal-mediated Chemistry, University of Utrecht, Padualaan 8, 3584 CH Utrecht, The Netherlands

Dr. J. T. B. H. Jastrzebski
Debye Institute, Department of Metal-mediated Chemistry, University of Utrecht, Padualaan 8, 3584 CH Utrecht, The Netherlands

Professor P. O'Brien
Department of Chemistry, Imperial College of Science, Technology and Medicine, South Kensington, London, SW7 2AY, UK

Professor H. Schmidbaur
Technische Universität München, Lichtenbergstrasse 4, D-85747 Garching, Germany

Professor G. van Koten
Debye Institute, Department of Metal-mediated Chemistry, University of Utrecht, Padualaan 8, 3584 CH Utrecht, The Netherlands

Dr. J. L. Wardell
Department of Chemistry, University of Aberdeen, Meston Walk, Old Aberdeen, AB9 2UE, UK

Abbreviations

The abbreviations used throughout 'Comprehensive Organometallic Chemistry II' are consistent with those used in 'Comprehensive Organometallic Chemistry' and with other standard texts in this area. The abbreviations in some instances may differ from those commonly used in other branches of chemistry.

Ac	acetyl
acac	acetylacetonate
AIBN	2,2'-azobisisobutyronitrile
Ar	aryl
arphos	1-(diphenylphosphino)-2-(diphenylarsino)ethane
Azb	azobenzene
9-BBN	9-borabicyclo[3.3.1]nonyl
9-BBN-H	9-borabicyclo[3.3.1]nonane
BHT	2,6-di-t-butyl-4-methylphenol (butylated hydroxytoluene)
bipy	2,2'-bipyridyl
t-BOC	t-butoxycarbonyl
bsa	N,O-bis(trimethylsilyl)acetamide
bstfa	N,O-bis(trimethylsilyl)trifluoroacetamide
btaf	benzyltrimethylammonium fluoride
Bz	benzyl
can	ceric ammonium nitrate
cbd	cyclobutadiene
1,5,9-cdt	cyclododeca-1,5,9-triene
chd	cyclohexadiene
chpt	cycloheptatriene
[Co]	cobalamin
(Co)	cobaloxime [Co(DMG)$_2$] derivative
cod	1,5-cyclooctadiene
cot	cyclooctatetraene
Cp	η^5-cyclopentadienyl
Cp*	pentamethylcyclopentadienyl
18-crown-6	1,4,7,10,13,16-hexaoxacyclooctadecane
CSA	camphorsulfonic acid
csi	chlorosulfonyl isocyanate
Cy	cyclohexyl
dabco	1,4-diazabicyclo[2.2.2]octane
dba	dibenzylideneacetone
dbn	1,5-diazabicyclo[4.3.0]non-5-ene
dbu	1,8-diazabicyclo[5.4.0]undec-7-ene
dcc	dicyclohexylcarbodiimide
dcpe	1,2-bis(dicyclohexylphosphino)ethane
ddq	2,3-dichloro-5,6-dicyano-1,4-benzoquinone
deac	diethylaluminum chloride
dead	diethyl azodicarboxylate
depe	1,2-bis(diethylphosphino)ethane
depm	1,2-bis(diethylphosphino)methane
det	diethyl tartrate (+ or −)

DHP	dihydropyran
diars	1,2-bis(dimethylarsino)benzene
dibal-H	diisobutylaluminum hydride
dien	diethylenetriamine
DIGLYME	bis(2-methoxyethyl)ether
diop	2,3-*O*-isopropylidene-2,3-dihydroxy-1,4-bis(diphenylphosphino)butane
dipt	diisopropyl tartrate (+ or −)
dma	dimethylacetamide
dmac	dimethylaluminum chloride
DMAD	dimethyl acetylenedicarboxylate
dmap	4-dimethylaminopyridine
DME	dimethoxyethane
DMF	*N*,*N*'-dimethylformamide
DMG	dimethylglyoximate
DMI	*N*,*N*'-dimethylimidazalone
dmpe	1,2-bis(dimethylphosphino)ethane
dmpm	bis(dimethylphosphino)methane
DMSO	dimethyl sulfoxide
dmtsf	dimethyl(methylthio)sulfonium fluoroborate
dpam	bis(diphenylarsino)methane
dppb	1,4-bis(diphenylphosphino)butane
dppe	1,2-bis(diphenylphosphino)ethane
dppf	1,1'-bis(diphenylphosphino)ferrocene
dpph	1,6-bis(diphenylphosphino)hexane
dppm	bis(diphenylphosphino)methane
dppp	1,3-bis(diphenylphosphino)propane

eadc	ethylaluminum dichloride
edta	ethylenediaminetetraacetate
eedq	*N*-ethoxycarbonyl-2-ethoxy-1,2-dihydroquinoline
en	ethylene-1,2-diamine
Et_2O	diethyl ether

F_6 acac	hexafluoroacetylacetonate
Fc	ferrocenyl
Fp	$Fe(CO)_2Cp$

HFA	hexafluoroacetone
hfacac	hexafluoroacetylacetonate
hfb	hexafluorobut-2-yne
HMPA	hexamethylphosphoramide
hobt	hydroxybenzotriazole

$IpcBH_2$	isopinocampheylborane
Ipc_2BH	diisopinocampheylborane

kapa	potassium 3-aminopropylamide
K-selectride	potassium tri-*s*-butylborohydride

LAH	lithium aluminum hydride
LDA	lithium diisopropylamide
LICA	lithium isopropylcyclohexylamide
LITMP	lithium tetramethylpiperidide
L-selectride	lithium tri-*s*-butylborohydride
LTA	lead tetraacetate

mcpba	*m*-chloroperbenzoic acid
MeCN	acetonitrile
MEM	methoxyethoxymethyl
MEM-Cl	β-methoxyethoxymethyl chloride

Mes	mesityl
mma	methyl methacrylate
mmc	methylmagnesium carbonate
MOM	methoxymethyl
Ms	methanesulfonyl
MSA	methanesulfonic acid
MsCl	methanesulfonyl chloride
nap	1-naphthyl
nbd	norbornadiene
NBS	*N*-bromosuccinimide
NCS	*N*-chlorosuccinimide
nmo	*N*-methylmorpholine *N*-oxide
NMP	*N*-methyl-2-pyrrolidone
Nu⁻	nucleophile
ox	oxalate
pcc	pyridinium chlorochromate
pdc	pyridinium dichromate
phen	1,10-phenanthroline
phth	phthaloyl
ppa	polyphosphoric acid
ppe	polyphosphate ester
[PPN]⁺	[(Ph₃P)₂N]⁺
ppts	pyridinium *p*-toluenesulfonate
py	pyridine
pz	pyrazolyl
Red-Al	sodium bis(2-methoxyethoxy)aluminum dihydride
sal	salicylaldehyde
salen	*N,N'*-bis(salicylaldehydo)ethylenediamine
SEM	β-trimethylsilylethoxymethyl
tas	tris(diethylamino)sulfonium
tasf	tris(diethylamino)sulfonium difluorotrimethylsilicate
tbaf	tetra-*n*-butylammonium fluoride
TBDMS	*t*-butyldimethylsilyl
TBDMS-Cl	*t*-butyldimethylsilyl chloride
TBDPS	*t*-butyldiphenylsilyl
tbhp	*t*-butyl hydroperoxide
TCE	2,2,2-trichloroethanol
TCNE	tetracyanoethene
TCNQ	7,7,8,8-tetracyanoquinodimethane
terpy	2,2':6',2"-terpyridyl
tes	triethylsilyl
Tf	triflyl (trifluoromethanesulfonyl)
TFA	trifluoracetic acid
TFAA	trifluoroacetic anhydride
tfacac	trifluoroacetylacetonate
THF	tetrahydrofuran
THP	tetrahydropyranyl
tipbs-Cl	2,4,6-triisopropylbenzenesulfonyl chloride
tips-Cl	1,3-dichloro-1,1,3,3-tetraisopropyldisiloxane
TMEDA	tetramethylethylenediamine [1,2-bis(dimethylamino)ethane]
TMS	trimethylsilyl
TMS-Cl	trimethylsilyl chloride
TMS-CN	trimethylsilyl cyanide
Tol	tolyl
tpp	*meso*-tetraphenylporphyrin

Tr	trityl (triphenylmethyl)
tren	2,2',2"-triaminotriethylamine
trien	triethylenetetraamine
triphos	1,1,1-tris(diphenylphosphinomethyl)ethane
Ts	tosyl
TsMIC	tosylmethyl isocyanide
ttfa	thallium trifluoroacetate

Contents of All Volumes

Volume 1   Lithium, Beryllium, and Boron Groups
1   Alkali Metals
2   Beryllium
3   Magnesium, Calcium, Strontium and Barium
4   Compounds with Three- or Four-coordinate Boron, Emphasizing Cyclic Systems
5   Boron Rings Ligated to Metals
6   Polyhedral Carbaboranes
7   Main-group Heteroboranes
8   Metallaboranes
9   Transition Metal Metallacarbaboranes
10   Aluminum
11   Gallium, Indium and Thallium, Excluding Transition Metal Derivatives
12   Transition Metal Complexes of Aluminum, Gallium, Indium and Thallium
Author Index
Subject Index

Volume 2   Silicon Group, Arsenic, Antimony, and Bismuth
1   Organosilanes
2   Carbacyclic Silanes
3   Organopolysilanes
4   Silicones
5   Germanium
6   Tin
7   Lead
8   Arsenic, Antimony and Bismuth
Author Index
Subject Index

Volume 3   Copper and Zinc Groups
1   Gold
2   Copper and Silver
3   Mercury
4   Cadmium and Zinc
Author Index
Subject Index

Volume 4   Scandium, Yttrium, Lanthanides and Actinides, and Titanium Group
1   Zero Oxidation State Complexes of Scandium, Yttrium and the Lanthanide Elements
2   Scandium, Yttrium, and the Lanthanide and Actinide Elements, Excluding their Zero Oxidation State Complexes
3   Titanium Complexes in Oxidation States Zero and Below
4   Titanium Complexes in Oxidation States +2 and +3
5   Titanium Complexes in Oxidation State +4
6   Zirconium and Hafnium Complexes in Oxidation States Zero and Below
7   Metallocene(II) Complexes of Zirconium and Hafnium
8   Zirconium and Hafnium Compounds in Oxidation State +3
9   Bis(cyclopentadienyl)zirconium and -hafnium Halide Complexes in Oxidation State +4
10   Bis(cyclopentadienyl) Metal(IV) Compounds with Si, Ge, Sn, N, P, As, Sb, O, S, Se, Te or Transition Metal-centred Ligands
11   Zirconium and Hafnium Complexes in Oxidation State +4

12   Cationic Organozirconium and Organohafnium Complexes
13   Cyclooctatetraene Complexes of Zirconium and Hafnium
Author Index
Subject Index

Volume 5   Vanadium and Chromium Groups
1   Vanadium
2   Niobium and Tantalum
3   Hexacarbonyls and Carbonyl Complexes of Carbon σ-Bonded Ligands of Chromium,
      Molybdenum and Tungsten
4   Carbonyl Complexes of Noncarbon σ-Bonded Ligands of Chromium, Molybdenum and
      Tungsten
5   Organometallic Complexes of Chromium, Molybdenum and Tungsten without Carbonyl
      Ligands
6   π-Complexes of Chromium, Molybdenum and Tungsten, Excluding those of Cyclopentadienyls
      and Arenes
7   Cyclopentadienyl Complexes of Chromium, Molybdenum and Tungsten
8   Arene and Heteroarene Complexes of Chromium, Molybdenum and Tungsten
Author Index
Subject Index

Volume 6   Manganese Group
1   Manganese Carbonyls and Manganese Carbonyl Halides
2   Manganese Alkyls and Hydrides
3   Manganese Complexes Containing Nonmetallic Elements
4   Manganese Hydrocarbon Complexes Excluding Cyclopentadienyl
5   Cyclopentadienyl Manganese Complexes
6   Manganese Nitrosyl and Isonitrile Complexes
7   High-valent Organomanganese Compounds
8   Technetium
9   Low-valent Organorhenium Compounds
10   High-valent Organorhenium Compounds
Author Index
Subject Index

Volume 7   Iron, Ruthenium, and Osmium
1   Iron Compounds without Hydrocarbon Ligands
2   Mononuclear Iron Compounds with η^1–η^6 Hydrocarbon Ligands
3   Dinuclear Iron Compounds with Hydrocarbon Ligands
4   Polynuclear Iron Compounds with Hydrocarbon Ligands
5   Introduction to Organoruthenium and Organoosmium Chemistry
6   Mononuclear Complexes of Ruthenium and Osmium Containing η^1 Carbon Ligands
7   Complexes of Ruthenium and Osmium Containing η^2–η^6 Hydrocarbon Ligands: (i) Complexes
      not Containing Cyclobutadiene, Cyclopentadienyl or η-Arene Coligands
8   Complexes of Ruthenium and Osmium Containing η^2–η^6 Hydrocarbon Ligands: (ii) Complexes
      Containing Four- and Five-membered Rings (Including MCp(arene) Complexes)
9   Complexes of Ruthenium and Osmium Containing η^2–η^6 Hydrocarbon Ligands: (iii)
      Complexes Containing Six-, Seven- and Eight-membered Rings
10   Ruthenocenes and Osmocenes
11   Binuclear Complexes of Ruthenium and Osmium Containing Metal–Metal Bonds
12   Trinuclear Clusters of Ruthenium and Osmium: (i) Introduction and Simple Neutral, Anionic
      and Hydrido Clusters
13   Trinuclear Clusters of Ruthenium and Osmium: (ii) Hydrocarbon Ligands on Metal Clusters
14   Trinuclear Clusters of Ruthenium and Osmium: (iii) Clusters with Metal–Carbon Bonds to
      Heteroatom Ligands
15   Tetranuclear Clusters of Ruthenium and Osmium
16   Medium- and High-nuclearity Clusters of Ruthenium and Osmium
Author Index
Subject Index

Volume 8   Cobalt, Rhodium, and Iridium
1   Cobalt
2   Rhodium
3   Iridium
4   Cluster Complexes of Cobalt, Rhodium, and Iridium
Author Index
Subject Index

Volume 9   Nickel, Palladium, and Platinum
1   Nickel Complexes with Carbonyl, Isocyanide, and Carbene Ligands
2   Nickel–Carbon σ-Bonded Complexes
3   Nickel–Carbon π-Bonded Complexes
4   Palladium Complexes with Carbonyl, Isocyanide and Carbene Ligands
5   Palladium–Carbon σ-Bonded Complexes
6   Palladium–Carbon π-Bonded Complexes
7   Platinum Complexes with Carbonyl, Isocyanide and Carbene Ligands
8   Platinum–Carbon σ-Bonded Complexes
9   Platinum–Carbon π-Bonded Complexes
Author Index
Subject Index

Volume 10   Heteronuclear Metal–Metal Bonds
1   Synthesis of Compounds Containing Heteronuclear Metal–Metal Bonds
2   Heterodinuclear Compounds
3   Heteronuclear Clusters Containing C_1, C_2, C_3, ..., C_n Acyclic Hydrocarbyl Ligands
4   Binary Carbonyls, Carbonyls plus Hydrides, Carbonyls plus Phosphines, Cyclic Hydrocarbyls and Main-group Ligands without Acyclic Hydrocarbyls
5   Cluster Complexes with Bonds Between Transition Elements and Copper, Silver and Gold
6   Cluster Complexes with Bonds Between Transition Elements and Zinc, Cadmium, and Mercury
7   Catalysis and Related Reactions with Compounds Containing Heteronuclear Metal–Metal Bonds
Author Index
Subject Index

Volume 11   Main-group Metal Organometallics in Organic Synthesis
1   Lithium
2   Sodium and Potassium
3   Magnesium
4   Zinc and Cadmium
5   Boron
6   Aluminum
7   Silicon
8   Tin
9   Mercury
10   Thallium
11   Lead
12   Antimony and Bismuth
13   Selenium
14   Tellurium
Author Index
Subject Index

Volume 12   Transition Metal Organometallics in Organic Synthesis
1   Introduction and Fundamentals
2   Transition Metal Hydrides: Hydrocarboxylation, Hydroformylation, and Asymmetric Hydrogenation
3.1 Transition Metal Alkyl Complexes from Hydrometallation
3.2 Transition Metal Alkyl Complexes from RLi and CuX
3.3 Transition Metal Alkyl Complexes: Main-group Transmetallation and Insertion Chemistry
3.4 Transition Metal Alkyl Complexes: Oxidative Addition and Transmetallation

3.5   Transition Metal Alkyl Complexes: Oxidative Addition and Insertion
3.6   Transition Metal Alkyl Complexes: Multiple Insertion Cascades
3.7   Transition Metal Alkyl Complexes: Reductive Dimerization of Alkenes and Alkynes
4   Transition Metal Carbonyl Complexes
5.1   Transition Metal Carbene Complexes: Cyclopropanation
5.2   Transition Metal Carbene Complexes: Diazodecomposition, Ylide, and Insertion
5.3   Transition Metal Carbene Complexes: Alkyne and Vinyl Ketene Chemistry
5.4   Transition Metal Carbene Complexes: Photochemical Reactions of Carbene Complexes
5.5   Transition Metal Carbene Complexes: Tebbe's Reagent and Related Nucleophilic Alkylidenes
6.1   Transition Metal Alkene, Diene, and Dienyl Complexes: Nucleophilic Attack on Alkene
       Complexes
6.2   Transition Metal Alkene, Diene, and Dienyl Complexes: Complexation of Dienes for Protection
6.3   Transition Metal Alkene, Diene, and Dienyl Complexes: Nucleophilic Attack on Diene and
       Dienyl Complexes
7.1   Transition Metal Alkyne Complexes: Transition Metal-stabilized Propargyl Systems
7.2   Transition Metal Alkyne Complexes: Pauson–Khand Reaction
7.3   Transition Metal Alkyne Complexes: Transition Metal-catalyzed Cyclotrimerization
7.4   Transition Metal Alkyne Complexes: Zirconium–Benzyne Complexes
8.1   Transition Metal Allyl Complexes: Telomerization of Dienes
8.2   Transition Metal Allyl Complexes: Pd, W, Mo-assisted Nucleophilic Attack
8.3   Transition Metal Allyl Complexes: Intramolecular Alkene and Alkyne Insertions
8.4   Transition Metal Allyl Complexes: Trimethylene Methane Complexes
8.5   Transition Metal Allyl Complexes: π-Allylnickel Halides and Other π-Allyl Complexes
       Excluding Palladium
9.1   Transition Metal Arene Complexes: Nucleophilic Addition
9.2   Transition Metal Arene Complexes: Ring Lithiation
9.3   Transition Metal Arene Complexes: Side-chain Activation and Control of Stereochemistry
10   Synthetically Useful Coupling Reactions Promoted by Ti, V, Nb, W, Mo Reagents
11.1   Transition Metal-catalyzed Oxidations: Asymmetric Epoxidation
11.2   Transition Metal-catalyzed Oxidations: Asymmetric Hydroxylation
11.3   Transition Metal-catalyzed Oxidations: Other Oxidations
12.1   Transition Metals in Polymer Synthesis: Ziegler–Natta Reaction
12.2   Transition Metals in Polymer Synthesis: Ring-opening Metathesis Polymerization and Other
       Transition Metal Polymerization Techniques
Author Index
Subject Index

Volume 13   Structure Index
Structures of Organometallic Compounds Determined by Diffraction Methods

Volume 14   Cumulative Indexes
Cumulative Formula Index
Cumulative Subject Index

1
Gold

ANDREAS GROHMANN and HUBERT SCHMIDBAUR
Technische Universität München, Garching, Germany

1.1 INTRODUCTION	1
1.2 SYNTHESIS, STRUCTURE AND REACTIONS OF σ-BONDED ORGANOGOLD COMPLEXES	3
1.2.1 Complexes of the Type [RAuL]	3
1.2.2 Gold(I) Complexes Containing the Structural Building Block C(AuL)$_n$, n ≥ 2	12
1.2.3 Gold(I) Complexes with Two Gold–Carbon Bonds	17
1.2.4 Complexes of Gold(I), Gold(II) and Gold(III) with Ylide Ligands	19
1.2.4.1 Mononuclear ylide complexes	20
1.2.4.2 Di- and polynuclear ylide complexes	23
1.2.5 Gold(III) Complexes with One Gold–Carbon Bond	31
1.2.6 Gold(III) Complexes with Two Gold–Carbon Bonds	33
1.2.6.1 Diorganylgold(III) halides and pseudohalides	33
1.2.6.2 Diorganylgold(III) complexes with group 16 donor ligands	34
1.2.6.3 Diorganylgold(III) complexes with group 15 donor ligands	36
1.2.7 Gold(III) Complexes with Three Gold–Carbon Bonds	39
1.2.8 Gold(III) Complexes with Four Gold–Carbon Bonds	42
1.3 HOMO- AND HETEROMETALLIC GOLD CLUSTERS CONTAINING GOLD–CARBON BONDS	43
1.4 SYNTHESIS AND PROPERTIES OF ALKENE, ALKYNE AND RELATED COMPLEXES OF GOLD	44
1.5 SYNTHESIS, PROPERTIES AND REACTIONS OF CARBENE COMPLEXES OF GOLD	44
1.6 SYNTHESIS AND PROPERTIES OF CARBONYL COMPLEXES OF GOLD	47
1.7 SYNTHESIS AND PROPERTIES OF ISOCYANIDE COMPLEXES OF GOLD	49
1.8 REFERENCES	49

1.1 INTRODUCTION

In terms of its chemistry, gold is no longer the Sleeping Beauty among the precious metals. A current inventory of research on gold, its compounds and their applications turns out to be a diverse list. While established areas of research[1] have been continuously expanded, since the early 1980s gold chemistry has, above all, been characterized by innovative approaches and an unprecedented diversification of research interests, and the organometallic chemistry of gold is no exception.[2]

New ways of reclaiming the metal are being explored, which are of particular relevance in view of the environmental hazard posed by the traditional amalgam and cyanide processes. The use of nontoxic and biologically degradable thiourea for the recovery of gold from electronic scrap is an example in point.[3] Biomass, prepared from algal cells, has been shown to be an efficient binding medium for gold salts in aqueous solution, and this biosorption process may provide a lucrative alternative for the recovery of gold from mining effluents.[4] On a similar note, a more traditional process in the extractive metallurgy of gold, which is based on the adsorption of gold complexes on activated carbon and subsequent elution (the 'carbon-in-pulp' process), is still being actively explored.[5]

Novel applications for metallic gold as well as its 'inorganic' complexes are being established in fields as diverse as electrochemistry,[6] nonlinear optics,[7] catalysis[8–12] and bioinorganic chemistry.[13] A particularly interesting example for a catalytic application is the use of luminescent gold(I) complexes as photocatalysts.[14,15] Developments with relevance to bioinorganic chemistry include the emergence of self-assembled thiolate monolayers on gold as attractive models of organic interfaces,[16] as well as the process of labelling proteins with water-soluble gold clusters and subsequent visualization by electron microscopy, which has proved to be an indispensable tool for determining the locations of functional sites in biological macromolecules.[17] By the same token, while the position of gold drugs as being the most effective therapeutic agents in the treatment of rheumatoid arthritis has remained unchallenged,[18,22] new gold complexes are being studied as part of the continuing search for substances which can be used in the therapy of cancer.[18,20,21] Finally, ultrafine particles of gold as well as cluster compounds (in which an array of gold atoms of an oxidation state intermediate between 0 and +1 is surrounded by ligands) have also constituted active areas of research in their own right. The synthesis, bonding, chemical and catalytic reactivities of such species have been investigated in depth,[22] and a number of excellent reviews has appeared (see Mingos *et al.*[23–5] and Steggerda *et al.*,[26] and references cited therein).

In step with these developments, the pace of research on the organic chemistry of gold has also quickened. Major advances have been made not only in the characterization of unusual new compounds but also in the application of these discoveries to practical purposes, such as surface coating and chemical vapour deposition.[27–9] Perhaps most importantly, however, the repeated confirmation of the existence of attractive gold–gold interactions in such compounds (and the observation of their often striking consequences) has proved highly stimulating in the quest for a sound theoretical description of these phenomena.[30,31]

Gold commonly occurs in its compounds in the oxidation states +1 and +3, corresponding to the electron configurations [Xe] $4f^{14} 5d^{10} 6s^0 6p^0$ and [Xe] $4f^{14} 5d^8 6s^0 6p^0$, respectively. Other oxidation states n (-1, $0 \leq n \leq +1$ (see above), +2, +5) are much less common,[32–4] and examples of compounds with formal oxidation state +4 have been reported only recently.[35] The vast majority of organometallic compounds contain either gold(I) or gold(III), although organogold(II) compounds (usually dimeric species containing bridging ligands) are known and have constituted one of the major areas of research in organogold chemistry since the 1980s. Gold(I) complexes are usually two-coordinate, linear, diamagnetic 14-electron species. Three-coordinate trigonal-planar complexes and tetrahedrally four-coordinate complexes of monovalent gold have been characterized but are not as numerous.[36,37] Gold(III) complexes are almost always four-coordinate 16-electron species with square-planar stereochemistry, and hence are diamagnetic.[38] For the purpose of this review, complexes of monovalent and tervalent gold may be assumed to possess linear and square-planar stereochemistry, respectively, unless particular mention is made otherwise.

As has been noted above, extensive evidence has emerged from structural and spectroscopic studies of organogold complexes and gold compounds in general for the existence of closer-than-normal Au⋯Au distances, indicating an attractive interaction between the metal centres.[39–45] The contacts occur in the range 0.25–0.32 nm and are thus shorter than the van der Waals contact (0.332 nm)[46] or, in many cases, the interatomic distance in the metal (0.289 nm).[47] Gold–gold bonding of this nature is most commonly encountered in the coordination chemistry of monovalent gold, where it occurs perpendicular to the principal axis of the linearly two-coordinate gold(I) atoms. It often has a decisive influence on solid-state structures. Intermolecular Au⋯Au bonding may lead to extended structures characterized by catenated or layered aggregation of the molecules,[48] while intramolecular association of gold atoms has been found to 'fix' the conformation of the ligand backbone in certain dinuclear gold(I) complexes.[49] This latter observation has allowed an estimate to be made of the energy of the Au⋯Au interaction; the energy is of the order of 29–33 kJ mol^{-1} and thus of a magnitude comparable with the energy involved in hydrogen bonding.[50] In keeping with this similarity, there is also evidence that Au⋯Au bonding may persist in solution[48] and even in the gas phase.[51]

The most striking effect of intramolecular Au⋯Au bonding is observed in complexes in which a central atom (usually carbon or another first- or second-row main group element) acts as a bridging ligand for several gold(I) units. The accumulation of Au⋯Au interactions in such systems is accompanied by a considerable lowering in overall energy, and polyaurated compounds thus show a strong propensity to further auration. This synergistic effect, which has been graphically denoted 'aurophilicity', allows the preparation of complexes with three, four, five or even six gold(I) units clustering about a common central atom, the latter species representing apparent violations of the classical octet rule. Important examples include homoleptic species such as [E(AuL)$_3$]$^+$ (E = O, S or Se), [E(AuL)$_4$]$^{n+}$ (E = C, n = 0; E = N, P or As, n = 1), the hypercoordinated cations [E(AuL)$_5$]$^{n+}$ (E = C, n = 1; E = N or P, n = 2) and [E(AuL)$_6$]$^{n+}$ (E = C, n = 2; E = P, n = 3),[52] as well as a boron-centred cluster cation of the type [B(AuL)$_4$(L)]$^+$ (L = PR$_3$).[53] Other polynuclear gold complexes in which the central atom (E = C, N or P) carries an additional nonmetallic ligand have also been prepared (see Schmidbaur *et al.*,[54] and references cited therein). Given the isolobal relationship of LAu$^+$ and Li$^+$ or H$^+$,[55] there exists an interesting analogy between gold clusters of this type and hyperlithiated or 'hyperprotonated' species such as [ELi$_x$]$^{y+}$ and [EH$_x$]$^{y+}$, respectively,[56] which have been the subject of experimental as well as theoretical investigations.[57]

The striking manifestations of secondary Au$\cdots$Au bonding have invited a detailed theoretical analysis of its underlying causes. The electronic structures of heavy atoms such as gold and, to a lesser extent, platinum or mercury are characterized by the operation of relativistic effects which induce considerable hybridization of the 6s and 5d orbitals, thus mobilizing the d-electrons for chemical bonding.[58] In the case of gold, the phenomenon is most pronounced for gold(I) but has also been discussed for other oxidation states.[59] Theoretical work on gold complexes which incorporates relativistic effects includes molecular orbital calculations on the heteroatom-centred clusters mentioned above,[30,60] as well as *ab initio* calculations on the hypothetical dimer [AuCl(PH$_3$)]$_2$.[31]

The presentation of the material in this review largely follows an order of increasing oxidation states, except where a classification by ligand type (as in the case of the ylide complexes) facilitates a generalized treatment. Within sections, attempts have usually been made to preserve the arrangement of material adopted in *COMC-I*, as this is hoped to aid the reader when searching for older references. Reviews which have been published since the early 1980s include surveys of general organogold chemistry,[61,62] as well as more specialized accounts[63] on topics such as structural chemistry (x-ray/Mössbauer data),[39] coordination chemistry,[64–8] and arylgold chemistry.[69] For further information, the reader is also referred to the '*Dictionary of Organometallic Compounds*' as another highly useful compilation of data concerning the organic chemistry of gold.[70]

1.2 SYNTHESIS, STRUCTURE AND REACTIONS OF σ-BONDED ORGANOGOLD COMPLEXES

1.2.1 Complexes of the Type [RAuL]

Complexes of the type [RAuL] may contain a variety of organic residues R, such as alkyl, vinyl, alkynyl or aryl. The neutral donor group L is most commonly a tertiary phosphine or an isocyanide ligand, and dinuclear complexes with bridging difunctional donor ligands are also known.[61,71,72] From a practical point of view, the series of organogold(I) isocyanide complexes is of particular interest as some of its members hold promise as precursors for the chemical vapour deposition of gold films.[29] A selection of these and other compounds of the type [RAuL] is presented in Tables 1–3.

Alkylgold(I) phosphine complexes are usually synthesized by the reaction of an alkyllithium or a Grignard reagent with a complex gold(I) halide, and the corresponding isocyanide complexes are accessible by analogous routes. Typical examples of the preparation of these types of complex are given in Equations (1)[73] and (2).[29] In a modified synthesis of alkylgold(I) phosphine complexes, tris(triphenylphosphinegold(I))oxonium tetrafluoroborate is used instead of phosphinegold(I) halides, and considerably shorter reaction times and generally higher yields of organogold compounds have been reported (Equations (3) and (4), e.g., R = Bz). In addition, the separation of starting material and final products is facilitated by the low solubility of the oxonium salt in common solvents such as ether or tetrahydrofuran.[74,75]

$$\text{[AuCl(SMe}_2\text{)]} + \text{MeMgCl} + \text{MeNC} \xrightarrow{-\text{MgCl}_2, -\text{SMe}_2} \text{[AuMe(CNMe)]} \qquad (2)$$

$$\text{[(Ph}_3\text{PAu)}_3\text{O][BF}_4\text{]} \xrightarrow[\text{ii, H}_2\text{O, } -\text{LiX}]{\text{i, RLi}} \text{[AuR(PPh}_3\text{)]} \qquad (3)$$

$$\text{[(Ph}_3\text{PAu)}_3\text{O][BF}_4\text{]} \xrightarrow[\text{ii, H}_2\text{O, } -\text{MgX}_2]{\text{i, RMgCl}} \text{[AuR(PPh}_3\text{)]} \qquad (4)$$

Table 1   Some recent examples of gold(I) complexes of the type [RAuL].

Complex	m.p. (°C)	Ref.
R = alkyl		
[{AuMe}₂{(Ph₂P)₂$\overline{\text{CCH}_2\text{CH}_2}$}]	203 (d)	73
[AuMe(AsPh₃)]	–	76
[AuMe(CNMe)]	95 (d)	29
[AuMe(CNPr^i)]	<40 (d)	29
[AuMe(CNC₆H₁₁)]	88 (d)	29
[AuMe(CNPh)]	102 (d)	29
R = functionalized alkyl and related ligands		
[Au(CCl₃)(PPh₃)]	>200 (d)	77
[Au(CF₃)(PMe₃)]	191–193 (d)	78,79
[Au(CH₂CN)(PPh₃)]	141–142	80
[Au(CCl₂CN)(PPh₃)]	172–173	80
[Au{CH(CN)(CO₂Et)}(PPh₃)]	147–149	81,82
[Au{C[=N(C₆H₃Me₂-2,6)](CH₂COPh)}]	160 (d)	83
[Au₂{(CH₂)₃PPh₂}₂]		84
[Au₂{2-C(TMS)₂(C₅H₄N)}₂]	122 (d)	85
[Au{2-(1-Ph-*closo*-1,2-C₂B₁₀H₁₀)}(PPh₃)]		76
[Au{2-(1-Ph-*closo*-1,2-C₂B₁₀H₁₀)}(AsPh₃)]		76
R = vinyl		
[Au(HC=CH₂)(CNMe)]		29
[Au(HC=CH₂)(CNBu^t)]		29

The nature of the bonding in [AuMe(PMe₃)] and similar alkylgold(I) complexes has been the subject of theoretical and spectroscopic work, in an effort to ascertain the involvement of the filled gold $5d$ orbitals. The evaluation of photoelectron spectra in combination with SCF-Xα-SW (self-consistent field-Xα-scattered wave) calculations indicates only a small $5d$ contribution to the overall bonding,[86] while a discussion on the basis of SCF-LCAO-MO calculations (including relativistic effects) suggests that the net contribution of the gold $5d$ electrons to the overall bonding in [AuMe(PMe₃)] is appreciable.[87] A structural study (by gas electron diffraction) of [AuMe(PMe₃)] has also appeared.[88] The molecule displays C_{3v} symmetry, and both the gold–carbon and the gold–phosphorus bond distances have been found to lie within the range usually observed for alkyl(phosphine)gold(I) complexes.[42–5]

Halogenoalkylgold(I) complexes can be prepared by photochemical insertion of fluoroalkenes into the gold–carbon bond of methylgold(I) complexes, and this and related processes involving other metals have been reviewed.[89] In addition, several new routes to halogenoalkylgold(I) complexes have been established.[90,91] These include the direct auration of halocarbons such as chloroform with tris(triphenylphosphinegold(I))oxonium tetrafluoroborate (see also earlier) to give the halogenomethyl-gold(I) derivatives,[77,80] and the reaction of bis(trifluoromethyl)cadmium with phosphinegold(I) halides[78,79] (Equations (5) and (6)). The auration product of chloroform is unstable with respect to decomposition into triphenylphosphinegold(I) chloride and dichlorocarbene, as has been ascertained by scavenging experiments with alkenes.[77] An unusual reaction is reported for the chloromethyl complex [Au(CH₂Cl)(PPh₃)]: interaction with [PtMe₂(2,2'-bipy)] results in oxidative addition to give [PtClMe₂(CH₂AuPPh₃)(2,2'-bipy)], a rare example of a heteronuclear μ-methylene complex that does not contain either a metal–metal bond or an additional bridging ligand.[92]

$$\text{CHCl}_3 \xrightarrow[\text{NaH}]{[(\text{Ph}_3\text{PAu})_3\text{O}][\text{BF}_4]} [\text{Au}(\text{CCl}_3)\text{PPh}_3] \tag{5}$$

$$2\,[\text{AuCl}(\text{PEt}_3)] + (\text{CF}_3)_2\text{Cd}\cdot\text{DME} \longrightarrow 2\,[\text{Au}(\text{CF}_3)(\text{PEt}_3)] + \text{CdCl}_2 \tag{6}$$
$$\text{DME} = \text{MeOCH}_2\text{CH}_2\text{OMe}$$

Alkylgold(I) complexes in which the alkyl group carries a donor functionality L represent a special class of compounds in that they often form cyclic di- or oligonuclear arrays. Pertinent examples include the complex [Ph₂P(CH₂)₃Au]₂ which is prepared according to Equation (7), and for which a dimeric structure is proposed.[84] The structure of the compound [{2-C(TMS)₂(C₅H₄N)}Au]₂ has been established by x-ray crystallography: an eight-membered ring contains the two gold(I) centres in close proximity (d(Au···Au) = 0.267 2(1) nm).[85] By contrast, the complex $\overline{\text{Pt}[\text{C}_6\text{H}_4\{2\text{-CH}(\text{AuPPh}_3)\text{PPh}_2\}]_2}$ represents a heterobimetallic species in which gold(I) is not part of a cyclic arrangement.[93]

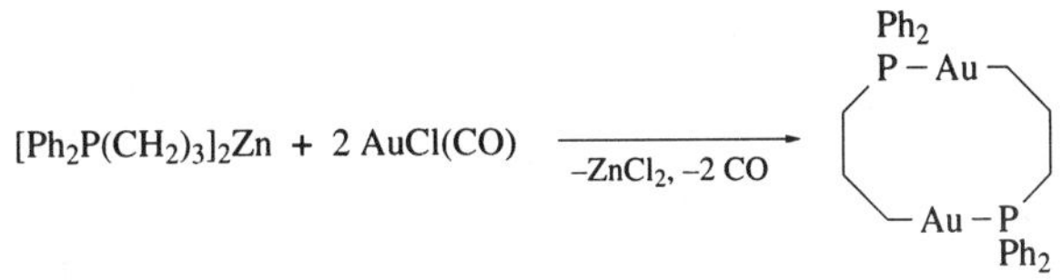

$$[Ph_2P(CH_2)_3]_2Zn + 2\,AuCl(CO) \xrightarrow[-ZnCl_2,\,-2\,CO]{} \quad \text{(structure)} \qquad (7)$$

Several alkyl ligands derived from organic molecules in which functional groups induce α-hydrogen acidity have been used to prepare the corresponding gold(I) complexes. In principle, ligands such as enolates are ambidentate and can coordinate to gold through either carbon or oxygen. Recently reported examples include various ketone and sulfonic acid derivatives, and carbon coordination is observed in all cases. An appraisal of gold(I)–carbon stretching vibration frequencies of such compounds has appeared.[94] The syntheses have been carried out by one of the following four methods.

(i)   The original 'alkyllithium route' (cf. Equation (1)).[85,95,96]
(ii) The modified 'alkyllithium route' (cf. Equation (3)).[97]
(iii) Deprotonation and subsequent auration with stable gold(I) alkoxides (Equation (8)).[98]
(iv) Reaction of silyl enol ethers with complex phosphinegold(I) halides in the presence of caesium fluoride (Equation (9)) [83] or of enol ethers with $[(Ph_3PAu)_3O][BF_4]$.[82] A variant consists in the use of silylcyclopropyl ethers for the preparation of phosphinegold(I) homoenolate complexes (Equation (10)).[99]

Less general reactions include the carbon-auration of pyrazolone derivatives with phosphinegold(I) halides in the presence of aqueous base,[100] and the auration of a palladium(II) phosphino enolate complex with $[AuCl(PPh_3)]$ in the presence of $AgBF_4$, resulting in the metallation of the C_α carbon atom of the former enolate moiety (Equation (11)).[101] A selection of compounds with their various methods of preparation is listed in Table 2.

$$[Au\{OCH(CF_3)_2\}PPh_3] + (CN)CH_2CO_2Et \xrightarrow{} [Au\{CH(CN)CO_2Et\}PPh_3] + (CF_3)_2CHOH \qquad (8)$$

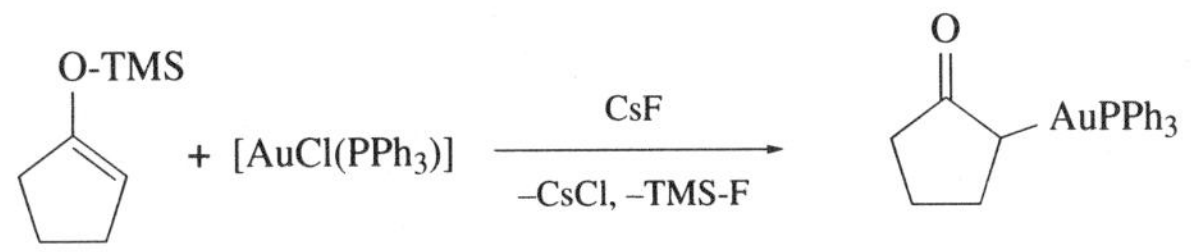

$$\text{(structure, O-TMS cyclopentenyl)} + [AuCl(PPh_3)] \xrightarrow[-CsCl,\,-TMS\text{-}F]{CsF} \text{(structure, cyclopentanone-AuPPh}_3) \qquad (9)$$

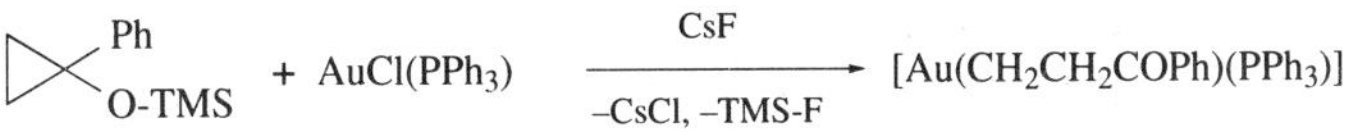

$$\text{(structure, Ph, O-TMS cyclopropyl)} + AuCl(PPh_3) \xrightarrow[-CsCl,\,-TMS\text{-}F]{CsF} [Au(CH_2CH_2COPh)(PPh_3)] \qquad (10)$$

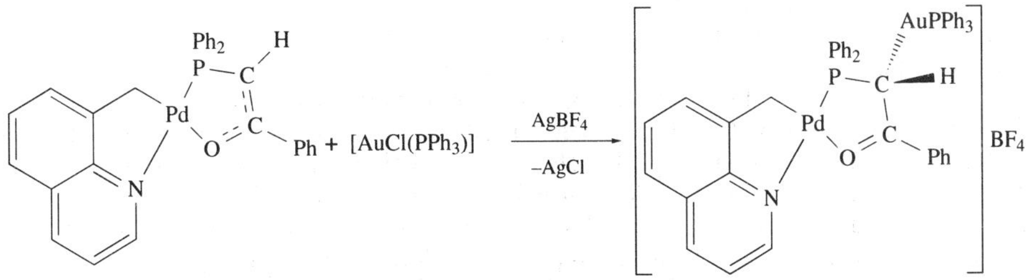

$$\text{(structure)} + [AuCl(PPh_3)] \xrightarrow[-AgCl]{AgBF_4} \text{(structure)} \; BF_4 \qquad (11)$$

Table 2 Alkylgold(I) complexes with carbonyl of sulfonyl functional groups.

Complex	Method of preparation[a]	Ref.
$[Au(CHPh-SO_2-Bu^t)(PPh_3)]$	(i)	95
$[Au\{(CCH_2CH_2)CO(CHCH_2CH_2)\}(PPh_3)]$	(i)	96
$[Au\{CH_2COC_5H_4Mn(CO)_3\}(PPh_3)]$	(ii)	97
$[Au\{CH(CN)_2\}(PPh_3)]$	(iii)	98
$[Au(CH_2COCH_2CH_2CH=CH_2)(PPh_3)]$	(iv)	99

[a] See text.

Finally, there are numerous complexes in which one or both protons in the methylene bridge of geminal diphosphines such as bis(diphenylphosphino)methane have been substituted with a phosphinegold(I) unit, as illustrated by the reaction products in Equation (12).[102] Deprotonation and subsequent auration can be accomplished by using gold(I) complexes containing basic ligands, such as [Au(acac)(PPh$_3$)][102] or [(Ph$_3$PAu)$_3$O][BF$_4$],[103] which are displaced as the conjugate acid in the process. Since the diphosphine ligands offer the possibility for further gold coordination at the phosphorus centres, they are versatile building blocks for the preparation of polynuclear gold complexes, and a large variety of structural types has been prepared.[104,105]

$$H_2C \underbrace{\left(\begin{smallmatrix} S \\ \parallel \\ PPh \end{smallmatrix} \right)}_2 \xrightarrow[-\ acacH]{[Au(acac)(PPh_3)]} \begin{smallmatrix} Ph_2(S=)P \diagdown \diagup H \\ Ph_2(S=)P \diagup \diagdown AuPPh_3 \end{smallmatrix} \ + \ \begin{smallmatrix} Ph_2(S=)P \diagdown \diagup AuPPh_3 \\ Ph_2(S=)P \diagup \diagdown AuPPh_3 \end{smallmatrix} \qquad (12)$$

There has been considerable activity concerning the synthesis and exploration of possible applications[106] of alkynylgold(I) compounds in recent years, and a number of x-ray structural (see below) as well as Mössbauer[107] studies have been reported. Previous routes to such complexes generally started with HAuCl$_4$, which is reduced by SO$_2$ in the presence of acetate, followed by addition of the terminal alkyne. In this way polymeric gold(I) acetylides [Au(C≡CR)]$_n$ are obtained which are versatile reagents in several ways: they have been used for the introduction of the acetylide ligand into heterobimetallic complexes of the late transition metals;[108–10] they react with electron-rich metal complexes by incorporation of the [Au(C≡CR)] unit to form neutral gold–metal clusters;[111,112] and they can be transformed into alkynylgold(I) complexes of the type [Au(C≡CR)(L)] simply by addition of suitable donor ligands, such as amines, tertiary phosphines or isonitriles (see also Bruce *et al.*[113]).[29,71,114] This group of ligands has been extended to include halide ions, giving anionic species of the general type [Au(C≡CR)(X)]$^-$.[115,116]

However, the preparation of complexes of the type [Au(C≡CR)(L)], as outlined earlier, often fails because the initially formed [Au(C≡CR)]$_n$ compounds may decompose readily, depending on the nature of the acetylide ligand. Several novel synthetic routes to alkynylgold(I) compounds that circumvent this problem have been established. Complex gold(I) chlorides containing a variety of tertiary phosphines have been found to react with a wide range of terminal alkynes, either in diethylamine in the presence of copper(I) halides,[113] or in alcoholic solution in the presence of sodium alkoxide,[113,117,118] to afford the corresponding alkynylgold(I) complexes in good yield. These transformations are equally applicable to unsubstituted ethyne, which gives dinuclear gold(I) acetylides [Au$_2$(C≡C)(PR$_3$)$_2$].[117,118] Examples are shown in Equations (13) and (14).

$$[AuCl(PPh_3)] \ + \ H\text{---}\!\!\equiv\!\!\text{---}C_6F_5 \xrightarrow[-\ [Et_2NH_2]Cl]{Et_2NH/Cu^ICl} [Au(\equiv\!\!\text{---}C_6F_5)(PPh_3)] \qquad (13)$$

$$[AuCl(P\{C_6H_4Me\text{-}4\}_3)] \ + \ H\text{---}\!\!\equiv\!\!\text{---} \xrightarrow{Et_2OH/NaOEt} [Au(\equiv\!\!\text{---})(P\{C_6H_4Me\text{-}4\}_3)] \qquad (14)$$

An alternative high-yield method for preparing gold(I) phenylacetylides involves the electrochemical oxidation of gold metal in an acetonitrile solution of the alkyne,[119] with the target compounds precipitating during electrolysis. The measured current efficiency of ~1 mol faraday^{-1} implies the following simple electrode processes:

Cathode: e$^-$ + PhC≡CH → (PhC≡C)$^-$ + 1/2 H$_2$ (g.)
Anode: Au → Au$^+$ + e$^-$
Overall reaction: Au$^+$ + (PhC≡C)$^-$ → (PhC≡C)Au

The simplicity of the procedure and the fact that the products can be isolated directly in a state of high purity make this a particularly attractive method for the small-scale preparation of alkynylgold(I) complexes.

Other routes that have led to alkynylgold(I) complexes have employed alkylgold(I) complexes[83] and *N*-substituted phosphinegold(I) imidazoles,[120] respectively. The anionic ligands in these reagents are sufficiently basic to deprotonate the alkyne moiety, thus forming acetylide complexes such as those shown in Figure 1.[120] A collection of other alkynylgold complexes is listed in Table 3. The alkynylgold(I) complexes of the type [Au(C≡CBu^t)(R$_2$PCH$_2$PR$_2$)] contain monodentate diphosphine ligands and can undergo interesting ligand exchange reactions involving a three-coordinate gold(I) centre.[121]

Alkynylgold(I) halide anion complexes have been prepared from suitable gold(I) acetylide precursors by addition of halide ion, as illustrated in Equation (15),[116] and a ^{13}C NMR study of such complexes has been reported.[115]

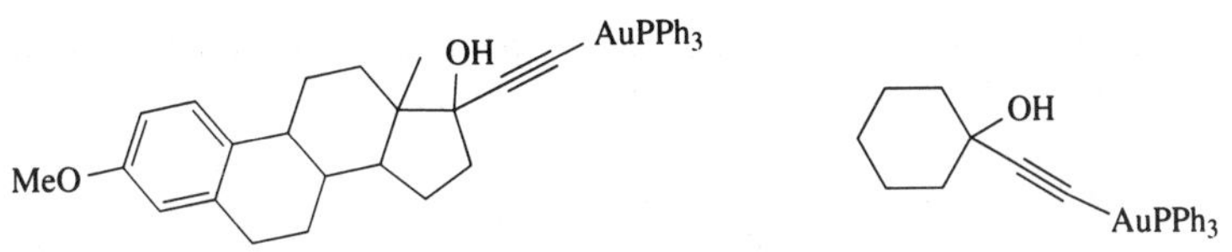

Figure 1 Alkynylgold(I) complexes (source *J. Organomet. Chem.*, 1988, **344**, 119).

Table 3 Alkynylgold(I) complexes.

Complex	m.p. (°C)	Ref.
[Au(C≡CPh)]$_n$	180 (d)	119
[Au(C≡CH)(PPh$_3$)]		117
[Au(C≡CMe)(PMe$_3$)]	123 (d)	29,113
[Au(C≡CMe)(PPh$_3$)]	173–175	113
[Au(C≡CCF$_3$)(PPh$_3$)]	155	117
[Au(C≡CEt)(PPh$_3$)]	154–155	117
[Au(C≡CPr^n)(PPh$_3$)]	256–259	113
[Au(C≡CBu^n)(PPh$_3$)]	279–281	113
[Au(C≡CBu^t)(PMe$_3$)]	183–186	113
[Au(C≡CBu^t)(PPh$_3$)]	181–182	113
[Au(C≡CCH$_2$OH)(PPh$_3$)]	234–235	113
[Au(C≡CCH$_2$OMe)(P{C$_6$H$_4$Me-4}$_3$)]		127
[Au(C≡CCH$_2$CH$_2$OH)(PPh$_3$)]	248–250	113
[Au(C≡CCO$_2$Me)(PPh$_3$)]	239–240	113
[Au(C≡CPh)(PMe$_3$)]	196–198	113
[Au(C≡CPh)(PPh{OMe}$_2$)]	212–215	113
[Au(C≡CPh)(PPh$_3$)]	163–165	117,119
[Au(C≡CPh)(P{C$_6$H$_4$Me-4}$_3$)]	146–148	117
[Au(C≡CPh)(phen)]	95–97	119
[Au(C≡CC$_6$F$_5$)(PPh$_3$)]	235–236	113
[Au(C≡CBu^t)(Me$_2$PCH$_2$PMe$_2$)]	165 (d)	121
[Au(C≡CBu^t)(Ph$_2$PCH$_2$PPh$_2$)]	99 (d)	121
[Au(C≡CPPh$_2$)]$_x$		122
[Au$_2$(C≡C)(PPh$_3$)$_2$]		117
[Au$_2$(C≡C){P(C$_6$H$_4$Me-4)$_3$}$_2$]	110–115	117,128
[Au$_2$(C≡C){P[C$_6$H$_4$(OMe-4)]$_3$}$_2$]	125	117,128
[Au$_2$(C≡C–(C$_6$H$_4$)$_n$–C≡C)(PMe$_3$)$_2$], n = 1, 2		123
[Au$_2$(C≡CPh)$_2$(Ph$_2$PC$_6$H$_4$PPh$_2$)]		123
[Au$_2$(C≡CPh)$_2$(Ph$_2$PCH$_2$CH$_2$PPh$_2$)]		106
[Au(C≡CMe)(CNMe)]	140 (d)	29
[Au(C≡CBu^t)(CNMe)]	180–181 (d)	29
[Au(C≡CPh)Cl]$^-$		116
[Au(C≡CPh)Br]$^-$		116
[Au(C≡CPh)I]$^-$		116

$$1/n \, [\text{Au(C≡CPh)}]_n + [\text{NEt}_4]\text{Cl} \longrightarrow [\text{NEt}_4][\text{Au(C≡CPh)Cl}] \qquad (15)$$

Bifunctional acetylene ligands such as R$_2$PC≡CH or C≡N–C$_6$H$_4$–C≡CH have been employed in the synthesis of a novel class of metal-containing polymers. As the polymer backbone contains both unsaturated organic fragments and inorganic elements, such as gold or a combination of gold and phosphorus, with valence *d* orbitals available for π-bonding, these materials may possess interesting properties such as electrical conductivity or optical nonlinearity.[122–5] The modes of preparation of such compounds are summarized in Equation (16) and Scheme 1. The structures of the phosphinoacetylide-containing polymers (oligomeric vs. polymeric, as inferred from solubility properties) have been found to vary with the nature of the substituents on phosphorus. A part of the polymer chain is shown in Figure 2.

$$[\text{AuCl(Ph}_2\text{P}{=}{=}\text{H)}] + \text{NaOMe} \longrightarrow [\{\text{Au(Ph}_2\text{PC≡C-)}\}_x] \qquad (16)$$

Polymers have also been prepared from diphosphines and diacetylides, both of which act as bridging ligands and alternate along the polymer chain (Figure 2). The synthesis starts from digold(I) diacetylides [(AuC≡CArC≡CAu)$_x$] (e.g., Ar = 1,4-C$_6$H$_4$), prepared by reaction of the dialkyne with [AuCl(SMe$_2$)] in the presence of sodium acetate as base. Reaction of the starting material with diphosphines (PP) will

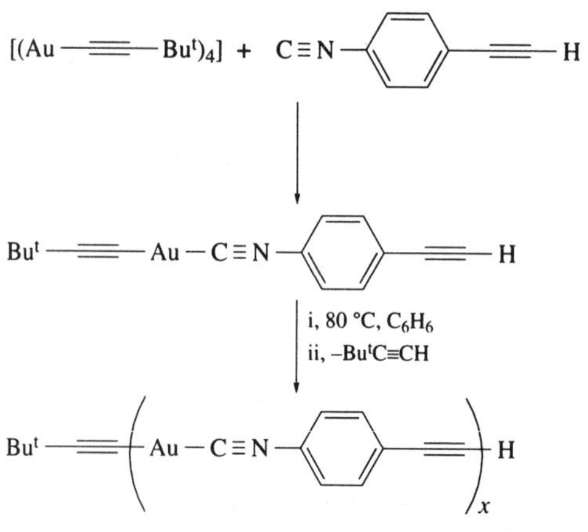

Scheme 1

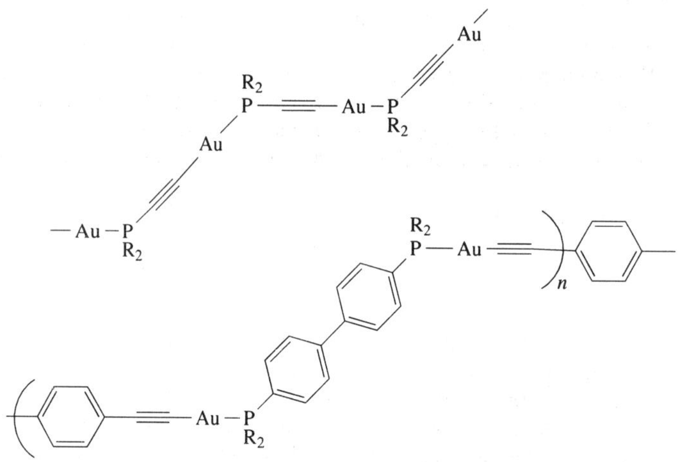

Figure 2 Suggested structures of $[\{Au(Ph_2PC\equiv C-)\}]_x$ and $[(C\equiv CArC\equiv C-Au-PP-Au-)_x]$.

give polymers of the general formula $[(C\equiv CArC\equiv C-Au-PP-Au-)_x]$. These are usually insoluble in common organic solvents, but the solubility can be improved by bulky alkyl substituents on phosphorus.

The crystal structures of several gold(I) acetylide complexes have been determined, and intermolecular Au⋯Au bonding interactions are observed in several cases (Figure 3).[106,125,126] The phosphine complexes[106,113,123,126–8] have Au–P bond lengths which are significantly elongated compared with the standard value of ~ 0.223 nm,[42–5] indicating a strong *trans* influence of the alkynyl group. A further important structural feature is the formation of clathrate complexes in the case of the diacetylide $[Au_2(C\equiv C)\{P(C_6H_4Me-4)_3\}_2]\cdot C_6H_6$, where a weak interaction between solvent and complex is suggested by the x-ray structure.[128] Finally, fast-atom bombardment (FAB) mass spectra of complexes such as $[Au_2(C\equiv C)\{P(C_6H_4Me-4)_3\}_2]$ suggest that Au⋯Au interactions are responsible for the observation of polynuclear ion aggregates in the gas phase.[51]

The chemistry of arylgold(I) complexes [ArAuL] has developed rapidly in recent years, and a review has appeared.[69] The available synthetic methods allow the preparation of complexes with a diverse range of ligands L, including tertiary phosphines, amines, isonitriles, tetrahydrothiophene and tertiary arsines. Difunctional phosphines such as bis(diphenylphosphino)amine give the corresponding dinuclear arylgold(I) complexes,[129,130] and a dimeric orthometallated phenylphosphine complex of formula $[Au_2\{C_6H_4(PPh_2)-2\}_2]$ has also been prepared.[131] Both types of complex undergo two-centre oxidative addition reactions reminiscent of those described for the structurally related gold(I) ylide complexes (cf. Section 1.2.4.2). Among the arylgold(I) complexes, halogenoaryl species have been found to be particularly stable thermally, and the strongly electron-withdrawing nature of the halogenoaryl group allows the isolation of organogold compounds with both neutral and anionic donor ligands. It has been noted that the structures of the pentahalogenophenyl complexes (as determined by x-ray diffraction) show distortions at the *ipso* carbon atoms, which require particular caution during refinement.[132]

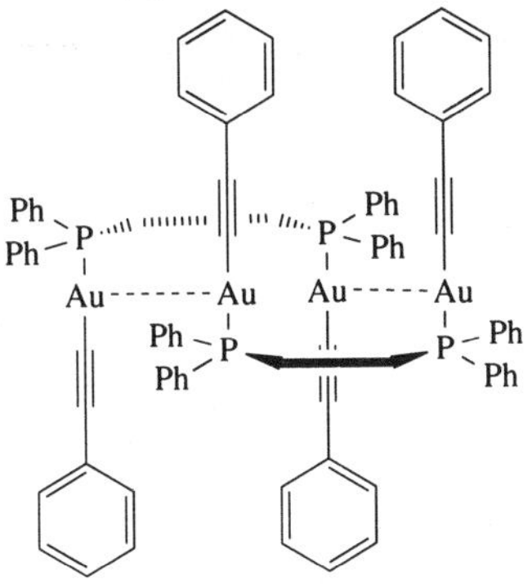

Figure 3 The spatial orientation of the two independent molecules in $[Au_2(C \equiv CPh)_2(Ph_2PCH_2CH_2PPh_2)]$.

In analogy to the synthesis of the alkyl derivatives, arylgold compounds [ArAuL] (L being a neutral donor ligand) can be prepared from complex gold(I) halides via the organolithium or the Grignard route.[133] [ArAuL] complexes can also be prepared with ligands L that are only weakly coordinating (such as tetrahydrothiophene, SC_4H_8) (Equation (17));[134] derivatives may then be obtained through displacement of this ligand by a variety of other more strongly coordinating donors (Equation (18)).[135] These may include unusual species, such as alkylidyne complexes [W($\equiv$CR)(CO)$_2$-(Me$_2$PCH$_2$CH$_2$PMe$_2$)Cl], which react to give heteronuclear complexes with bridging carbyne ligands between tungsten and gold.[136] A selection of arylgold(I) complexes that have been recently prepared is presented in Table 4. Complexes of the type [ArAuL] readily undergo oxidative addition of halogen, acyl halides or similar reagents to give gold(III) complexes of the type [ArAu(X)$_2$L].[137-9]

$$[Au(SC_4H_8)_2]ClO_4 + [Bu_4N][Au(C_6H_2F_3)_2] \longrightarrow 2 [Au(C_6H_2F_3)(SC_4H_8)] + [Bu_4N]ClO_4 \qquad (17)$$

$$[Au(C_6F_5)(SC_4H_8)] + H_2NCH_2CH_2NH_2 \longrightarrow [Au(C_6F_5)(H_2NCH_2CH_2NH_2)] \qquad (18)$$

Anionic complexes [ArAuL]$^-$, in which L is a halide or pseudohalide donor ligand, have so far only been isolated if the aryl group is strongly electron withdrawing, as in nitrophenyl or pentafluorophenyl derivatives.[69,152] Two methods of preparation have been reported: one relies on the displacement of tetrahydrothiophene, SC_4H_8, by halide or pseudohalide ion,[153] while the other involves the arylation of dihalogenoaurate ion with diarylmercury compounds.[141] Examples of these two reactions are presented in Equations (19) and (20), respectively.

$$[Au(C_6H_2F_3-2,4,6)(SC_4H_8)] + [N(PPh_3)_2]SCN \longrightarrow [N(PPh_3)_2][Au(C_6H_2F_3-2,4,6)(SCN)] + SC_4H_8 \qquad (19)$$

$$[PPh_3(CH_2Ph)][AuCl_2] + [Hg\{C_6H_2(NO_2)_3-2,4,6\}_2] \longrightarrow [PPh_3(CH_2Ph)][Au\{C_6H_2(NO_2)_3-2,4,6\}Cl] + \qquad (20)$$
$$[Hg\{C_6H_2(NO_2)_3-2,4,6\}Cl]$$

Phosphinegold(I) complexes containing a cyclopentadienyl or related ligand have been the subject of a number of studies, the major question being the degree of fluxionality in such molecules. Several cyclopentadienyl complexes have been prepared (Table 5), either by the organolithium method[154] or by direct auration of the parent cyclopentadiene with tris(triphenylphosphinegold(I))oxonium tetrafluoroborate[155,156] (Equations (21) and (22)). A unique reaction is that of pentakis(methoxy-carbonyl)cyclopentadiene (1),[157] which is a strong acid and will displace weaker organic acids upon interaction with their metal salts (Equation (23)).[158]

$$[AuCl(PPr^i_3)] + LiCp \longrightarrow [AuCp(PPr^i_3)] + LiCl \qquad (21)$$

$$[(Ph_3PAu)_3O][BF_4] + 3 C_5Ph_4H_2 \longrightarrow 3 [Au(C_5Ph_4H)(PPh_3)] + [H_3O][BF_4] \qquad (22)$$

$$[Au(O_2CMe)(PPh_3)] + C_5(CO_2Me)_5H \longrightarrow [Au\{C_5(CO_2Me)_5\}(PPh_3)] + MeCO_2H \qquad (23)$$

The dynamic behaviour of such complexes (with regard to the NMR timescale) has been found to depend on the nature of the substituents R on the five-membered ring. For R = H,[154] Me,[154] Bz[159] or CO_2Me,[158] and for R$_5$ = Me(CO$_2$Me)$_4$,[160] the cyclopentadienylgold(I) complexes are fluxional in solution, with rapid migration of the Au(PR$_3$) group around the η^1-C$_5$ ring even at low temperature (−80 °C). The

Table 4 Arylgold(I) complexes [ArAuL].

Complex	m.p. (°C)	Ref.
[AuPh(CNMe)]		29
[Au{C$_6$H$_5$(η^6-Cr(CO)$_3$)}(PPh$_3$)]		140
[Au(C$_6$H$_4${NO$_2$}-2)(AsPh$_3$)]		139
[Au(C$_6$H$_4${NO$_2$}-2)(SbPh$_3$)$_2$]	120 (d)	139
[Au(C$_6$H$_4${NO$_2$}-2)(phen)]	100 (d)	139
[Au(C$_6$H$_2${NO$_2$}$_3$-2,4,6)(dmphen)]^a		141
[Au$_2$(C$_6$H$_4${PPh$_2$}-2)$_2$]		131
[Au$_2$(C$_6$H$_4${AsPh$_2$}-2)$_2$]		142
[Au(C$_6$H$_2$F$_3$)(SC$_4$H$_8$)]	73 (d)	134
[Au$_2$(C$_6$F$_5$)$_2$(H$_2$C{PPh$_2$}$_2$)]		143
[Au$_2$(C$_6$F$_5$)$_2$(H$_2$C{AsPh$_2$}$_2$)]	189 (d)	138
[Au$_2$(C$_6$F$_5$)$_2$(Fe{C$_5$H$_4$PPh$_2$}$_2$)]	245 (d)	144
[Au$_2$(C$_6$F$_5$)$_2$(Ph$_2$PC(Ph$_2$PAuPPh$_2$)$_2$CPPh$_2$)]	170	145
[Au$_2$(C$_6$F$_5$)$_2$(Au{(PPh$_2$)CH(PPh$_2$Me)}$_2$)]$^+$		146
[Au$_2$(C$_6$F$_5$)$_2${(Ph$_2$P)$_2$–Mn(CO)$_4$}]$^-$		147
[Au$_2$(C$_6$F$_5$)$_2$(HN{PPh$_2$}$_2$)]	205 (d)	130
[Au$_2$(C$_6$Cl$_5$)$_2$(HN{PPh$_2$}$_2$)]	144 (d)	130
[Au$_2$(C$_6$F$_5$)$_2$(μ-S$_2$C–PEt$_3$)]	144	148
[Au$_2$(C$_6$F$_5$)$_2$(SP(Ph$_2$)CH$_2$P(Ph$_2$)S)]	185	149
[Au$_2$(C$_6$F$_5$)$_2$(C$_3$H$_4$NS$_2$)]^b	195	150
[Au$_2$(C$_6$F$_5$)$_2$(C$_5$H$_4$NS)]^b	143	150
[Au$_3$(C$_6$F$_5$)$_3$(HC{PPh$_2$}$_3$)]	220 (d)	145
[Au(C$_6$F$_5$)([W]{μ-C–C$_6$H$_4$Me-4})]^c		136
[(AuC$_3$N$_2$Me)$_3$]^d	272–275 (d)	151
[Au(C$_6$F$_5$)Cl]$^-$		152
[Au(C$_6$H$_2$F$_3$-2,4,6)(Br)]$^-$		153
[Au(C$_6$H$_2$F$_3$-2,4,6)(SCN)]$^-$		153
[Au{C$_6$H$_2$(NO$_2$)$_3$-2,4,6}Cl]$^-$		141

^{a} dmphen = 2,9-dimethyl-1,10-phenanthroline. ^{b} HL: C$_3$H$_5$NS$_2$, C$_5$H$_5$NS: heterocyclic thiones. ^{c} [W] = [W(CO)$_2$(Me$_2$PCH$_2$CH$_2$PMe$_2$)Cl]. ^{d} C$_3$N$_2$ = C^2-imidazolyl

Table 5 Cyclopentadienyl and related gold(I) complexes.

Complex	m.p. (°C)	Ref.
[Au(Cp)PMe$_3$]	87 (d)	154
[Au(Cp)PPr$^i_3$]	105 (d)	154
[Au(Cp*)PPr$^i_3$]	92 (d)	154
[Au(Cp*)PPh$_3$]	83 (d)	154
[Au(C$_5$HPh$_4$)PPh$_3$]	159–160 (d)	156
[Au(C$_5$Bz$_5$)PPh$_3$]		159
[Au(C$_5$Me{CO$_2$Me}$_4$)PPh$_3$]	157–159	160
[Au(C$_5${CO$_2$Me}$_5$)PPh$_3$]	145	158
[Au(C$_{13}$H$_9$)PPh$_3$]^a	135–137 (d)	161
[Au$_2$(C$_5$H$_4$FeC$_5$H$_4$)(PPh$_3$)$_2$]	105–106 (d)	79
[Au(C{NMe}CHCHN)]$_3$		151
[Au({C$_3$B$_2$HMe$_4$}Co{Cp})(PPh$_3$)]^b	190 (d)	162

^{a} C$_{13}$H$_9$ = 9-Fluorenyl. ^{b} C$_3$B$_2$HMe$_4$ = 2,3,4,5-Tetramethyl-2,3-dihydro-1*H*-1,3-diborol-2-yl.

^{13}C chemical shift of the ring carbon atoms is significantly different from the value observed for pentahapto coordination in other metal complexes. By contrast, the tetraphenylcyclopentadienyl complex has been found to be stereochemically rigid.[155]

Low-temperature x-ray crystallographic studies of most of the complexes have been carried out. In all cases, an unusual structural feature is found in the interaction of the gold atom with the cyclopentadienyl ring, which shows a 'slip distortion' toward a η^3-mode of coordination. In addition to the gold–carbon σ-bond (0.215 nm), two Au···C contacts (0.27 nm) with the Cp carbon atoms adjacent to the *ipso* position are observed.

Cyclopentadienyl- and pentamethylcyclopentadienylgold(I) complexes react with terminal alkynes such as HC≡CH, HC≡CPh and HC≡CCO$_2$Me to give the corresponding alkynylgold(I) compounds.[154] A most unusual reactivity towards aurating agents such as triphenylphosphinegold(I) tetrafluoroborate is observed with the tetraphenylcyclopentadienylgold(I) phosphine complex, which can add one or even two triphenylphosphinegold(I) units to form compounds containing three-centre, two-electron bonds.[155,163] Such derivatives are described in Section 1.2.2.

Gold(I) complexes of Cp-related ligands have also been prepared. 9-Fluorenyl(triphenylphosphine)-gold(I) (**2**)[161] can be obtained via the oxonium salt route, but the reduced C–H acidity of fluorene requires the use of an additional base, such as sodium hydride, to deprotonate the hydrocarbon. The phosphinegold moiety coordinates in η^1 fashion, and the molecule is found to be stereochemically rigid in solution. Genuine σ-coordination of the phosphinegold unit is also observed in the solid state, and there are no close contacts with carbon atoms adjacent to the *ipso* position of the fluorenyl ring.

Ferrocenylgold compounds have been prepared by the organolithium method, and auration of one or both of the cyclopentadienyl rings can be achieved.[81] Similar to tetraphenylcyclopentadienylgold(I) phosphine complexes, these compounds can incorporate additional phosphinegold(I) units upon interaction with triphenylphosphinegold(I) tetrafluoroborate to give electron-deficient species featuring three-centre, two-electron bonds, as described below.

Reaction of *C*-imidazolyllithium derivatives with triphenylphosphinegold(I) chloride results in the formation of *C*-aurated imidazolylgold(I) compounds.[137,151,164] In these, the imidazole ring can act as a donor ligand towards other imidazolylgold(I) units, and displacement of phosphine leads to the formation of cyclic trimers. Structure (**3**) illustrates the trinuclear complex [Au(C{NMe}CHCHN)]₃.

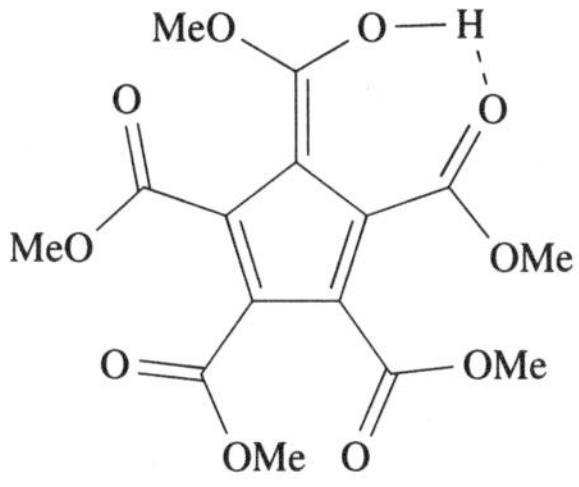

(1)

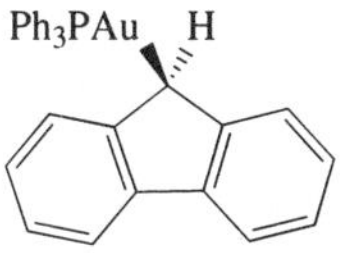

(2)

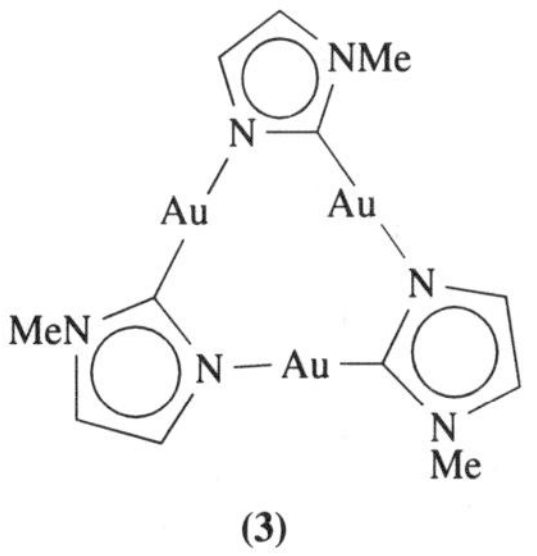

(3)

Several metallocene complexes of cobalt and rhodium have been studied in which one cyclopentadienyl ring is replaced by a diborole moiety.[162] Under suitable conditions, auration of the C-2 carbon atom in the diborolyl ligand can be effected, and a species containing a pentacoordinate carbon atom is obtained (Equation (24)). Similar to the structural characteristics of related cyclopentadienyl-gold(I) complexes, the gold atom is bonded to the diborolyl moiety in an η^1 fashion. However, the distances between gold and the two boron atoms adjacent to C-2 are in a range where a bonding interaction ('slip distortion') cannot be ruled out.

Iminoalkyl complexes of gold(I) can be prepared by the simultaneous treatment of a phosphinegold halide [AuX(PR₃)] with isocyanides and alcohols in the presence of alkali (Equation (25)).[165] In some cases, an intermolecular reaction similar to that observed for imidazolylgold(I) compounds results in the displacement of the phosphine ligands by the iminoalkyl moiety to form cyclic oligomers. The presence of bulky phosphine ligands seems to favour the formation of monomeric species, for one of which an x-ray crystallographic study has been reported.[166]

A related compound is obtained from an unusual tungsten carbene thioketene adduct upon auration, as shown in Equation (26).[167]

Gold

$$(24)$$

$$[AuCl(PBu^t_3)] + TsCH_2NC + MeOH + KOH \longrightarrow [Au\{C(OMe)=NCH_2Ts\}(PBu^t_3)] + KCl + H_2O \quad (25)$$

$$(26)$$

1.2.2 Gold(I) Complexes Containing the Structural Building Block $C(AuL)_n$, $n \geq 2$

Many solid-state structures of organogold(I) compounds are characterized by close contacts between the metal atoms. Examples from previous sections include the mono- and dinuclear alkynylgold(I) complexes $[Au(C\equiv CPh)(PPh_3)]$[126] and $[Au_2(C\equiv CPh)_2(Ph_2PCH_2CH_2PPh_2)]$,[106] with intermolecular Au···Au distances of 0.338 nm and 0.315 nm, respectively, and the dinuclear compound $[Au_2(C_5H_4N\{C(TMS)_2\}-2)_2]$, in which the intramolecular gold–gold contact is as short as 0.267 nm.[85]

Given the interatomic distance of 0.289 nm in metallic gold,[47] such close contacts clearly indicate an attractive interaction. As was discussed in Section 1.1, such bonding interactions are a recurrent motif in the coordination chemistry of monovalent gold, and they often play a decisive role in determining the molecular conformation and/or the solid-state structure of its compounds. Bridging ligands which allow the coordination of several gold(I) centres in close proximity accentuate this phenomenon as they introduce the possibility of intramolecular bonding. The smallest possible organic bridging ligand is a single carbon atom. Whereas compounds containing the dinuclear fragment $C(AuL)_2$ have been known for some time (and hence will be described first), species with more highly aurated carbon atoms have become accessible only recently. These polyauriomethanes comprise a new class of compounds with highly unusual properties, in large measure due to the presence of an increased number of intramolecular gold–gold bonding interactions.

Historically, the first compounds to be characterized in which a single carbon atom acts as a bridging ligand between two gold(I) units are the tolyl and ferrocenyl derivatives $[Tol(AuPPh_3)_2][BF_4]$ and $[\{(C_5H_5)Fe(C_5H_4)\}(AuPPh_3)_2][BF_4]$, respectively, prepared as shown in Equation (27).[168]

$$2 [Tol(AuPPh_3)] + H[BF_4] \longrightarrow [Tol(AuPPh_3)_2][BF_4] + PhMe \quad (27)$$

Protonation of the aryl moiety in the starting material liberates the reactive species $[Au(PPh_3)]^+$, which subsequently attacks a second molecule of the arylgold(I) complex in an electrophilic fashion. Structural studies of the resulting complexes gave the unexpected result that in either case two $[Au(PPh_3)]^+$ units share a carbon atom of the aromatic ring as a common bonding partner, each gold atom being linearly two-coordinate. The bonding arrangement has been described as a three-centre, two-electron bond, but resonance structures in which the aromaticity of the ring is lost and the positive charge delocalized over the ring are also conceivable. The short gold–gold distance (0.277 nm in the ferrocenyl complex) and the correspondingly small Au–C–Au bond angle of 78° clearly show a bonding interaction between the gold centres.

Similar complexes have subsequently been prepared. One example is the tolyl complex $[Tol\{Au_2(dppb)\}][BF_4]$, in which the two independent PPh_3 groups have been replaced by the bidentate chelating ligand 1,4-bis(diphenylphosphino)butane (dppb). This complex has been studied by ^{13}C NMR as well as Mössbauer spectroscopy, and the evidence obtained shows both triphenylphosphinegold(I) units to be in an identical coordination environment.[169] The reaction leading to such species is obviously not affected by the accumulation of positive charges within the same molecule, as indicated by the facile preparation of stable $[\{(C_5H_4(AuPPh_3)_2)\}Fe\{C_5H_4(AuPPh_3)_2\}][BF_4]_2$ (Equation (28)).[81] The syntheses of the unusual trinuclear complexes $[Ph_4C_5(AuPPh_3)_3][BF_4]$[155,163] and $[(CN)_2C(AuPPh_3)_3][BF_4]$[82] along a similar route provide another case in point.

$$5\,[AuCl(CO)] + 2\,[Ph_3PAu]^+[BF_4]^- \longrightarrow \left[\right]^{2+} [BF_4]^-_2 \quad (28)$$

There has also been a report of dinuclear gold(I) alkynyl complexes, $[R-C\equiv C(AuPPh_3)_2][BF_4]$ (R = Ph or Pr^i), which are prepared in the same way. Infrared spectroscopic data indicate the presence of a bridging carbon atom rather than direct interaction of the second $[Au(PPh_3)]^+$ cation with the triple bond.[170] The distances between the gold atoms in a number of these compounds have been studied by means of the radial distribution function method, which allows the direct determination of such parameters in amorphous or polycrystalline samples.[171,172]

A related group of compounds has aryl groups bridging gold atoms or gold and a second metal, thus giving cyclic oligomers. The tetrameric nature of $[Au(C_6F_3H_2)]_4$ (prepared by the interaction of the diarylaurate with $H[PF_6]$)[173] has been inferred from molecular-weight determinations, while the structure of pentameric mesitylgold(I) has been confirmed by x-ray diffraction.[174,175] A beautiful star-shaped molecule is found, with a 10-membered ring of alternating carbon and gold(I) atoms, the latter showing significant deviations from the expected linear coordination (C–Au–C bond angles 148–153°). The gold centres approach each other so closely (0.269–0.271 nm) that the compound can be considered a pentanuclear gold(I) cluster. The values for the Au–C–Au bond angles are correspondingly low (76–78°). The synthesis of this compound is detailed in Equation (29), and the structure is depicted in Figure 4. $[Au(C_6H_2Me_3-2,4,6)]_5$ will react with mono- and bidentate phosphine ligands to give the corresponding mesitylgold(I) phosphine complexes.

$$5\,[AuCl(CO)] + 5\,[MgBr(C_6H_2Me_3-2,4,6)] \xrightarrow[-\,CO,\,-MgBrCl]{} [Au(C_6H_2Me-2,4,6)]_5 \quad (29)$$

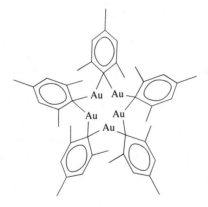

Figure 4 The structure of $[Au(C_6H_2Me_3-2,4,6)]_5$.

There exists a certain degree of analogy between the coordination behaviour of monovalent gold and alkali metals such as lithium, and the complex $[Au_2Li_2(C_6H_4\{CH_2NMe_2-2\})_4]$ is an example in point. Its x-ray structure[176] shows both gold and lithium atoms in a square-planar cyclic arrangement, with asymmetric aryl bridges between the metal atoms. The asymmetry is induced by the dimethylamino groups which participate in the overall stabilization of the structure by forming dative bonds to the lithium atoms. Again, the coordination environment of gold(I) deviates significantly from linearity, with the C–Au–C angles having values of 167°. A detailed description of the chemistry of this and similar complexes containing copper and zinc was presented in *COMC-I*.

Complexes in which an aliphatic carbon atom acts as a bridging ligand between two gold(I) units are now also well represented.[177] Several can be considered to be derived from the simplest hydrocarbon (methane) by formal substitution of two hydrogen atoms with isolobal phosphinegold(I) units. A few examples of these diaurio(I)methanes and related complexes are shown in Figure 5, and Table 6 provides a more comprehensive listing. $(NC)_2C(AuPPh_3)_2$ can be prepared by auration of C–H-acidic maleodinitrile with tris(triphenylphosphinegold(I))oxonium tetrafluoroborate $[(Ph_3PAu)_3O][BF_4]$.[82] In a similar way, the methylene group of *N*-alkylated barbituric acid can be doubly aurated by reaction with sodium methoxide/methanol in the presence of $[AuCl(PPh_3)]$,[178,179] and the bis(ylide) derivative

[(Me$_3$P)$_2$C(AuMe)$_2$] has been prepared by reaction of hexamethylcarbodiphosphorane with two equivalents of methyl(trimethylphosphine)gold(I).[180]

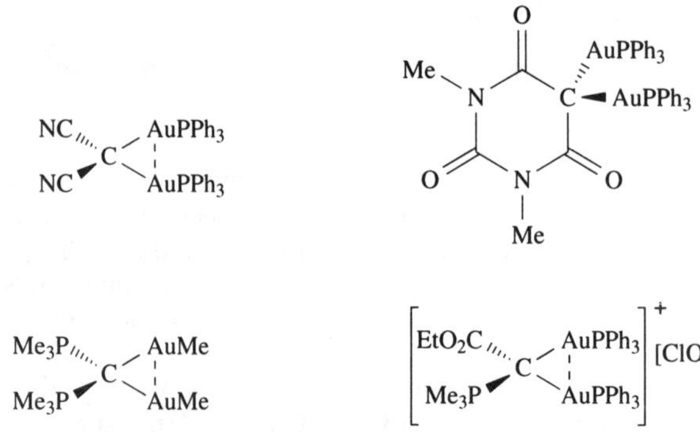

Figure 5   The structures of some diaurio(I)methanes and related complexes.

Table 6   Compounds containing the structural building block C(AuL)$_2$.

Complex	d(Au···Au) (nm)	Ref.
[(Ph$_3$PAu)$_2$(μ-C{CN}$_2$)]	0.291	82
[(Ph$_3$PAu)$_2$(μ-C·CO·NMe·CO·NMe·CO)]		178
[(Ph$_3$PAu)$_2$(μ-C·CMe·N·NPh·CO)]		100,184
[(MeAu)$_2${μ-C(PMe$_3$)$_2$}]		180
[(Ph$_3$PAu)$_2${μ-C(PMePh$_2$)$_2$}][Br]$_2$		185
[(Ph$_3$PAu)$_2${μ-C(PPh$_3$)(CO·NMe$_2$)}]	0.294	182
[(Ph$_3$PAu)$_2${μ-C·B(NPr$^i_2$)·C(TMS)$_2$·B(NPr$^i_2$)}]		186
[(Ph$_3$PAu)$_2$(μ-C·PPh$_2$·{Fe(CNPh)$_3$Cl}·PPh$_2$)]$^+$	0.289	187
[{(PhMe$_2$PAu)$_2$(μ-C(PPh$_3$))}$_2$C=O]	0.295–0.334	188
[(Ph$_3$PAu)$_2$(μ-C(CO$_2$Et))·PPh$_2$·CH(CO$_2$Et)(AuPPh$_3$)]$^+$		189

A number of other diaurated ylide derivatives have been obtained from the interaction of functionalized phosphonium salts such as [Ph$_3$PCH$_2$(C$_5$H$_4$N-2)][ClO$_4$] with two equivalents of triphenylphosphinegold(I) acetylacetonate, which induce double deprotonation and subsequent auration of the methylene group.[181-3] The resulting complexes are particularly interesting in that addition of a silver salt leads to the formation of a triangular Au$_2$Ag arrangement of metal atoms, which is held together by bonding interactions (ca. 0.3 nm) between the gold and silver centres. Furthermore, the anion-coordinated silver cation also coordinates to the pyridyl residue, thereby inducing a change in conformation with respect to the diaurated starting material, as has been verified by x-ray crystallographic studies. The overall process is depicted in Scheme 2. Exchange of coordinated silver nitrate for a third [Au(PPh$_3$)]$^+$ to form a homonuclear species is also possible.

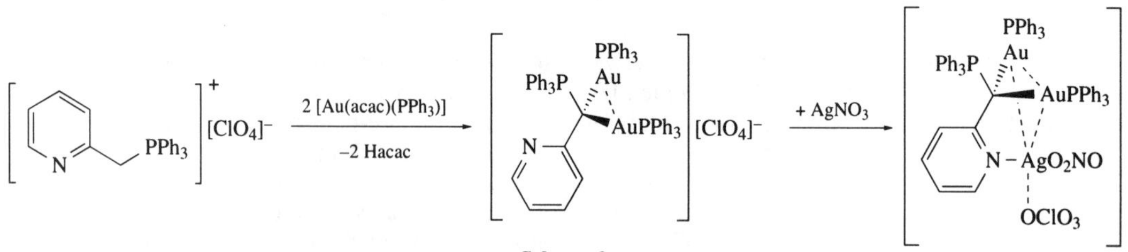

Scheme 2

As is evident from Table 6, there are numerous examples of dinuclear organogold(I) compounds in which a single carbon atom acts as a bridging ligand between the two metal centres, thereby enabling the gold atoms to enter into an intramolecular bonding relationship. The scene is now set for a description of polynuclear complexes in which the degree of auration at carbon is three or higher. The preparation of such compounds has proved surprisingly facile, due to the increasing number of gold–gold bonding interactions. Peripheral Au···Au bonding not only favours the formation of these complexes, but is the underlying reason why in some cases the coordination number of carbon can be increased beyond four, culminating in a series of most unusual, stable, hypercoordinate polyauriomethanium cations. In the following, the order of presentation is by increasing nuclearity.

The first specific synthesis of a triauriomethane to be reported was that of the trinuclear compound $[Me_3PC(AuPPh_3)_3]Cl$. It is formed in good yield by reaction of the silylated ylide $Me_3P=CH(TMS)$ with $[AuCl(PPh_3)]$ in the presence of caesium fluoride, as shown in Equation (30).[185] The colourless reaction product is of remarkable stability, and does not deteriorate upon exposure to air, light or water. The structural equivalence of the three $[Au(PPh_3)]$ units, as inferred from the Mössbauer spectrum, is corroborated by crystallographic analysis. The three gold atoms are about equidistant from the common carbon atom, and gold–gold bonding interactions are implied by the close distance between the metal atoms, with an average value of 0.318 nm. The solid-state structure is also remarkable in that the cations show no signs of oligomerization via Au···Au contacts, which is observed for other triaurioelement compounds of the type $[E(AuL)_3]^+[X]^-$ (E = O, S or Se).[190–2]

Another triauriomethane derivative is a particularly illustrative example of the striking tendency of gold(I) units to cluster around a central carbon atom, in so far as it represents a hypercoordinate carbon species. Such species are clearly stabilized through peripheral gold–gold bonding and represent an unusual, but nevertheless firmly established, class of compounds. Bis(trimethylsilyl)methyl triphenylphosphinegold(I) reacts with the oxonium salt $[(Ph_3PAu)_3O][BF_4]$ to give a mixture of two products (Equation (31)).[193]

$$(TMS)_2CHAuPPh_3 \xrightarrow{[(Ph_3PAu)_3O][BF_4]} (TMS)_2C(AuPPh_3)_2 + [(TMS)_2C(AuPPh_3)_3][BF_4] \qquad (31)$$

One product has been identified as a disilyl diauriomethane, while the major product is found to be the stable trinuclear complex $[(Ph_3PAu)_3\{\mu\text{-}C(TMS)_2\}][BF_4]$. Clearly, this compound forms from the diauriomethane precursor by incorporation of yet another $[Au(PPh_3)]^+$ unit, even though the result is a positively charged species with a hypercoordinate carbon atom. The x-ray structure analysis shows carbon to be present in a highly distorted trigonal-bipyramidal coordination environment, with the two silyl groups in equatorial positions and the three $[Au(PPh_3)]$ units in the two axial and the remaining equatorial positions. The axis Au_{ax}–C–Au_{ax} (152°) deviates strongly from linearity, with the two axial gold atoms tilted towards the equatorial atom, leading to an arrangement in which two remarkably close gold–gold contacts (0.272 nm) are observed. The Au–C and Si–C bonds are within the range established for such distances, although an interesting variation is observed for the Si–C bonds. The Si–C(centre) bonds are significantly longer than the Si–C bonds within the trimethylsilyl groups, probably reflecting the different hybridization states of the tetra- and pentacoordinate carbon atoms involved.

A different triauriomethane derivative is an important intermediate in an unusual transformation which provides access to gold–carbon compounds of even higher nuclearity, containing the structural fragment $[RC(AuL)_4]^+$. The overall reaction is depicted in Scheme 3.[194] The synthesis starts from 2,4,4-trimethyl-4,5-dihydrooxazole, a compound with marked C–H-acidity in the exocyclic methyl group. Triple lithiation followed by derivatization with chlorotrimethylsilane affords a triply silylated, open-chain ketenimine which reacts with $[AuCl(PPh_3)]$ in the presence of caesium fluoride to give a dimeric, octanuclear gold(I) cluster in close to quantitative yield. The reaction proceeds beyond the stage of a triauriomethyl oxazolinyl derivative which can only be observed as a side product. The crystal structure of the dicationic cluster consists of a C_2 symmetric arrangement of two such triauriomethyl oxazolinyl moieties, linked via two gold(I) atoms (cf. Scheme 3). The two pentacoordinate carbon atoms are coordinated to the oxazolinyl residue as well as to four gold(I) units. The compound can therefore be considered to contain two hypercoordinate tetraauriomethane fragments sharing a common gold(I) unit.

Hypercoordinate, tetraaurated carbon centres in which all gold(I) atoms carry a phosphine ligand as a second bonding partner have been realized in ethane-derived complexes of the type $[Me–C(AuPR_3)_4]^+[BF_4]^-$.[195] The common starting material for this series of compounds is 1,1,1-tris(dimethoxyboryl)ethane, which reacts with complex phosphinegold(I) halides in the presence of caesium fluoride to give the pentacoordinate organogold(I) clusters as air-stable substances (Equation (32)). Complexes containing four monodentate phosphine ligands PR_3 (R = Ph or C_6H_{11}) or two intramolecularly bridging bidentate phosphine ligands $1,2\text{-}(dppb)C_6H_4$ have been prepared.

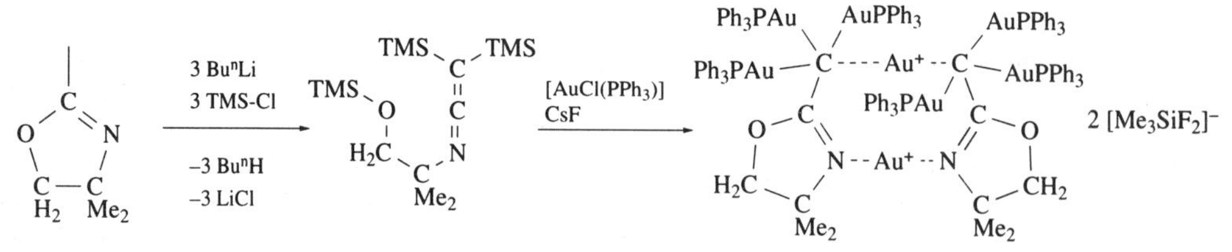

Scheme 3

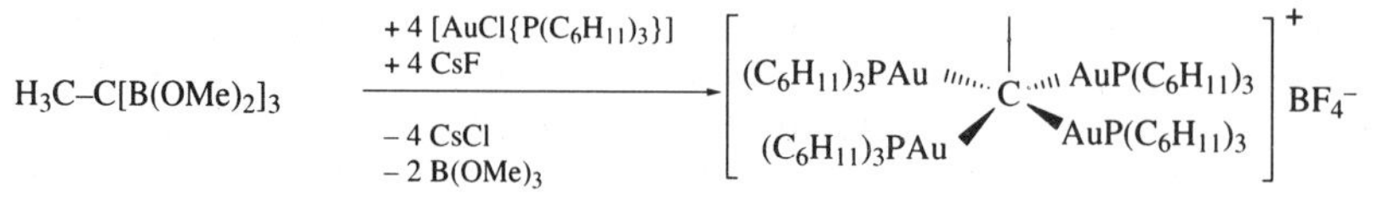

$$H_3C-C[B(OMe)_2]_3 \xrightarrow[\substack{-4\,CsCl \\ -2\,B(OMe)_3}]{\substack{+4\,[AuCl\{P(C_6H_{11})_3\}] \\ +4\,CsF}} \left[\begin{array}{c} (C_6H_{11})_3PAu\,\cdots\,C\,\cdots\,AuP(C_6H_{11})_3 \\ (C_6H_{11})_3PAu \qquad AuP(C_6H_{11})_3 \end{array} \right]^+ BF_4^- \qquad (32)$$

The equivalence of the phosphinegold units has been established in all cases by the presence of only one singlet resonance in the ^{31}P NMR spectra. Likewise, the proton NMR spectra are characterized by symmetrical 1:4:6:4:1 quintets for the methyl hydrogen atoms, with $^4J(HP)$ coupling constants of ca. 5 Hz. An x-ray study has revealed a slightly distorted tetragonal-pyramidal structure for the cation in the cyclohexylphosphine derivative, with the [Me–C] fragment capping a square of phosphinegold(I) ligands. Again, bonding interactions between neighbouring gold atoms (average $d(Au–Au) = 0.28$ nm) are responsible for the stabilization of the system. By contrast, the carbon–carbon bond distance in the ethane-derived organic fragment is unexceptional, and corresponds to a standard C–C single bond.

In attempts to synthesize homoleptic polyauriomethanes $C(AuPR_3)_4$, the preparation has only been successful when phosphine ligands of sufficient steric bulk were employed. Tetrakis[(tricyclohexylphosphine)gold(I)]methane is obtained from a mixture of $C[B(OMe)_2]_4$, $[AuCl\{P(c-C_6H_{11})_3\}]$ and CsF as a colourless crystalline solid with the expected tetrahedral array of gold atoms around the central carbon atom.[196] The ^{13}C resonance of the central atom has been detected in a ^{13}C-enriched sample as a quintet signal (intensity ratio 1:4:6:4:1) at $\delta = 99.3$ ppm ($^2J(PC) = 61$ Hz).

The presence of sterically less demanding phosphine ligands L (such as PPh_3 or $P(Tol-4)_3$) on gold(I) changes the outcome of the reactions dramatically (Equation (33)).[197] The reactions now proceed beyond the stage of tetraauration at carbon, and homoleptic pentaaurated cations $[C(AuPR_3)_5]^+$ are produced in high yield. The isolated tetrafluoroborate salts are colourless, diamagnetic, quite stable solid materials. Whereas NMR experiments indicate the equivalence of the phosphinegold(I) units in solution even at low temperature, the x-ray crystallographic analysis (for L = PPh_3) reveals a solid-state structure for the cation in which a trigonal-bipyramidal array of gold atoms is centred by a pentacoordinate interstitial carbon atom. The framework is characterized by close axial–equatorial Au···Au contacts of 0.29–0.30 nm, whereas the distances between the gold atoms in the equatorial plane are necessarily longer (0.36 nm) and thus do not fall in the range of bonding interactions.

$$H_2C[B(OMe)_2]_2 \xrightarrow[\substack{[P(O)(NMe_2)_3]}]{[AuCl(PPh_3)],\ CsF} [C(AuPPh_3)_5]^+[BF_4]^- \qquad (33)$$

These comparatively long distances do, however, indicate that there is still room in the structures of the pentacoordinate species to accommodate yet another $[LAu]^+$ unit, and six-fold auration of carbon is indeed observed under suitable conditions.[198] The dications $[C(AuPR_3)_6]^{2+}$ with a large variety of phosphine ligands have been characterized (R = Et, Pr^i, Ph, Tol-4 or C_6H_4X-4, with X = Cl, Br, NMe_2 or OMe), and the isolation of species containing tailor-made chelating phosphine ligands is equally straightforward.[199–201] The salts are stable, colourless, crystalline substances. They are accessible in high yields and have been found to be diamagnetic in all cases. A typical reaction is given in Equation (34).

$$C[B(OMe)_2]_4 + 6\,[AuCl(PPh_3)] + 6\,CsF \longrightarrow [C(AuPPh_3)_6]^{2+}\{[MeOBF_3]^-\}_2 + 6\,CsCl + 2\,B(OMe)_3 \qquad (34)$$

The dication of the triphenylphosphine derivative has been shown by x-ray structure analysis to consist of an only slightly distorted octahedron of gold atoms, with a carbon atom occupying the crystallographic inversion centre. The gold–carbon distances are unexceptional (0.212 nm), and an average value of 0.30 nm is found for the 12 Au···Au contacts in the cluster. Similar structures have also been determined for other derivatives. The equivalence of the phosphinegold(I) units in solution is indicated by the presence of a singlet in the ^{31}P NMR spectra (at both ambient and low temperature), as well as by the observation of a septet for the interstitial carbon atom in the ^{13}C NMR spectra of isotopically enriched material (e.g., R = Pr^i, $\delta = 154.6$ ppm, $^2J(PC) = 52$ Hz). Conventional as well as

[252]Cf plasma desorption mass spectrometric studies have also been reported.[202] The experimental evidence obtained for these carbon-centred hexanuclear gold(I) clusters sheds new light on the octahedral cluster $[\{AuP(Tol-4)_3\}_6]^{2+}$, whose crystal structure was reported in 1972.[203,204] Attempts to reproduce its synthesis have failed, and theoretical work predicted the structure of $[(AuL)_6]^{2+}$ aggregates to be that of an edge-sharing double tetrahedron,[205] as has been verified in separate experiments.[206,207] The original x-ray data for the hexanuclear tolylphosphinegold(I) cluster show considerable residual electron density at the centre of the gold octahedron, and the otherwise inconsistent structure can be rationalized if a central carbon atom is assumed to be present in this cluster as well.

The synthesis of polyauriomethanes and hypercoordinate carbon–gold clusters has led to a resurgence of interest in the theoretical description of these and related systems. A certain analogy between the coordination behaviour of monovalent gold and that of alkali metals such as lithium towards carbon was noted.[198] *Ab initio* calculations have predicted species such as $[CLi_5]^+$ and $[CLi_6]^{2+}$ to be energetically favourable, and the former has been detected in mass spectrometric studies.[208] On a similar note, given the isolobal relationship between $[LAu]^+$, Li^+ and H^+,[55] the analogy between the cations $[C(AuL)_5]^+$ and $[C(AuL)_6]^{2+}$, on the one hand, and species such as CH_5^+ and (as yet hypothetical) CH_6^{2+}, on the other, is obvious.[209]

It is particularly noteworthy that the existence of stable, carbon-centred octahedral gold(I) clusters was predicted as early as 1976. On the basis of molecular orbital calculations, it was argued that the octahedral cluster $[\{AuP(Tol-4)_3\}_6]^{2+}$ as such should be unstable, and that a much more stable ion would result if the gold cluster contained an interstitial carbido ligand.[210] More specific theoretical studies have been undertaken, and molecular orbital analyses of gold cluster compounds containing interstitial main-group atoms[30,60,211] or an interstitial dicarbido fragment[212] have been carried out. While relativistic effects have been confirmed to play a dominant role in the bonding of such systems (see Section 1.1), a qualitative picture of the bonding situation can be obtained from simple molecular orbital interaction diagrams. One such diagram (for the carbon-centred octahedral cluster $[CAu_6]^{2+}$) is presented in Figure 6.[30] Only gold orbitals pertinent to the bonding with the central carbon atom are shown. Under the local symmetry O_h, the contributions of four valence electrons each from the cluster $[Au_6]^{2+}$ and from the central carbon atom result in the formation of four bonding states (a_{1g} and t_{1u}). The diagram thus offers an initial explanation for the stability of the cluster as well as its diamagnetic nature.

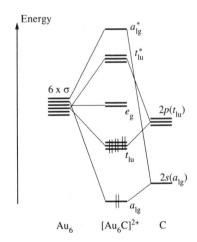

Figure 6 Schematic orbital-interaction diagram for the carbon-centred cluster $[CAu_6]^{2+}$ (reproduced by permission of the American Chemical Society from *Inorg. Chem.*, 1991, **30**, 3986).

1.2.3 Gold(I) Complexes with Two Gold–Carbon Bonds

Simple dialkylaurates(I) are still rarely encountered in the literature. They are best prepared by the interaction of an alkylgold(I) phosphine adduct with an organolithium reagent.[71] This method has also been used for the preparation of complexes containing functionalized alkyl groups, as shown in Equation (35).[74] Treatment of the product of Equation (35) with potassium carbonate affords the analogous potassium bis(diethylcarbamoylcyclopropyl)aurate, which possesses a dimeric structure in the solid state. Two gold and two potassium atoms are arranged in the shape of a distorted tetrahedron ($d(Au\cdots Au) = 0.360$ nm), with the organic ligands bridging the cluster edges. The structure thus resembles that of those aryl aurates in which an aryl group functions as a bridging ligand between gold and lithium, as discussed in Section 1.2.2. The carbamoyl derivatives are remarkably stable towards oxygen and moisture, quite unlike their unfunctionalized dialkyl analogues.

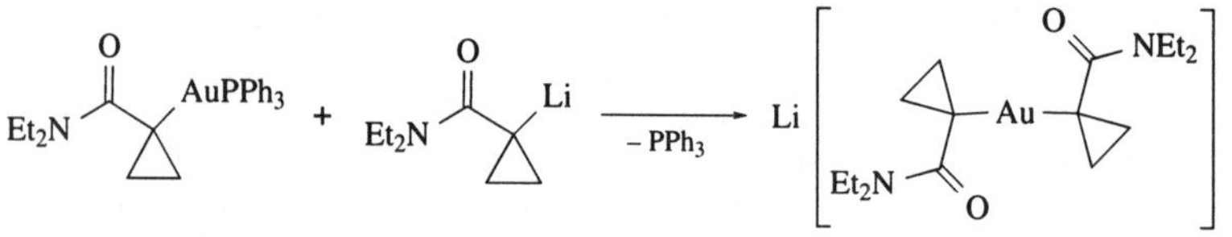

$$\text{Et}_2\text{N}\overset{O}{-}\text{C}\overset{}{-}\text{AuPPh}_3 \; + \; \text{Et}_2\text{N}\overset{O}{-}\text{C}\overset{}{-}\text{Li} \; \xrightarrow[-\text{PPh}_3]{} \; \text{Li}\left[\begin{array}{c} \text{Et}_2\text{N}\overset{O}{-}\text{C} \\ \text{CH}-\text{Au}-\text{CH} \\ \text{C}\overset{O}{-}\text{NEt}_2 \end{array} \right] \tag{35}$$

The stabilization of dialkylaurates(I) by means of functionalized alkyl groups appears to be a general phenomenon. Another example is provided by the complex [Ph$_3$P=N=PPh$_3$][Au(acac)$_2$] which, in addition, is a versatile precursor for other dialkylaurate derivatives and related compounds.[213] Some relevant reactions are summarized in Scheme 4. Similar products have also been obtained from complex gold(I) halides by reaction with carbanionic ligands such as bis(diphenylphosphino)methanide.[146]

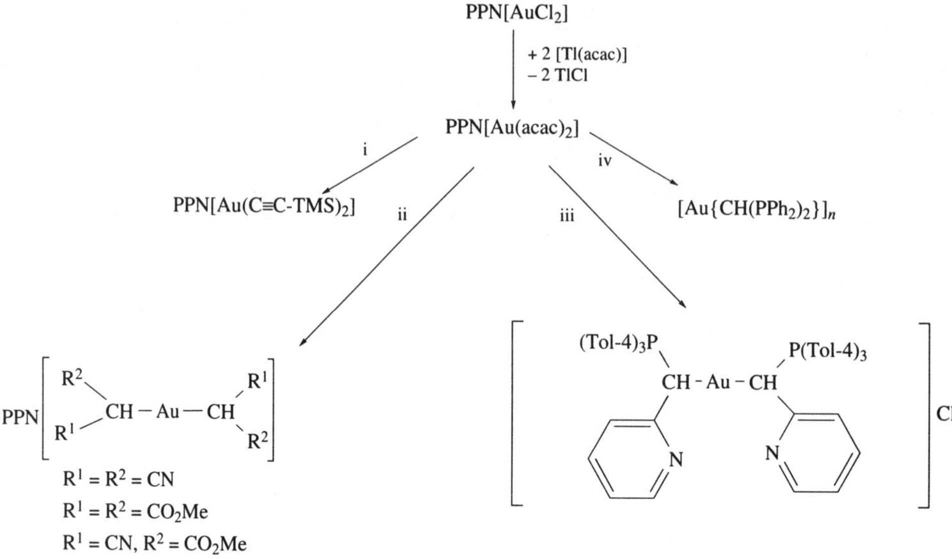

PPN[AuCl$_2$]

$+ 2$ [Tl(acac)]
$- 2$ TlCl

PPN[Au(acac)$_2$]

i → PPN[Au(C≡C-TMS)$_2$]

iv → [Au{CH(PPh$_2$)$_2$}]$_n$

ii / iii

$$\text{PPN}\left[\begin{array}{c} R^2 \qquad\quad R^1 \\ \diagdown\;\;\;\;\;\diagup \\ \text{CH}-\text{Au}-\text{CH} \\ \diagup\;\;\;\;\;\diagdown \\ R^1 \qquad\quad R^2 \end{array} \right]$$

$R^1 = R^2 = $ CN
$R^1 = R^2 = $ CO$_2$Me
$R^1 = $ CN, $R^2 = $ CO$_2$Me

$$\left[\begin{array}{c} (\text{Tol-4})_3\text{P} \qquad\qquad \text{P(Tol-4)}_3 \\ \diagdown\;\;\;\;\;\;\;\;\;\;\;\;\diagup \\ \text{CH}-\text{Au}-\text{CH} \\ \end{array} \right] \text{ClO}_4$$

i, HC≡C-TMS, −acacH; ii, CH$_2$R^{1}R^2, −acacH; iii, [(Tol-4)$_3$P(CH$_2$\{2-C$_5$H$_4$N\})]ClO$_4$, −acacH, −[PPN]ClO$_4$
iv, CH$_2$(PPh$_2$)$_2$, −acacH, −[PPN]acac

Scheme 4

A related type of compound is that of the dialkynylaurates(I), some of which are accessible from dialkylaurates by treatment with terminal alkynes (Scheme 4). Alternatively, the bis(phenylacetylide) complex PPN[Au(C≡CPh)$_2$] may be prepared by metathesis of the organocuprate PPN[Cu(C≡CPh)$_2$] with various gold(I) complexes such as [Au(C≡CPh)(PPh$_3$)], [AuCl(PPh$_3$)], [AuCl(C≡CPh)]$^-$ and [Au(C≡CPh)]$_n$.[214] A less straightforward synthesis of dialkynylaurates(I) starts from gold(I) iodide and potassium acetylide, requiring liquid ammonia as the reaction medium.[215] Although no structural data are to date (1994) available for these compounds, polymeric association of the linearly two-coordinate aurate anions in the solid state (via Au⋯Au interactions) may be assumed. Close metal–metal contacts are the dominating feature in the solid-state structures of a series of related (in most cases heterometallic) organogold compounds, prepared from the alkynylgold(I) complexes PPN[Au(C≡CPh)$_2$] or [Au(C≡CPh)(PPh$_3$)] by reaction with the coinage metal acetylides [Cu(C≡CR)]$_n$, [Ag(C≡CR)]$_n$ and [Au(C≡CR)]$_n$ (R = Ph or Tol-4).[216–22] Some of these clusters have been structurally characterized. They consist of an array of bis(alkynyl)aurate moieties into which the heterometallic fragments are incorporated via metal–metal and π-bonds to the acetylide residues. An example is given in Figure 7.[221]

Diarylaurate(I) complexes have been prepared by a variety of methods, depending on the nature of the aromatic ligands.[69,223] While diarylaurates as such are generally more stable than their unfunctionalized dialkyl analogues, this stability is particularly enhanced in the case of aryl groups bearing a number of strongly electron-withdrawing ligands. Polyhalogenophenylaurates(I) have been much studied recently, and their preparation and properties have been the subject of a review.[69] The most versatile method for the preparation of such compounds is the arylation of a complex gold(I) halide with two equivalents of aryllithium (Equation (36)).

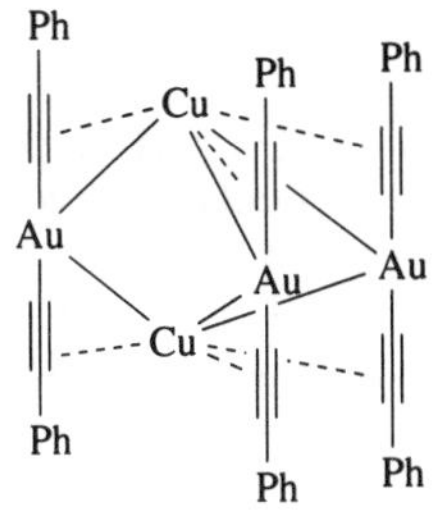

Figure 7 The structure of the pentanuclear cluster $[Au_3Cu_2(C\equiv CPh)_6]^-$.

$$[AuCl(SC_4H_8)] + 2\,Li(C_6H_2F_3\text{-}2,4,6) \xrightarrow{\text{Bu}_4\text{NCl}} Bu_4N[Au(C_6H_2F_3\text{-}2,4,6)_2] + 2\,LiCl + SC_4H_8 \qquad (36)$$

Similar to the bis(alkynyl)aurates(I) discussed earlier, the bis(polyhalogenophenyl)aurate(I) complexes obtained according to Equation (36) show a propensity for metal–metal aggregation when treated with heterometallic reagents such as $AgClO_4$. Upon addition of neutral donor ligands L, polymeric chains are formed which contain the repeat unit $R_2Au(\mu\text{-}AgL)_2AuR_2$, linked through Au$\cdots$Au interactions in the range 0.29–0.30 nm (Equation (37)). Interestingly, the complexes possess direct Au–Ag bonds unsupported by bridging ligands. The structure of polymeric $[AuAg(C_6F_5)_2(SC_4H_8)]_n$ is shown in Figure 8.[224,225]

$$Bu_4N[AuR_2] + AgClO_4 \xrightarrow[-Bu_4NClO_4]{L} 1/n\,[AuR_2AgL]_n \qquad (37)$$

$$R = Ph,\ C_6Cl_5,\ C_6F_3H_2$$
$$L = N,\ P,\ O,\ S\ donors,\ alkene,\ alkyne,\ arene$$

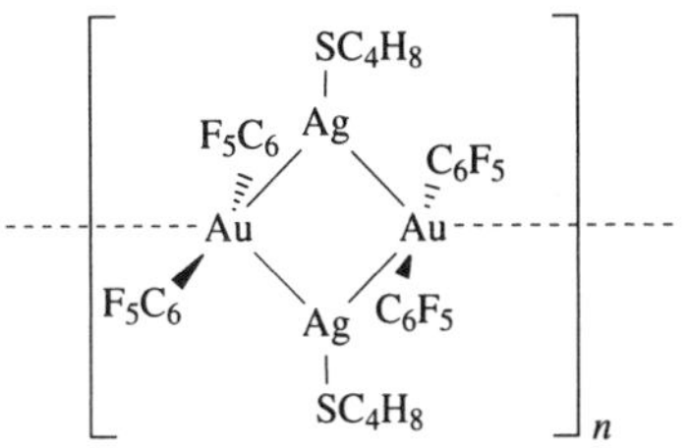

Figure 8 Structure of polymeric $[AuAg(C_6F_5)_2(SC_4H_8)]_n$; $d(Au–Ag) = 0.272$ nm, $d(Au\cdots Au) = 0.289$ nm.

Finally, there are a number of compounds in which an arylgold(I) moiety is bonded to a bis(diphenylphosphino)methanide group, which in turn acts as a bridging ligand in a polynuclear gold or gold–silver complex.[69] An example is shown in Scheme 5.[226,227]

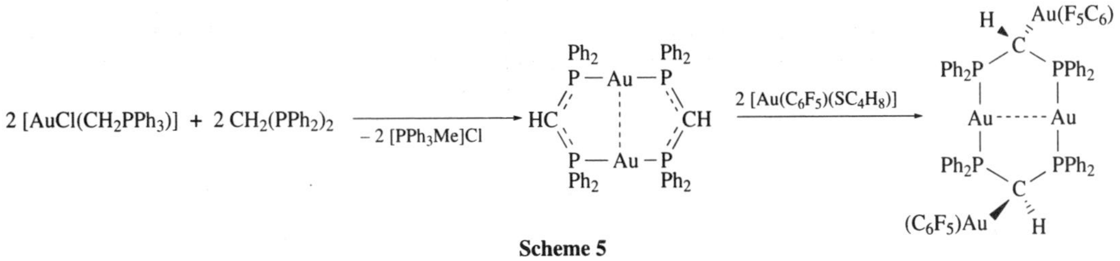

Scheme 5

1.2.4 Complexes of Gold(I), Gold(II) and Gold(III) with Ylide Ligands

There now exists a large number of methods for the synthesis of organogold complexes with various ylide ligands. Owing to the dipolar nature of ylides, such complexes are characterized by exceptionally strong metal–carbon bonds which often impart high thermal stability and also inertness with respect to oxygen and water.[228,229] The complexes maybe of a mono-, di- or polynuclear constitution, and the metal may be present as either mono-, di- or tervalent gold, but mixed-valent gold(I)–gold(III) compounds are also known. There is an equally extensive reaction chemistry, and successive oxidation of the gold centres ($Au^I \rightarrow Au^{II} \rightarrow Au^{III}$) can often be accomplished. Ylide complexes of gold have therefore been summarized in the following sections, regardless of oxidation state.

1.2.4.1 Mononuclear ylide complexes

Mononuclear ylide complexes of gold(I) and gold(III) have been prepared in a variety of ways,[228,229] as detailed below. Gold(I) complexes are particularly diverse, existing as cations or as uncharged molecules, with a neutral donor ligand, another ylide moiety, a halide or pseudohalide, a carbanionic group, or a carbonyl metalate as the second ligand. The largest number of complexes of both gold(I) and gold(III) has been prepared with phosphorus ylides, but species containing arsenic or sulfoxonium ylides have also been reported. Some prototypical examples are summarized in Table 7.

Table 7 Mononuclear ylide complexes of gold(I) and gold(III).

Complex	m.p. (°C)	Ref.
Gold(I)		
$[AuCl(CH_2PPh_2Me)]$	148 (d)	230
$[AuCl\{CH(PPh_3)C(=O)Me\}]$	154	231
$[AuCl\{CH(PPh_3)CO_2Me\}]$	180	232
$[AuCl(Ph_3P=C=PPh_3)]$	250 (d)	233
$[AuBr(CH_2PPh_3)]$	164 (d)	230
$[Au(CN)(CH_2PPh_3)]$	185–188	234
$[Au(C_6F_5)(CH_2PPh_3)]$	157	235
$[Au(C_6F_5)(CH_2AsPh_3)]$	144	235
$[Au(C{\equiv}CPh)(CH_2PPh_3)]$		236
$[Au\{Co(CO)_4\}(CH_2PPh_3)]$		236
$[Au(CH_2PPh_3)(SC_4H_8)]ClO_4$		236
$[Au(CH_2PPh_3)(SbPh_3)]ClO_4$		236
$[Au(CH_2PPh_3)(phen)]ClO_4$		236
$[Au(CH_2AsPh_3)(SC_4H_8)]ClO_4$		236
$[Au\{CH(PPh_3)C(=O)Me\}(PPh_3)]ClO_4$	125	231
$[Au\{CH(PPh_3)CO_2Me\}(PPh_3)]ClO_4$	97	232
$[Au(CH_2PMe_3)_2]BF_4$	214 (d)	230
$[Au(CH_2PMe_3)_2][Au^{III}(1,2-Me_2C_2B_9H_9)_2]$		237
$[Au\{CH(PPh_3)C(=O)Me\}_2]ClO_4$	128	231
$[Au\{CH(PPh_3)CO_2Me\}_2]ClO_4$	125	232
$[Au(CH_2S(=O)Me_2)_2]Cl$		238
$[Ph_3P=N=PPh_3][Au\{CH_2P(S)Ph_2\}_2]$		239
Gold(III)		
$[AuCl_3\{CH(PPh_3)C(=O)Me\}]$	120	231
$[AuCl_3\{CH(PPh_3)CO_2Me\}]$	161	232
$[AuCl_2\{CH(PPh_3)C(=O)Me\}_2]ClO_4$	130	231
$[AuCl_2\{CH(PPh_3)CO_2Me\}_2]ClO_4$	150	232
$[AuBr\{CH_2P(S)Ph_2\}_2]$		240
$[Au(S_2CNEt_2)_2\{CH_2P(S)Ph_2\}]$	156	240
$[AuI_3(CH_2PPh_3)]$	133 (d)	241
trans-$[AuI_2(CN)(CH_2PPh_3)]$	165–170	234
$[Au(SCN)_3(CH_2PPh_3)]$	140 (d)	241
$[AuMe_3(CH_2PPh_3)]$		242
$[AuMe_3\{CH_2S(=O)Me_2\}]$		242
$[AuMe_2\{\eta^2-CH_2PMe_2BH_2PMe_2CH_2\}]$		243
$[Au(C_6F_5)_3(CH_2PPh_3)]$	252	235
$[Au(C_6F_5)_3(CH_2PPh_2Me)]$		244
$[Au(C_6F_5)_3(CH_2AsPh_3)]$	231	235
cis-$[Au(C_6F_5)_2Cl(CH_2PPh_3)]$	214 (d)	245
trans-$[Au(C_6F_5)Cl_2(CH_2PPh_3)]$	135 (d)	241

The reaction of alkylgold(I) phosphine complexes with ylides leads to alkylgold(I) monoylide species with concomitant liberation of phosphine, as illustrated in a general form in Equation (38).[61,71]

$$[AuR^1(PR^4_3)] + R^3_3P=CR^2_2 \xrightarrow[-PR^4_3]{} [AuR^1(CR^2_2PR^3_3)]$$ (38)

$$R^1 = alkyl$$
$$R^2 = H, alkyl, aryl$$
$$R^3, R^4 = alkyl, aryl$$

Halogenoarylgold(I) ylide complexes have been generated from suitable precursors via an indirect route: the interaction of phosphonium halides with $[Au(C_6F_5)(SC_4H_8)]$ leads to phosphonium organo(halogeno)aurates which can subsequently be reacted with base to give the respective ylide derivatives (Scheme 6). Gold(III) ylide complexes of the type [Au(C$_6$F$_5$)$_3$(ylide)] have been prepared analogously from [Au(C$_6$F$_5$)$_3$(SC$_4$H$_8$)].[235]

$$[Au(C_6F_5)(SC_4H_8)] + [MePPh_3]Cl \xrightarrow[-SC_4H_8]{} [MePPh_3][AuCl(C_6F_5)] \xrightarrow[-H_2, -NaCl]{NaH} [Au(C_6F_5)(CH_2PPh_3)]$$

Scheme 6

The reaction of phosphinegold(I) halides with ylides of the type R$^1_2$C=PR$^2_3$ proceeds in two steps: initial displacement of halide gives an isolable monoylide species, which can be transformed into a bis(ylide)gold(I) complex if an excess of ylide is employed (Scheme 7).[61,71] In some instances, however, the reaction produces the bis(ylide) species directly, regardless of the relative ratios of gold(I) halide and ylide employed, as in the case of the bis(cyclopropylide) complex [Au{C(CH$_2$)$_2$PPh$_3$}$_2$]Cl.[246] Bis(ylide) complexes form particularly easily with a chelating bis(ylide) ligand, an example being the compound 1,3-propanediylbis[(9-fluorenylidene)diphenylphosphorane]gold(I) chloride (Figure 9).[247] Facile double substitution also occurs with amino-substituted phosphorus ylides and sulfoxonium ylides to give complexes such as [Au{CH$_2$P(NMe$_2$)$_3$}$_2$]Cl[248] and [Au{CH$_2$S(=O)Me$_2$}$_2$]Cl.[238] NMR[247,248] and x-ray photoelectron spectroscopic studies[238,249] of such bis(ylide)gold(I) complexes indicate the equivalence of the ylide moieties in solution and a trigonal arrangement of ylide and halide ligands in the solid state.

$$[AuCl(PR^3_3)] + R^2_3P=CR^1_2 \longrightarrow [AuCR^1_2PR^2_3(PR^3_3)]Cl \xrightarrow[-PR^3_3]{R^2_3P=CR^1_2} [Au(CR^1_2PR^2_3)_2]Cl$$

R^1 = H, alkyl, aryl

R^2, R^3 = alkyl, aryl

Scheme 7

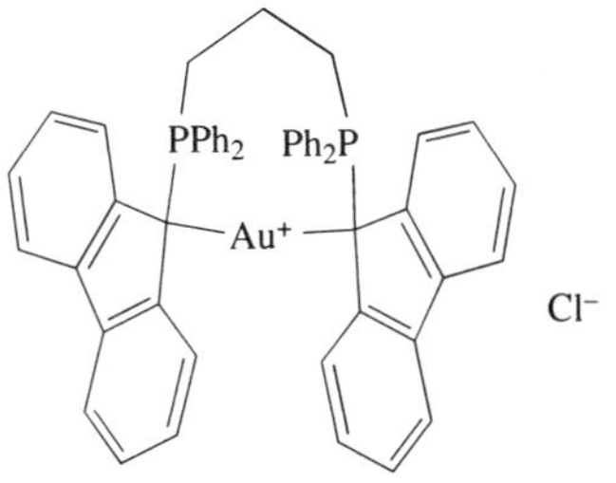

Figure 9 A chelated bis(ylide)gold(I) complex.

Carbonyl-stabilized phosphorus ylides are less nucleophilic and hence do not react with phosphinegold(I) halides, but their gold(I) complexes can be generated from precursors such as [Au(acac)PPh$_3$] or [AuCl(SC$_4$H$_8$] by reaction with phosphonium salts and ylides, respectively, and again both mono- and bis(ylide) complexes have been obtained (Equation (39) and Scheme 8).[231,232] Although the ylide carbon atoms in these complexes are asymmetric, no stereospecificity is to be expected in the coordination process, and racemic mixtures are formed throughout.

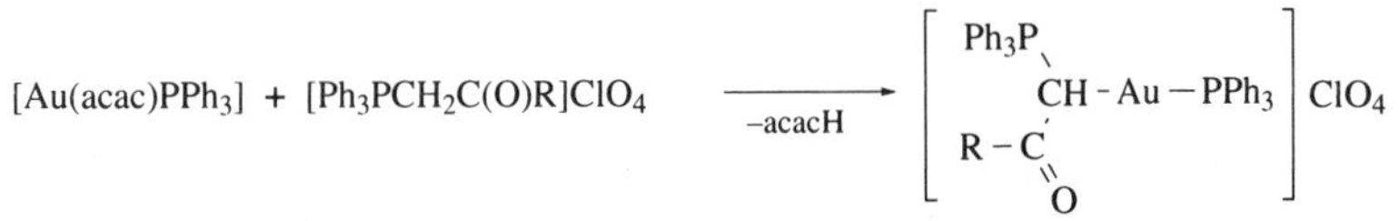

$$[Au(acac)PPh_3] + [Ph_3PCH_2C(O)R]ClO_4 \xrightarrow{-acacH} \begin{bmatrix} Ph_3P \\ \quad CH-Au-PPh_3 \\ R-C \\ \quad\quad O \end{bmatrix} ClO_4 \quad (39)$$

R = Me, Ph, OMe, OEt

$$[AuCl(SC_4H_8)] + Ph_3PCHC(O)R \longrightarrow \begin{matrix} PPh_3 \\ Cl-Au-CH \\ \quad\quad C-R \\ \quad\quad O \end{matrix} \xrightarrow[-NaCl]{+ Ph_3PCHC(O)R, + NaClO_4} \begin{bmatrix} O \\ R-C \quad\quad PPh_3 \\ \quad CH-Au-CH \\ Ph_3P \quad\quad C-R \\ \quad\quad\quad O \end{bmatrix} ClO_4$$

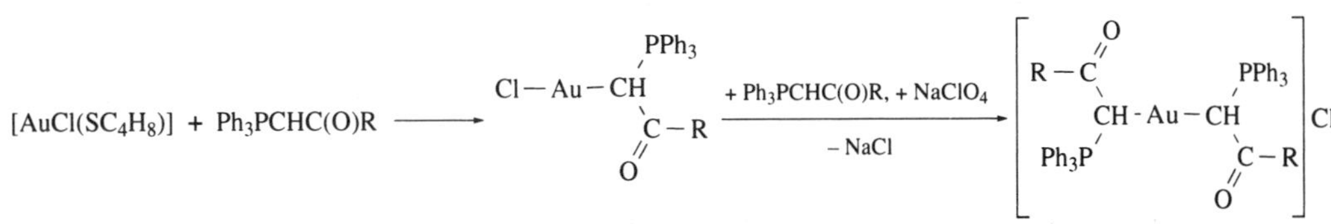

Scheme 8

R = Me, Ph, OMe, OEt

Gold

An alternative strategy for the preparation of mononuclear ylide complexes is to start from gold(I) precursors which already contain a ylide ligand. Displacement in such complexes of tetrahydrothiophene (SC_4H_8) by neutral or anionic ligands (including polyfunctional phosphines, acetylides and carbonyl metallates) leads to a variety of mono- and dinuclear compounds of remarkable stability (Scheme 9).[144,236]

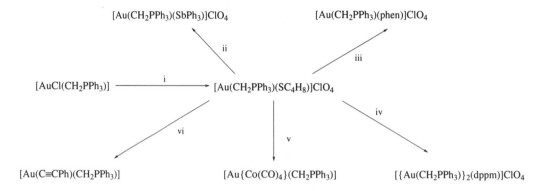

i, $[Ag(ClO_4)(SC_4H_8)]$; ii, $SbPh_3$; iii, phen, 1, 10-phenanthroline, iv, dppm; v, $[N(PPh_3)_2][Co(CO)_4]$; vi, $PhC\equiv CH$, KOH

Scheme 9

Interaction of gold(I) ylide dimers such as $[Au(CH_2)_2PPh_2]_2$[250,251] or gold(I) ylide monomers such as $[Au(C_6F_5)(CH_2PPh_3)]$[230] with HCl, HBr or acetyl bromide gives gold(I) mono(ylide) complexes in which halide is the second ligand, while acids with weakly coordinating anions X (BF_4^- or ClO_4^-) convert the halogenoarylgold(I) ylide complexes into bis(ylide) species of the type $[Au(ylide)_2]X$.[230]

Conversely, certain mono(ylide) complexes are accessible from the corresponding bis(ylide) compounds. Reaction between complexes of the type $[Au(ylide)_2]ClO_4$ and aurates(I) such as $NR_4[AuX_2]$ (X = Cl, Br, C_6F_5 or C_6Cl_5) affords neutral halogenogold(I) and (halogenoaryl)gold(I) ylide species, respectively (Equation (40)). Related gold(III) complexes have been prepared in a similar manner.[245]

$$[Au(CH_2PPh_3)_2]ClO_4 + Bu^n_4N[Au(C_6F_5)_2] \xrightarrow[-Bu^n_4NClO_4]{} 2[Au(C_6F_5)(CH_2PPh_3)] \qquad (40)$$

A few mono(ylide) complexes of gold(I) have been obtained from the interaction of phosphorus ylides with [Au(CO)Cl] and AuCN. The complex $[AuCl(Ph_3P=C=PPh_3)]$ forms readily when carbonylgold(I) chloride is reacted with the double ylide hexaphenylcarbodiphosphorane. It is stable to oxygen and water and also thermally stable up to 250 °C.[233] Similarly, the reaction of gold(I) cyanide with triphenylphosphoniomethanide yields the complex $[Au(CN)(CH_2PPh_3)]$ which, in addition to its high thermal stability, is characterized by its lack of reactivity towards hydrogen chloride. By contrast, HCl will readily cleave the Au–C bond in alkylgold(I) complexes. However, the complex does undergo oxidative addition reactions with halogens X_2 to give the gold(III) complexes *trans*-$[AuX_2(CN)(CH_2PPh_3)]$ (X = Cl, Br or I).[234]

A number of mononuclear gold(III) ylide complexes may be similarly prepared by oxidative addition of halogens to suitable gold(I) ylide precursors. This approach is exemplified by complexes of the general types $[AuX_3(ylide)]$,[241] *trans*-$[Au(C_6F_5)X_2(ylide)]$[241] and *trans*-$[AuX_2(ylide)_2]ClO_4$ (X = Cl or Br),[231,232] some of which may be further converted into the related iodo or thiocyanato complexes by reaction with potassium iodide or potassium thiocyanate.[241] Additionally, a few methods for the direct synthesis of gold(III) ylide complexes have been described. Thus, interaction of trimethylgold(III) phosphine complexes with phosphorus or sulfoxonium ylides results in displacement of phosphine to yield complexes such as $[AuMe_3(CH_2PPh_3)]$ and $[AuMe_3\{CH_2S(=O)Me_2\}]$, respectively, which again have remarkable stability.[242] A gold(III) complex containing a chelating bis(ylide) ligand is represented by the species $[AuMe_2\{\eta^2\text{-}CH_2PMe_2BH_2PMe_2CH_2\}]$, obtained from the reaction of dimethylgold(III) chloride with lithium boranato-bis(dimethylphosphonium methylide), as shown in Equation (41).[243]

The reaction chemistry of both mono- and bis(ylide) complexes of gold(I) has been explored in several instances. In the case of carbonyl-stabilized ylide ligands, the acidic proton on the α-methylene group can be replaced with a further gold(I) unit, which leads to polynuclear species. Some of these have solid-state structures characterized by close intramolecular Au⋯Au contacts of ~0.29 nm and

$$\underset{\text{Li}}{\overset{\text{H}_2}{\underset{\text{Me}_2\text{P}\diagdown\text{B}\diagup\text{PMe}_2}{}}} + 1/2\ \underset{\text{Me}}{\overset{\text{Me}}{}}\text{Au}\underset{\text{Cl}}{\overset{\text{Cl}}{}}\text{Au}\underset{\text{Me}}{\overset{\text{Me}}{}} \xrightarrow{-\text{LiCl}} \underset{\text{Au}(\text{Me})(\text{Me})}{\overset{\text{H}_2\text{B}}{\underset{}{\text{Me}_2\text{P}\diagup\diagdown\text{PMe}_2}}} \qquad (41)$$

belong to the family of polyauriomethane complexes discussed previously (Equations (42) and (43)).[231,232]

$$\left[\underset{\text{MeO}_2\text{C}}{\overset{\text{Ph}_3\text{P}}{}}\text{CH}-\text{AuPPh}_3\right]\text{ClO}_4 \xrightarrow[-\text{acacH}]{+ [\text{Au}(\text{acac})\text{PPh}_3]} \left[\underset{\text{MeO}_2\text{C}}{\overset{\text{Ph}_3\text{P}}{}}\text{C}\underset{\text{Au}\,\text{PPh}_3}{\overset{\text{PPh}_3\,\text{Au}}{}}\right]\text{ClO}_4 \qquad (42)$$

$$\left[\underset{\text{Ph}_3\text{P}}{\overset{\text{EtO}_2\text{C}}{}}\text{CH}-\text{Au}-\text{CH}\underset{\text{CO}_2\text{Et}}{\overset{\text{PPh}_3}{}}\right]\text{ClO}_4 \xrightarrow[-\,2\ \text{acacH}]{+\,2\ [\text{Au}(\text{acac})\text{PPh}_3]} \left[\underset{\text{Ph}_3\text{PAu}}{\overset{\text{EtO}_2\text{C}}{}}\text{Ph}_3\text{P}-\text{C}-\text{Au}-\text{C}\underset{\text{AuPPh}_3}{\overset{\text{PPh}_3}{}}\text{CO}_2\text{Et}\right]\text{ClO}_4 \qquad (43)$$

The halogenoarylgold(I) ylide complexes have also been investigated in their reactivity towards chlorobis(pentafluorophenyl)thallium(III), and partial oxidation of gold(I) to give chloro-substituted halogenoarylgold(III) ylide complexes is observed, together with the formation of bis(ylide) complexes of monovalent gold (Equation (44)).[252]

$$[\text{Tl}(\text{C}_6\text{F}_5)_2\text{Cl}] + [\text{Au}(\text{C}_6\text{F}_5)(\text{CH}_2\text{PPh}_3)] \longrightarrow \textit{trans}\text{-}[\text{Au}(\text{C}_6\text{F}_5)_2\text{Cl}(\text{CH}_2\text{PPh}_3)] + [\text{Au}(\text{CH}_2\text{PPh}_3)_2][\text{Au}(\text{C}_6\text{F}_5)_4] + \qquad (44)$$
$$\text{TlCl} + [\text{Tl}(\text{C}_6\text{F}_5)_3]$$

Of particular interest are bimetallic gold–silver complexes which may be prepared from gold(I)bis(ylide) species such as $[\text{Au}(\text{CH}_2\text{PPh}_3)_2]\text{ClO}_4$ and AgClO_4 (leading to neutral clusters, Equation (45)) or $[\text{Ag}(\text{ClO}_4)\text{PPh}_3]$ (leading to dicationic species).[253] The complexes contain Au_2Ag_2 rings in which the Au–Ag bonds are unsupported by bridging ligands. In contrast with the polymeric gold–silver clusters containing bis(halogenoaryl)gold(I) units, no Au⋯Au interactions are observed in this case, most likely because of the steric demands of the PPh_3 groups in the ylide ligands. The complexes are air and moisture stable in the solid state, but decompose in solution to regenerate the starting materials. The x-ray structure of one of the neutral complexes has been established, which is characterized by an Au–Ag bond distance of 0.278 nm.

$$2\ [\text{Au}(\text{CH}_2\text{PPh}_3)_2]\text{ClO}_4 + 2\ \text{AgClO}_4 \longrightarrow [(\text{Ph}_3\text{PCH}_2)_2\text{Au}\{\mu\text{-Ag}(\text{OClO}_3)_2\}_2\text{Au}(\text{CH}_2\text{PPh}_3)_2] \qquad (45)$$

1.2.4.2 Di- and polynuclear ylide complexes

In addition to the mononuclear gold ylide complexes described earlier, a large number of dinuclear compounds has been prepared in which the ylide moiety functions as a bridging ligand. Complexes with two bridging ylide groups and two gold atoms in a metallacyclic array are by far the most numerous, but variants containing one ylide and one diphosphine bridge as well as open-chain compounds with only one bridging ylide ligand have also been reported (structures (4)–(11)). All complexes are remarkably inert with respect to hydrolysis and oxidative or thermal decomposition, representing some of the most stable organometallic compounds of gold.[177,188,189,228,229,243,254-6]

Three general methods for the preparation of such complexes have been established, their applicability depending on the nature of the ylide ligand. Complexes such as (4)–(6) are accessible by transylidation of the corresponding mononuclear bis(ylide) compounds (Equation (46)), while species of type (7) and (8) have been prepared from complex gold chlorides by reaction with the lithium salt of the appropriate ligand. The third method employs symmetric diphosphine precursors, from which the asymmetric complexes (9) and (10) can be made by reaction with one equivalent of ylide ligand. A less general route has been used for the preparation of (11), starting from the appropriate diphosphonium salt and [AuCl(ylide)]. All complexes are similar, in that the metal atoms are held in close proximity by the

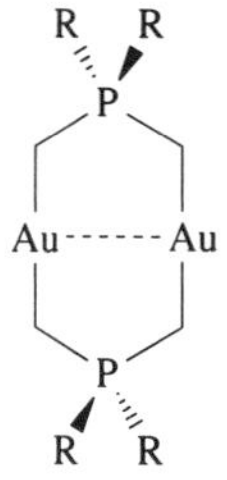

(4) R = alkyl
(5) R = phenyl

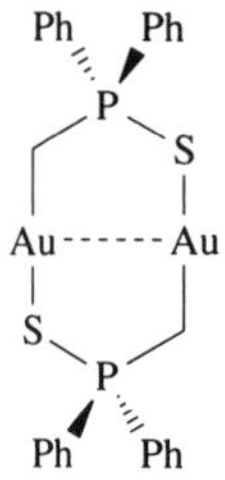

(6)

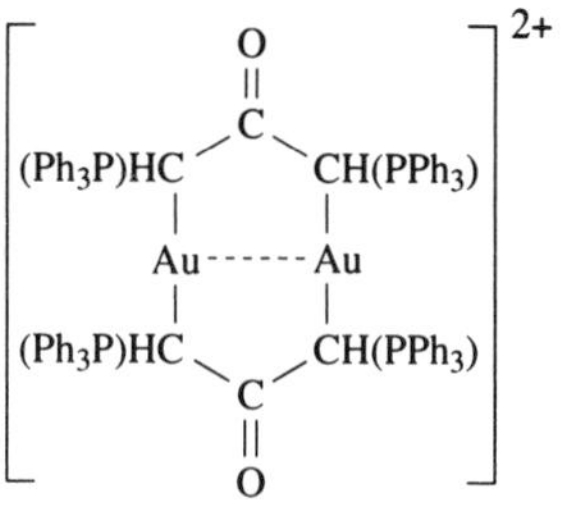

(7)

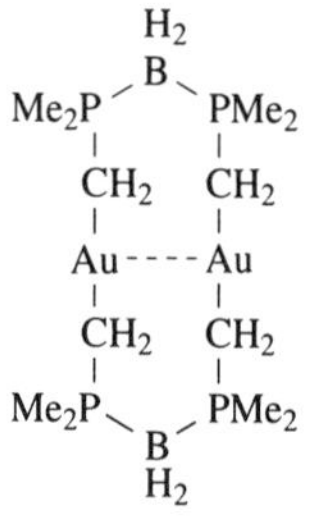

(8)

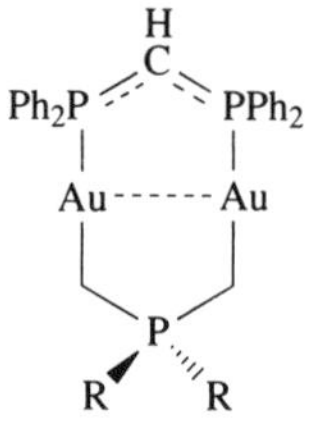

(9)
R = alkyl

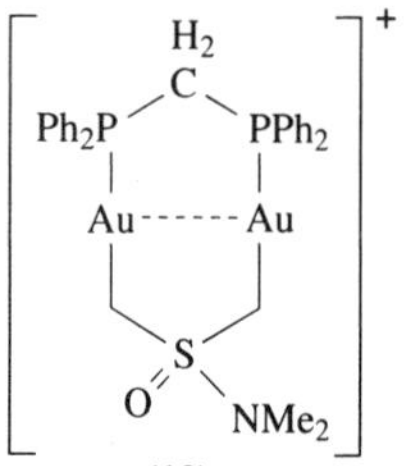

(10)

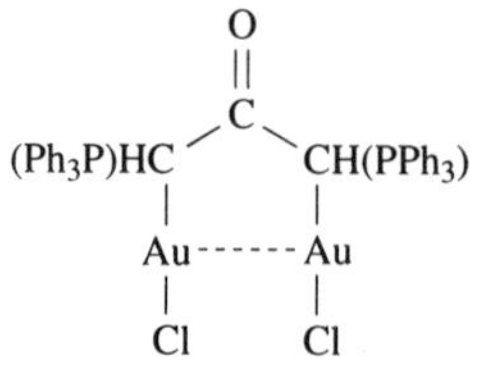

(11)

bridging ligands. Structural as well as theoretical studies suggest the existence of an attractive Au···Au interaction between the gold(I) centres, with metal–metal distances typically in the vicinity of 0.30 nm.[257] This structural similarity gives rise to similar physical and chemical properties which, moreover, resemble those of related species containing bridging diphosphine ligands (such as bis(diphenyl-phospino)methane[258] or bis(diphenylphosphino)amine[130]). Several comparative studies of the spectral characteristics of such compounds (including electronic absorption and luminescence data) have appeared.[259–61]

$$2\ [Au(CH_2PMeR_2)_2]X \longrightarrow 2\ [PMe_2R_2]X + \ \begin{matrix} \\ R_2P^+ \end{matrix} \begin{matrix} -Au^- \\ -Au^- \end{matrix} \begin{matrix} +PR_2 \\ \end{matrix} \qquad (46)$$

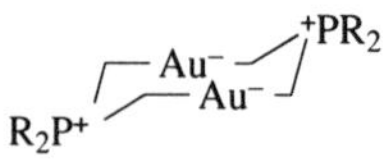

Such dinuclear bis(ylide) complexes are characterized by a rich reaction chemistry which involves the [Au···Au] structural unit and, to a lesser extent, the bridging ylide ligands. The attractive Au···Au interaction facilitates oxidative addition to the metal centres which, depending on the type of substrate, may lead to bicyclic gold(II) compounds containing a discrete transannular Au–Au bond, conventional gold(III) complexes as well as gold(III) 'A-frame' species containing bridging methylene groups, and mixed-valent gold(I)–gold(III) complexes.[262] In some cases, further derivatization of the resulting compounds by ligand exchange is also possible. An overview of these reactions is given in Schemes 10–13 for the example of (5), and Tables 8–10 contain examples of the various types of product thus obtained. Conversions involving the bridging ylide ligands include isomerizations,[263] acid-induced cleavage reactions to give mononuclear species,[264] oxidative cleavage to give ring-opened dimeric complexes,[265] and replacement of acidic protons with further gold(I) units to give polynuclear complexes reminiscent of the polyauriomethanes.[177,188,189] Some features of this chemistry are discussed in more detail in the following.

Some substrates (e.g., halogens) react with dimeric gold(I) ylide complexes by two-centre, two-electron oxidative addition to give the corresponding gold(II) complexes as isolable species (Scheme

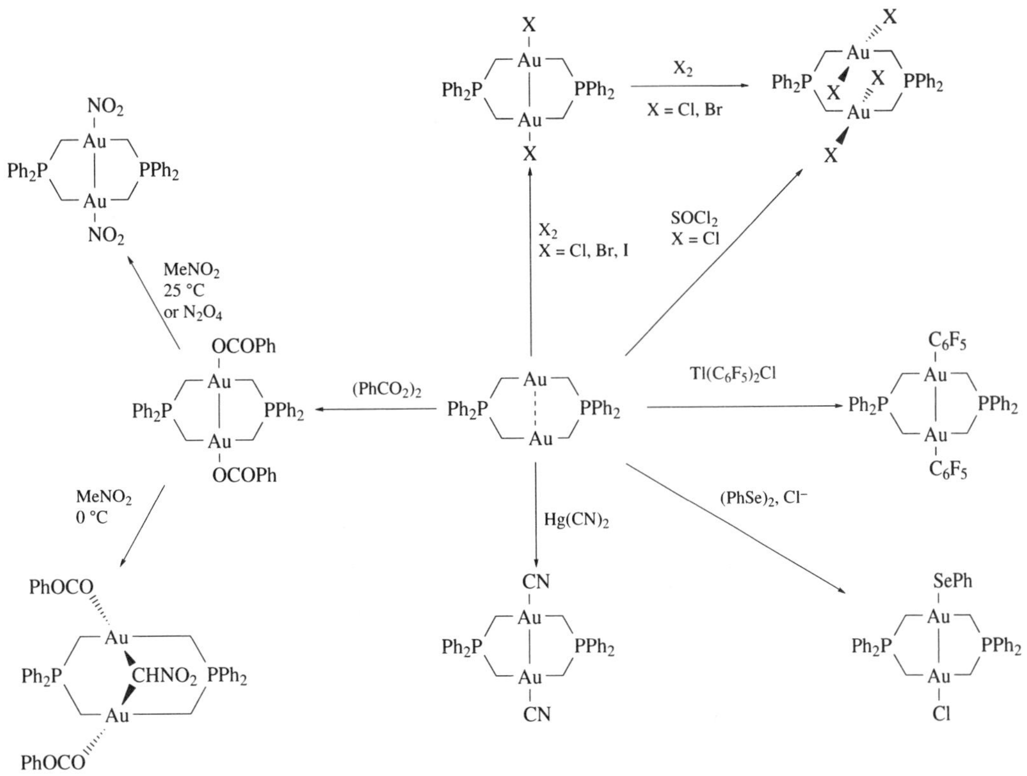

Scheme 10

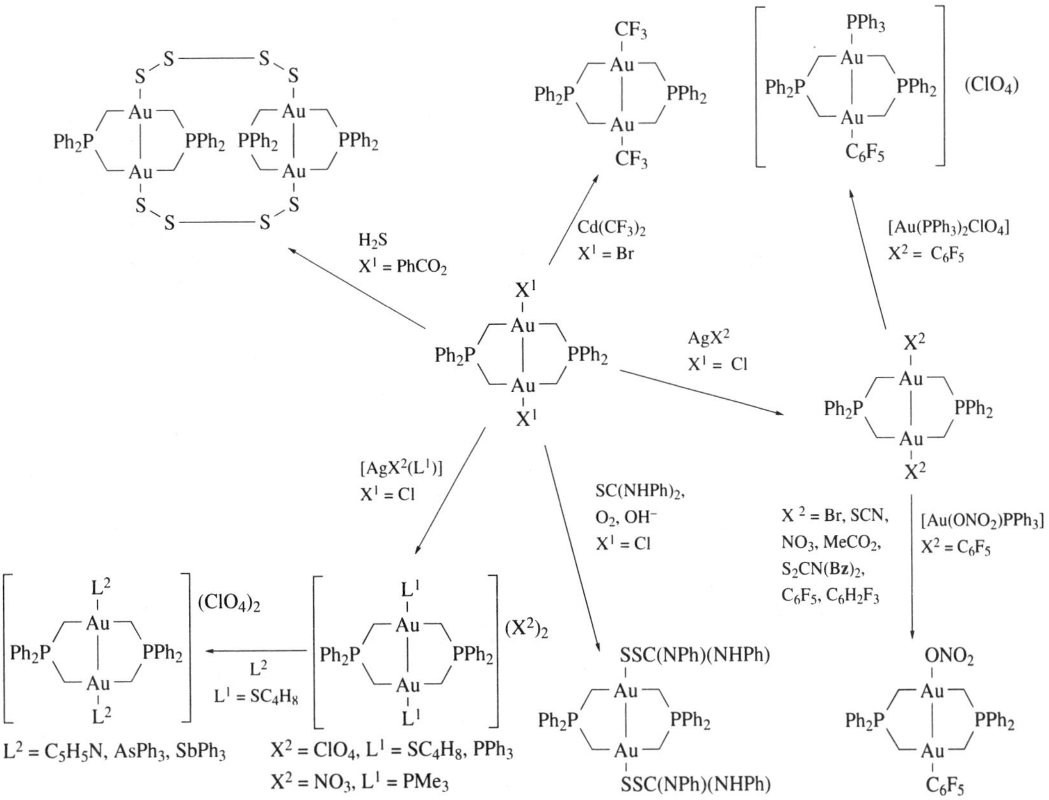

Scheme 11

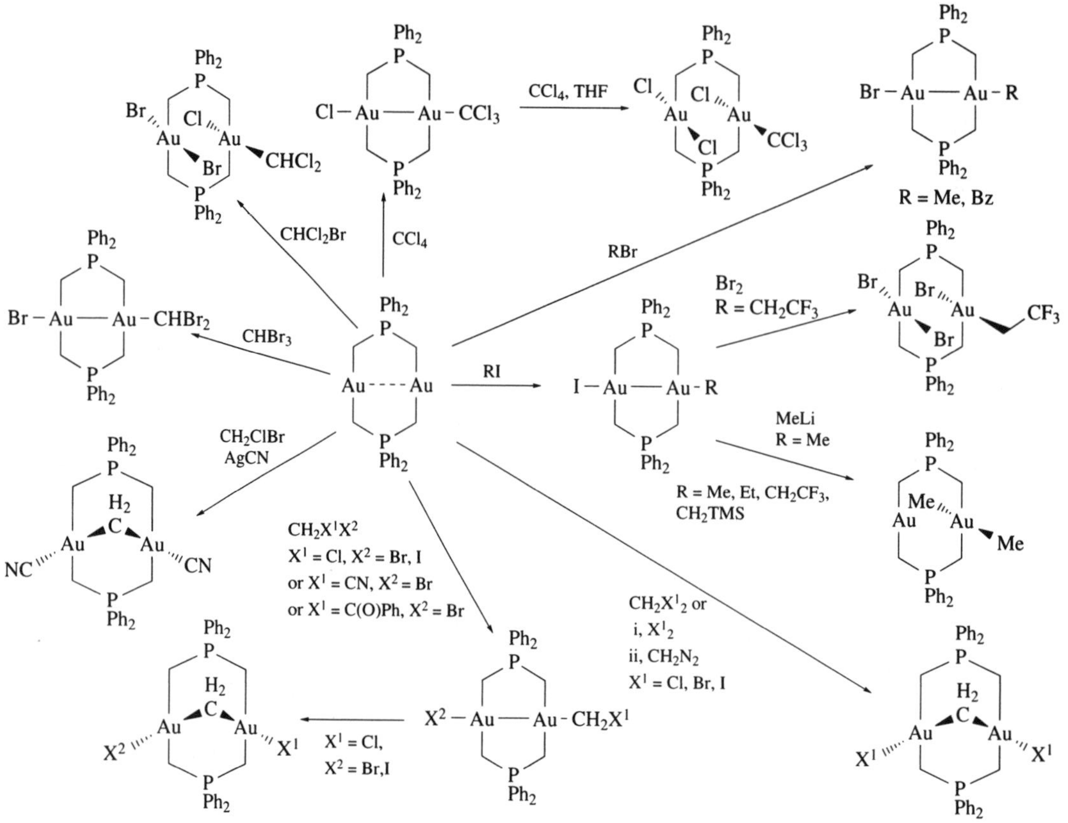

Scheme 12

10). These compounds are diamagnetic and are characterized by a discrete metal–metal bond, the presence of which has been confirmed by various structural techniques, including x-ray diffraction,[228] Mössbauer spectroscopy[228] and Raman spectroscopy.[300] The eight-membered ring may adopt a chair or a boat conformation, with square-planar coordination at gold and a linear arrangement of atoms in the X–Au–Au–X subunit. Further oxidative addition of halogens to the dihalide complexes cleaves the metal–metal bond, leading to dimeric gold(III) ylide complexes, and geometrical isomers with ligands in a *trans/trans*,[288,290] *cis/trans*[263] or *cis/cis*[291] geometry have all been structurally characterized (Figure 10). Interestingly, some of these species interconvert in solution upon prolonged standing of samples.[263] As a further point of note, several isomers feature unusually close contacts between the metal centres, thus providing rare examples for the operation of relativistic effects in the coordination chemistry of gold(III).[290]

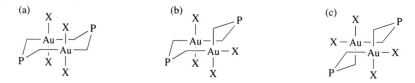

Figure 10 The possible geometrical isomers in the case of gold(III) ylide dimers: (a) *trans,trans*, (b) *cis,trans*, and (c) *cis,cis*.

With benzoyl peroxide, the dimer [Au$_2${(CH$_2$)$_2$PPh$_2$}$_2$] gives a dinuclear gold(II) ylide complex containing oxygen-bound carboxylate ligands.[269] This compound shows interesting reactivity in the presence of nitromethane as co-solvent. If kept at 0 °C, a dibenzoate complex (with A-frame geometry) can be isolated wherein a CHNO$_2$ group bridges two gold(III) centres,[296] while at ambient temperature and on prolonged standing, the reaction gives a nitritogold(II) complex, as illustrated in Scheme 10. The latter compound may also be prepared directly by reacting [Au$_2${(CH$_2$)$_2$PPh$_2$}$_2$] with N$_2$O$_4$.[269]

As can be seen in Scheme 11, various procedures have been established for the further derivatization of dinuclear gold(II) ylide complexes by ligand substitution. Halide ligands can be substituted with neutral donor ligands such as tetrahydrothiophene or pyridine (giving the corresponding cationic

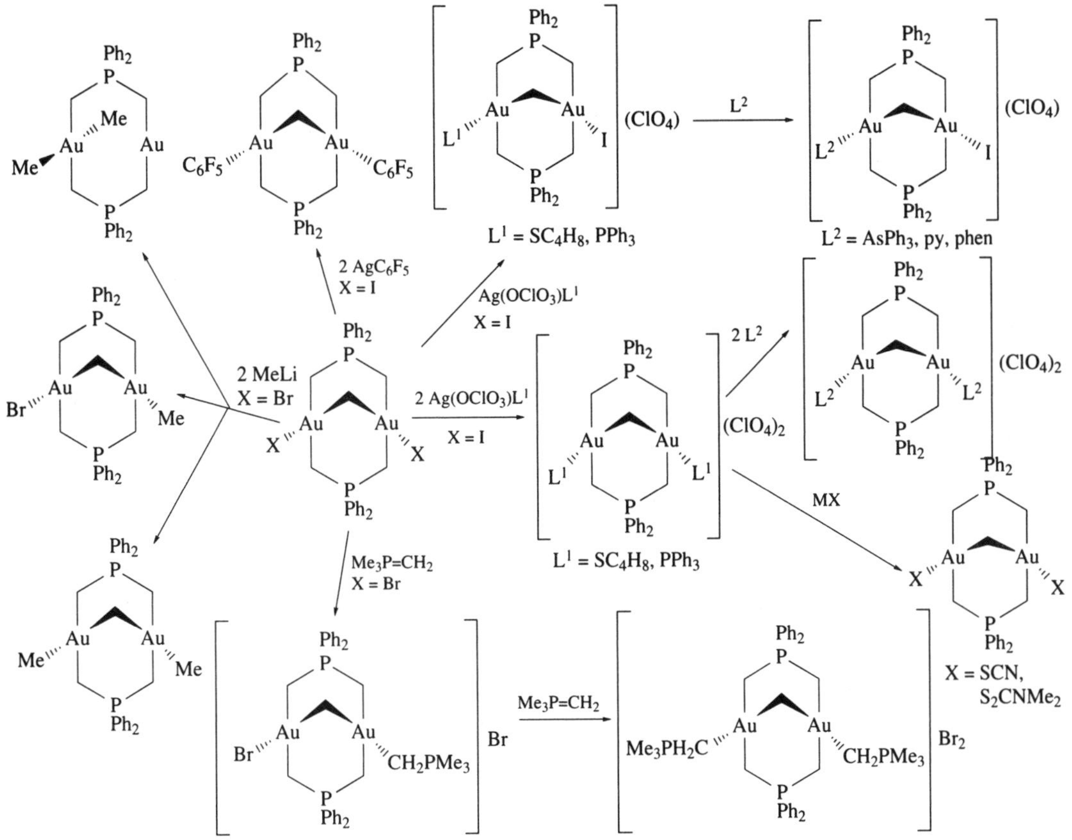

Scheme 13

complexes),[268] or with other anionic ligands (such as pseudohalides or carboxylates).[268] The digold(II) dibenzoate complex reacts with hydrogen sulfide in THF to give one of several interesting macrocyclic complexes in which two $[Au_2\{(CH_2)PPh_2\}_2]$ units are bridged by two polysulfide chains.[272] The introduction of carbanionic groups is also straightforward, leading to organogold(II) complexes with exocyclic trifluoromethyl,[274] pentafluorophenyl[268,274] or phosphorus ylide ligands.[278,287] While substitution usually occurs at both gold(II) centres simultaneously, some asymmetric derivatives have been prepared from the pentafluorophenyl-containing complexes.[278]

Considerable effort has been expended to elucidate the course of oxidative addition reactions of halogenoalkanes to dinuclear gold(I) bis(ylide) complexes,[90] experimental evidence supporting the notion that the order of reactivity of such substrates is inversely proportional to the carbon–halogen bond dissociation energies.[286] As shown in Scheme 12, $[Au_2\{(CH_2)_2PPh_2\}_2]$ reacts with alkyl halides RX (X = Br or I) to give asymmetrically substituted bicyclic digold(II) complexes.[279–81,301] The reaction is reversible if R bears no electron-withdrawing substituents,[281] NMR spectroscopic data indicating that in solution the gold(II) alkyl halide complexes are in equilibrium with the gold(I) complex and the halocarbon.[279] In the solid state, the gold(II) methyl bromide and gold(II) methyl iodide complexes undergo a thermally induced reductive elimination giving $[Au_2\{(CH_2)_2PPh_2\}_2]$ and MeBr or MeI, respectively.[279] Furthermore, $[Au_2\{(CH_2)_2PPh_2\}_2]$ catalyzes halogen exchange between MeBr and CD_3I in a process which is thought to involve an S_N2 reaction between the free alkyl halide and a postulated intermediate $[Au_2\{(CH_2)_2PPh_2\}_2\{C(H \text{ or } D)_3\}]^+X^-$.[279,301] There is no indication that alkyl halides RX add to $[Au_2\{(CH_2)_2PPh_2\}_2]$ by two-electron oxidation of a single gold centre, although asymmetric gold(I)–gold(III) complexes do form when complexes such as $[Au_2(Me)(I)\{(CH_2)_2PPh_2\}_2]$ are reacted with MeLi (Scheme 12).[280] As maybe expected, digold(II) halogenoalkyl halide complexes are susceptible to further oxidative addition when treated with halogen. The reaction, which can be used to prepare digold(III) halogenoalkyl trihalide complexes (Scheme 12),[292] apparently proceeds via an intermediate cationic μ-halogeno A-frame species, as suggested by the serendipitous isolation of small quantities of $[Au_2(\mu\text{-Br})\{(CH_2)_2PPh_2\}_2][IBr_2]$ from the mixture of $[Au_2(CH_2CF_3)(I)\{(CH_2)_2PPh_2\}_2]$ and Br_2.[289] Several geometrical isomers of the digold(III) halogenoalkyl trihalide complexes have been studied using x-ray techniques.[292]

Table 8 Dinuclear gold(II) complexes with bridging ylide ligands.

Complex	R^1	X^1	X^2	R^2	L	$d(Au–Au)$ (nm)	Ref.
$[Au_2X^1{}_2\{(CH_2)_2PR^1{}_2\}_2]$	Me	Cl					266
	Me	Br					266
	Ph	Cl				0.260	267
	Ph	Br				0.261	268
	Ph	SCN					268
	Ph	NO_2				0.260	269
	Ph	NO_3					268
	Ph	$MeCO_2$				0.259	268,270
	Ph	$PhCO_2$				0.258	271
	Ph	$S_2CN(Bz)_2$					268
	Ph	S_4				0.266	272
	Ph	SSC(NPh)(NHPh)				0.265	273
	Ph	CF_3				0.268	274
	Ph	$C_6H_2F_3$				–	268
	Ph	C_6F_5				0.268	268,274
	Ph	CN				0.264	275
$[Au_2X^1{}_2\{(CH_2)P(S)Ph_2\}_2]$		Cl				0.255	276
		I				0.261	254
$[Au_2X^1X^2\{(CH_2)_2PPh_2\}_2]$		C_6H_5Se	Cl			0.266	277
		S_4	S_5			0.265	272
		C_6F_5	NO_3				278
		C_6F_5	SCN				278
		C_6F_5	$MeCO_2$				278
$[Au_2R^2X^1\{(CH_2)_2PR^1{}_2\}_2]$	Me	Br		Bz			279
	Me	I		Me		0.270	279,280
	Me	I		CH_2TMS			279
	Ph	Br		Me		0.267	279
	Ph	Br		Bz			279
	Ph	I		Me			279
	Ph	I		Et		0.268	281
	Ph	I		CH_2TMS			279
	Ph	I		CH_2CF_3		0.268	281
	Ph	Br		CH_2Cl		0.265	282,283
	Ph	Br		CH_2CN		0.269	284
	Ph	Br		$CH_2C(O)Ph$		0.269	284
	Ph	I		CH_2Cl		0.268	285
	Ph	Br		$CHBr_2$			286
	Ph	Cl		CCl_3		0.265	267
$[Au_2X^1L\{(CH_2)_2PPh_2\}_2][X^2]$		I	I		CH_2PMe_3	0.268	287
		C_6F_5	ClO_4		PPh_3	0.266	278
		C_6F_5	ClO_4		SC_4H_8		278
$[Au_2L_2\{(CH_2)_2PPh_2\}_2][X^2]_2$			ClO_4		PPh_3	0.258	268
			I		CH_2PMe_3		287
			ClO_4		CH_2PPh_3		278
			ClO_4		SC_4H_8		268
			ClO_4		C_5H_5N		268
			ClO_4		PPh_3		268
			ClO_4		$AsPh_3$		268
			ClO_4		$SbPh_3$		268
			NO_3		PMe_3		268

With methylene dihalides CH_2X_2 (X = Cl, Br or I), dimeric gold(I) ylide complexes undergo a two-centre, four-electron oxidative addition reaction to give gold(III) A-frame complexes in which a methylene group bridges two gold centres (Scheme 13).[266,293] Alternatively, such complexes have been prepared by the reaction of $[Au_2\{(CH_2)_2PPh_2\}_2]$ first with a halogen and then with diazomethane.[285] While no dinuclear gold(II) halogenoalkyl halide complexes are isolated from the reactions of symmetrical methylene dihalides, the intermediacy of such species is documented for asymmetric substrates, as exemplified by the reactions of $[Au_2\{(CH_2)_2PPh_2\}_2]$ with CH_2ClBr,[282] CH_2ClI,[285] $CH_2Br(CN)$,[284] or $CH_2Br\{C(=O)Ph\}$[284] (Scheme 12). A remarkable feature of the crystal structure of $[Au_2(CH_2Cl)(Br)\{(CH_2)_2PPh_2\}_2]$ is the unusually close contact between one gold centre and the chlorine atom of the coordinated chloromethyl group ($d(Au–Cl) = 0.2895$ nm), the Au–C–Cl bond angle being only 96.2° compared with the normal value of 109.5° for an sp^3 carbon centre. This finding has been interpreted as being evidence for the formation of a carbene intermediate preceding the formation of an

Table 9 Dinuclear gold(III) complexes with bridging ylide ligands.

Complex	R	X_2^1	Y	X^1	X^2	L	$d(Au\text{-}Au)$ (nm)	*Ref.*
$[Au_2(X^1{}_2)_2\{(CH_2)_2PR_2\}_2]$								
(−)	Me	Cl_2						266
(*cis/trans*)	Ph	Cl_2					0.309	263
(*trans/trans*)	Ph	Cl_2					0.309	288
(*cis/cis*)	Ph	Br_2					0.450	289
(*trans/trans*)	Ph	Br_2					0.307	263,290
(*cis/cis*)	Ph	$C_6H_4S_2$					0.440	291
$[Au_2(X^1{}_2)(RY)\{(CH_2)_2PPh_2\}_2]$								
(*trans/trans*)	CCl_3	Cl_2	Cl				0.309	267
(*cis/trans*)	CH_2CF_3	Br_2	Br				0.310	292
(*trans/trans*)	$CHCl_2$	Br_2	Cl				0.307	292
$[Au_2(\mu\text{-}CH_2)X^1{}_2\{(CH_2)_2PR_2\}_2]$	Me			Cl				293
	Me			Br				293
	Me			I				293
	Bu^t			Cl				266
	Ph			Br			0.314	282,283
	Ph			I				294
	Ph			CN			0.317	275,285
	Ph			$PhCO_2$				285
	Ph			SCN				285,294
	Ph			S_2CNMe_2				294
	Ph			Me			0.312	295
	Ph			C_6F_5			0.311	294
$[Au_2(\mu\text{-}CHY)X^1{}_2\{(CH_2)_2PPh_2\}_2]$			NO_2	$OC(O)Ph$			0.307	269,296
$[Au_2(\mu\text{-}CH_2)X^1X^2\{(CH_2)_2PR_2\}_2]$	Ph			Cl	Br			285
	Ph			Cl	I			285
	Ph			Br	Me			295
$[Au_2(\mu\text{-}CH_2)LX^1\{(CH_2)_2PPh_2\}_2][X^2]$				Br	Br	CH_2PMe_3		297
				I	ClO_4	PPh_3		294
				I	ClO_4	$AsPh_3$		294
				I	ClO_4	C_5H_5N		294
				I	ClO_4	SC_4H_8		294
				I	ClO_4	phen		294
$[Au_2(\mu\text{-}CH_2)L_2\{(CH_2)_2PPh_2\}_2][X^1]_2$				Br		CH_2PMe_3		297
				ClO_4		PPh_3		294
				ClO_4		$AsPh_3$		294
				ClO_4		C_5H_5N		294
				ClO_4		SC_4H_8		294
				ClO_4		phen		294

Table 10 Dinuclear mixed-valent gold(I)–gold(III) complexes with bridging ylide ligands.

	X	Z	$d(Au\cdots Au)$ (nm)	*Ref.*
$[Au(AuX_2)\{(CH_2)(Z)PPh_2\}_2]$	Br	CH_2	0.306	298
	Me	CH_2		295
	Cl	S		240
	I	S	0.305	254,299

A-frame complex containing a bridging methylene group.[282,283] By contrast, the Au–C–Cl bond angle in the related complex $[Au_2(CH_2Cl)(I)\{(CH_2)_2PPh_2\}_2]$ is 115.6°, and hence in the expected range.[285] In both reaction sequences, the formulation of the final product as the mixed halogenated μ-methylene A-frame species is tentative (Scheme 12); at least in the case of CH_2ClBr, further halide exchange with solvent molecules leads to the formation of the symmetrically disubstituted μ-methylene dibromo complex $[Au_2(\mu\text{-}CH_2)Br_2\{(CH_2)_2PPh_2\}_2]$, which has been structurally characterized.[282,283] A related reaction is the treatment of $[Au_2\{(CH_2)_2PPh_2\}_2]$ first with CH_2ClBr and then with silver salts such as AgCN, AgSCN or $Ag(O_2CPh)$ to give the symmetrically disubstituted μ-methylene complexes,[275,285] and derivatizations of digold(III) μ-methylene dihalide complexes with other reagents to introduce carbanionic ligands,[294,295] pseudohalides,[294] phosphorus ylides[297] or neutral donor ligands[294] have also been reported (Scheme 13). The treatment with organolithium compounds gives several well-defined products, including a partially alkylated intermediate and a mixed-valent Au^I–Au^{III} species. The doubly methylated complex [Au$_2$Me$_2${(CH$_2$)$_2$PPh$_2$}$_2$] shows interesting reactivity in that it undergoes thermally induced reductive elimination of propane, the dinuclear gold(I) bis(ylide) complex being regenerated in the process.[295]

The reaction of [Au$_2${(CH$_2$)$_2$PPh$_2$}$_2$] with trihalogenomethanes may give two products (Scheme 12): initially, a digold(II) halogenomethyl halide species is formed, as exemplified by the isolation of [Au$_2$(CHBr$_2$)(Br){(CH$_2$)$_2$PPh$_2$}$_2$] from a mixture of [Au$_2${(CH$_2$)$_2$PPh$_2$}$_2$] and CHBr$_3$.[286] In a second step, renewed oxidative addition leads to the formation of digold(III) halogenomethyl trihalide complexes, as shown in Scheme 12 for the substrate CHCl$_2$Br.[292] Tetrahalogenomethanes react in an analogous way, the interaction of [Au$_2${(CH$_2$)$_2$PPh$_2$}$_2$] with CCl$_4$ having been studied in some detail.[267] Oxidative addition of one equivalent of substrate leads to the isolable digold(II) halogenomethyl halide species [Au$_2$(CCl$_3$)(Cl){(CH$_2$)$_2$PPh$_2$}$_2$] which, in the presence of THF, reacts a second time with CCl$_4$ to give the digold(III) trichloromethyl trichloride complex [Au$_2$(CCl$_3$)Cl$_3${(CH$_2$)$_2$PPh$_2$}$_2$]. By contrast, prolonged interaction of [Au$_2${(CH$_2$)$_2$PPh$_2$}$_2$] with CCl$_4$ in the absence of THF generates the digold(II) dichloride complex [Au$_2$Cl$_2${(CH$_2$)$_2$PPh$_2$}$_2$], and it has been noted that the two products do not interconvert in the presence of CCl$_4$. Reactions in the CCl$_4$/THF solvent mixture are accompanied by the formation of chloroform, suggesting a radical pathway wherein THF acts as a hydrogen atom source for CCl$_3$ radicals stabilized by complexation to a gold centre.

As is apparent from the foregoing discussion, two-electron oxidative addition of substrates to dinuclear gold(I) ylide complexes usually involves both gold centres to yield gold(II)–gold(II) complexes. Remarkably, both gold(I)–gold(III) and gold(II)–gold(II) complexes have been obtained from the reaction of dimeric gold(I) methylenethiophosphinate complexes with iodine. As illustrated in Scheme 14, the isovalent isomer contains both chelating ligands in a geometry which is *trans* with respect to the Au^{II}–Au^{II} bond, while the mixed-valent isomer exhibits a *cis* chelate ligand configuration with a linear S–Au^I–S and a square-planar *trans*-[Au^{III}I$_2$C$_2$] unit.[254] A related mixed-valent dichloride derivative is also known.[240]

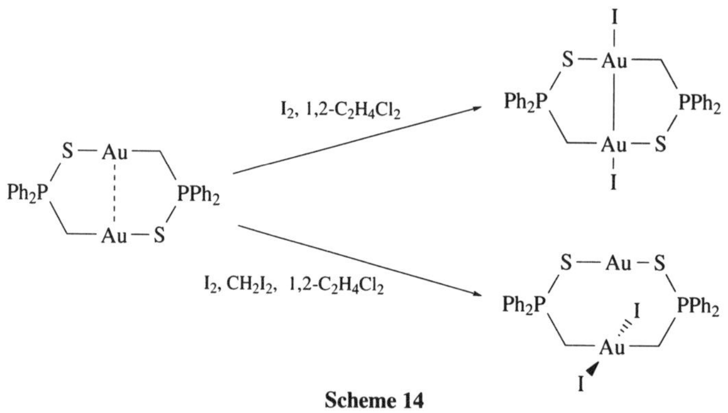

Scheme 14

Methylene thiophosphinate ligands have also been used to synthesize dinuclear gold–mercury complexes which show similar isomerism: depending on the synthetic procedure, species containing S–Au–C/S–Hg–C or S–Au–S/C–Hg–C structural units are obtained.[302] The latter isomer undergoes an interesting ring expansion reaction on treatment with diazomethane, in the course of which one methylene group inserts into each gold–sulfur bond to give a 10-membered heterobimetallic ring.[303] The trinuclear complex [Au$_2$Pt{CH$_2$P(S)Ph$_2$}$_4$] contains a linear [Au···Pt···Au] axis, with two pairs of ligands forming mutually perpendicular eight-membered rings and sharing a square-planar [PtS$_4$] fragment. Oxidative addition of chlorine gives the symmetrical complex containing a metal–metal bonded [Cl–Au–Pt–Au–Cl] unit, rather than a mixed-valent species.[239]

A number of other procedures have led to mixed-valent gold(I)–gold(III) ylide complexes, but these do not involve oxidative addition in the key step. Thus, reaction of digold(II) bis(ylide) dihalide complexes such as [Au$_2$I$_2${(CH$_2$)$_2$PPh$_2$}$_2$] with various phosphorus ylides in a 1:2 molar ratio leads to the formation of dicationic complexes with a gold(I)–gold(III) constitution,[287] as shown by Mössbauer spectroscopy.[304] This result is in agreement with the finding that oxidative addition of MeI to [Au$_2${(CH$_2$)$_2$PPh$_2$}$_2$] yields the asymmetrically substituted digold(II) complex, whereas attempts to carry out further alkylation with MeLi afford only a mixed-valent isomer (Scheme 12). Apparently, the strong *trans* influence of two alkyl groups on the Au–Au bond results in destabilization of this structural unit and induces disproportionation (Scheme 15).

In a single-centre reduction process, treatment of the digold(III) complex [Au$_2$Br$_4${(CH$_2$)$_2$PPh$_2$}$_2$] with AgCN under suitable conditions has been reported to produce the mixed-valent complex

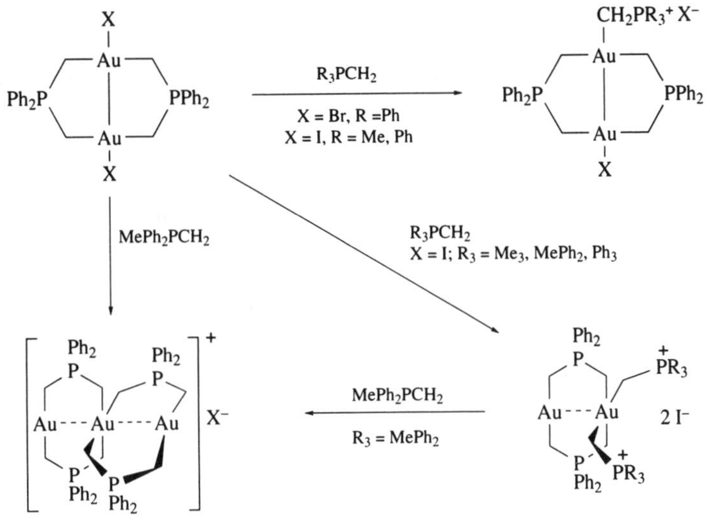

Scheme 15

[Au(AuBr$_2$){(CH$_2$)$_2$PPh$_2$}$_2$], characterized by x-ray crystallography and photoelectron spectroscopy.[298] An open-chain dinuclear gold(I)–gold(III) complex with one ylide ligand bridging the two gold atoms and the other in a chelating mode can be obtained by disproportionation of the digold(II) complex [Au$_2$Cl$_2${(CH$_2$)$_2$PPh$_2$}$_2$] in polar solvents such as nitromethane (Equation (47)).[299]

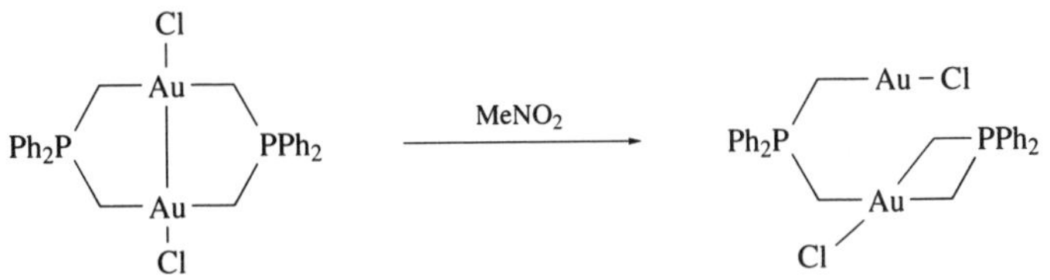

(47)

Finally, complexes containing direct formal bonds between gold(I) and gold(III) or gold(II) which are unsupported by other ligands have been prepared in straightforward reactions (Equations (48)[305] and (49)[306]). It has been noted, however, that the assignment of oxidation states to the gold atoms in these oligonuclear complexes is a matter of conjecture, pending the results of a Mössbauer spectroscopic study. X-ray structural studies have revealed Au–Au distances in these molecules which fall in the range 0.257–0.277 nm.

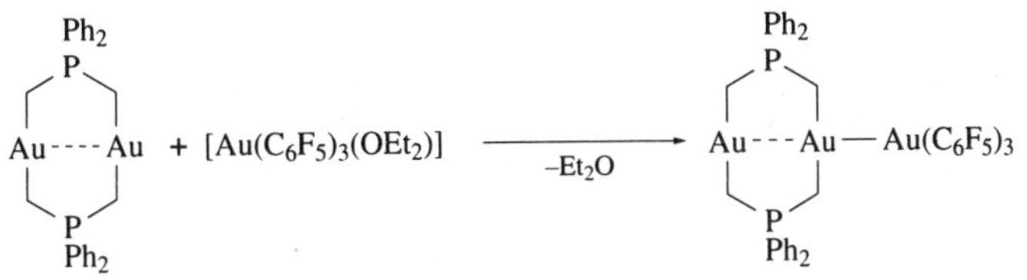

(48)

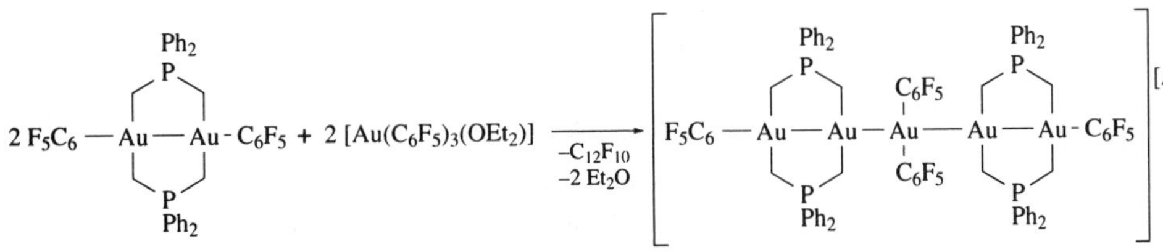

(49)

1.2.5 Gold(III) Complexes with One Gold–Carbon Bond

Alkylgold(III) complexes containing one gold–carbon bond are still rare,[71,307] and only a few trifluoromethyl compounds of the type [Au(CF$_3$)X$_2$(PR$_3$)] (X = Br or I; R = Me, Et or Ph) have been reported. They are formed by oxidative addition of halogen to the respective trifluoromethyl(phosphine)gold(I) complex.[76] By contrast, various methods now exist for the synthesis of monoarylgold(III) species. One of the earliest established procedures involves the electrophilic substitution by gold(III) at the aromatic ring of benzene derivatives (auration).[71] Systems such as the trichloro(phenyl)aurate anion have been investigated with respect to their ligand substitution kinetics, and the *trans* influence of the aryl ligand has been assessed.[308]

Another synthesis of monoarylgold(III) compounds starts from pentahalogenoarylgold(I) complexes, which are sufficiently stable to undergo oxidative addition of halogen without cleavage of the gold–carbon bond (Equation (50)).[69] In these products, the thiophene ligand may then be substituted by other neutral or anionic ligands (e.g., phosphine, isocyanide, 1,10-phenanthroline, halide), and a wide variety of derivatives has been thus obtained, examples of which are given in Table 11.[69] Gold-197 Mössbauer spectra of $[Au(C_6F_5)Cl_2(phen)]$ support the existence of a weak interaction of gold(III) with the second donor atom of the ligand, approaching pentacoordination. This type of compound may be further derivatized by treatment with $AgClO_4$, giving four-coordinate cationic complexes of the types $[Au(C_6F_5)X(phen)]ClO_4$ (X = Cl or Br) and $[Au(C_6F_5)(PPh_3)(phen)](ClO_4)_2$.[309]

$$[Au(C_6F_5)(SC_4H_8)] + Cl_2 \longrightarrow [Au(C_6F_5)Cl_2(SC_4H_8)] \qquad (50)$$

Table 11 Some monoarylgold(III) complexes.

Complex	m.p. (°C)	Ref.
$[Au(C_6F_5)Cl_2(SC_4H_8)]$		69
$[Au(C_6Cl_5)Cl_2(SC_4H_8)]$		69
$[Au(C_6F_5)Br_2(CNC_6H_4Me-4)]$	148 (d)	69
$[Au(C_6F_5)Cl_2(phen)]$	149 (d)	309
$[Au(C_6F_5)Cl_2(pdma)]^a$	100	309
$[Ph_3PCH_2C_6H_5][Au(C_6F_5)Br_2Cl]$		69
$[Au(C_6F_5)Cl(phen)]ClO_4$	195 (d)	309
$[Au(C_6F_5)(PPh_3)(phen)](ClO_4)_2$	147 (d)	309
$[Au(C_6H_4\{N=NPh\}-2)Cl_2]$	218 (d)	310
$[Au(C_6H_4\{N=NPh\}-2)(MeCO_2)_2]$	140 (d)	311
$[Au(C_6H_4\{N=NPh\}-2)(O_2C–CO_2)]$	129 (d)	311
$[Au(C_6H_4\{N=NPh\}-2)Cl_2(PPh_3)]$	181 (d)	310
$[\{Au(C_6H_4\{N=NPh\}-2)Cl_2\}_2\{Ph_2P(CH_2)_2PPh_2\}]$	169	310
$[Ph_3PCH_2C_6H_5][Au(C_6H_4\{N=NPh\}-2)Cl_3]$	83	310
$[Au(C_6H_4\{N=NPh\}-2)Cl(PPh_3)_2]ClO_4$	136 (d)	310
$[Au(C_6H_4\{N=NPh\}-2)(C_5H_5N)_2](ClO_4)_2$	210 (d)	311
$[\{Au(C_6H_4\{N=NPh\}-2)(PPh_3)_2\}_2](ClO_4)_4$	185 (d)	312
$[Au(C_6H_4\{CH_2NMe_2\}-2)Cl_2]$	185 (d)	313
$[Au(C_6H_4\{CH_2NMe_2\}-2)I_2]$	105 (d)	313
$[Au(C_6H_4\{CH_2NMe_2\}-2)Cl(PPh_3)]Cl$	174	313
$[Au(C_6H_4\{CH_2NMe_2\}-2)(phen)](ClO_4)_2$		69
$[Au(C_6H_4\{CH_2NMe_2\}-2)(CN)(phen)]BF_4$	125 (d)	314
$[Au(C_6H_4\{CH_2NMe_2\}-2)(phen)(PPh_3)](BF_4)_2$	145	314
$[Au(C_6H_4\{2-C_5H_4N\}-2)Cl_2]$		315

^{a} pdma = Phenylene-1,2-bis(dimethylarsine).

Arylgold(III) complexes with the stoichiometry $[Au(Ar)X_2L]$ may also be obtained by arylation of $[AuCl_3(SC_4H_8)]$ or $[AuCl_4]^-$ with arylmercury(II) compounds, and two examples for this conversion are presented in Equations (51) and (52).[310,313] The derivatization of these complexes (by substitution of the chloride ligands or displacement of the nitrogen-donor centre) is straightforward, and examples of products obtained from reactions with neutral or anionic ligands are listed in Table 11. These may again be transformed in subsequent conversions. Specifically, $[Au(C_6H_4\{CH_2NMe_2\}-2)(phen)](ClO_4)_2$ reacts with triphenylphosphine to give the pentacoordinated complex $[Au(C_6H_4\{CH_2NMe_2\}-2)-(phen)(PPh_3)](BF_4)_2$, which has been shown by x-ray crystallography to contain gold(III) in a distorted square-pyramidal coordination environment.[314]

$$2\,[AuCl_3(SC_4H_8)] + 2\,[Hg(C_6H_4\{N=NPh\}-2)Cl] \xrightarrow{2\,NMe_4Cl} 2 \quad \text{(structure)} + (NMe_4)_2[Hg_2Cl_6] + 2\,SC_4H_8 \qquad (51)$$

$$4 [AuCl_4]^- + 2 [Hg(C_6H_4\{CH_2NMe_2\}-2)_2] \longrightarrow 4 \underset{\underset{Me_2}{|}}{\overset{\overset{Cl}{|}}{Au-Cl}} + [Hg_2Cl_6]^{2-} + 2\,Cl^- \qquad (52)$$

Several closely related aryl–gold(III) complexes have also been formed in the course of cyclometallation reactions. The reaction of substituted pyridine ligands such as 2-phenylpyridine (denoted by HL) with Na[AuCl₄] at ambient temperature yields the nonmetallated complex [Au(HL)Cl₃] which, upon heating, is transformed into the metallated compound [Au(L)Cl₂].[315] This process is illustrated in Equation (53). The product has also been prepared by the transmetallation of the appropriate arylmercury(II) chloride with [AuCl₄]⁻, a reaction resembling that depicted in Equation (51). The ligand 6-(2-thienyl)-2,2'-bipyridine gives identical reactions with Na[AuCl₄], and an x-ray crystallographic study of the cyclometallated product has revealed its interesting dimeric structure (Figure 11).[316–18]

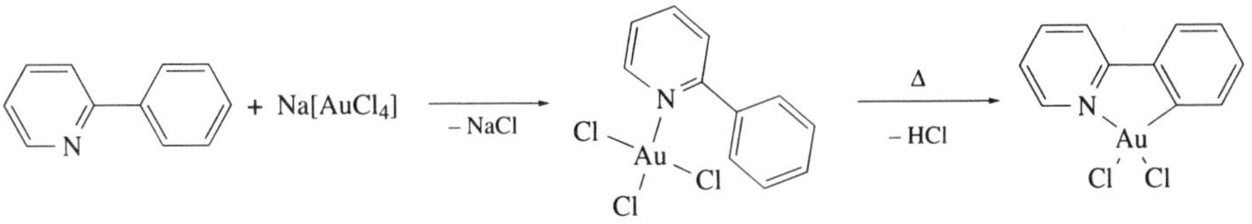

$$\text{(53)}$$

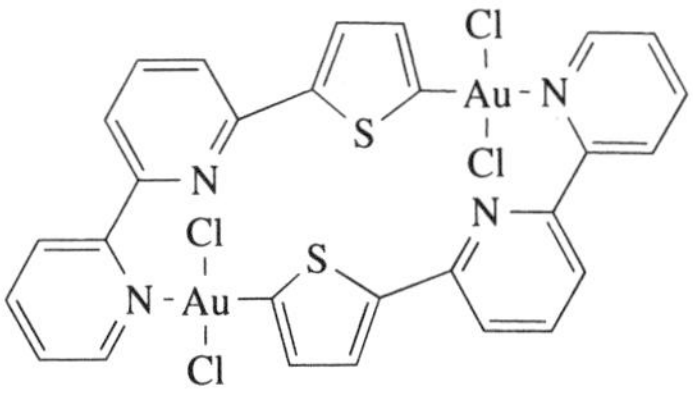

Figure 11 Structure of the dimeric complex [{Au(thbipy)Cl₂}₂] (thbipy = 6-(2-thienyl)-2,2'-bipyridine).

1.2.6 Gold(III) Complexes with Two Gold–Carbon Bonds

1.2.6.1 Diorganylgold(III) halides and pseudohalides

Compounds of the type [AuR₂X]₂ have been known for a long time, the organic ligands being unsubstituted alkyl groups in the majority of cases.[71] One of the simplest representatives is dimethylchlorogold(III), [Au₂(μ-Cl)₂Me₄], whose interaction with nucleosides has been studied as part of an investigation aimed at the identification of biologically active organometallic compounds.[319] The emphasis of recent synthetic work, however, lies on gold(III) species with trifluoromethyl, aryl or alkenyl ligands.

Trifluoromethyl-containing gold complexes are potentially useful precursors in chemical vapour deposition studies as they are expected to show enhanced volatility.[320] This is observed for the dimeric complexes [Au₂(μ-Br)₂(CF₃)₄] and [Au₂(μ-I)₂(CF₃)₄], which are accessible by cocondensation of gold with CF₃Br and CF₃I, respectively, and may be sublimed at room temperature (10⁻² torr). The solid-state structure of the iodide complex has been determined but is unexceptional, resembling that of other nonfluorinated alkylgold halides.[321] Another trifluoromethylgold(III) complex is [*cis*-Au(CF₃)₂-(I)(PMe₃)], which is formed in quantitative yield by the oxidative addition of trifluoromethyl iodide to [Au(CF₃)(PMe₃)], the reaction being assumed to proceed via a radical intermediate.[78,79]

Diarylaurate(III) complexes of the type [AuR₂X₂]⁻ (R = C₆F₅ or C₆F₃H₂-2,4,6; X = Cl, Br or I) were first prepared by oxidative addition of halogens to the corresponding organogold(I) complexes [AuR₂]⁻.[69,71] Reaction of these compounds with AgClO₄ results in the abstraction of halide to give the dimeric diarylgold(III) halide complexes [Au₂(μ-X)₂(C₆F₅)₄] (X = Cl or Br) which, in turn, react with sodium azide or potassium thiocyanate to give the corresponding dinuclear pseudohalide derivatives. Treatment of the dimeric chloride complex with Tl(acac) results in the formation of monomeric [Au(C₆F₅)₂(acac)], while the cleavage of the chloro bridges with neutral ligands yields complexes of the types [Au(C₆F₅)₂(L–L)][Au(C₆F₅)₂Cl₂] (L–L = 1,10-phenanthroline and related ligands) or [Au(C₆F₅)₂Cl(L)] (L = pyridine).[322]

The preparation of diarylaurate(III) complexes of the type $[AuR_2X_2]^-$ by oxidative addition of halogens as described earlier is dependent on the particular stability of polyhalogenoarylgold complexes and is not generally applicable, as less stable organogold(I) derivatives usually react with cleavage of the gold–carbon bonds under the given conditions. An alternative synthesis of diarylgold(III) halides uses organomercurials as arylation reagents, and compounds of the type $[Me_4N][cis\text{-}AuR_2Cl_2]$ can be prepared where R = $\{C_6H_4(NO_2)\text{-}2\}$ or $\{C_6H_3Me\text{-}2(NO_2)\text{-}6\}$ (Equation (54)).[323] The complexes may then be derivatized by substitution of one or both halide ligands with various reagents. Thus, reaction with KCN results in substitution of both halide ligands by cyanide, while treatment with neutral mono- or bidentate ligands leads to neutral or cationic complexes (Equations (55)–(57)).[323]

$$2\,[Hg\{C_6H_4(NO_2)\text{-}2\}_2] + [AuCl_4]^- \longrightarrow [Au\{C_6H_4(NO_2)\text{-}2\}_2Cl_2]^- + 2\,[Hg\{C_6H_4(NO_2)\text{-}2\}Cl] \quad (54)$$

$$[Me_4N][Au\{C_6H_4(NO_2)\text{-}2\}_2Cl_2] + 2\,KCN \xrightarrow[-2\,KCl]{} [Me_4N][Au\{C_6H_4(NO_2)\text{-}2\}_2(CN)_2] \quad (55)$$

$$[Me_4N][Au\{C_6H_4(NO_2)\text{-}2\}_2Cl_2] + PPh_3 \xrightarrow[-[Me_4N]Cl]{} [Au\{C_6H_4(NO_2)\text{-}2\}_2Cl(PPh_3)] \quad (56)$$

$$[Me_4N][Au\{C_6H_4(NO_2)\text{-}2\}_2Cl_2] + Ph_2PCH_2CH_2PPh_2 \xrightarrow[\substack{-[Me_4N]Cl \\ -NaCl}]{NaClO_4} [Au\{C_6H_4(NO_2)\text{-}2\}_2(\eta^2\text{-}Ph_2PCH_2CH_2PPh_2)]ClO_4 \quad (57)$$

A diorganylgold(III) halide complex in which gold is incorporated in a carbocyclic ring is the auracyclopentadiene derivative $[Au_2(\mu\text{-}Cl)_2(C_4Ph_4)_2]$. The reaction with various anionic and neutral ligands again leads to cleavage of the chloride bridges, and some of the resulting products are illustrated in Scheme 16. For the complex $[AuCl(C_4Ph_4)(phen)]$, x-ray structural data indicate a distorted square-pyramidal coordination environment of the metal centre.[324]

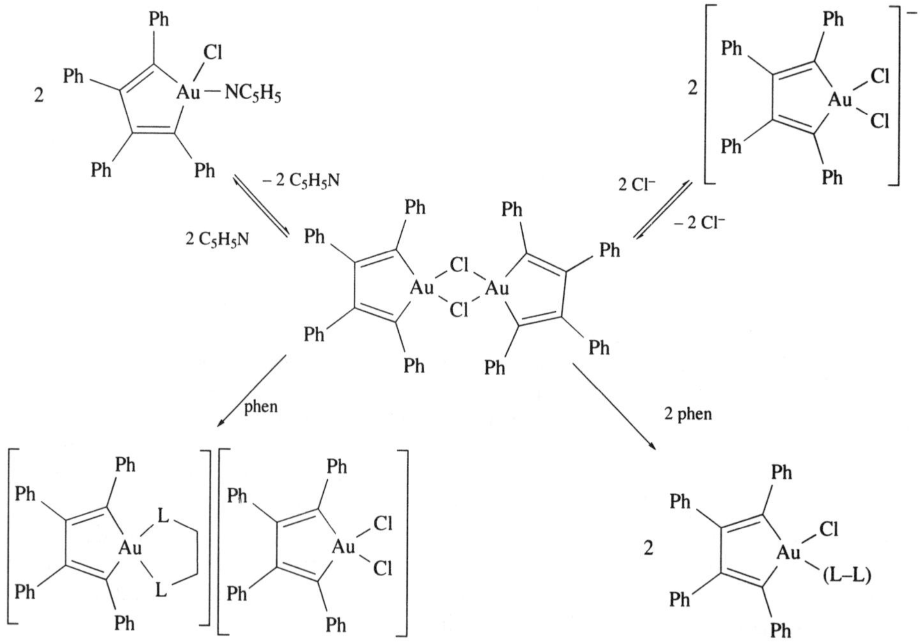

Scheme 16

1.2.6.2 Diorganylgold(III) complexes with group 16 donor ligands

Dimethyl(acetylacetonato)gold(III), $[AuMe_2(acac)]$, is of great interest as a precursor for the chemical vapour deposition of gold. A gas-phase electron diffraction study of the complex indicates a square-planar structure in which the acetylacetonate ligand is bonded to gold(III) via both its oxygen atoms, with $d(Au\text{-}O) = 0.208\,5(7)$ nm, $d(Au\text{-}C) = 0.205\,4(5)$ nm, essentially right angles at gold ($\angle(O\text{-}Au\text{-}O) = 90.9(6)°$, $\angle(C\text{-}Au\text{-}C) = 93(2)°$) and the normal parameters of an acac ligand.[325] The mechanisms of thermal and photochemical decomposition of this complex (12) and its dimethyl-d_6 congener (13) have been studied by UV–visible and proton NMR spectroscopy in solution. Thermal decomposition is very solvent dependent and is not observed in noncoordinating, nonpolar solvents such as cyclohexane. It proceeds with reductive elimination of ethane and protonation of the acetylacetonate ligand. The generation of free radicals is suggested by the observation of cross-over products such as ethane-d_3 in the decomposition of a 50:50 mixture of (12) and (13), but the predominant mode of decomposition appears to be concerted reductive elimination. This contrasts with the photochemical decomposition mechanism, where the formation of free radicals predominates.[27]

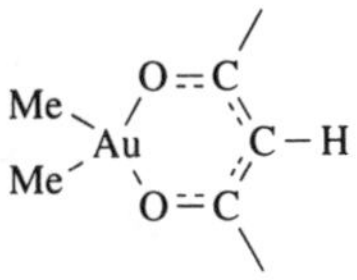

(12) 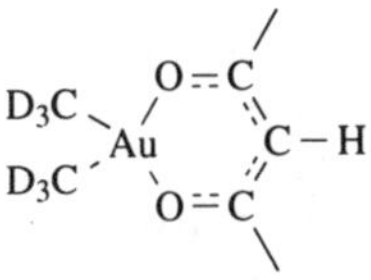   (13)

The related complex [Au(C$_4$Ph$_4$)(acac)] can be prepared from the auracyclopentadiene species [AuCl(C$_4$Ph$_4$)(SC$_4$H$_8$)] or [Au$_2$(μ-Cl)$_2$(C$_4$Ph$_4$)$_2$] by reaction with Tl(acac), and has been used as the starting material in the synthesis of several other complexes with oxygen donor ligands (Scheme 17).[326]

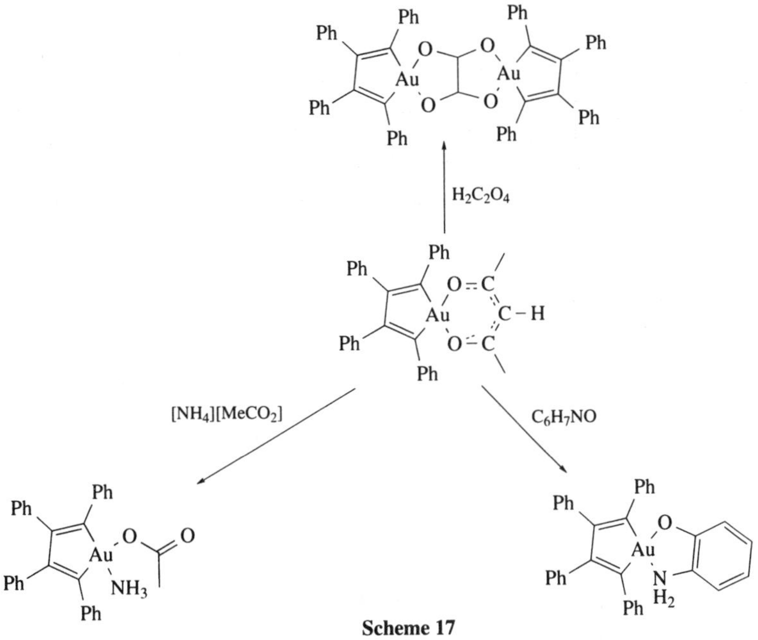

Scheme 17

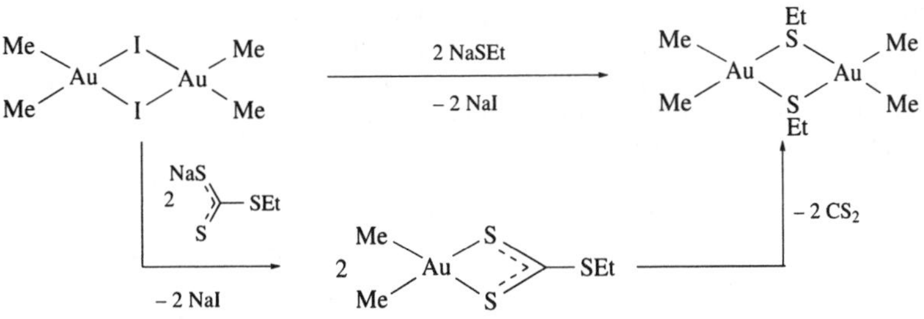

Scheme 18

Sulfur donor ligands have also been employed in the synthesis of several types of dialkyl- and diarylgold(III) complexes. A dimeric dimethylgold(III) thiolate may be prepared from [Au$_2$(μ-I)$_2$Me$_4$] by reaction with either sodium thiolate or sodium thioxanthate, the latter reaction proceeding through an intermediate which decomposes with liberation of carbon disulfide (Scheme 18).[327] With other chelating S,S donor ligands, [Au$_2$(μ-I)$_2$Me$_4$] gives a variety of stable mononuclear, neutral dimethylgold(III) complexes, examples being the compounds [AuMe$_2${S$_2$PMe$_2$}] and [AuMe$_2${Me$_2$P(S)NP(S)Me$_2$}].[327] An analogous reaction is that of [Au$_2$(μ-Cl)$_2$(C$_6$F$_5$)$_4$] with dialkyldithiocarbamates, leading to complexes such as [Au(C$_6$F$_5$)$_2$(S$_2$CNMe$_2$)].[328] Diorganylgold(III) dialkyldithiocarbamates have also been obtained together with other products from the oxidative addition of tetraalkylthiuram disulfides to organogold(I) complexes RAuPPh$_3$ (R = alkyl, phenyl or ferrocenyl) (Equation (58)).[329]

$$[Au(Fc)(PPh_3)] + Me_2NC(=S)S-SC(=S)NMe_2 \longrightarrow [Au(Fc)\{SC(=S)NMe_2\}_2] + [Au(Fc)_2\{SC(=S)NMe_2\}] + \cdots \quad (58)$$

Several other compounds with chelating sulfur-containing ligands are known.[330] Treatment of *cis*-[Au(C$_6$F$_5$)$_2$Cl(η^1-Ph$_2$PCH$_2$PPh$_2$)] with elemental sulfur leads to oxidation of the noncoordinated phosphorus centre to give the phosphine sulfide, which adopts a chelating bonding mode upon further reaction of the complex with AgClO$_4$. The resulting compound can be deprotonated with sodium hydride to give the methanide derivative, the crystal structure of which has been determined. The reactions of this product with a variety of gold and silver compounds readily give polynuclear complexes containing methylene carbon–metal bonds (Scheme 19).[331]

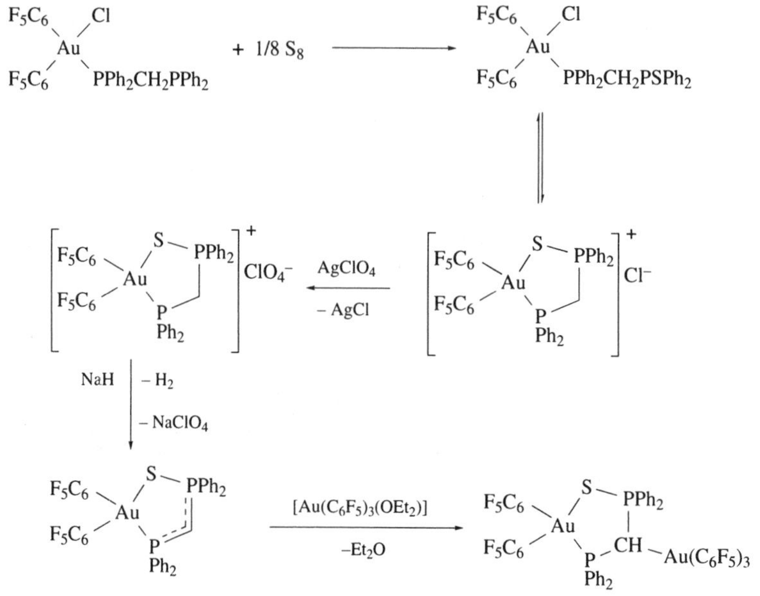

Scheme 19

1.2.6.3 Diorganylgold(III) complexes with group 15 donor ligands

The simplest complexes with nitrogen ligands are the oligomeric dimethylgold(III) amides which may be prepared by the reaction of [Au$_2$(μ-I)$_2$Me$_4$] with alkali metal amides. Depending on the reaction conditions, Au–N heterocycles of varying ring size are obtained (Scheme 20). All compounds are colourless solids, stable towards oxygen and moisture, but decompose upon exposure to light. The crystal structures show (Au–N)$_x$ rings with symmetrical amido bridges, but while the Au$_2$N$_2$ ring in [Au$_2$(μ-NHMe)$_2$Me$_4$] is puckered (d(Au–N)$_{av}$ = 0.214 nm; ∠(N–Au–N) = 77.4(2)°, ∠(Au–N–Au) = 92.5(2)°),[332] that in [Au$_2$(μ-NMe$_2$)$_2$Me$_4$] is planar, apparently owing to increased repulsion of the methyl groups (d(Au–N) = 0.214 0(5) nm; ∠(N–Au–N) = 82.0(2)°, ∠(Au–N–Au) = 98.0(2)°).[333] In the trimeric complex [Au$_3$(μ-NH$_2$)$_3$Me$_6$], the six-membered Au–N heterocycle adopts a chair conformation.[334]

Dimethylgold(III) nitrate forms mononuclear complexes with a wide range of polydentate nitrogen-donor ligands (L) containing imidazole,[335,336] pyridine,[335–7] pyrazole[335,337,338] and other functionalities. The gold(III) centre is invariably present in a square-planar *cis*-C$_2$AuN$_2$ coordination environment, with additional weak axial Au···N interactions in some cases. In solution, some of the compounds show fluxional behaviour involving pentacoordinate intermediates, as has been inferred from variable-temperature NMR data. Several complexes have also been studied by mass spectrometry.[339] Two examples of complexes of the general composition [AuMe$_2$L]NO$_3$ and [AuMe$_2$L] are given by Structures (**14**) and (**15**). By contrast, simple pyrazole ligands (pzH) react with dimethylgold(III) nitrate in water to give dimeric complexes of the type [AuMe$_2$(pz)]$_2$ (**16**).[340]

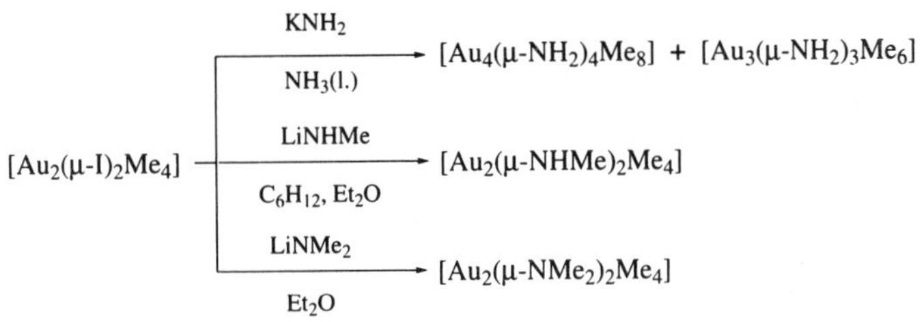

Scheme 20

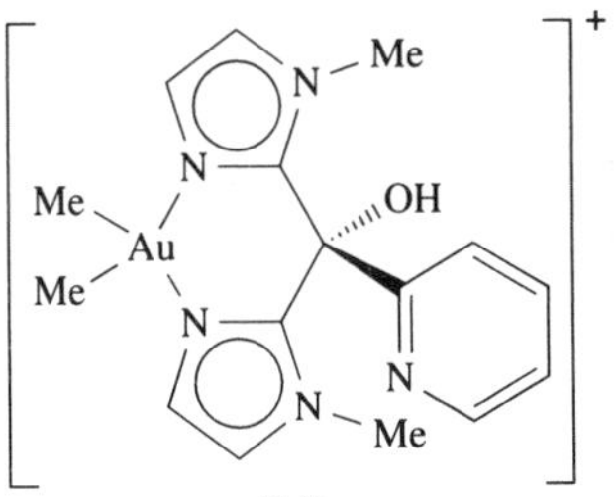

(14)

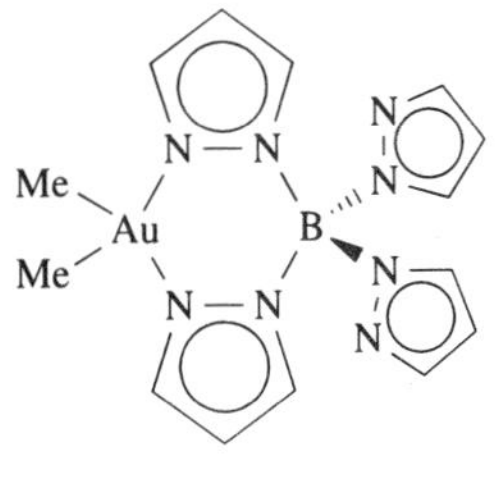

(15)

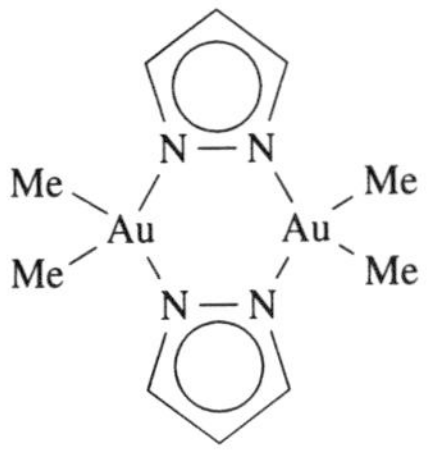

(16)

Structurally related complexes can also be prepared by the reaction of gold(III) dihalide precursors with Grignard reagents (Equation (59))[341] and by treatment of the auracyclopentadiene derivative [Au(C$_4$Ph$_4$)(acac)] with protonated nitrogen ligands, similar to the reactions depicted in Scheme 17.[326] The complexes prepared according to Equation (59) show interesting luminescence properties.[341]

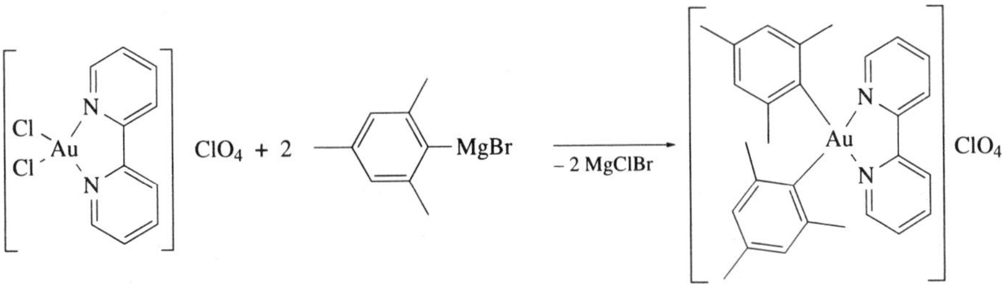

(59)

Several types of diarylgold(III) complex with chelating C,N donor ligands are known which can be prepared by arylation of suitable gold(III) precursors with organomercury(II) reagents.[69] One reaction is shown in Scheme 21,[342] and the analogous complexes with Ar = η^2-{C$_6$H$_4$(CH$_2$NMe$_2$)-2} have also been prepared.[343] Sequential arylation allows the synthesis of complexes with two different aryl ligands (Scheme 22).[344] The doubly C,N-chelated cationic complexes can be derivatized further by reactions with anionic (halide, cyanide, acetate) or neutral ligands (pyridine), which proceed by displacement of one of the intramolecular nitrogen donor groups to give the respective neutral or cationic products.

Sequential arylation has also been used to prepare diarylgold(III) complexes in which the second aryl ligand does not carry a potential donor group, and complexes such as (17) and (18) have been obtained from reactions analogous to the first step in Scheme 22.[345-8] The chloride ligand in these complexes is readily substituted, and bromide, iodide and acetate derivatives as well as a cationic pyridine adduct have been synthesized, together with other related species.

Upon treatment with phosphine or chloride, some of the C,N-chelated diarylgold(III) chloride complexes undergo reductive elimination to give biaryls. The high-yield reaction proceeds at room temperature and can be used for the synthesis of both symmetrical and unsymmetrical products. An example is given in Equation (60).[349]

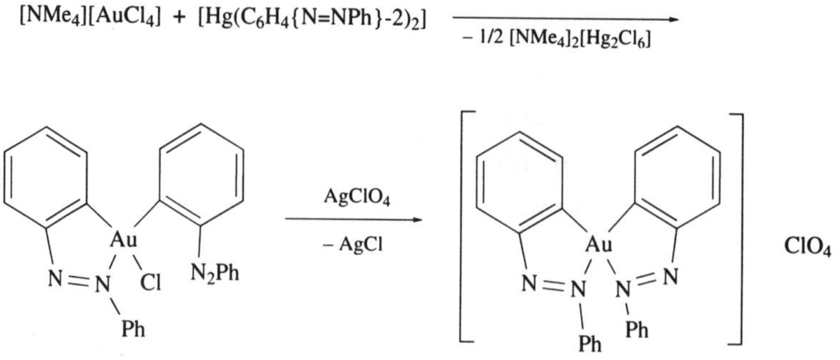

$$[NMe_4][AuCl_4] + [Hg(C_6H_4\{N=NPh\}-2)_2] \xrightarrow[-1/2\ [NMe_4]_2[Hg_2Cl_6]]{}$$

Scheme 21

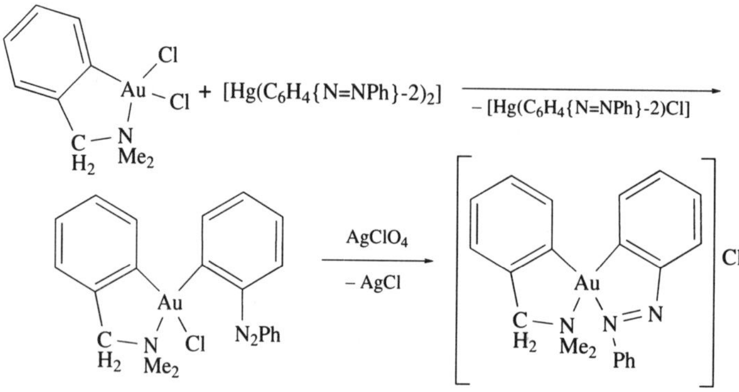

Scheme 22

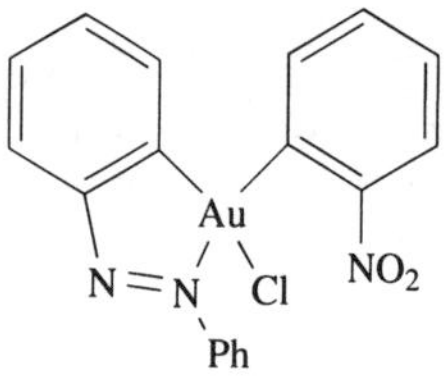

(17)

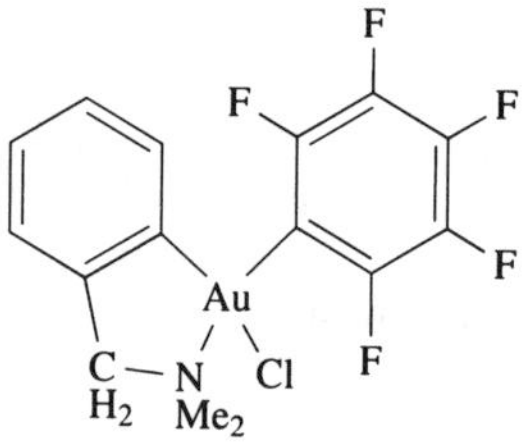

(18)

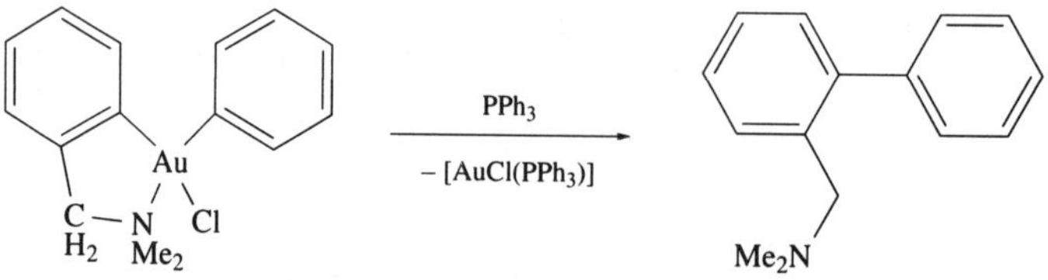

(60)

Gold(III) complexes with one aryl and one acetonyl ligand form as products of an unusual C–H activation of acetone,[350] which occurs by an intramolecular cooperative process which involves the metal centre and a phenyl(azophenyl) ligand attached to it. Thus, [Au{η^2-C$_6$H$_4$(N=NPh)-2}Cl$_2$], when treated in acetone with various reagents such as Tl(acac), KCN, AgClO$_4$, 1,10-phenanthroline or diarylmercury(II) compounds (e.g., [Hg(C$_6$F$_5$)$_2$]), gives [Au{η^2-C$_6$H$_4$(N=NPh)-2}(η^1-CH$_2$COMe)Cl] as the final product.[312] Some intermediates in this process have been isolated (Scheme 23), and the C–H activation of other ketones has also been studied.[351] Like the diaryl complexes described earlier, the aryl acetonyl complex can be derivatized in various ways, either by substitution of the chloride ligand or by cleavage of the gold–nitrogen bond with other donor ligands.[352]

Several diarylgold(III) complexes with monodentate alkyldiphenylphosphonio(diphenylphosphino)methane ligands have been prepared. These compounds react with sodium hydride to give the corresponding methanide complexes, which in turn may be transformed into polynuclear gold(III)–gold(I) complexes upon treatment with gold(I) reagents containing labile ligands. An example of this reaction sequence is given in Scheme 24.[330]

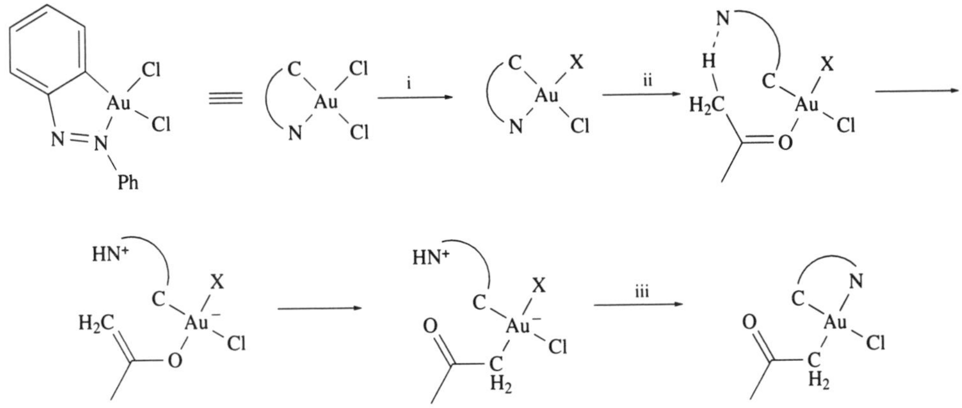

i, X = acac-C, CN, $C_6H_4(N=NPh)$-2, C_6F_5, or $C_6H_4(NO_2)$-2; ii, + Me_2CO; iii, – HX

Scheme 23

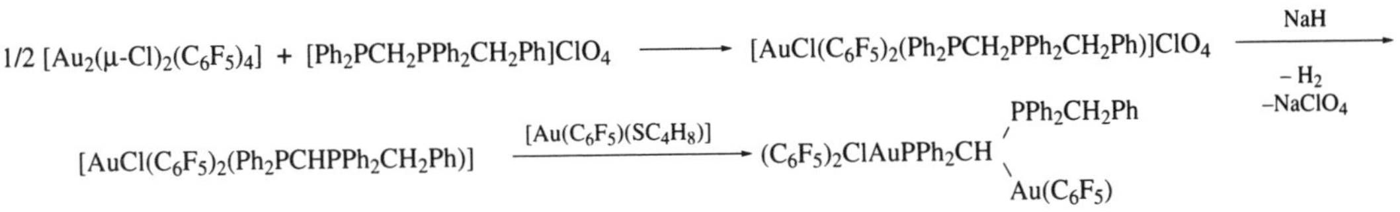

Scheme 24

Most other reported diorganylgold(III) complexes with phosphorus donor ligands contain difunctional phosphines such as dppm or closely related species.[258,330] Various complexes can be prepared by a sequence of derivatizations.[104,353,354] Dimeric $[Au_2(\mu\text{-}Cl)_2(C_6F_5)_4]$ may be cleaved with dppm to give $[Au(C_6F_5)_2Cl(dppm)]$, which on reaction with $AgClO_4$ forms the cationic complex $[Au(C_6F_5)_2(dppm)]ClO_4$. This is transformed into the deprotonated complex $[Au(C_6F_5)_2(Ph_2PCHPPh_2)]$ by treatment with sodium hydride, and the methanide derivative can then react further with gold(I) complexes to yield di- or trinuclear complexes bridged by diphenylphosphinomethanido groups, such as $[(C_6F_5)_2Au\{(Ph_2P)_2CHAu(C_6F_5)\}]$ or $[[(C_6F_5)_2Au\{(Ph_2P)_2CH\}]_2Au]ClO_4$. The preparation of the latter complex is summarized in Scheme 25. Renewed deprotonation and metallation to give polynuclear methanediide derivatives is also possible. Again, there are few limitations to the range of substituents in the various reagents which have been used. A comprehensive review of these complexes has been published.[330]

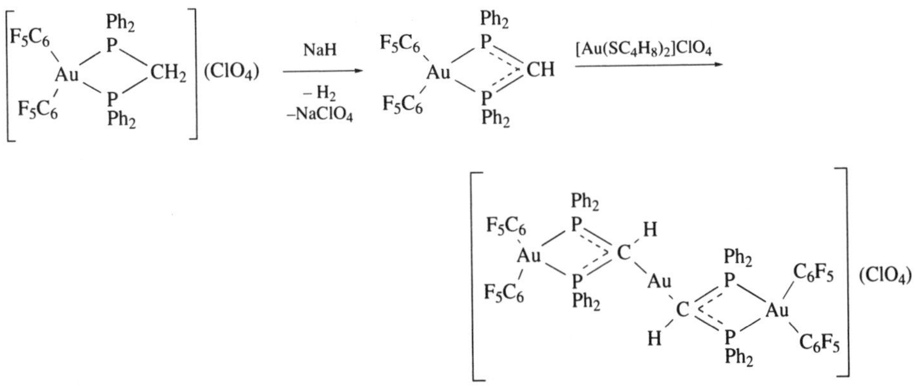

Scheme 25

1.2.7 Gold(III) Complexes with Three Gold–Carbon Bonds

The principal methods available for the synthesis of compounds containing alkyl and/or aryl ligands are the reactions of suitable gold(III) halide complexes with organolithium or Grignard reagents and the oxidative arylation of gold(I) complexes with organothallium(III) compounds.[69,71] Specific examples of these reactions are given in Equations (61),[355] (62)[356] and (63),[357] and a selection of recently prepared complexes is presented in Table 12.

$$Au_2Br_6 \xrightarrow[\text{ii, 2 PMe}_3]{\text{i, 6 MeLi, } - \text{6 LiBr}} 2\ [AuMe_3(PMe_3)] \tag{61}$$

$$cis\text{-}[AuIMe_2(PPh_3)] + PhMgBr \xrightarrow[-\text{ MgBrI}]{} cis\text{-}[AuMe_2Ph(PPh_3)] \tag{62}$$

$$[Au(C_6F_5)(SC_4H_8)] + [Tl(C_6F_5)_2Cl] \xrightarrow[-\text{ TlCl}]{} [Au(C_6F_5)_3(SC_4H_8)] \tag{63}$$

Table 12 Some gold(III) complexes of the types $[AuR^1_3L]$ and $[AuR^1_2R^2L]$.

Complex	m.p. (°C)	Ref.
$[AuMe_3(PMe_3)]$		355
$[Au(CF_3)_3(PMe_3)]$		77
$[Au(CF_3)_3(PEt_3)_3]$		78
$cis\text{-}[AuMe_2Et(PPh_3)]$		358
$cis\text{-}[AuMe_2(trans\text{-}MeCH=CH)(PPh_3)]$		358
$cis\text{-}[AuMe_2(C{\equiv}CPh)(PPh_3)]$		358
$cis\text{-}[AuMe_2Ph(PPh_3)]$	101 (d)	358
$cis\text{-}[AuMe_2Ph(P\{C_6H_4(OMe)\text{-}4\}_3)]$	113 (d)	356
$cis\text{-}[AuMe_2Ph(P\{C_6H_4F\text{-}4\}_3)]$	115 (d)	356
$cis\text{-}[AuMe_2(CO_2Me)(PPh_3)]$	124 (d)	359
$cis\text{-}[AuMe_2(CO_2Et)(PPh_3)]$	130 (d)	359
$cis\text{-}[AuMe_2(CO_2Pr^i)(PPh_3)]$	120 (d)	359
$[Au(C_6F_5)_3(SC_4H_8)]$	190	357
$[Au(C_6F_5)_3(OEt_2)]$		360
$[Au(C_6F_5)_3(O{=}CMe_2)]$		360
$[Au(C_6F_5)_3(OAsPh_3)]$	157	360
$[Au(C_6F_5)_3(\eta^1\text{-}Ph_2AsCH_2AsPh_2)]$	197	138
$[Au(C_6F_5)_3(\eta^1\text{-}SPh_2PCH_2PPh_2Me)]ClO_4$	138	255
$[Au_2(C_6F_5)_6(\mu\text{-}Ph_2P\text{-}\{Fc\}\text{-}PPh_2)]^a$	260	144
$[NBu_4][Au_2(C_6F_5)_6(\mu\text{-}S_2CNEt_2)]$	203	328
$[Au_2(C_6F_5)_6(\mu\text{-}S_2CPBu_3)]$	152	148
$[Au(C_6Cl_5)_3(SC_4H_8)]$	150 (d)	361
$[N(PPh_3)_2][Au(C_6Cl_5)_3Cl]$	145 (d)	361

^{a} Fc = 1,1'-Ferrocenediyl

A study of the bonding situation in a series of trimethylgold(III) phosphine complexes has appeared, in which the role played by the gold $5d$ orbitals was evaluated on the basis of photoelectron spectra. While the Au–C bonds apparently have both gold $5d$ and gold $6s/6p$ character, the Au–P bond has mainly gold $5d$ and gold $6s$ character, the dominating contribution being that of the gold $5d_{x^2-y^2}$ orbital.[355]

Some tris(trifluoromethyl)gold(III) complexes are also known. $[Au(CF_3)_3(PMe_3)]$ can be synthesized in high yield by treatment of $[AuI(CF_3)_2(PMe_3)]$ with $[Cd(CF_3)_2]\cdot DME$ in the presence of excess trifluoromethyliodide (Equation (64)).[78] The complex has also been obtained by donor ligand stabilization of $[Au(CF_3)_3]$, which can be generated by co-condensation of gold atoms with CF_3 radicals.[362] The related complex $[Au(CF_3)_3(PEt_3)]$ is reported to form when CF_3I and $[Au(CF_3)(PEt_3)]$ are reacted over extended periods of time (>10 d).[79]

$$2\ [AuI(CF_3)_2(PMe_3)] + [Cd(CF_3)_2]\cdot DME \xrightarrow[-\text{ CdI}_2]{CF_3I} 2\ [Au(CF_3)_3(PMe_3)] \tag{64}$$

The mixed compounds of the type $[AuMe_2R^1(PR^2_3)]$ (R^1 = alkenyl, alkynyl or aryl; R^2 = alkyl or aryl) often have *cis* stereochemistry, as indicated by NMR data.[356,363,364] When heated, both the trimethyl- and the mixed *cis*-dimethyl(organyl)(phosphine)gold(III) complexes undergo reductive elimination of two organic groups after dissociation of the phosphine ligand, a reaction which is important for the catalytic coupling between alkyllithium reagents and alkyl halides.[355] As illustrated in Scheme 26, two types of product may form in the case of the mixed complexes. While reductive elimination of R–Me is predominant when R = alkenyl or aryl, the formation of ethane is favoured when R is an alkynyl or an electron-withdrawing alkyl group.[358] In addition to electronic effects, the selectivity of the process also depends on the steric demand of both the phosphine ligand and R.[356,363] In another type of reaction, dimethylarylgold(III) complexes undergo selective cleavage of the gold–aryl bond when treated with electrophiles such as HCl, HgCl$_2$ or [PtI$_2$(PMe$_2$Ph)$_2$], the only gold-containing products being complexes of the type *cis*-[AuXMe$_2$(PPh$_3$)] (X = Cl or I).[364]

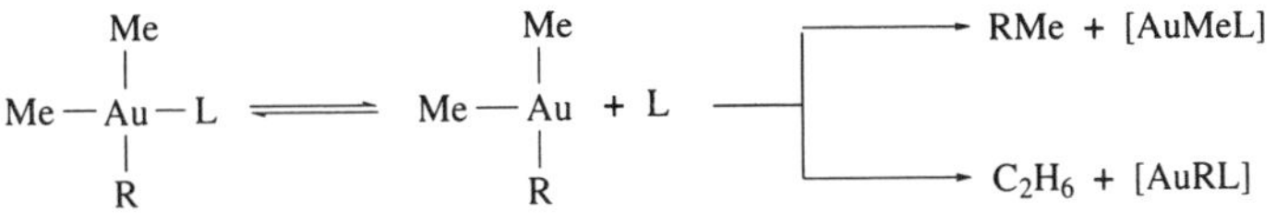

$$\begin{array}{c} \text{Me} \\ | \\ \text{Me}-\text{Au}-\text{L} \\ | \\ \text{R} \end{array} \rightleftharpoons \begin{array}{c} \text{Me} \\ | \\ \text{Me}-\text{Au} \\ | \\ \text{R} \end{array} + \text{L} \longrightarrow \begin{cases} \text{RMe} + [\text{AuMeL}] \\ \\ \text{C}_2\text{H}_6 + [\text{AuRL}] \end{cases}$$

Scheme 26

A related group of compounds is that of the gold(III) dimethyl(alkoxycarbonyl) complexes, accessible by the reaction of carbon monoxide with dimethyl(alkoxy)(triphenylphosphine)gold(III), which is prepared *in situ* from *cis*-[AuIMe$_2$(PPh$_3$)] and sodium alkoxide in methanol (Equation (65)).[359,365] Thermolysis of the methoxycarbonyl complex in benzene leads to reductive elimination of methyl acetate and ethane, indicating competition between the two modes of decomposition illustrated in Scheme 26. The reaction of the same complex with electrophiles such as hydrogen chloride proceeds with liberation of carbon monoxide and methanol (Equation (66)).

$$cis\text{-}[\text{AuIMe}_2(\text{PPh}_3)] + \text{NaOMe} + \text{CO} \xrightarrow{\text{MeOH}} cis\text{-}[\text{AuMe}_2(\text{CO}_2\text{Me})(\text{PPh}_3)] \qquad (65)$$

$$cis\text{-}[\text{AuMe}_2(\text{CO}_2\text{Me})(\text{PPh}_3)] + \text{HCl} \longrightarrow \text{MeOH} + \text{CO} + cis\text{-}[\text{AuClMe}_2(\text{PPh}_3)] \qquad (66)$$

As regards triarylgold(III) complexes, only polyhalogenophenyl derivatives are known. [Au(C$_6$F$_5$)$_3$(SC$_4$H$_8$)] is a useful precursor for the preparation of a wide variety of other compounds by ligand substitution. It reacts with neutral (L = NH$_3$, C$_5$H$_5$N, P(OPh)$_3$, AsPh$_3$, SbPh$_3$ or CNC$_6$H$_4$Me-4)[357] or anionic ligands (X = Cl, Br, I, SCN or N$_3$)[357,361] to give complexes of the types [Au(C$_6$F$_5$)$_3$L] and [Au(C$_6$F$_5$)$_3$X]$^-$, and dinuclear derivatives (e.g., (**19**)[328] and (**20**)[148]) have been similarly obtained from the reaction with bridging ligands.[138,144,148,357] The complex [Au(C$_6$F$_5$)$_3$(OEt$_2$)],[360] prepared according to Equation (67), is another useful starting material as the ether molecule is displaced even more readily, allowing the synthesis of compounds which are inaccesible from the thiophene adduct (Equation (68)).[150,255]

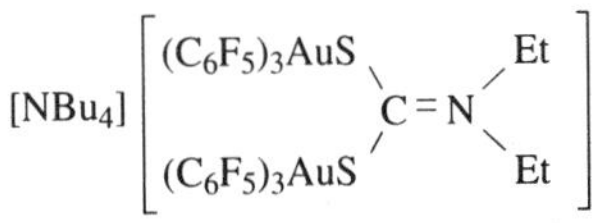

$$[\text{NBu}_4] \begin{bmatrix} (\text{C}_6\text{F}_5)_3\text{AuS} \\ \diagdown \\ \quad\quad\quad \text{C}=\text{N} \\ \diagup \\ (\text{C}_6\text{F}_5)_3\text{AuS} \end{bmatrix} \begin{matrix} \text{Et} \\ \diagup \\ \diagdown \\ \text{Et} \end{matrix}$$

(**19**)

$$\begin{array}{c} (\text{C}_6\text{F}_5)_3\text{AuS} \\ \diagdown \\ \quad\quad\quad \text{C}-\text{PEt}_3 \\ \diagup \\ (\text{C}_6\text{F}_5)_3\text{AuS} \end{array}$$

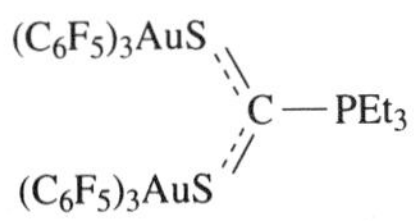

(**20**)

$$[\text{NBu}_4][\text{Au}(\text{C}_6\text{F}_5)_3\text{Br}] + \text{AgClO}_4 \xrightarrow[-\text{AgBr}, -[\text{NBu}_4]\text{ClO}_4]{\text{Et}_2\text{O}} [\text{Au}(\text{C}_6\text{F}_5)_3(\text{OEt}_2)] \qquad (67)$$

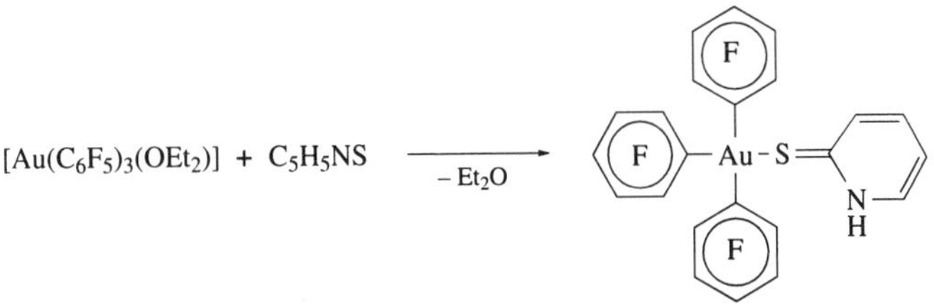

$$[\text{Au}(\text{C}_6\text{F}_5)_3(\text{OEt}_2)] + \text{C}_5\text{H}_5\text{NS} \xrightarrow{-\text{Et}_2\text{O}} \text{[structure]} \qquad (68)$$

Gold(III) complexes with two aryl ligands and one methanide ligand have also been prepared, as illustrated in Scheme 27.[330,366,367] The synthesis starts from diarylgold(III) precursors containing substituted phosphinophosphonium ligands, which adopt a chelating bonding mode on deprotonation. Depending on the reaction conditions, chelate rings of varying size may be obtained. Thus, Na$_2$CO$_3$ or AgClO$_4$ as deprotonating agents effect single deprotonation and simultaneous halide abstraction, leading to cationic four- or five-membered auracycles, while NaH deprotonates both CH$_2$ groups to give neutral complexes containing five- or six-membered rings, whose methanide carbon atoms can serve as electron donors to other metal centres, thus forming polynuclear derivatives. Sequential deprotonation and metallation of the same methylene group is also possible, giving five-membered methanediide auracycles with a [>C(AuPPh_3)_2] structural unit which resembles that found in certain diauriomethanes.[368] Analogous reactions have also been carried out with gold(III) complexes of bis(diphenylphosphino)methane disulfide.[149]

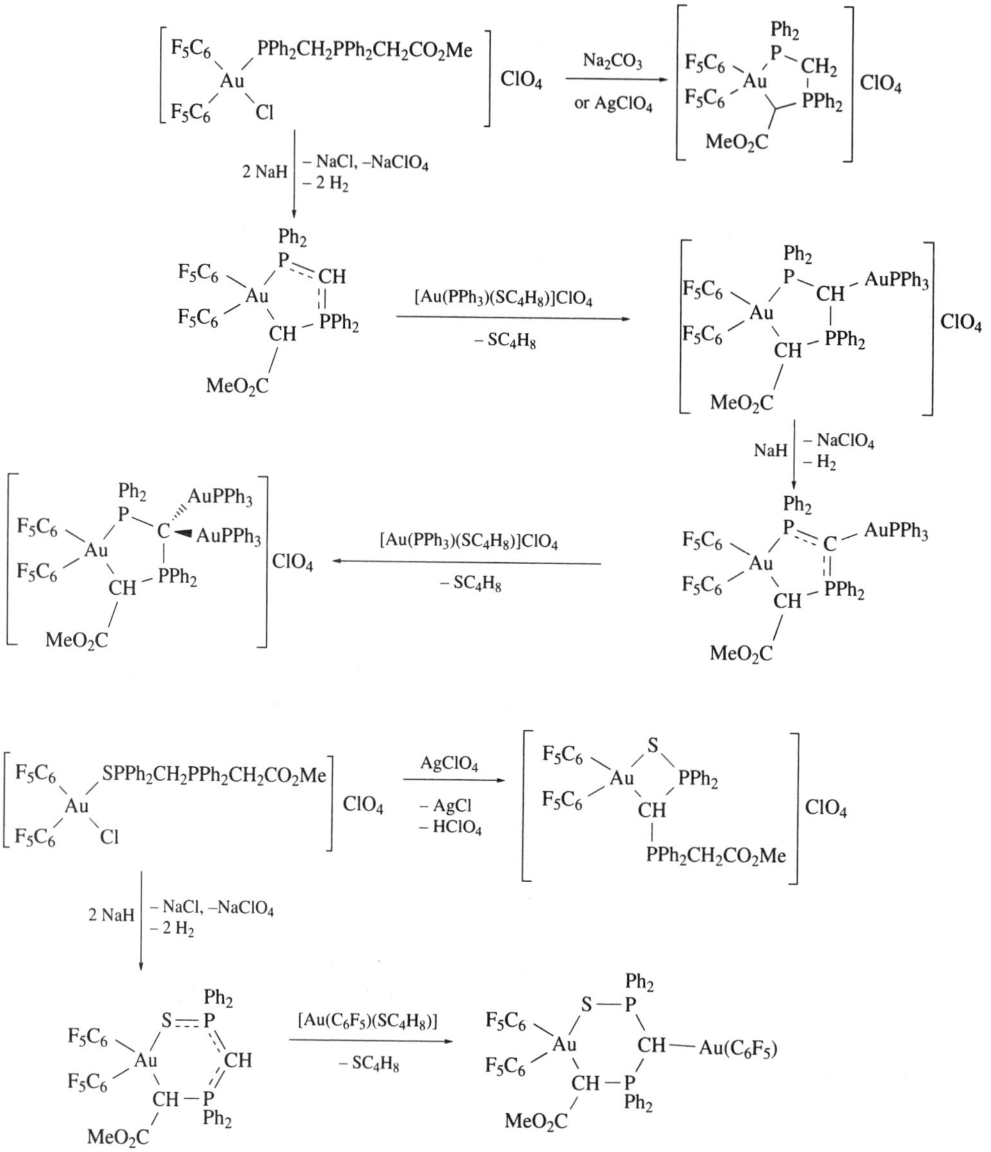

Scheme 27

1.2.8 Gold(III) Complexes with Four Gold–Carbon Bonds

The simplest gold(III) complexes with four gold–carbon bonds are the tetraalkylaurates(III), prepared by the reaction of trialkyl(triphenylphosphine)gold(III) complexes with alkyllithiums, which proceeds by phosphine displacement as illustrated in Equation (69). Recent studies have shown the reaction to be stereoselective, with *cis-* (or *trans-*) dimethylalkylgold(III) complexes giving square-planar *cis-* (or *trans-*) tetraalkylaurates, and an associative mechanism involving a pentacoordinate gold(III) intermediate has been postulated.[369]

Several tetraarylaurates(III) are known, and the crystal structures of both the tetraphenyl and the tetrakis(pentafluorophenyl) derivatives have been reported. In [NBu_4][AuPh_4], the anion contains four

$$cis\text{-}[AuMe_2Et(PPh_3)] + LiEt \xrightarrow[-PPh_3]{} Li\{cis\text{-}[AuMe_2Et_2]\} \qquad (69)$$

phenyl rings surrounding the central gold(III) atom in a square-planar arrangement, with phenyl groups *trans* to each other lying in the same plane. These planes are tilted so that two pairs of *cis* phenyl rings have *ortho* carbon atoms which approach each other, and the anion does not have a fourfold axis normal to the coordination plane defined by gold and the four *ipso* carbon atoms.[370] The structure is thus at variance with that of the salt $[N(PPh_3)_2][Au(C_6F_5)_4]$, in which the anions show a propeller-like arrangement of ligands (D_4 symmetry).[274] Tetraarylaurate(III) complexes containing polyhalogenophenyl ligands[69] are best prepared by further arylation of triarylhalogenoaurate(III) with $Ag(C_6F_5)$ (Equation (70)).[357]

$$[N(PPh_3)_2][Au(C_6F_5)_3Cl] + Ag(C_6F_5) \xrightarrow[-AgCl]{} [N(PPh_3)_2][Au(C_6F_5)_4] \qquad (70)$$

A very interesting mixed complex is the stable hydrogen dimethylbis(2-pyridyl)aurate(III), prepared by the reaction of *cis*-dimethyliodo(triphenylphosphine)gold(III) with 2-pyridyllithium at low temperature and subsequent hydrolysis.[371] NMR data and data obtained from thermolysis experiments in the presence of protic and aprotic deuterated solvents suggest the presence of an intramolecularly bound proton, as illustrated in Equation (71). The compound is remarkably stable towards oxygen and moisture, and can be handled in air.

$$(71)$$

Finally, a number of gold(III) carbene and ylide complexes are known in which the metal forms four gold–carbon bonds. These are treated in the appropriate sections.

1.3 HOMO- AND HETEROMETALLIC GOLD CLUSTERS CONTAINING GOLD–CARBON BONDS

Of the many gold cluster compounds in which the metal has an oxidation state between 0 and +1, only a few also feature bonds between gold and carbon.[61,71] A recent addition to this group of compounds is the neutral cluster $[Au_{10}(C_6F_5)_4(PPh_3)_5]$, prepared selectively by the reaction of $[Au_9(PPh_3)_8](NO_3)_3$ with $NBu_4[Au(C_6F_5)_2]$ in a 1:3 molar ratio. The cluster skeleton contains a hexagonal ring of six edge- and face-sharing $Au(AuL)_3$ tetrahedra with a common central gold atom, and the triphenylphosphine and pentafluorophenyl ligands are arranged around the cluster core in such a way that the overall symmetry is approximately C_{2v}.[372] Several other gold phosphine clusters containing Au–C bonds have isonitriles as peripheral ligands. Species such as $[Au_8(PPh_3)_7(CN\text{-}Bu^t)](NO_3)_2$ and $[Au_9(PPh_3)_6(CNBu^t)_2](NO_3)_3$ have been prepared by the addition of one equivalent of isonitrile to the starting material $[Au_8(PPh_3)_7](NO_3)_2$ and by isonitrile-induced displacement of phosphine ligands in $[Au_9(PPh_3)_8](NO_3)_3$, respectively, and the reactivity of the products towards halide ion and amines has also been studied.[373] Similarly, ligand displacement in related platinum–gold clusters has led to species such as $[PtAu_8(PPh_3)_7(CNBu^t)_2](NO_3)_2$.[374]

Heterometallic cluster compounds of gold can often be prepared by treating suitable substrates with aurating agents such as $[AuCl(PPh_3)]$ or $[O(AuPPh_3)_3][BF_4]$.[375] The reaction is important because of the isolobal analogy between H^+ and gold(I)–ligand fragments, and has served to determine the extent of structural analogies between the corresponding hydrido and gold-containing clusters.[55] A variety of substrates has been studied in this context, some of them containing organic groups such as alkyne[376] or carbido ligands.[377] In the case of such compounds, auration may involve the metal framework and/or the organic ligand, and several mixed-metal gold clusters containing gold–carbon bonds have been prepared in this way.[378] In a related reaction, it has been possible to replace an agostic hydrogen atom of a metal-coordinated allyl group by an isolobal gold ligand.[379] Examples of the various kinds of compound which have been prepared are collected in Figure 12.

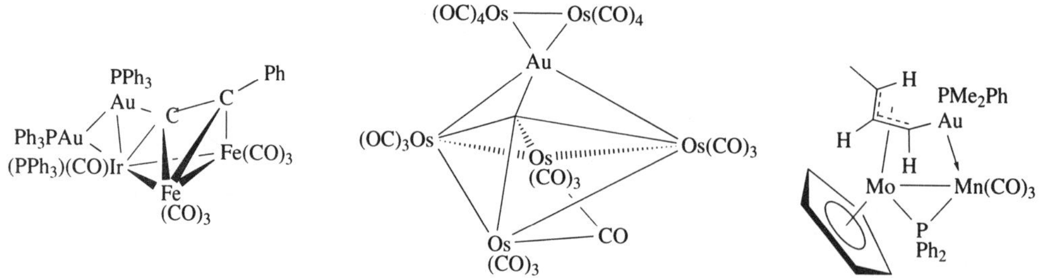

Figure 12 Heterometallic cluster compounds of monovalent gold containing gold–carbon bonds.[376,377,379]

1.4 SYNTHESIS AND PROPERTIES OF ALKENE, ALKYNE AND RELATED COMPLEXES OF GOLD

Complexes of zerovalent gold (in the form of atoms or the surface of the bulk metal) and unsaturated molecules such as alkenes,[380,381] alkynes,[381,382] allenes,[383] arenes[384] or fullerenes[385] have been the subject of a number of experimental studies. Molecular species, which are formed by low-temperature cocondensation of gold atoms with unsaturated organic substrates, have been characterized by their infrared/Raman, UV–visible absorption and ESR spectra, but the interpretation of these data is not without contradictions. In the case of ethene and propene,[380,381] analysis of the ESR spectra has suggested the presence of π-coordinated gold(0)–mono(alkene) and gold(0)–bis(alkene) complexes, the latter featuring the metal atom flanked by two ligand molecules oriented parallel to each other. With ethyne, only the existence of the π-bonded mono(alkyne) complex has been established.[381] Both kinds of complex have been suggested to be stabilized by the dative interactions of the Dewar–Chatt–Duncanson scheme, with donation of π-electron density to the metal and back-donation from the metal *d*-orbitals to the molecular π*-orbitals. By contrast, another ESR study of the gold(0)–ethyne system has presented evidence for the formation of the σ-bonded vinyl species $[AuCH=(C\cdot)H]$ and $[Au(C\cdot)=CH_2]$, while no mono(alkyne) complex of the type $[Au(HC\equiv CH)]$ appears to be present.[382] Radical species with unpaired electrons on carbon have similarly been postulated as the products of the reaction of gold atoms with allene, as monitored by ESR spectroscopy.[383] The bonding situation in gold(0)–alkene complexes has been the subject of a recent theoretical study, which takes into account UV–visible spectroscopic data and summarizes the findings of earlier experimental work.[386]

Alkene complexes of gold(I) can be prepared either by the reaction of tetrachloroauric acid with the alkene or directly from the alkene and gold(I) halide.[71] While such species usually have poor thermal stability and decompose in solution at room temperature, the *cis*-cyclooctene and norbornene complexes of AuCl (obtained from [AuCl(CO)] and the corresponding alkene by CO displacement) have been found to be less prone to decomposition, and an x-ray study of the former has been reported.[387] The structure of $[AuCl(cis\text{-}C_8H_{14})]$ consists of discrete AuCl(alkene) molecules, with nonbonding intermolecular Au···Au distances of 0.376 nm. The coordination geometry at gold is essentially linear, the angle between the Au–Cl bond (0.226 6(4) nm) and the vector connecting the gold atom with the centre of the C=C double bond (0.207(2) nm) being 179.2(6)°. The C=C bond distance is 0.138(2) nm and thus lies within the range of values normally encountered in alkene complexes of univalent cations.

It has been noted that metal–carbon triple bonds behave in a similar way to alkynes with respect to π-complex formation,[71] and various heterometallic gold complexes with bridging carbyne ligands have been synthesized on this basis. The alkylidyne complexes $[W(\equiv CR)(CO)_2Cp]$, $[W(\equiv CR^1)(CO)_2(\eta^5\text{-}C_2B_9H_9R^2_2)]^-$ (R^1 = Me or Tol-4; R^2 = H or Me), and analogues containing other isolobal ligands, readily react with gold(I) chlorides such as $[AuCl(PPh_3)]$ or $[AuCl(SC_4H_8)]$ to give dimetallacyclopropene species, formed by addition of a gold(I) fragment to the metal–ligand multiple bond (Figure 13).[388,389] Trimetal compounds in which an alkylidyne ligand caps a triangle of metal atoms have been obtained in an analogous way.[390] A review of these gold complexes and related copper species has recently been published.[391]

1.5 SYNTHESIS, PROPERTIES AND REACTIONS OF CARBENE COMPLEXES OF GOLD

Carbene complexes of gold are accessible by a variety of transformations,[71] the most general being the reaction of gold isocyanide complexes with alcohols[392] or amines.[393] Thus, both mono- and bis(carbene) species have been prepared. While the former can be reacted with base to give

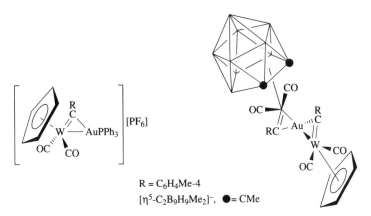

R = C$_6$H$_4$Me-4
[η^5-C$_2$B$_9$H$_9$Me$_2$]$^-$, ● = CMe

Figure 13 Dimetallacyclopropene complexes derived from addition of gold(I) fragments to metal–alkylidyne complexes.

iminomethylgold(I) complexes (Section 1.2.1), oxidation of the latter by Br$_2$ or I$_2$ generates carbene complexes of gold(III), without cleavage of the Au–C bonds (Scheme 28). Specific studies concerning the reactions of gold carbene complexes under mass spectrometric conditions have also been reported.[394,395]

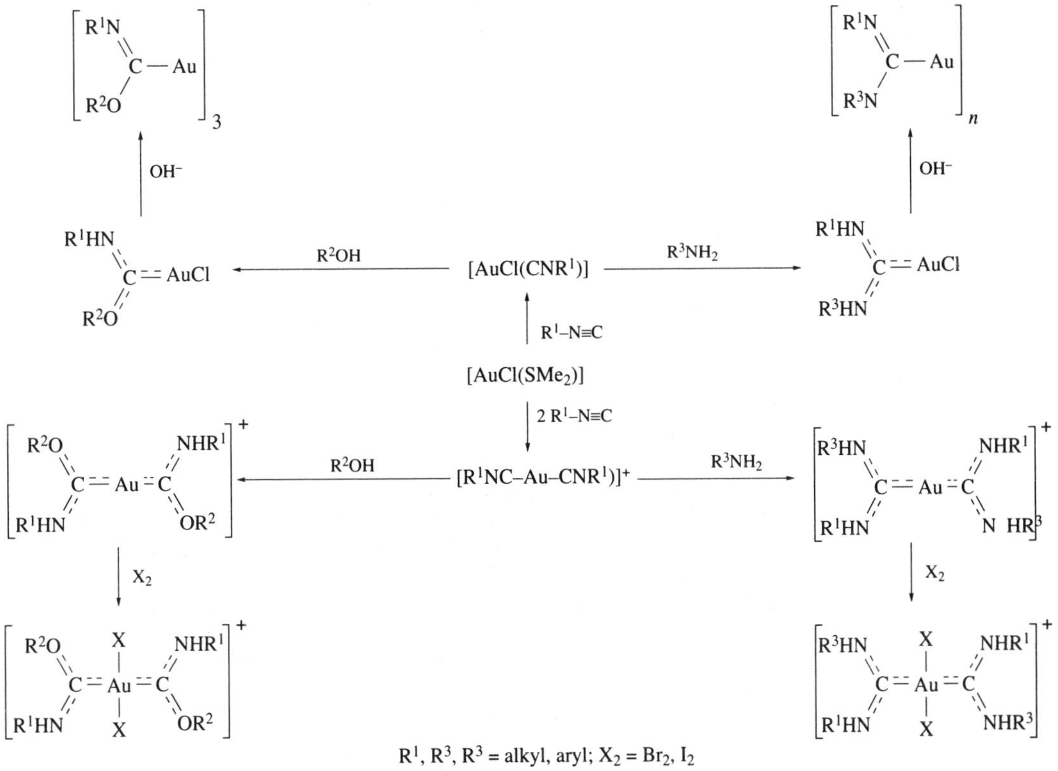

R^1, R^3, R^3 = alkyl, aryl; X$_2$ = Br$_2$, I$_2$

Scheme 28

Complexes containing chelating or bridging bis(carbene) ligands have been similarly prepared, either by the reaction of suitable (isocyanide)gold compounds with difunctional amines[396] or by the reaction of (diamine)gold species with isocyanides.[135] In some of these complexes of gold(I) or gold(III), the gold atom is surrounded solely by carbon-coordinated ligands. In the case of difunctional isocyanide ligands with the isocyano groups in close proximity, the reaction of their dinuclear gold complexes with amines is characterized by an interesting neighbouring group participation, leading to NC coupling of the two isocyano functions to give a benzimidazole ring system (Scheme 29).[397]

In the mono- and bis(carbene) complexes, the C–N/C–O and Au–C bonds have partial double-bond character, and rotation about these bonds is therefore restricted, giving rise to geometrical isomers.

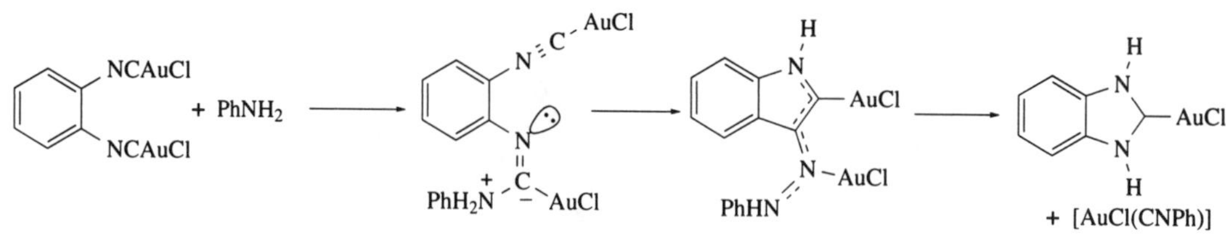

Scheme 29

These can be identified as separate species in solution, although slow interconversion is observed in some cases. Some of the possible canonical structures are summarized in Structures (**21**)–(**26**).[393]

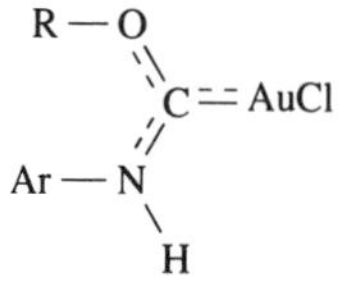

cis, cis
(**21**)

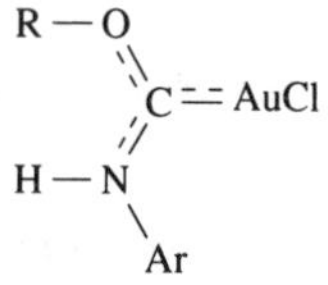

cis, trans
(**22**)

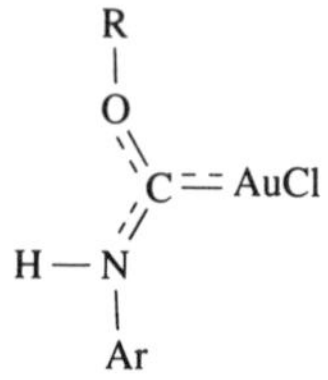

trans, trans
(**23**)

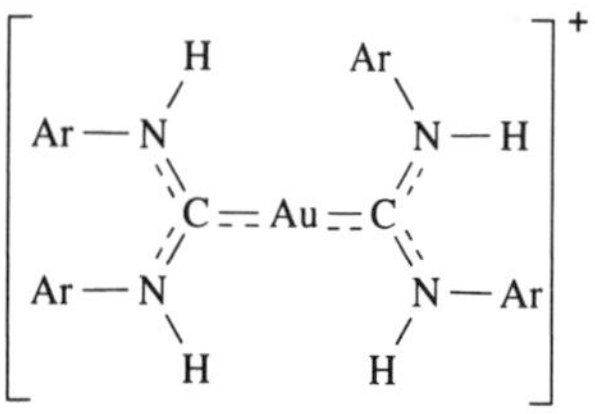

(*cis, cis*)$_2$
(**24**)

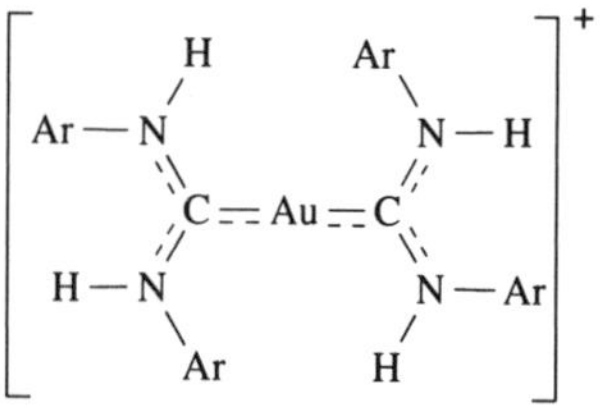

(*cis, trans*)$_2$
(**25**)

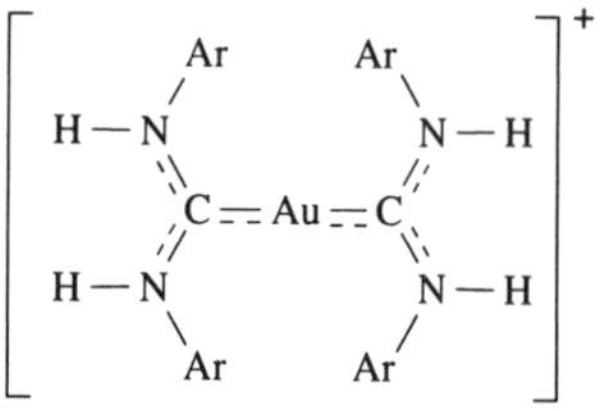

(*trans, trans*)$_2$
(**26**)

While the isomers of a given complex usually show identical Mössbauer spectra,[393] a number of structures have been unequivocally established by x-ray methods. The complex [AuCl{C(NHTol-4)(OEt)}] has the substituents in the *cis,trans* configuration, (**22**), and the arrangement of the molecules in the solid state is such that a weak intermolecular Au···Au interaction with *d*(Au···Au) = 0.333 nm can be observed.[393] Another structure of interest is that of the bis(carbene) complex [Au{C(NHTol-4)}$_2$]BF_4, where both carbene ligands are in the *trans,trans* configuration, (**26**). This arrangement, albeit unexpected for reasons of steric hindrance, allows the formation of two hydrogen bonds between each carbene ligand and the BF_4^- counterions.[398] Similar hydrogen bonding is found for the *cis,cis* isomer of the bis(carbene) complex [Au{C(NHTol-4)(OEt)}$_2$]ClO_4. The *trans,trans* isomer has also been isolated (from the same solution as its *cis,cis* congener), but no structural peculiarities are observed in this case.[393]

Another versatile method for the preparation of gold–carbene complexes involves the use of suitable chromium, molybdenum or tungsten pentacarbonyl compounds as carbene ligand transfer agents, and a number of carbene complexes of both gold(I) and gold(III) have been prepared in this way from $HAuCl_4$.[399–401] By way of halogen exchange, carbenegold(III) trichloride may subsequently be transformed into the carbenegold(III) tribromide and triiodide complexes through the action of BBr_3 and BI_3, respectively. An overview of relevant reactions is given in Scheme 30. Dinuclear gold complexes with bridging bis(carbene) ligands can be similarly prepared from the corresponding dinuclear tungsten bis(carbene) precursors.[402]

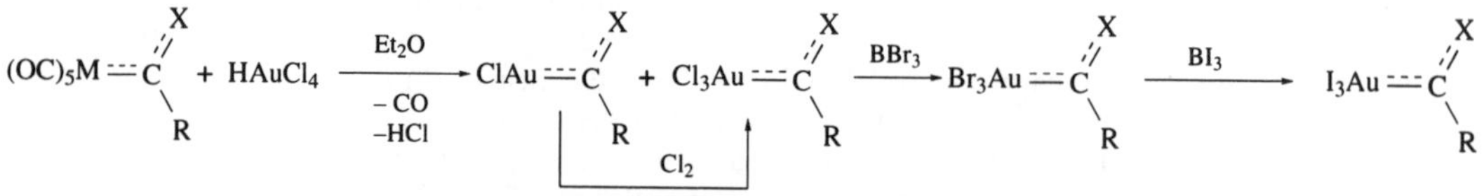

M = Cr, Mo, W; X = OMe, NH_2, NHMe, NMe_2; R = alkyl, aryl

Scheme 30

It is interesting to note that transfer of the carbene ligand generally proceeds with retention of configuration at the C–X bond, as has been concluded from NMR spectroscopic data.[401] This excludes a reaction mechanism in which the carbene carbon atom temporarily assumes sp^3 character, as would be the case in a dimetallacyclopropane intermediate. Instead, the gold chloride moiety is likely to insert into the metal–carbene bond of the carbene transfer agent, thereby maintaining the partial double-bond character of the bonds within the carbene ligand at all times. Reductive elimination of a (carbonyl)metal halide then concludes the reaction sequence to give a gold(I)–carbene complex (Scheme 31).

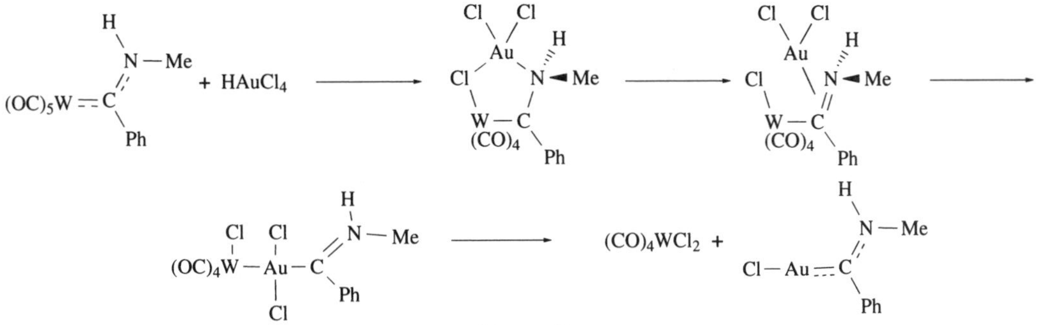

Scheme 31

Various gold carbene complexes with heterocyclic carbene ligands, derived from imidazoles[137,151,164,403] or thiazoles,[404] have also been reported. Most commonly, such complexes are obtained on addition of the carbon-lithiated heterocycle to gold(I) chloride compounds and subsequent protonation or alkylation of the products formed. In this manner, bis(carbene) species have been prepared, as well as carbene complexes which contain halide, phosphine or pentahalogenophenyl groups as a second ligand, as shown in Scheme 32.

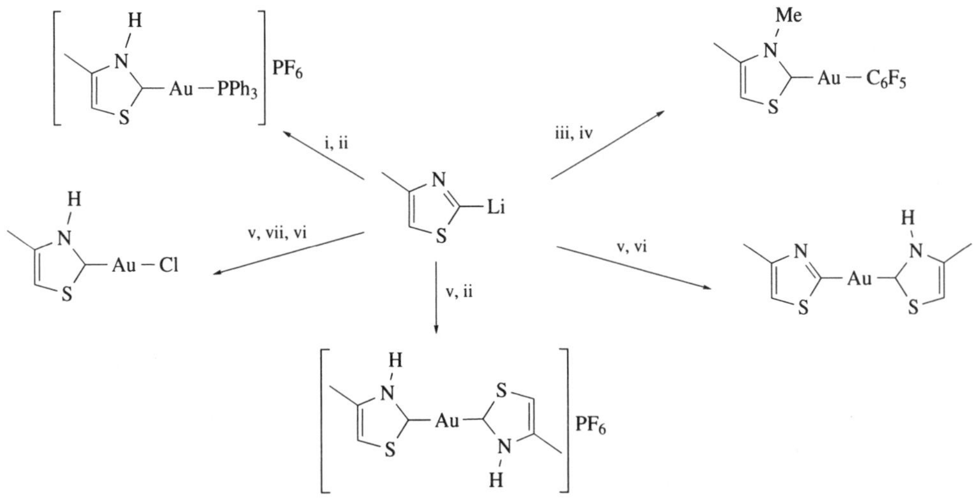

i, $[AuCl(PPh_3)]$; ii, HPF_6, iii, $[Au(C_6F_5)(SC_4H_8)]$; iv, MeOTf; v, $[AuCl(SC_4H_8)]$; vi, HCl, vii, MeLi

Scheme 32

1.6 SYNTHESIS AND PROPERTIES OF CARBONYL COMPLEXES OF GOLD

Co-condensation of gold atoms with carbon monoxide has yielded several gold(0) carbonyls, among them $Au(CO)$[405] and $Au(CO)_2$,[406] which have been trapped in inert-gas matrices at 6–10 K and studied by ESR, IR and UV spectroscopy. Similar experimental conditions have also allowed the observation of dinuclear $Au_2(CO)_4$ and, possibly, isocarbonyl(carbonyl)gold(0), $Au(CO)(OC)$, while there is no indication of the existence of a tricarbonylgold(0) complex $Au(CO)_3$.[406] All the observed species exhibit extremely low thermal stability, and their vibrational spectra show no resemblance to carbonylgold(I) derivatives. For example, in matrix-isolated $Au(CO)_2$, $v(CO)$ is reduced from the value of free CO at 2134 cm^{-1}, while in the $[Au(CO)_2]^+$ cation $v(CO)$ is raised.[407] A series of molecular orbital calculations for $Au(CO)$, $Au(CO)_2$ and the linkage isomer $Au(CO)(OC)$ has addressed the distinct vibrational and optical properties of these species.[408]

The most important carbonyl derivative of monovalent gold is carbonylchlorogold(I), [AuCl(CO)], which is a useful starting material for the preparation of other gold(I) compounds by CO displacement. The related bromide, [AuBr(CO)], has been detected in solution but is very unstable, while attempts to prepare the iodide have been unsuccessful altogether. Methods of synthesis for these compounds have been surveyed in a recent review.[409] Other carbonyl derivatives of gold(I) include carbonyl(fluorosulfato)gold(I), [AuOSO$_2$F(CO)], and the bis(carbonyl)gold(I) cation present in the uranate and antimonate salts [Au(CO)$_2$][UF$_6$] and [Au(CO)$_2$][Sb$_2$F$_{11}$], respectively. In addition to these mononuclear carbonyl complexes of gold(I), a dinuclear compound of the composition [Au$_2$Cl$_4$(CO)] is also known.[71]

[AuCl(CO)] is best prepared by reductive carbonylation of anhydrous gold(III) chloride with carbon monoxide at ambient temperature and pressure, [Au$_2$Cl$_6$] being generated *in situ* by dehydration of tetrachloroauric acid with thionyl chloride (Equations (72) and (73)).[410] Carbonylchlorogold(I) can be isolated as a colourless crystalline material which is extremely moisture-sensitive. The compound is found to be monomeric both in the solid state and in solution, the very high CO stretching vibration frequency (in solution 2152–2162 cm^{-1}, slightly solvent dependent)[410] indicating a high bond order for the carbonyl group and little π back-donation from the metal to CO. The crystal structure (orthorhombic, space group *CmCm*)[411] consists of discrete, exactly linear molecules (by symmetry), and the shortest Au$\cdots$Au contacts between molecules are 0.338 nm. The Au–Cl bond distance (0.226 1(6) nm) falls within the usual range, whereas the Au–C bond is rather short (0.193(2) nm; this value, however, may not be entirely reliable due to libration).

$$2 \{[H_3O]^+[AuCl_4]^-\} + 2\,SOCl_2 \longrightarrow 2\,SO_2 + 6\,HCl + [Au_2Cl_6] \qquad (72)$$

$$[Au_2Cl_6] + 4\,CO \longrightarrow 2\,[AuCl(CO)] + 2\,COCl_2 \qquad (73)$$

Carbonylchlorogold(I) has also been generated on certain zeolite supports with a view to investigating catalytic applications, and evidence has been presented that such material effects the catalytic reduction of NO with CO at 200–373 K (giving a mixture of N$_2$O and CO$_2$).[412]

Treatment of [AuCl(CO)] with [Au$_2$Cl$_6$] in thionyl chloride affords the complex [Au$_2$Cl$_4$(CO)], formulated as a mixed-valence compound [OCAu(μ-Cl)AuCl$_3$] (ν(CO) = 2180 cm^{-1} in thionyl chloride). The compound can be isolated at $-30\,°C$, but decomposes when warmed to room temperature.[413]

Carbonylbromogold(I), [AuBr(CO)], has been obtained in solutions of halogenated hydrocarbons by reductive carbonylation of [Au$_2$Br$_6$] or carbonylation of [AuBr], but its thermal instability prevents its isolation.[413,414] While CO stretching vibration frequencies (2151–2159 cm^{-1}, depending on the solvent) differ little from those determined for [AuCl(CO)], complex formation is indicated by the 1:1 (Au:CO) molar ratio observed in the course of gas-volumetric measurements of the CO absorbed by AuBr in solution. The existence of [AuBr(CO)] is further corroborated by low-temperature ^{13}C NMR spectroscopy (CD$_2$Cl$_2$, $-70\,°C$: $\delta(CO) = 173.8$ ppm, compared with $\delta(CO) = 171.5$ ppm for [AuCl(CO)]).

Carbonyl(fluorosulfato)gold(I) is formed by reduction of gold tris(fluorosulfate) with CO in fluorosulfuric acid, followed by thermal decomposition of the resulting bis(carbonyl) gold(I) fluorosulfate (Equations (74) and (75)).[407] [AuOSO$_2$F(CO)] is a white, moisture-sensitive solid which melts at 49–50 °C, can be sublimed from the melt at 80 °C, and is thermally stable up to 190 °C, when decomposition sets in. The CO stretching frequency is observed at 2198 cm^{-1} and 2195 cm^{-1} in the Raman and IR spectra, respectively, which is notably higher than the value of 2143 cm^{-1} reported for free CO. This suggests that the σ-acceptor ability of gold(I) increases in the presence of an anionic ligand which is only weakly nucleophilic (such as SO$_3$F$^-$), and that there is virtually no π back donation from the metal.[415] The cation [Au(CO)$_2$]$^+$ is found to be linearly two-coordinate; it is stable in solution and is also characterized by very high CO stretching frequencies at 2251 (Raman) and 2211 (IR) cm^{-1}, which reflect the presence of a positive charge on the metal, in addition to the effects described earlier.

$$Au(OSO_2F)_3 + 3\,CO \xrightarrow[\text{HSO}_3\text{F}]{25\,°C} [Au(CO)_2]^+ + CO_2 + S_2O_5F_2 + SO_3F^- \qquad (74)$$

$$[Au(CO)_2]^+ + [SO_3F]^- \xrightarrow[\text{HSO}_3\text{F}]{80\,°C} [AuOSO_2F(CO)] + CO \qquad (75)$$

The [Au(CO)$_2$]$^+$ cation is also present in the salts [Au(CO)$_2$][UF$_6$][416] and [Au(CO)$_2$][Sb$_2$F$_{11}$],[417] synthesized from elemental gold and UF$_6$ in anhydrous HF in the presence of CO as a complexing agent, and by solvolysis of [AuOSO$_2$F(CO)] in liquid SbF$_5$ in the presence of CO, respectively (Equations (76) and (77)). While $[Au(CO)_2][UF_6]$ is of low thermal stability, $[Au(CO)_2][Sb_2F_{11}]$ is thermally stable up to 130 °C. In both cases, the near absence of gold-to-carbon π back-donation is reflected in very high CO stretching frequencies in the IR spectra, with values of 2200 cm^{-1} and 2217 cm^{-1}, respectively. In the ^{13}C NMR spectrum of $[Au(CO)_2][Sb_2F_{11}]$, a single-line resonance at 174 ppm is attributed to $[Au(^{13}CO)_2]^+$, and there is evidence for the presence of the monocarbonyl cation $[Au(^{13}CO)]^+$ as well.

$$Au + 2CO + UF_6 \xrightarrow{\text{HF}} [Au(CO)_2]^+[UF_6]^- \qquad (76)$$

$$[AuOSO_2F(CO)] + CO + 4SbF_5 \xrightarrow{\text{SbF}_5} [Au(CO)_2][Sb_2F_{11}] + Sb_2F_9SO_3F \qquad (77)$$

The ability of carbon monoxide to act as a reducing agent has also been used in the preparation of subvalent gold phosphine cluster complexes from gold(I) compounds of the type $[AuX(PPh_3)]$, where X is a weakly coordinated ligand such as NO_3^- or ClO_4^-. The intermediacy of a species formulated as $[Au(CO)(PPh_3)]^+$ has been postulated, and a gold(I)–fulminate complex, $[Au(CNO)(PPh_3)]$, is formed as a side product.[418,419]

1.7 SYNTHESIS AND PROPERTIES OF ISOCYANIDE COMPLEXES OF GOLD

Two types of gold isocyanide complex have already been mentioned: the organogold(I) species $[RAu(C\equiv NR)]$ (Section 1.2.1) and the bis(isocyanide) complexes $[Au(C\equiv NR)_2]X$, which have been used as precursors in the synthesis of bis(carbene) complexes of gold (Section 1.5). Related compounds are those of the type $[AuX(C\equiv NR)]$ (X = Cl or CN) and the dicyanoaurates $M[Au(CN)_2]$. These complexes have been studied extensively in connection with the investigation of the electronic structure of linearly two-coordinate compounds, and the participation of the gold(I) orbitals in σ-bonding and metal-to-ligand π-bonding has been compared for $[Au(CN)_2]^-$ and the isocyanide complexes $[Au(CNEt)_2]^+$ and $[Au(CN)(CNMe)]$.[420]

(Isocyanide)gold(I) chloride complexes may be prepared by the reaction of tetrachloroauric acid with two equivalents of isocyanide.[421,422] The reduction of gold(III) to gold(I) is accompanied by the conversion of one equivalent of isocyanide into an alkyl carbamate, whose identity has been established by x-ray crystallography in one case.[423] For alkyl isocyanides $RN\equiv C$ containing a tertiary alkyl group R, an alternative mode of reaction involves dealkylation with concomitant formation of cyanide ion, which then acts as a ligand towards gold(I) to give the corresponding (isocyanide)gold(I) cyanide complexes. Both types of compound can be synthesized from *t*-butyl isocyanide, as illustrated in Equations (78) and (79).

$$HAuCl_4 \cdot H_2O + 2Bu^tN\equiv C + EtOH \xrightarrow{\text{EtOH}} [AuCl(C\equiv NBu^t)] + Bu^tNH - \overset{\displaystyle O}{\underset{\displaystyle OEt}{\overset{\|}{C}}} + 3HCl \qquad (78)$$

$$KAuCl_4 \cdot H_2O + 3Bu^tN\equiv C + MeOH \xrightarrow{\text{MeOH}} [Au(CN)(C\equiv NBu^t)] + Bu^tNH - \overset{\displaystyle O}{\underset{\displaystyle OMe}{\overset{\|}{C}}} + 3HCl + KCl + \overset{\diagdown}{\underset{\diagup}{C}} = CH_2 \qquad (79)$$

Dinuclear complexes containing bridging diisocyanide ligands have been prepared in a similar manner.[421,423,424] They show interesting solid-state photoluminescence which appears to be associated with weak inter- or intramolecular Au⋯Au contacts in the range 0.33–0.35 nm, as determined by x-ray crystallography.

1.8 REFERENCES

1. H. Schmidbaur, *Angew. Chem., Int. Ed. Engl.*, 1976, **15**, 728.
2. W. S. Rapson, *Gold Bull.*, 1988, **21**, 10.
3. G. Mehringer and J. Simon, *Metall (Berlin)*, 1989, **43**, 624.
4. J. W. Watkins, II, R. C. Elder, B. Greene and D. W. Darnall, *Inorg. Chem.*, 1987, **26**, 1147.
5. G. J. McDougall and R. D. Hancock, *Gold Bull.*, 1981, **14**, 138.
6. Y. Hori, A. Murata, K. Kikuchi and S. Suzuki, *J. Chem. Soc., Chem. Commun.*, 1987, 728.
7. A. W. Olsen and Z. H. Kafafi, *J. Am. Chem. Soc.*, 1991, **113**, 7758.
8. T. Aida, R. Higuchi and H. Niiyama, *Chem. Lett.*, 1990, 2247.
9. B. Nkosi, N. J. Coville and G. J. Hutchings, *J. Chem. Soc., Chem. Commun.*, 1988, 71.

10. J. Schwank, *Gold Bull.*, 1985, **18**, 2.
11. F. Gasparrini, M. Giovannoli, D. Misiti, G. Natile, G. Palmieri and L. Maresca, *J. Am. Chem. Soc.*, 1993, **115**, 4401.
12. A. Togni and S. D. Pastor, *J. Org. Chem.*, 1990, **55**, 1649.
13. H. Yang and P. A. Frey, *Biochemistry*, 1984, **23**, 3863.
14. D. Li, C.-M. Che, H.-L. Kwong and V. W.-W. Yam, *J. Chem. Soc., Dalton Trans.*, 1992, 3325.
15. C. King, M. N. I. Khan, R. J. Staples and J. P. Fackler, Jr., *Inorg. Chem.*, 1992, **31**, 3236.
16. G. M. Whitesides and P. E. Laibinis, *Langmuir*, 1990, **6**, 87.
17. W. Jahn, *Z. Naturforsch., Teil B*, 1989, **44**, 1313.
18. P. J. Sadler, *Adv. Inorg. Chem.*, 1991, **36**, 1.
19. G. G. Graham, J. B. Ziegler and G. D. Champion, *Agents Actions Suppl.*, 1993, **44**, 209.
20. G. Stocco, F. Gattuso, A. A. Isab and C. F. Shaw, III, *Inorg. Chim. Acta*, 1993, **209**, 129.
21. O. M. NiDubhghaill and P. J. Sadler, in 'Metal Complexes in Cancer Chemotherapy', ed. B. K. Keppler, VCH, Weinheim, 1993, p. 221.
22. D. G. Duff, A. Baiker and P. P. Edwards, *J. Chem. Soc., Chem. Commun.*, 1993, 96.
23. D. M. P. Mingos and M. J. Watson, *Adv. Inorg. Chem.*, 1992, **39**, 327.
24. D. M. P. Mingos, T. Slee and L. Zhenyang, *Chem. Rev.*, 1990, **90**, 383.
25. D. M. P. Mingos, *Spec. Publ. R. Soc. Chem.*, 1993, **131**, 189.
26. J. J. Steggerda, J. J. Bour and J. W. A. van der Velden, *Recl. Trav. Chim. Pays-Bas*, 1982, **101**, 164.
27. R. B. Klassen and T. H. Baum, *Organometallics*, 1989, **8**, 2477.
28. K. L. Kompa, *Angew. Chem., Int. Ed. Engl.*, 1988, **27**, 1314.
29. R. J. Puddephatt and I. Treurnicht, *J. Organomet. Chem.*, 1987, **319**, 129.
30. A. Görling, N. Rösch, D. E. Ellis and H. Schmidbaur, *Inorg. Chem.*, 1991, **30**, 3986.
31. P. Pyykkö and Y. Zhao, *Angew. Chem., Int. Ed. Engl.*, 1991, **30**, 604.
32. A. P. Koley, R. Nirmala, L. S. Prasad, S. Ghosh and P. T. Manoharan, *Inorg. Chem.*, 1992, **31**, 1764.
33. B. G. Müller, *Angew. Chem., Int. Ed. Engl.*, 1987, **26**, 1081.
34. H. Schmidbaur and K. C. Dash, *Adv. Inorg. Chem. Radiochem.*, 1982, **25**, 239.
35. G. Rindorf, N. Thorup, T. Bjørnholm and K. Bechgaard, *Acta Crystallogr., Sect. C*, 1990, **46**, 1437.
36. O. Crespo, M. C. Gimeno, A. Laguna and P. G. Jones, *J. Chem. Soc., Dalton Trans.*, 1992, 1601.
37. A. L. Balch and E. Y. Fung, *Inorg. Chem.*, 1990, **29**, 4764.
38. J. E. Huheey, in 'Inorganic Chemistry: Principles of Structure and Reactivity', 3rd edn., Harper & Row, New York, 1983, p. 408.
39. S. S. Pathaneni and G. R. Desiraju, *J. Chem. Soc., Dalton Trans.*, 1993, 319.
40. R. V. Parish, *Hyperfine Interact.*, 1988, **40**, 159.
44. M. Melník and R. V. Parish, *Coord. Chem. Rev.*, 1986, **70**, 157.
45. P. G. Jones, *Gold Bull.*, 1981, **14**, 102.
46. P. G. Jones, *Gold Bull.*, 1981, **14**, 159.
47. P. G. Jones, *Gold Bull.*, 1983, **16**, 114.
48. P. G. Jones, *Gold Bull.*, 1986, **19**, 46.
49. A. Bondi, *J. Phys. Chem.*, 1964, **68**, 441.
50. W. B. Pearson, 'Lattice Spacings and Structures of Metals and Alloys', Pergamon, Oxford, 1957.
51. H. Schmidbaur, *Gold Bull.*, 1990, **23**, 11.
52. H. Schmidbaur, K. Dziwok, A. Grohmann and G. Müller, *Chem. Ber.*, 1989, **122**, 893.
53. H. Schmidbaur, W. Graf and G. Müller, *Angew. Chem., Int. Ed. Engl.*, 1988, **27**, 417.
54. M. I. Bruce and M. J. Liddell, *J. Organomet. Chem.*, 1992, **427**, 263.
55. H. Schmidbaur, *Pure Appl. Chem.*, 1993, **65**, 691.
56. A. Blumenthal, H. Beruda and H. Schmidbaur, *J. Chem. Soc., Chem. Commun.*, 1993, 1005.
54. H. Schmidbaur, E. Zeller and J. Ohshita, *Inorg. Chem.*, 1993, **32**, 4524.
55. J. W. Lauher and K. Wald, *J. Am. Chem. Soc.*, 1981, **103**, 7648.
56. A. Grohmann, J. Riede and H. Schmidbaur, *Nature*, 1990, **345**, 140.
57. H. Kudo, *Nature*, 1992, **355**, 432.
58. P. Pyykkö, *Chem. Rev.*, 1988, **88**, 563.
59. P. Schwerdtfeger, P. D. W. Boyd, S. Brienne and A. K. Burrell, *Inorg. Chem.*, 1992, **31**, 3411.
60. P. Pyykkö and Y. Zhao, *Chem. Phys. Lett.*, 1991, **177**, 103.
61. G. K. Anderson, *Adv. Organomet. Chem.*, 1982, **20**, 39.
62. K. I. Grandberg and V. P. Dyadchenko, *J. Organomet. Chem.*, 1994, **474**, 1.
63. S. Komiya, *Trends Inorg. Chem.*, 1990, **1**, 15.
64. C. E. Housecroft, *Coord. Chem. Rev.*, 1993, **127**, 187.
65. C. E. Housecroft, *Coord. Chem. Rev.*, 1992, **115**, 117.
66. W. E. Smith, *Coord. Chem. Rev.*, 1985, **67**, 311.
67. W. E. Smith, *Coord. Chem. Rev.*, 1982, **45**, 319.
68. W. E. Smith, *Coord. Chem. Rev.*, 1981, **35**, 259.
69. R. Usón and A. Laguna, *Coord. Chem. Rev.*, 1986, **70**, 1.
70. J. Buckingham and J. E. Macintyre (eds.), 'Dictionary of Organometallic Compounds', Chapman & Hall, London, 1984–1990.
71. R. J. Puddephatt, in 'COMC-I', vol. 2, p. 765.
72. H. Schmidbaur, in 'Gmelin Handbuch der Anorganischen Chemie. Organogold Compounds', Springer, Berlin, 1980.
73. H. Schmidbaur *et al.*, *Organometallics*, 1986, **5**, 566.
74. E. G. Perevalova *et al.*, *J. Organomet. Chem.*, 1989, **369**, 267.
75. E. G. Perevalova, K. I. Grandberg, E. I. Smyslova and E. S. Kalyuzhnaya, *Bull. Acad. Sci. USSR, Chem. Ser.*, 1985, 191.
76. B. D. Reid and A. J. Welch, *J. Organomet. Chem.*, 1992, **438**, 371.
77. E. G. Perevalova, E. I. Smyslova and K. I. Grandberg, *Bull. Acad. Sci. USSR, Chem. Ser.*, 1982, 2506.
78. R. D. Sanner, J. H. Satcher, Jr. and M. W. Droege, *Organometallics*, 1989, **8**, 1498.
79. H. K. Nair and J. A. Morrison, *J. Organomet. Chem.*, 1989, **376**, 149.

80. E. G. Perevalova, Y. T. Struchkov, V. P. Dyadchenko, E. I. Smyslova, Y. L. Slovokhotov and K. I. Grandberg, *Bull. Acad. Sci. USSR, Chem. Ser.*, 1983, 2529.
81. E. G. Perevalova, T. V. Baukova, M. M. Sazonenko and K. I. Grandberg, *Bull. Acad. Sci. USSR, Chem. Ser.*, 1985, 1726.
82. E. I. Smyslova, E. G. Perevalova, V. P. Dyadchenko, K. I. Grandberg, Y. L. Slovokhotov and Y. T. Struchkov, *J. Organomet. Chem.*, 1981, **215**, 269.
83. M. Murakami, M. Inouye, M. Suginome and Y. Ito, *Bull. Chem. Soc. Jpn.*, 1988, **61**, 3649.
84. J. Dekker, J. W. Münninghoff, J. Boersma and A. L. Spek, *Organometallics*, 1987, **6**, 1236.
85. R. I. Papasergio, C. L. Raston and A. H. White, *J. Chem. Soc., Dalton Trans.*, 1987, 3085.
86. G. M. Bancroft, T. Chan, R. J. Puddephatt and J. S. Tse, *Inorg. Chem.*, 1982, **21**, 2946.
87. R. L. DeKock, E. J. Baerends, P. M. Boerrigter and R. Hengelmolen, *J. Am. Chem. Soc.*, 1984, **106**, 3387.
88. A. Haaland, J. Hougen, H. V. Volden and R. J. Puddephatt, *J. Organomet. Chem.*, 1987, **325**, 311.
89. J. Sýkora and J. Šima, *Coord. Chem. Rev.*, 1990, **107**, 1.
90. H. B. Friedrich and J. R. Moss, *Adv. Organomet. Chem.*, 1991, **33**, 235.
91. J. A. Morrison, *Adv. Inorg. Chem. Radiochem.*, 1983, **27**, 293.
92. G. J. Arsenault, M. Crespo and R. J. Puddephatt, *Organometallics*, 1987, **6**, 2255.
93. H.-P. Abicht, P. Lehniger and K. Issleib, *J. Organomet. Chem.*, 1983, **250**, 609.
94. S. Bruni, A. L. Bandini, F. Cariati and F. Speroni, *Inorg. Chim. Acta*, 1993, **203**, 127.
95. H.-J. Kneuper, K. Harms and G. Boche, *J. Organomet. Chem.*, 1989, **364**, 275.
96. E. G. Perevalova *et al.*, *J. Organomet. Chem.*, 1985, **286**, 129.
97. E. G. Perevalova, M. D. Reshetova and G. M. Kokhanyuk, *J. Gen. Chem. USSR*, 1984, 2424.
98. S. Komiya, M. Iwata, T. Sone and A. Fukuoka, *J. Chem. Soc., Chem. Commun.*, 1992, 1109.
99. Y. Ito, M. Inouye, M. Suginome and M. Murakami, *J. Organomet. Chem.*, 1988, **342**, C41.
100. F. Bonati, A. Burini, B. R. Pietroni and M. Felici, *J. Organomet. Chem.*, 1984, **273**, 275.
101. P. Veya, C. Floriani, A. Chiesi-Villa, C. Guastini, A. Dedieu, F. Ingold and P. Braunstein, *Organometallics*, 1993, **12**, 4359.
102. M. C. Gimeno, A. Laguna, M. Laguna, F. Sanmartín and P. G. Jones, *Organometallics*, 1993, **12**, 3984.
103. M. I. Bruce, P. J. Low, B. W. Skelton and A. H. White, *J. Chem. Soc., Dalton Trans.*, 1993, 3145.
104. E. J. Fernández, M. C. Gimeno, P. G. Jones, A. Laguna, M. Laguna and J. M. López-de-Luzuriaga, *J. Chem. Soc., Dalton Trans.*, 1992, 3365.
105. N. C. Payne, R. Ramachandran, I. Treurnicht and R. J. Puddephatt, *Organometallics*, 1990, **9**, 880.
106. D. Li, X. Hong, C.-M. Che, W.-C. Lo and S.-M. Peng, *J. Chem. Soc., Dalton Trans.*, 1993, 2929.
107. F. Bonati, A. Burini, B. R. Pietroni, E. Torregiani, S. Calogero and F. E. Wagner, *J. Organomet. Chem.*, 1991, **408**, 125.
108. X. L. R. Fontaine, S. J. Higgins, C. R. Langrick and B. L. Shaw, *J. Chem. Soc., Dalton Trans.*, 1987, 777.
109. A. J. Deeming, S. Donovan-Mtunzi and K. Hardcastle, *J. Chem. Soc., Dalton Trans.*, 1986, 543.
110. A. T. Hutton, P. G. Pringle and B. L. Shaw, *Organometallics*, 1983, **2**, 1889.
111. G. J. Arsenault, L. Manojlović-Muir, K. W. Muir, R. J. Puddephatt and I. Treurnicht, *Angew. Chem., Int. Ed. Engl.*, 1987, **26**, 86.
112. L. Manojlović-Muir, K. W. Muir, I. Treurnicht and R. J. Puddephatt, *Inorg. Chem.*, 1987, **26**, 2418.
113. M. I. Bruce, E. Horn, J. G. Matisons and M. R. Snow, *Aust. J. Chem.*, 1984, **37**, 1163.
114. R. Nast, *Coord. Chem. Rev.*, 1982, **47**, 89.
115. O. M. Abu-Salah, A.-R. A. Al-Ohaly, S. S. Al-Showiman and I. M. Al-Najjar, *Transition Met. Chem.*, 1985, **10**, 207.
116. O. M. Abu-Salah and A. R. Al-Ohaly, *Inorg. Chim. Acta*, 1983, **77**, L159.
117. R. J. Cross and M. F. Davidson, *J. Chem. Soc., Dalton Trans.*, 1986, 411.
118. R. J. Cross M. F. Davidson and A. J. McLennan, *J. Organomet. Chem.*, 1984, **265**, C37.
119. A. T. Casey and A. M. Vecchio, *Appl. Organomet. Chem.*, 1990, **4**, 513.
120. F. Bonati, A. Burini, B. R. Pietroni, E. Giorgini and B. Bovio, *J. Organomet. Chem.*, 1988, **344**, 119.
121. N. C. Payne, R. J. Puddephatt, R. Ravindranath and I. Treurnicht, *Can. J. Chem.*, 1988, **66**, 3176.
122. G. Jia, R. J. Puddephatt and J. J. Vittal, *J. Organomet. Chem.*, 1993, **449**, 211.
123. G. Jia, R. J. Puddephatt, J. D. Scott and J. J. Vittal, *Organometallics*, 1993, **12**, 3565.
124. G. Jia, R. J. Puddephatt, J. J. Vittal and N. C. Payne, *Organometallics*, 1993, **12**, 263.
125. G. Jia, N. C. Payne, J. J. Vittal and R. J. Puddephatt, *Organometallics*, 1993, **12**, 4771.
126. M. I. Bruce and D. N. Duffy, *Aust. J. Chem.*, 1986, **39**, 1697.
127. G. A. Carriedo, V. Riera, X. Soláns and J. Soláns, *Acta Crystallogr., Sect. C*, 1988, **44**, 978.
128. M. I. Bruce, K. R. Grundy, M. J. Liddell, M. R. Snow and E. R. T. Tiekink, *J. Organomet. Chem.*, 1988, **344**, C49.
129. L. S. Moore, R. V. Parish, R. Usón, A. Laguna, M. Laguna and M. N. Fraile, *J. Chem. Soc., Dalton Trans.*, 1988, 23.
130. R. Usón, A. Laguna, M. Laguna, M. N. Fraile, P. G. Jones and G. M. Sheldrick, *J. Chem. Soc., Dalton Trans.*, 1986, 291.
131. M. A. Bennett, S. K. Bhargava, K. D. Griffiths and G. B. Robertson, *Angew. Chem., Int. Ed. Engl.*, 1987, **26**, 260.
132. P. G. Jones, *J. Organomet. Chem.*, 1988, **345**, 405.
133. R. Usón, A. Laguna, M. Laguna, I. Colera and E. de Jesús, *J. Organomet. Chem.*, 1984, **263**, 121.
134. R. Usón, A. Laguna, A. Navarro, R. V. Parish and L. S. Moore, *Inorg. Chim. Acta*, 1986, **112**, 205.
135. R. Usón, A. Laguna and M. D. Villacampa, *Inorg. Chim. Acta*, 1984, **81**, 25.
136. G. A. Carriedo, G. Sánchez, V. Riera, C. Bois, Y. Jeannin and D. Miguel, *J. Chem. Soc., Dalton Trans.*, 1990, 3355.
137. F. Bonati, A. Burini, B. R. Pietroni and B. Bovio, *J. Organomet. Chem.*, 1991, **408**, 271.
138. A. Laguna, M. Laguna, J. Fañanas, P. G. Jones and C. Fittschen, *Inorg. Chim. Acta*, 1986, **121**, 39.
139. J. Vicente, A. Arcas, M. Mora, X. Soláns and M. Font-Altaba, *J. Organomet. Chem.*, 1986, **309**, 369.
140. P. H. van Rooyen, M. Schindehutte and S. Lotz, *Organometallics*, 1992, **11**, 1104.
141. J. Vicente, A. Arcas, P. G. Jones and J. Lautner, *J. Chem. Soc., Dalton Trans.*, 1990, 451.
142. M. A. Bennett, S. K. Bhargava, K. D. Griffiths, G. B. Robertson, W. A. Wickramasinghe and A. C. Willis, *Angew. Chem., Int. Ed. Engl.*, 1987, **26**, 258; 1988, **27**, 589.
143. P. G. Jones and C. Thöne, *Acta Crystallogr., Sect. C*, 1992, **48**, 1312.
144. M. C. Gimeno, A. Laguna, C. Sarroca and P. G. Jones, *Inorg. Chem.*, 1993, **32**, 5926.
145. E. J. Fernández, M. C. Gimeno, P. G. Jones, A. Laguna, M. Laguna and J. M. López-de-Luzuriaga, *J. Chem. Soc., Dalton Trans.*, 1993, 3401.
146. R. Usón *et al.*, *J. Chem. Soc., Dalton Trans.*, 1986, 669.

147. G. A. Carriedo, V. Riera, M. L. Rodríguez, P. G. Jones and J. Lautner, *J. Chem. Soc., Dalton Trans.*, 1989, 639.
148. R. Usón, A. Laguna, M. Laguna, M. L. Castilla, P. G. Jones and C. Fittschen, *J. Chem. Soc., Dalton Trans.*, 1987, 3017.
149. A. Laguna, M. Laguna, A. Rojo and M. N. Fraile, *J. Organomet. Chem.*, 1986, **315**, 269.
150. R. Usón *et al.*, *J. Chem. Soc., Dalton Trans.*, 1990, 3457.
151. F. Bonati, A. Burini, B. R. Pietroni and B. Bovio, *J. Organomet. Chem.*, 1989, **375**, 147.
152. D. A. Briggs, R. G. Raptis and J. P. Fackler, Jr., *Acta Crystallogr., Sect. C*, 1988, **44**, 1313.
153. A. Laguna, M. Laguna, J. Jiménez and A. J. Fumanal, *J. Organomet. Chem.*, 1990, **396**, 121.
154. H. Werner, H. Otto, T. Ngo-Khac and C. Burschka, *J. Organomet. Chem.*, 1984, **262**, 123.
155. E. G. Perevalova, K. I. Grandberg, V. P. Dyadchenko and T. V. Baukova, *J. Organomet. Chem.*, 1981, **217**, 403.
156. T. V. Baukova, Y. L. Slovokhotov and Y. T. Struchkov, *J. Organomet. Chem.*, 1981, **220**, 125.
157. M. I. Bruce and A. H. White, *Aust. J. Chem.*, 1990, **43**, 949.
158. M. I. Bruce, J. K. Walton, B. W. Skelton and A. H. White, *J. Chem. Soc., Dalton Trans.*, 1983, 809.
159. H. Schumann, F. H. Görlitz and A. Dietrich, *Chem. Ber.*, 1989, **122**, 1423.
160. M. I. Bruce, P. A. Humphrey, M. L. Williams, B. W. Skelton and A. H. White, *Aust. J. Chem.*, 1989, **42**, 1847.
161. Y. T. Struchkov, Y. L. Slovokhotov, D. N. Kravtsov, T. V. Baukova, E. G. Perevalova and K. I. Grandberg, *J. Organomet. Chem.*, 1988, **338**, 269.
162. K. Geilich, K. Stumpf, H. Pritzkow and W. Siebert, *Chem. Ber.*, 1987, **120**, 911.
163. T. V. Baukova, Y. L. Slovokhotov and Y. T. Struchkov, *J. Organomet. Chem.*, 1981, **221**, 375.
164. B. Bovio, A. Burini and B. R. Pietroni, *J. Organomet. Chem.*, 1993, **452**, 287.
165. G. Banditelli, A. L. Bandini, F. Bonati, R. G. Goel and G. Minghetti, *Gazz. Chim. Ital.*, 1982, **112**, 539.
166. M. Lanfranchi, M. A. Pellinghelli, A. Tiripicchio and F. Bonati, *Acta Crystallogr., Sect. C*, 1985, **41**, 52.
167. H. G. Raubenheimer, G. J. Kruger, C. F. Marais, J. T. Z. Hattingh, L. Linford and P. H. van Rooyen, *J. Organomet. Chem.*, 1988, **355**, 337.
168. A. N. Nesmeyanov, E. G. Perevalova, K. I. Grandberg, D. A. Lemenovskii, T. V. Baukova and O. B. Afanassova, *J. Organomet. Chem.*, 1974, **65**, 131.
169. H. Schmidbaur and Y. Inoguchi, *Chem. Ber.*, 1980, **113**, 1646.
170. E. G. Perevalova, E. I. Smyslova, V. P. Dyadchenko and K. I. Grandberg, *Bull. Acad. Sci. USSR, Chem. Ser.*, 1984, 883.
171. V. I. Korsunsky, *J. Organomet. Chem.*, 1986, **311**, 357.
172. V. I. Korsunsky, K. I. Grandberg, E. I. Smyslova and T. V. Baukova, *J. Organomet. Chem.*, 1987, **335**, 277.
173. R. Usón, A. Laguna and P. Brun, *J. Organomet. Chem.*, 1980, **197**, 369.
174. S. Gambarotta, C. Floriani, A. Chiesi-Villa and C. Guastini, *J. Chem. Soc., Chem. Commun.*, 1983, 1304.
175. E. M. Meyer, S. Gambarotta, C. Floriani, A. Chiesi-Villa and C. Guastini, *Organometallics*, 1989, **8**, 1067.
176. G. van Koten, J. T. B. H. Jastrzebski, C. H. Stam and N. C. Niemann, *J. Am. Chem. Soc.*, 1984, **106**, 1880.
177. I. J. B. Lin, C. W. Liu, L.-K. Liu and Y.-S. Wen, *Organometallics*, 1992, **11**, 1447.
178. F. Bonati, A. Burini, B. R. Pietroni and B. Bovio, *J. Organomet. Chem.*, 1986, **317**, 121.
179. F. Bonati, A. Burini, B. R. Pietroni and E. Giorgini, *Inorg. Chim. Acta*, 1987, **137**, 81.
180. H. Schmidbaur and O. Gasser, *Angew. Chem., Int. Ed. Engl.*, 1976, **15**, 502.
181. J. Vicente, M.-T. Chicote, M.-C. Lagunas and P. G. Jones, *J. Chem. Soc., Chem. Commun.*, 1991, 1730.
182. J. Vicente, M.-T. Chicote, M.-C. Lagunas and P. G. Jones, *J. Chem. Soc., Dalton Trans.*, 1991, 2579.
183. J. Vicente, M.-T. Chicote and M.-C. Lagunas, *Inorg. Chem.*, 1993, **32**, 3748.
184. F. Bonati, A. Burini, M. Felici and B. R. Pietroni, *Gazz. Chim. Ital.*, 1983, **113**, 105.
185. H. Schmidbaur, F. Scherbaum, B. Huber and G. Müller, *Angew. Chem., Int. Ed. Engl.*, 1988, **27**, 419.
186. G. Karger, P. Hornbach, A. Krämer, H. Pritzkow and W. Siebert, *Chem. Ber.*, 1989, **122**, 1881.
187. V. Riera, J. Ruiz, X. Soláns and E. Tauler, *J. Chem. Soc., Dalton Trans.*, 1990, 1607.
188. J. Vicente, M.-T. Chicote, I. Saura-Llamas, P. G. Jones, K. Meyer-Bäse and C. Freire Erdbrügger, *Organometallics*, 1988, **7**, 997.
189. J. Vicente, M.-T. Chicote and I. Saura-Llamas, *J. Chem. Soc., Dalton Trans.*, 1990, 1941.
190. A. N. Nesmeyanov, E. G. Perevalova, Y. T. Struchkov, M. Y. Antipin, K. I. Grandberg and V. P. Dyadchenko, *J. Organomet. Chem.*, 1980, **201**, 343.
191. C. Lensch, P. G. Jones and G. M. Sheldrick, *Z. Naturforsch., Teil B*, 1982, **37**, 944.
192. P. G. Jones, G. M. Sheldrick and E. Hädicke, *Acta Crystallogr., Sect. B*, 1980, **36**, 2777.
193. N. Dufour, A. Schier and H. Schmidbaur, *Organometallics*, 1993, **12**, 2408.
194. F. Scherbaum, B. Huber, G. Müller and H. Schmidbaur, *Angew. Chem., Int. Ed. Engl.*, 1988, **27**, 1542.
195. O. Steigelmann, P. Bissinger and H. Schmidbaur, *Z. Naturforsch., Teil B*, 1993, **48**, 72.
196. H. Schmidbaur and O. Steigelmann, *Z. Naturforsch., Teil B*, 1992, **47**, 1721.
197. F. Scherbaum, A. Grohmann, G. Müller and H. Schmidbaur, *Angew. Chem., Int. Ed. Engl.*, 1989, **28**, 463.
198. F. Scherbaum, A. Grohmann, B. Huber, C. Krüger and H. Schmidbaur, *Angew. Chem., Int. Ed. Engl.*, 1988, **27**, 1544.
199. O. Steigelmann, P. Bissinger and H. Schmidbaur, *Angew. Chem., Int. Ed. Engl.*, 1990, **29**, 1399.
200. H. Schmidbaur, B. Brachthäuser and O. Steigelmann, *Angew. Chem., Int. Ed. Engl.*, 1991, **30**, 1488.
201. H. Schmidbaur, B. Brachthäuser, O. Steigelmann and H. Beruda, *Chem. Ber.*, 1992, **125**, 2705.
202. C. J. McNeal *et al.*, *Inorg. Chem.*, 1993, **32**, 5582.
203. P. L. Bellon, M. Manassero, L. Naldini and M. Sansoni, *J. Chem. Soc., Chem. Commun.*, 1972, 1035.
204. P. L. Bellon, M. Manassero and M. Sansoni, *J. Chem. Soc., Dalton Trans.*, 1973, 2423.
205. D. G. Evans and D. M. P. Mingos, *J. Organomet. Chem.*, 1982, **232**, 171.
206. C. E. Briant, K. P. Hall and D. M. P. Mingos, *J. Organomet. Chem.*, 1983, **254**, C18.
207. C. E. Briant, K. P. Hall, D. M. P. Mingos and A. C. Wheeler, *J. Chem. Soc., Dalton Trans.*, 1986, 687.
208. E. D. Jemmis *et al.*, *J. Am. Chem. Soc.*, 1982, **104**, 4275.
209. D. M. P. Mingos, *Nature*, 1990, **345**, 113.
210. D. M. P. Mingos, *J. Chem. Soc., Dalton Trans.*, 1976, 1163.
211. N. Rösch, A. Görling, D. E. Ellis and H. Schmidbaur, *Angew. Chem., Int. Ed. Engl.*, 1989, **28**, 1357.
212. D. M. P. Mingos and R. P. F. Kanters, *J. Organomet. Chem.*, 1990, **384**, 405.
213. J. Vicente, M.-T. Chicote, I. Saura-Llamas and M.-C. Lagunas, *J. Chem. Soc., Chem. Commun.*, 1992, 915.
214. O. M. Abu-Salah and A. R. Al-Ohaly, *J. Organomet. Chem.*, 1983, **255**, C39.

215. R. Nast, P. Schneller and A. Hengefeld, *J. Organomet. Chem.*, 1981, **214**, 273.
216. O. M. Abu-Salah, A. R. A. Al-Ohaly and Z. F. Mutter, *J. Organomet. Chem.*, 1990, **391**, 267.
217. O. M. Abu-Salah, A. R. A. Al-Ohaly and Z. F. Mutter, *J. Organomet. Chem.*, 1990, **389**, 427.
218. O. M. Abu-Salah, *J. Organomet. Chem.*, 1990, **387**, 123.
219. O. M. Abu-Salah and A. R. A. Al-Ohaly, *J. Chem. Soc., Dalton Trans.*, 1988, 2297.
220. O. M. Abu-Salah and C. B. Knobler, *J. Organomet. Chem.*, 1986, **302**, C10.
221. O. M. Abu-Salah, A.-R. A. Al-Ohaly and C. B. Knobler, *J. Chem. Soc., Chem. Commun.*, 1985, 1502.
222. O. M. Abu-Salah, *J. Organomet. Chem.*, 1984, **270**, C26.
223. R. Usón, A. Laguna, E. J. Fernández, M. E. Ruiz-Romero, P. G. Jones and J. Lautner, *J. Chem. Soc., Dalton Trans.*, 1989, 2127.
224. R. Usón, *J. Organomet. Chem.*, 1989, **372**, 171.
225. R. Usón, A. Laguna, M. Laguna, B. R. Manzano, P. G. Jones and G. M. Sheldrick, *J. Chem. Soc., Dalton Trans.*, 1984, 285.
226. R. Usón, A. Laguna, M. Laguna and M. C. Gimeno, *J. Chem. Soc., Dalton Trans.*, 1989, 1883.
227. R. Usón *et al.*, *J. Chem. Soc., Chem. Commun.*, 1986, 509.
228. H. Schmidbaur, *Angew. Chem., Int. Ed. Engl.*, 1983, **22**, 907.
229. W. C. Kaska, *Coord. Chem. Rev.*, 1983, **48**, 1.
230. R. Usón, A. Laguna, M. Laguna, A. Usón and M. C. Gimeno, *Inorg. Chim. Acta*, 1986, **114**, 91.
231. J. Vicente, M. T. Chicote, I. Saura-Llamas, J. Turpin and J. Fernández-Baeza, *J. Organomet. Chem.*, 1987, **333**, 129.
232. J. Vicente *et al.*, *J. Chem. Soc., Dalton Trans.*, 1985, 1163.
233. H. Schmidbaur, C. E. Zybill, G. Müller and C. Krüger, *Angew. Chem., Int. Ed. Engl.*, 1983, **22**, 729.
234. G. A. Bowmaker and H. Schmidbaur, *J. Chem. Soc., Dalton Trans.*, 1984, 2859.
235. R. Usón, A. Laguna, M. Laguna and A. Usón, *Inorg. Chim. Acta*, 1983, **73**, 63.
236. C. J. Aguirre, M. C. Gimeno, A. Laguna, M. Laguna, J. M. López de Luzuriaga and F. Puente, *Inorg. Chim. Acta*, 1993, **208**, 31.
237. M. C. Gimeno, A. Laguna, M. Laguna and P. G. Jones, *Inorg. Chim. Acta*, 1991, **189**, 117.
238. H. Konno and Y. Yamamoto, *Bull. Chem. Soc. Jpn.*, 1987, **60**, 2561.
239. H. H. Murray, D. A. Briggs, G. Garzón, R. G. Raptis, L. C. Porter and J. P. Fackler, Jr., *Organometallics*, 1987, **6**, 1992.
240. H. H. Murray, G. Garzón, R. G. Raptis, A. M. Mazany, L. C. Porter and J. P. Fackler, Jr., *Inorg. Chem.*, 1988, **27**, 836.
241. R. Usón *et al.*, *J. Organomet. Chem.*, 1987, **336**, 461.
242. J. Stein, J. P. Fackler, Jr., C. Paparizos and H.-W. Chen, *J. Am. Chem. Soc.*, 1981, **103**, 2192.
243. H. Schmidbaur, G. Müller, K. C. Dash and B. Milewski-Mahrla, *Chem. Ber.*, 1981, **114**, 441.
244. P. G. Jones, *Acta Crystallogr., Sect. C*, 1992, **48**, 1209.
245. R. Usón, A. Laguna, M. Laguna, A. Usón and M. C. Gimeno, *Organometallics*, 1987, **6**, 682.
246. A. Schier and H. Schmidbaur, *Z. Naturforsch., Teil B*, 1982, **37**, 1518.
247. N. Holy, U. Deschler and H. Schmidbaur, *Chem. Ber.*, 1982, **115**, 1379.
248. Y. Yamamoto, *Bull. Chem. Soc. Jpn.*, 1984, **57**, 43.
249. Y. Yamamoto and H. Konno, *Bull. Chem. Soc. Jpn.*, 1986, **59**, 1327.
250. L. C. Porter, H. Knachel and J. P. Fackler, Jr., *Acta Crystallogr., Sect. C*, 1987, **43**, 1833.
251. L. C. Porter, H. Knachel and J. P. Fackler, Jr., *Acta Crystallogr., Sect. C*, 1986, **42**, 1125.
252. R. Usón, A. Laguna, M. Laguna, A. Usón and M. C. Gimeno, *J. Chem. Soc., Dalton Trans.*, 1988, 701.
253. R. Usón, A. Laguna, M. Laguna, A. Usón, P. G. Jones and C. Freire Erdbrügger, *Organometallics*, 1987, **6**, 1778.
254. A. M. Mazany and J. P. Fackler, Jr., *J. Am. Chem. Soc.*, 1984, **106**, 801.
255. R. Usón *et al.*, *J. Chem. Soc., Dalton Trans.*, 1990, 333.
256. H. Schmidbaur, J. R. Mandl, J.-M. Bassett, G. Blaschke and B. Zimmer-Gasser, *Chem. Ber.*, 1981, **114**, 433.
257. Y. Jiang, S. Alvarez and R. Hoffmann, *Inorg. Chem.*, 1985, **24**, 749.
258. B. Chaudret, B. Delavaux and R. Poilblanc, *Coord. Chem. Rev.*, 1988, **86**, 191.
259. C. King, J.-C. Wang, M. N. I. Khan and J. P. Fackler, Jr., *Inorg. Chem.*, 1989, **28**, 2145.
260. H.-R. C. Jaw, M. M. Savas, R. D. Rogers and W. R. Mason, *Inorg. Chem.*, 1989, **28**, 1028.
261. W. Ludwig and W. Meyer, *Helv. Chim. Acta*, 1982, **65**, 934.
262. J. P. Fackler, Jr. and J. D. Basil, in 'Inorganic Chemistry: Towards the 21st Century', ACS Symposium Series No. 211, ed. M. H. Chisholm, American Chemical Society, Washington, DC, 1983, p. 201.
263. D. S. Dudis and J. P. Fackler, Jr., *Inorg. Chem.*, 1985, **24**, 3758.
264. H. C. Knachel, C. A. Dettorre, H. J. Galaska, T. A. Salupo, J. P. Fackler, Jr. and H. H. Murray, *Inorg. Chim. Acta*, 1987, **126**, 7.
265. R. G. Raptis, J. P. Fackler, Jr., J. D. Basil and D. S. Dudis, *Inorg. Chem.*, 1991, **30**, 3072.
266. H. Schmidbaur and P. Jandik, *Inorg. Chim. Acta*, 1983, **74**, 97.
267. H. H. Murray, III, J. P. Fackler, Jr., L. C. Porter and A. M. Mazany, *J. Chem. Soc., Chem. Commun.*, 1986, 321.
268. R. Usón, A. Laguna, M. Laguna, J. Jiménez and P. G. Jones, *J. Chem. Soc., Dalton Trans.*, 1991, 1361.
269. B. Trzcinska-Bancroft, M. N. I. Khan and J. P. Fackler, Jr., *Organometallics*, 1988, **7**, 993.
270. L. C. Porter and J. P. Fackler, Jr., *Acta Crystallogr., Sect. C*, 1986, **42**, 1646.
271. L. C. Porter and J. P. Fackler, Jr., *Acta Crystallogr., Sect. C*, 1986, **42**, 1128.
272. J. P. Fackler, Jr. and L. C. Porter, *J. Am. Chem. Soc.*, 1986, **108**, 2750.
273. D. D. Heinrich and J. P. Fackler, Jr., *J. Chem. Soc., Chem. Commun.*, 1987, 1260.
274. H. H. Murray, J. P. Fackler, Jr., L. C. Porter, D. A. Briggs, M. A. Guerra and R. J. Lagow, *Inorg. Chem.*, 1987, **26**, 357.
275. H. H. Murray, A. M. Mazany and J. P. Fackler, Jr., *Organometallics*, 1985, **4**, 154.
276. L. C. Porter and J. P. Fackler, Jr., *Acta Crystallogr., Sect. C*, 1987, **43**, 587.
277. L. C. Porter and J. P. Fackler, Jr., *Acta Crystallogr., Sect. C*, 1987, **43**, 29.
278. A. Laguna, M. Laguna, J. Jiménez, F. J. Lahoz and E. Olmos, *J. Organomet. Chem.*, 1992, **435**, 235.
279. J. D. Basil *et al.*, *J. Am. Chem. Soc.*, 1985, **107**, 6908.
280. J. P. Fackler, Jr. and J. D. Basil, *Organometallics*, 1982, **1**, 871.
281. H. H. Murray, J. P. Fackler, Jr. and B. Trzcinska-Bancroft, *Organometallics*, 1985, **4**, 1633.
282. H. H. Murray, III, J. P. Fackler, Jr. and D. A. Tocher, *J. Chem. Soc., Chem. Commun.*, 1985, 1278.

283. H. H. Murray, III, J. P. Fackler, Jr. and D. A. Tocher, *J. Chem. Soc., Chem. Commun.*, 1986, 580.
284. H. H. Murray, J. P. Fackler, Jr., A. M. Mazany, L. C. Porter, J. Shain and L. R. Falvello, *Inorg. Chim. Acta*, 1986, **114**, 171.
285. H. H. Murray, J. P. Fackler, Jr. and A. M. Mazany, *Organometallics*, 1984, **3**, 1310.
286. H. H. Murray and J. P. Fackler, Jr., *Inorg. Chim. Acta*, 1986, **115**, 207.
287. H. Schmidbaur, C. Hartmann, G. Reber and G. Müller, *Angew. Chem., Int. Ed. Engl.*, 1987, **26**, 1146.
288. L. C. Porter, H. H. Murray and J. P. Fackler, Jr., *Acta Crystallogr., Sect. C*, 1987, **43**, 877.
289. R. G. Raptis, H. H. Murray, R. J. Staples, L. C. Porter and J. P. Fackler, Jr., *Inorg. Chem.*, 1993, **32**, 5576.
290. R. G. Raptis, J. P. Fackler, Jr., H. H. Murray and L. C. Porter, *Inorg. Chem.*, 1989, **28**, 4057.
291. D. D. Heinrich and J. P. Fackler, Jr., *Inorg. Chem.*, 1990, **29**, 4402.
292. H. H. Murray, III, L. C. Porter, J. P. Fackler, Jr. and R. G. Raptis, *J. Chem. Soc., Dalton Trans.*, 1988, 2669.
293. P. Jandik, U. Schubert and H. Schmidbaur, *Angew. Chem., Int. Ed. Engl.*, 1982, **21**, 73.
294. M. Bardají, M. C. Gimeno, J. Jiménez, A. Laguna, M. Laguna and P. G. Jones, *J. Organomet. Chem.*, 1992, **441**, 339.
295. H. Schmidbaur, C. Hartmann, J. Riede, B. Huber and G. Müller, *Organometallics*, 1986, **5**, 1652.
296. H. C. Knachel, D. S. Dudis and J. P. Fackler, Jr., *Organometallics*, 1984, **3**, 1312.
297. H. Schmidbaur and C. Hartmann, *Angew. Chem., Int. Ed. Engl.*, 1986, **25**, 575.
298. R. G. Raptis, L. C. Porter, R. J. Emrich, H. H. Murray and J. P. Fackler, Jr., *Inorg. Chem.*, 1990, **29**, 4408.
299. J. P. Fackler, Jr. and B. Trzcinska-Bancroft, *Organometallics*, 1985, **4**, 1891.
300. R. J. H. Clark, J. H. Tocher, J. P. Fackler, Jr., R. Neira, H. H. Murray and H. Knackel, *J. Organomet. Chem.*, 1986, **303**, 437.
301. J. P. Fackler, Jr., H. H. Murray and J. D. Basil, *Organometallics*, 1984, **3**, 821.
302. S. Wang and J. P. Fackler, Jr., *Organometallics*, 1988, **7**, 2415.
303. S. Wang and J. P. Fackler, Jr., *Organometallics*, 1989, **8**, 1578.
304. H. Schmidbaur, C. Hartmann and F. E. Wagner, *Angew. Chem., Int. Ed. Engl.*, 1987, **26**, 1148.
305. R. Usón, A. Laguna, M. Laguna, M. T. Tartón and P. G. Jones, *J. Chem. Soc., Chem. Commun.*, 1988, 740.
306. R. Usón, A. Laguna, M. Laguna, J. Jiménez and P. G. Jones, *Angew. Chem., Int. Ed. Engl.*, 1991, **30**, 198.
307. M. A. Bennett, *J. Organomet. Chem.*, 1986, **300**, 7.
308. E. Ahmed, R. J. H. Clark, M. L. Tobe and L. Cattalini, *J. Chem. Soc., Dalton Trans.*, 1990, 2701.
309. R. Usón, A. Laguna, M. U. de la Orden, R. V. Parish and L. S. Moore, *J. Organomet. Chem.*, 1985, **282**, 145.
310. J. Vicente, M. T. Chicote and M. D. Bermúdez, *Inorg. Chim. Acta*, 1982, **63**, 35.
311. J. Vicente, M. T. Chicote, M. D. Bermúdez and M. Garcia-Garcia, *J. Organomet. Chem.*, 1985, **295**, 125.
312. J. Vicente, M.-D. Bermúdez, M.-T. Chicote and M.-J. Sánchez-Santano, *J. Chem. Soc., Dalton Trans.*, 1990, 1945.
313. J. Vicente, M. T. Chicote and M. D. Bermúdez, *J. Organomet. Chem.*, 1984, **268**, 191.
314. J. Vicente, M. T. Chicote, M. D. Bermúdez, P. G. Jones, C. Fittschen and G. M. Sheldrick, *J. Chem. Soc., Dalton Trans.*, 1986, 2361.
315. E. C. Constable and T. A. Leese, *J. Organomet. Chem.*, 1989, **363**, 419.
316. E. C. Constable, R. P. G. Henney, P. R. Raithby and L. R. Sousa, *J. Chem. Soc., Dalton Trans.*, 1992, 2251.
317. E. C. Constable, R. P. G. Henney, P. R. Raithby and L. R. Sousa, *Angew. Chem., Int. Ed. Engl.*, 1991, **30**, 1363.
318. E. C. Constable, R. P. G. Henney and T. A. Leese, *J. Organomet. Chem.*, 1989, **361**, 277.
319. Y. Mizuno and S. Komiya, *Inorg. Chim. Acta*, 1986, **125**, L13.
320. J. A. Morrison, *Adv. Organomet. Chem.*, 1993, **35**, 211.
321. J. L. Margrave, K. H. Whitmire, R. H. Hauge and N. T. Norem, *Inorg. Chem.*, 1990, **29**, 3252.
322. R. Usón, A. Laguna, M. Laguna and M. Abad, *J. Organomet. Chem.*, 1983, **249**, 437.
323. J. Vicente, M. T. Chicote, A. Arcas, M. Artigao and R. Jimenez, *J. Organomet. Chem.*, 1983, **247**, 123.
324. R. Usón, J. Vicente, M. T. Chicote, P. G. Jones and G. M. Sheldrick, *J. Chem. Soc., Dalton Trans.*, 1983, 1131.
325. S. Shibata, K. Iijima and T. H. Baum, *J. Chem. Soc., Dalton Trans.*, 1990, 1519.
326. R. Usón, J. Vicente and M. T. Chicote, *J. Organomet. Chem.*, 1981, **209**, 271.
327. H. W. Chen, C. Paparizos and J. P. Fackler, Jr., *Inorg. Chim. Acta*, 1985, **96**, 137.
328. R. Usón, A. Laguna, M. Laguna, M. L. Castilla, P. G. Jones and K. Meyer-Bäse, *J. Organomet. Chem.*, 1987, **336**, 453.
329. E. G. Perevalova, K. I. Grandberg, V. P. Dyadchenko and O. N. Kalinina, *J. Organomet. Chem.*, 1988, **352**, C37.
330. A. Laguna and M. Laguna, *J. Organomet. Chem.*, 1990, **394**, 743.
331. R. Usón, A. Laguna, M. Laguna, M. N. Fraile, P. G. Jones and C. Freire Erdbrügger, *J. Chem. Soc., Dalton Trans.*, 1989, 73.
332. U. Grässle, W. Hiller and J. Strähle, *Z. Anorg. Allg. Chem.*, 1985, **529**, 29.
333. H.-N. Adams, U. Grässle, W. Hiller and J. Strähle, *Z. Anorg. Allg. Chem.*, 1983, **504**, 7.
334. U. Grässle and J. Strähle, *Z. Anorg. Allg. Chem.*, 1985, **531**, 26.
335. P. K. Byers, A. J. Canty, N. J. Minchin, J. M. Patrick, B. W. Skelton and A. H. White, *J. Chem. Soc., Dalton Trans.*, 1985, 1183.
336. P. K. Byers, A. J. Canty, L. M. Engelhardt, J. M. Patrick and A. H. White, *J. Chem. Soc., Dalton Trans.*, 1985, 981.
337. A. J. Canty, N. J. Minchin, P. C. Healy and A. H. White, *J. Chem. Soc., Dalton Trans.*, 1982, 1795.
338. A. J. Canty, N. J. Minchin, J. M. Patrick and A. H. White, *Aust. J. Chem.*, 1983, **36**, 1107.
339. A. J. Canty, R. Colton and I. M. Thomas, *J. Organomet. Chem.*, 1993, **455**, 283.
340. P. K. Byers, A. J. Canty, K. Mills and L. Titcombe, *J. Organomet. Chem.*, 1985, **295**, 401.
341. V. W.-W. Yam, S. W.-K. Choi, T.-F. Lai and W.-K. Lee, *J. Chem. Soc., Dalton Trans.*, 1993, 1001.
342. J. Vicente, M.-T. Chicote, M.-D. Bermúdez, X. Soláns and M. Font-Altaba, *J. Chem. Soc., Dalton Trans.*, 1984, 557.
343. J. Vicente, M. D. Bermúdez, M. J. Sánchez-Santano and J. Payá, *Inorg. Chim. Acta*, 1990, **174**, 53.
344. J. Vicente *et al.*, *J. Organomet. Chem.*, 1986, **310**, 401.
345. J. Vicente, M. D. Bermúdez, J. Escribano, M. P. Carrillo and P. G. Jones, *J. Chem. Soc., Dalton Trans.*, 1990, 3083.
346. J. Vicente, M.-D. Bermúdez, M.-T. Chicote and M.-J. Sánchez-Santano, *J. Organomet. Chem.*, 1990, **381**, 285.
347. J. Vicente, M. D. Bermúdez, M. T. Chicote and M. J. Sánchez-Santano, *J. Organomet. Chem.*, 1989, **371**, 129.
348. J. Vicente, M. T. Chicote, M. D. Bermúdez, M. J. Sánchez-Santano and P. G. Jones, *J. Organomet. Chem.*, 1988, **354**, 381.
349. J. Vicente, M. D. Bermúdez and J. Escribano, *Organometallics*, 1991, **10**, 3380.
350. J. Vicente, M.-D. Bermúdez, M.-T. Chicote and M.-J. Sánchez-Santano, *J. Chem. Soc., Chem. Commun.*, 1989, 141.
351. J. Vicente, M.-D. Bermúdez, M.-P. Carrillo and P. G. Jones, *J. Chem. Soc., Dalton Trans.*, 1992, 1975.

352. J. Vicente, M.-D. Bermúdez, M.-P. Carrillo and P. G. Jones, *J. Organomet. Chem.*, 1993, **456**, 305.
353. R. Usón, A. Laguna, M. Laguna, B. R. Manzano, P. G. Jones and G. M. Sheldrick, *J. Chem. Soc., Dalton Trans.*, 1985, 2417.
354. R. Usón, A. Laguna, M. Laguna, B. R. Manzano, P. G. Jones and G. M. Sheldrick, *J. Chem. Soc., Dalton Trans.*, 1984, 839.
355. G. M. Bancroft, T. C. S. Chan and R. J. Puddephatt, *Inorg. Chem.*, 1983, **22**, 2133.
356. S. Komiya and A. Shibue, *Organometallics*, 1985, **4**, 684.
357. R. Usón, A. Laguna, M. Laguna, E. Fernández, P. G. Jones and G. M. Sheldrick, *J. Chem. Soc., Dalton Trans.*, 1982, 1971.
358. S. Komiya, S. Ozaki and A. Shibue, *J. Chem. Soc., Chem. Commun.*, 1986, 1555.
359. S. Komiya, T. Sone, S. Ozaki, M. Ishikawa and N. Kasuga, *J. Organomet. Chem.*, 1992, **428**, 303.
360. R. Usón, A. Laguna, M. Laguna, J. Jiménez and M. E. Durana, *Inorg. Chim. Acta*, 1990, **168**, 89.
361. R. Usón, A. Laguna and M. D. Villacampa, *Inorg. Chim. Acta*, 1986, **122**, 81.
362. M. A. Guerra, T. R. Bierschenk and R. J. Lagow, *J. Organomet. Chem.*, 1986, **307**, C58.
363. S. Komiya, A. Shibue and S. Ozaki, *J. Organomet. Chem.*, 1987, **319**, C31.
364. J. K. Jawad, R. J. Puddephatt and M. A. Stalteri, *Inorg. Chem.*, 1982, **21**, 332.
365. S. Komiya, M. Ishikawa and S. Ozaki, *Organometallics*, 1988, **7**, 2238.
366. R. Usón, A. Laguna, M. Laguna, I. Lázaro, P. G. Jones and C. Fittschen, *J. Chem. Soc., Dalton Trans.*, 1988, 2323.
367. R. Usón, A. Laguna, M. Laguna, I. Lázaro and P. G. Jones, *Organometallics*, 1987, **6**, 2326.
368. M. C. Gimeno, P. G. Jones, A. Laguna, M. Laguna and I. Lázaro, *J. Chem. Soc., Dalton Trans.*, 1993, 2223.
369. S. Komiya, S. Ozaki, I. Endo, K. Inoue, N. Kasuga and Y. Ishizaki, *J. Organomet. Chem.*, 1992, **433**, 337.
370. A. J. Markwell, *J. Organomet. Chem.*, 1985, **293**, 257.
371. S. Komiya, S. Meguro, A. Shibue and S. Ozaki, *J. Organomet. Chem.*, 1987, **328**, C40.
372. A. Laguna, M. Laguna, M. C. Gimeno and P. G. Jones, *Organometallics*, 1992, **11**, 2759.
373. W. Bos *et al.*, *J. Organomet. Chem.*, 1986, **307**, 385.
374. R. P. F. Kanters, P. P. J. Schlebos, J. J. Bour, J. Wijnhoven, E. van den Berg and J. J. Steggerda, *J. Organomet. Chem.*, 1990, **388**, 233.
375. A. M. Mueting, *et al.*, *New J. Chem.*, 1988, **12**, 505.
376. M. I. Bruce, G. A. Koutsantonis and E. R. T. Tiekink, *J. Organomet. Chem.*, 1991, **408**, 77.
377. C. M. Hay, B. F. G. Johnson, J. Lewis, N. D. Prior, P. R. Raithby and W. T. Wong, *J. Organomet. Chem.*, 1991, **401**, C20.
378. M. I. Bruce, *J. Organomet. Chem.*, 1990, **400**, 321.
379. A. D. Horton, M. J. Mays and M. McPartlin, *J. Chem. Soc., Chem. Commun.*, 1987, 424.
380. P. H. Kasai, *J. Am. Chem. Soc.*, 1984, **106**, 3069.
381. P. H. Kasai, *J. Am. Chem. Soc.*, 1983, **105**, 6704.
382. J. H. B. Chenier, J. A. Howard, B. Mile and R. Sutcliffe, *J. Am. Chem. Soc.*, 1983, **105**, 788.
383. J. H. B. Chenier, J. A. Howard and B. Mile, *J. Am. Chem. Soc.*, 1985, **107**, 4190.
384. A. J. Buck, B. Mile and J. A. Howard, *J. Am. Chem. Soc.*, 1983, **105**, 3381.
385. R. L. Garrell, T. M. Herne, C. A. Szafranski, F. Diederich, F. Ettl and R. L. Whetten, *J. Am. Chem. Soc.*, 1991, **113**, 6302.
386. G. Nicolas and F. Spiegelmann, *J. Am. Chem. Soc.*, 1990, **112**, 5410.
387. D. Belli Dell'Amico, F. Calderazzo, R. Dantona, J. Strähle and H. Weiss, *Organometallics*, 1987, **6**, 1207.
388. J. E. Goldberg, D. F. Mullica, E. L. Sappenfield and F. G. A. Stone, *J. Chem. Soc., Dalton Trans.*, 1992, 2495.
389. G. A. Carriedo, V. Riera, G. Sánchez, X. Soláns and M. Labrador, *J. Organomet. Chem.*, 1990, **391**, 431.
390. J. C. Jeffery, P. A. Jelliss and F. G. A. Stone, *J. Chem. Soc., Dalton Trans.*, 1993, 1073.
391. F. G. A. Stone, *Adv. Organomet. Chem.*, 1990, **31**, 53.
392. W. Beck, W. Weigand, U. Nagel and M. Schaal, *Angew. Chem., Int. Ed. Engl.*, 1984, **23**, 377.
393. G. Banditelli, F. Bonati, S. Calogero, G. Valle, F. E. Wagner and R. Wordel, *Organometallics*, 1986, **5**, 1346.
394. A. L. Bandini, G. Banditelli, G. Minghetti, B. Pelli and P. Traldi, *Organometallics*, 1989, **8**, 590.
395. A. K. Chowdhury and C. L. Wilkins, *J. Am. Chem. Soc.*, 1987, **109**, 5336.
396. R. Usón, A. Laguna, M. D. Villacampa, P. G. Jones and G. M. Sheldrick, *J. Chem. Soc., Dalton Trans.*, 1984, 2035.
397. W. P. Fehlhammer and W. Finck, *J. Organomet. Chem.*, 1991, **414**, 261.
398. G. Banditelli, F. Bonati, S. Calogero and G. Valle, *J. Organomet. Chem.*, 1984, **275**, 153.
399. E. O. Fischer and M. Böck, *J. Organomet. Chem.*, 1985, **287**, 279.
400. E. O. Fischer, M. Böck and R. Aumann, *Chem. Ber.*, 1983, **116**, 3618.
401. R. Aumann and E. O. Fischer, *Chem. Ber.*, 1981, **114**, 1853.
402. E. O. Fischer and M. Böck, *Monatsh. Chem.*, 1984, **115**, 1159.
403. J. F. Britten, C. J. L. Lock and Z. Wang, *Acta Crystallogr., Sect. C*, 1992, **48**, 1600.
404. H. G. Raubenheimer, F. Scott, M. Roos and R. Otte, *J. Chem. Soc., Chem. Commun.*, 1990, 1722.
405. P. H. Kasai and P. M. Jones, *J. Am. Chem. Soc.*, 1985, **107**, 6385.
406. J. H. B. Chenier, J. A. Howard, H. A. Joly, B. Mile and M. Tomietto, *Can. J. Chem.*, 1989, **67**, 655.
407. H. Willner and F. Aubke, *Inorg. Chem.*, 1990, **29**, 2195.
408. D. F. McIntosh, G. A. Ozin and R. P. Messmer, *Inorg. Chem.*, 1981, **20**, 3640.
409. F. Calderazzo, *J. Organomet. Chem.*, 1990, **400**, 303.
410. D. Belli Dell'Amico and F. Calderazzo, *Inorg. Synth.*, 1986, **24**, 236.
411. P. G. Jones, *Z. Naturforsch., Teil B*, 1982, **37**, 823.
412. S. Qiu, R. Ohnishi and M. Ichikawa, *J. Chem. Soc., Chem. Commun.*, 1992, 1425.
413. D. Belli Dell'Amico, F. Calderazzo, P. Robino and A. Segre, *J. Chem. Soc., Dalton Trans.*, 1991, 3017.
414. D. Belli Dell'Amico, F. Calderazzo, P. Robino and A. Segre, *Gazz. Chim. Ital.*, 1991, **121**, 51.
415. A. Veldkamp and G. Frenking, *Organometallics*, 1993, **12**, 4613.
416. M. Adelhelm, W. Bacher, E. G. Höhn and E. Jacob, *Chem. Ber.*, 1991, **124**, 1559.
417. H. Willner *et al.*, *J. Am. Chem. Soc.*, 1992, **114**, 8972.
418. W. P. Bosman, W. Bos, J. M. M. Smits, P. T. Beurskens, J. J. Bour and J. J. Steggerda, *Inorg. Chem.*, 1986, **25**, 2093.
419. W. Bos, J. J. Bour, J. J. Steggerda and L. H. Pignolet, *Inorg. Chem.*, 1985, **24**, 4298.
420. S. K. Chastain and W. R. Mason, *Inorg. Chem.*, 1982, **21**, 3717.
421. D. Perreault, M. Drouin, A. Michel and P. D. Harvey, *Inorg. Chem.*, 1991, **30**, 2.

422. D. S. Eggleston, D. F. Chodosh, R. L. Webb and L. L. Davis, *Acta Crystallogr., Sect. C*, 1986, **42**, 36.
423. C.-M. Che, H.-K. Yip, W.-T. Wong and T.-F. Lai, *Inorg. Chim. Acta*, 1992, **197**, 177.
424. C.-M. Che, W.-T. Wong, T.-F. Lai and H.-L. Kwong, *J. Chem. Soc., Chem. Commun.*, 1989, 243.

2

Copper and Silver

GERARD VAN KOTEN, STUART L. JAMES and JOHANN T. B. H. JASTRZEBSKI
University of Utrecht, The Netherlands

2.1 INTRODUCTION	57
2.2 PREPARATIVE ROUTES TO ORGANOCOPPER AND -SILVER COMPOUNDS	60
2.2.1 General Remarks	60
2.2.2 Synthesis of Organocopper Compounds	60
2.2.2.1 Via transmetallation reactions: reagents and copper salts	60
2.2.2.2 Via transmetallation reactions: complex formation with organometallics and metal salts	64
2.2.2.3 Via metallation reactions	67
2.2.2.4 Via interaggregate exchange/self-assembly	69
2.2.2.5 Miscellaneous preparative routes	70
2.2.3 Synthesis of Organosilver Compounds	71
2.3 STRUCTURE AND BONDING IN ORGANOCOPPER AND ORGANOSILVER COMPOUNDS	75
2.3.1 General Remarks: Thermal Stability	75
2.3.2 Structures in the Solid State Solved by X-ray Analysis	77
2.3.2.1 Alkylcopper and -silver compounds	77
2.3.2.2 Arylcopper and -silver compounds	82
2.3.2.3 Alkenylcopper compounds	95
2.3.2.4 Homo- and heteroleptic alkynylcopper and -silver compounds	96
2.3.2.5 Organocuprates and -argentates	100
2.3.2.6 Ligand stabilized neutral and cationic organocopper and -silver compounds	110
2.3.2.7 Miscellaneous organocopper and -silver compounds	115
2.3.3 Polymeric Structures (Including ^{63,65}Cu NQR Spectroscopy)	120
2.3.4 Bonding and Stereochemical Aspects (Including Theoretical Studies)	121
2.3.4.1 The nature of the copper– and silver–carbon bonds and of the metal–metal interaction	121
2.3.4.2 Structural aspects of organocopper and -silver compounds in solution (including NMR studies and EXAFS spectroscopic studies)	125
2.4 REFERENCES	129

2.1 INTRODUCTION

Of all the organometallic reagents used in organic synthesis, organocopper reagents are among the most widely applied. The great number and diversity of accessible organocopper reagents, each with its specific reactivity,[1] make them highly attractive to the synthetic organic chemist. These organocopper reagents are generally prepared *in situ* from the corresponding organomagnesium or lithium compounds by the simple addition of varying amounts (from catalytic to over stoichiometric) of a suitable copper(I) salt. As organomagnesium or -lithium reagents can only tolerate a limited range of functionalities, new transmetallation procedures have emerged, in which combinations of transition metal-based organometallic reagents and copper salts are used,[2] as well as reagents derived directly from various forms of reactive metallic copper and functionalized organic halides.[3] In contrast to the high degree of sophistication achieved in the application of (organo)copper reagents, which has allowed the development of numerous pathways for the production of economically important organic products, the understanding of the synthetic and structural aspects of the organometallic species involved has remained in its infancy. Perhaps this is not altogether very surprising. Although they are all based on monovalent copper, that is, copper in the formal oxidation state one, connected to a monoanionic organo group, the synthesis of most of the simple alkyl- and arylcopper compounds [CuR] in pure form was for some time seemingly impossible. The susceptibility of the copper–carbon bond to thermal decomposition (pure, solid alkylcoppers explode even below 0 °C) and to oxidation and hydrolysis, as well as the lack of solubility of most organocoppers in hydrocarbon solvents, has thwarted many attempts to isolate pure compounds. These aspects are also encountered in organosilver chemistry when dealing with organosilver compounds containing monovalent silver. The sometimes considerable light sensitivity of these compounds further adds to the experimental difficulties.

Not until 1923 was the first organocopper compound actually isolated, when Reich[4] reported the preparation of impure [CuPh] from the reaction of CuI with MgBrPh in diethyl ether. The first well-defined, pure compounds were halogenated organocoppers, for example, $[Cu(C_6F_5)]$ and $[Cu(C_6H_4CF_3-3)]$, reported by Cairncross and co-workers,[5] which appeared to exist in benzene solution as tetrameric and octameric species, respectively. The first reported structure in the solid state of an arylcopper was the *ortho*-substituted phenylcopper compound, $[Cu(C_6H_3CH_2NMe_2-2-Me-5)]$ (1).[6] This compound exists in the form of a tetrameric $[Cu_4R_4]$ aggregate. Subsequent to this came the observation that organocopper compounds can also form discrete aggregates with copper salts, namely the report of the structure of $[Cu_6Br_2(C_6H_4NMe_2-2)_4]$ (2), a $[Cu_6Br_2R_4]$ species, obtained from the reaction of the corresponding *ortho*-substituted phenyllithium reagent with an over-stoichiometric amount of CuBr.[7] In 1972 and 1973 came the first full structural analysis (^{1}H, ^{13}C, ^{7}Li and ^{107,109}Ag NMR and colligative measurements) of an organocopper and an organosilver species in solution, the neutral arylcuprate and -argentate compounds $[Li_2Cu_2(C_6H_4CH_2NMe_2-2)_4]$ (3)[8] and $[Li_2Ag_2(C_6H_4CH_2NMe_2-2)_4]$ (4),[9] which are both $[Li_2M_2R_4]$ (M = Cu or Ag) aggregates. In 1985, the structure of arylcuprate (3) was confirmed in the solid state with an x-ray structure determination.[10] In 1973 the first structure of an alkylcopper, $[Cu_4(CH_2TMS)_4]$ (5) (a $[Cu_4R_4]$ aggregate), was reported.[11] Except for the neutral cuprate (3), for which NMR and colligative measurements gave conclusive structural data, it is the evidence gained from x-ray structure determinations of organocopper and -silver species that provides new insights into the bonding and synthetic aspects of organocopper and -silver chemistry. The increasing importance of x-ray structure determinations for organocopper and -silver chemistry is illustrated in Table 1.

In the space available, this chapter cannot cover all the reported organocopper chemistry. Attention will be focused even more than in the previous review on discussion of the new insights into the nature and structure of organocopper and -silver compounds gained from x-ray structure determinations or solution spectroscopic studies performed on analytically pure samples. The preparation of organocopper and -silver compounds will be discussed with emphasis on preparative routes allowing the isolation of pure compounds. One new aspect is that seemingly complex compounds can now be straightforwardly prepared and isolated in the pure form by simply mixing together constituent species, for example, a stable tricopper species (11) with $[Cu_3RX_2L]$ stoichiometry is assembled quantitatively by mixing the mesitylcopper pentamer (CuR) and a copper arenethiolate trimer aggregate (CuX), with triphenylphosphine (L), in the appropriate solvent,[15] (Equation (1) and Scheme 6). Another important development has been the notion that various solvents, such as dimethyl sulfide (DMS), can aid the selective formation of desired organocopper and -silver compounds. The double role of DMS as a reaction medium and as a potential ligand is demonstrated by the synthesis in DMS of phenyl-[16] and *o*-tolylcopper,[17] which were ultimately isolated as DMS complexed arylcopper tetramers (12) and (13) respectively (Equations (2) and (3); see also Section 2.2.2.1). A further new aspect in organocopper chemistry is the increasing interest in mechanistic aspects of organic synthetic reactions in which an organocopper reagent takes part. These studies will be discussed in the light of the recent information gained on the structure and reactivity of the pure compounds.

The reader should note that this chapter is largely confined to organocopper and -silver compounds in which the organo group is bound to copper or silver as a monoanionic ligand. The vast area of alkene, alkyne, CO and isocyanide σ- and π-complexes of copper and silver is not treated here.[18] It is hoped that this chapter contains not only the key references that the reader may require but also serves as an introduction to a fascinating and still rapidly developing field of organometallic chemistry, not least because of its importance in organic synthesis.

Table 1 First examples and numbers of x-ray crystal structures of organocopper compounds in the periods up to 1982 and 1983–1994.

First examples	Numbers of structures in the period	
	up to 1982	1983–1994
C_{sp3}	3	9
$[Cu_4(CH_2TMS)_4]$ **(5)**[11]		
C_{sp2}		
Aryl	6	28

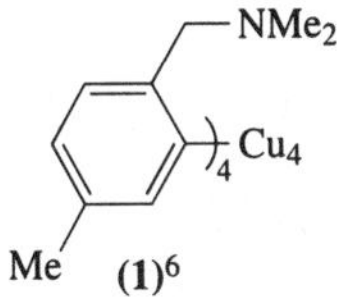

(1)[6]

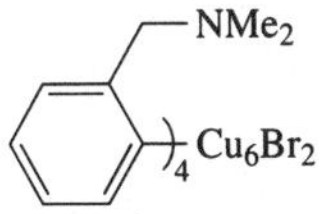

(2)[7]

Alkenyl	1	1

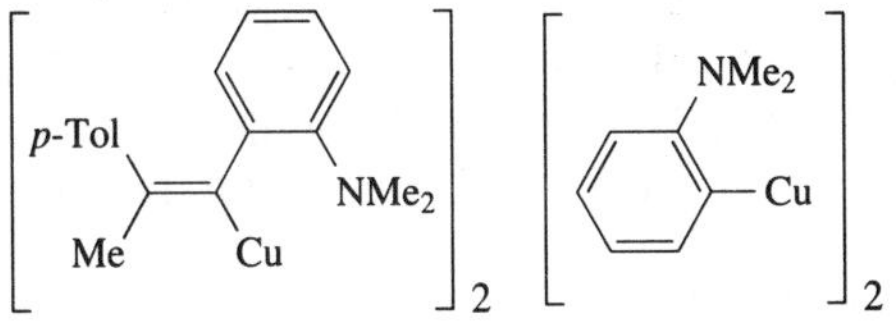

(6)[12]

C_{sp}	3	8
$[Cu_4(C{\equiv}CPh)_4(PMe_3)_4]$**(7)**[13]		
Cuprates (including neutral and ligated cuprates)	1	14

$[Li_2Cu_2(C_6H_4CH_2NMe_2\text{-}2)_4]$ **(3)**[8]
$[Li_2Cu_3Ph_6]^-$ **(8)**[14a]
$[Cu\{C(TMS)_3\}_2]^-$ **(9)**[14b]
$[Cu_5Ph_6]^-$ **(10)**[14]

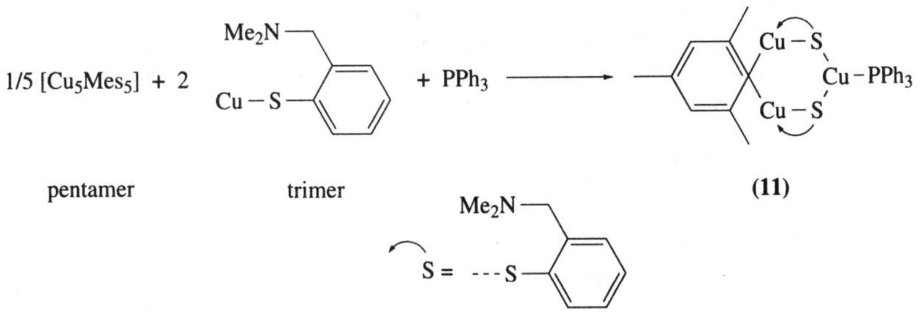

$1/5\ [Cu_5Mes_5]\ +\ 2$ (pentamer) + (trimer) $+\ PPh_3\ \longrightarrow$ **(11)** $\qquad$ (1)

$$4\ PhLi\ \xrightarrow[\text{DMS, } -4\ LiBr]{4\ CuBr}\ [Cu_4Ph_4(DMS)_2]\quad\textbf{(12)} \qquad (2)$$

$$Li(C_6H_4Me\text{-}2)\ \xrightarrow[\text{ii, crystallization from DMS}]{\text{i, [CuBr(DMS)] in Et}_2O}\ [Cu_4(C_6H_4Me\text{-}2)_4(DMS)_2]\quad\textbf{(13)} \qquad (3)$$

2.2 PREPARATIVE ROUTES TO ORGANOCOPPER AND -SILVER COMPOUNDS

2.2.1 General Remarks

Detailed accounts covering the literature up to the mid-1970s of the methods used for the preparation of organocopper and -silver compounds can be found in References 19–26. In these accounts all preparations are listed irrespective of whether or not a pure organocopper or -silver compound was actually isolated. *COMC-I*[19] treated exclusively the organocopper and -silver compounds which had been isolated in the pure state, or had been adequately characterized without actual isolation (see Tables 1 and 2 in *COMC-I*[19]). On the basis of the increased structural information now available on these compounds, which shows that they are for the most part aggregated species, synthetic methods have improved considerably in recent years. This greater knowledge indicated that a rational synthesis of organocopper and -silver compounds requires the selective use of (i) copper and silver starting materials, for example, salts, (ii) well-defined organometallic reagents (RLi, $R_2Mg/RMgX$, R_2Zn, etc.), and (iii) additional ligands (including solvents with or without coordinating properties). This review will concentrate on a discussion of the synthetic routes available to the (in)organic chemist to arrive at (analytically) pure products in acceptable yields. The *in situ* formation of compounds commonly employed when organocopper and -silver reagents for organic synthesis are prepared has been treated in detail.[1,19] With few exceptions, organocopper and -silver compounds are susceptible to hydrolysis and oxidation reactions and to reactions with CO, CO_2 and H_2. Moreover, their thermal (see Table 3 in Section 2.3.1) stability can be very limited (note: alkylcopper and -silver compounds can explode on heating). Therefore, an inert atmosphere must be maintained during the preparation, isolation and handling of most of these compounds. In addition, the organosilver compounds are often light sensitive, and during preparation and storage shielding from direct light is required. Consequently, the experience of the authors is that the successful synthesis of pure products requires the use of pure starting materials and careful selection of the reaction conditions: extensive purification of crude reaction mixtures to arrive eventually at pure organocopper or -silver compounds has commonly been found to be ineffective.

The compounds listed in Table 2 have been isolated in the pure form. Also included in this table are compounds for which the structure in solution could be determined on the basis of colligative and NMR spectroscopic data. In particular for several organosilver, argentate and cuprate compounds, on the basis of ^{107,109}Ag and/or ^{6,7}Li NMR data, detailed proposals for the structure in solution have been made (see Section 2.3.4.2(i)), which later proved correct when the structure in the solid state was obtained.

In the following section the most important synthetic routes to pure organo-copper and -silver compounds will be discussed. These are summarized in Table 2.

2.2.2 Synthesis of Organocopper Compounds

2.2.2.1 Via transmetallation reactions: reagents and copper salts

The most widely used method for the preparation of pure organocopper compounds of the type RCu involves the transmetallation reaction of a copper(I) salt with an organolithium, -magnesium or -zinc reagent using Et_2O or THF, aromatics (benzene or toluene) or DMS as a solvent. This method has been successfully used for the preparation and isolation in the pure state of alkyl- (**5**) (note: violent decomposition can occur),[11] alkenyl-,[78] alkynyl-,[20] aryl- (**14**),[20] ferrocenyl-[79] and cyclopentadienylcopper (**15**)[80,81] compounds (Equation (4)).

$$CuX + LiR \xrightarrow{\text{solvent}} CuR + LiX \qquad (4)$$

R = Me, CH$_2$TMS(**5**), Ph, C$_6$H$_4$CH$_2$NMe$_2$-2(**14**)
4-MeC$_6$H$_4$C(Me)C=C(C$_6$H$_4$NMe$_2$-2), C≡CPh,
(η-C$_5$H$_3$CH$_2$NMe$_2$-2)FeCp(**15**)

The nature of the anion in the copper(I) salt has a profound influence on the success of the synthesis of pure organocopper compounds. CuBr is preferred in most cases over CuCl and CuI. In addition, organic copper(I) salts may be used, for example, CuCN,[20] CuSCN,[20] CuOSO$_2$CF$_3$,[82] CuSPh and CuO$_2$CPh, or soluble copper halide complexes, for example, [CuBr(DMS)][83] and [CuI(PBu$_3$)]$_4$.[65] In particular [CuBr(DMS)], introduced by House *et al.*[84] and later extensively exploited by Bertz and Dabbagh,[83] is a convenient, soluble precursor for the preparation of pure organocoppers (often with

Table 2 Isolated organocopper and -silver compounds and reagents used.

Isolated compound	Reagents used	Ref.
Organocopper compounds		
C_{sp^3}		
$[\{2\text{-(TMS)}_2C(Cu)C_5H_4N\}_2]$	$CuCl + Li(2\text{-(TMS)}_2C_5H_4N)$	27
$[Cu_4(2\text{-CH(TMS)}C_5H_4N)_4]$	$CuCl + Li(2\text{-CH(TMS)}C_5H_4N)$	28
$[FeCp(\eta\text{-}C_5H_3CH_2NMe_2\text{-}2\text{-Cu})]_4$	$\{FeCp(\eta\text{-}C_5H_3CH_2NMe_2\text{-}2)\}\cdot CuI + Li\{FeCp$ $(\eta\text{-}C_5H_3CH_2NMe_2\text{-}2)\}$	29
$[(CF_3)_2Cu^{III}SC(S)NEt_2]$	$CdI^+[(CF_3)_2Cu]^- + (Et_2NC(S)S)_2$ $CuBr_2(SC(S)NEt_2) + 1/2Cd(CF_3)_2 + CF_3CdI$	30
$[Cu_4(CH_2TMS)_4]$	$CuI + LiCH_2TMS$	11
$[Cu_3\{CH(PPh_2)_2\}_3]$	$[CuAr] + Ph_2PCH_2PPh_2$	31
$[Cu(PMe_3)_4]^+[CuMe_2]^-$	$[Cu_2(O_2CMe)_4] + 2MgMe_2 + PMe_3$ (excess)	32
$[CuMe(PPh_3)_3]\cdot THF$	$[Cu(acac)_2] + 4AlMe_2(O^iPr) + 3.5\ PPh_3$	33
$[Cu_2\{(CH_2)_2PMe_2\}_2]$	$2CuCl + 4Me_3P{=}CH_2$	34,35
C_{sp^2}		
$[Cu_4Ph_4(SMe_2)_2]$	$CuBr + LiPh$	16
$[MeC(CH_2PPh_2)_3CuPh]$	$(MeC(CH_2PPh_2)_3)CuCl + LiPh$	36
$[Cu_5Mes_5]$	$CuCl + MesMgBr$	37,38
$[Cu_4(Mes_4)(C_4H_8S_2)]$	$[Cu_5Mes_5] + C_4H_8S$	37,38
$[Cu_3(Mes)(\mu\text{-}O_2CPh)_2]$	$[Cu_5Mes_5] + [Cu_4(O_2CPh)_4]$	39
$[Cu_2(SC_6H_4(CH_2NMe_2)\text{-}2)(Mes)]_2$	$[Cu(SC_6H_4(CH_2NMe_2)\text{-}2)]_3 + 0.6[Cu_5Mes_5]$	15
$[Cu_3(SC_6H_4(CH_2NMe_2)\text{-}2\text{-Cl-3})_2(Mes)(PPh_3)]$	$[Cu(SC_6H_4(CH_2NMe_2)\text{-}2\text{-Cl-3})]_3 + 0.3[Cu_5Mes_5] + 1.4PPh_3$	15
$[Cu_4(C_6H_4Me\text{-}2)_4(SMe_2)_2]$	$CuBr(DMS) + [Li(C_6H_4Me\text{-}2)]_n$	17
$[Cu_8(C_6H_4OMe\text{-}2)_8]$	$CuBr + LiC_6H_4OMe\text{-}2$	40
$[Cu_4(2,4,6\text{-Pr}^i_3C_6H_2)_4]$	$CuBr + (2,4,6\text{-Pr}^i_3C_6H_2Li)$	41
$[Cu(2,4,6\text{-Ph}_3C_6H_2)]^?_?$	$CuCl + (2,4,6\text{-Ph}_3C_6H_2MgBr)$	42
$[Cu_4\{C_6H_3(CH_2NMe_2)\text{-}2\text{-Me-5}\}_4]$	$CuBr + LiC_6H_3(CH_2NMe_2)\text{-}2\text{-Me-5}$	43
$[Cu_4(C_{10}H_6NMe_2\text{-}8)_4]$	$CuBr\{P(OMe)_3\} + Li(C_{10}H_6NMe_2\text{-}8)$	44
$[Cu_2(MeOxl)_2]$	$CuBr + Li(MeOxl)$	45
$[Cu_6Br_2(MeOxl)_4]$	$CuBr + 2[Cu(MeOxl)]$	45
$[Cu_6Br_2(C_6H_4NMe_2\text{-}2)_4]$	$3CuBr + 2LiC_6H_4NMe_2\text{-}2$	46,47
$[Cu_4Br_2\{C_6H_3(CH_2NMe_2)_2\text{-}o,o'\}_2]$	$2CuBr + Bu^nLi + o,o'\text{-}(Me_2NCH_2)_2C_6H_3Br$	48
$[Cu(dppe)_2]^+[Cu(Mes_2)]^-$	$[Cu_5Mes_5] + dppe$	49
$[Cu_4Vi_2(C_6H_4NMe_2\text{-}2)_2]^a$	$[Cu_4Vi_2Br_2] + Li(C_6H_4NMe_2\text{-}2)$	12,50
$[Cu_4Vi_2Br_2]^a$	$2CuBr + ViLi$	50
C_{sp}		
$[Cu_3\{SC_6H_4(CH_2NMe_2)\text{-}2\}_2(C{\equiv}CBu^t)]_2$	$[CuSC_6H_4(CH_2NMe_2)\text{-}2]_3 + LiC{\equiv}CBu^t$ $4/3[CuSC_6H_4(CH_2NMe_2)\text{-}2]_3 + 2CuC{\equiv}CBu^t$	51,52
$[Cu(C{\equiv}CPh)(Ph_2Ppy\text{-}P)]_4$	$[Cu_2(\mu\text{-}Ph_2Ppy)_2(MeCN)_2][BF_4]_2 + 2PhC{\equiv}CH + KOH$ (excess)	53
$[CuC{\equiv}CPh]_n$		54
$[Cu(PMe_3)C{\equiv}CPh]_4$		13
$[Cu(PPh_3)C{\equiv}CPh]_4$	$[Cu(PPh_3)_2BH_4] + PhC{\equiv}CH + KOH$	55
$[Cu_3(C{\equiv}CPh)(dppm)_3]^{2+}[BF_4]_2^-$	$3[Cu_2(\mu\text{-}dppm)_2(MeCN)_2][BF_4]_2 + 2LiC{\equiv}CPh$	56
$[Cu_4(L)_2(C{\equiv}CPh)]^{3+}[ClO_4]^-_3{}^b$	$Cu^{II}(L)(OH)_2(ClO_4)_2\cdot H_2O + PhC{\equiv}CH$	57
Mixed $sp^2\text{-}sp$		
$[Cu_6(C_6H_4NMe_2\text{-}2)_4(C{\equiv}CC_6H_4Me\text{-}4)_2]$	$[Cu_6Br_2(C_6H_4NMe_2\text{-}2)_4] + 2Li(C{\equiv}CC_6H_4Me\text{-}4)_2]$ $2CuC_6H_4NMe_2\text{-}2 + CuC{\equiv}CC_6H_4Me\text{-}4$ $6CuC_6H_4NMe_2\text{-}2 + HC{\equiv}CC_6H_4Me\text{-}4$	58,59 60
$[CuCp^*\{C(PPh_3)_2\}]$	$[CuCl\{C(PPh_3)_2\}] + [KCp^*]$	61
Mixed-metal compounds		
Cu–Li		
$[Cu_2Li_2(CH_2TMS)_4(DMS)_2]_\infty$	$CuBr + 2Li(CH_2TMS)$ in DMS	62
$[MeCuP(Bu^t)_2\{Li(THF)\}_3]$	$[Cu(PBu^t_2)]_4 + 4LiMe$	63
$[Li(THF)_4]^+[Cu\{C(TMS)_3\}_2]^-$	$2CuI + [Li(THF)_4]^+[Li\{C(TMS)_3\}_2]^-$	14b
$[Li(12\text{-crown-4})_2]^+[CuMe_2]^-$	$CuI + 2LiMe + 2(12\text{-crown-4})$	64
$[Li(12\text{-crown-4})_2]^+[Cu(Br)CH(TMS)_2]^-$	$CuI + 2Li(CH(TMS)_2) + 2(12\text{-crown-4})$	64
$[Li(12\text{-crown-4})_2]^+[CuPh_2]^-$	$CuI + 2LiPh + 2(12\text{-crown-4})$	64
$[Li(THF)_4]^+[Cu_5Ph_6]^-$	$CuBr + 1.3LiPh$	14c
$[Li(PMDTA)(THF)]^+[Cu_5Ph_6]^-$	$CuBr + 2LiPh + PMDTA$	14c
$[Li_2Cu_2Ph_4(OEt)_2]$	$2[Cu(PBu^n_3)I] + 4LiPh$	65
$[Li_4Cl_2(Et_2O)_{10}]^{2+}[Li_2Cu_3Ph_6]^-_2$	$CuCN + 2LiPh$	14a
$[Li(Et_2O)_4]^+[Cu_4LiPh_6]^-$	$CuBr + 2LiPh$	66
$[Li_2Cu_2Ph_4(SMe_2)_3]$	$CuBr + 2LiPh$	16
$[Li_3(CuPh_3)(CuPh_2)(SMe_2)_4]$	$CuBr + 3LiPh$	16
$[Li_5Cu_4Ph_9(SMe_2)_9]$	$CuBr + 3.3LiPh$	16
$[Cu_2Li_2(Ar)_4]^c$	$[Li_4(Ar)_4] + 2CuBr$ or $[Cu_4(Ar)_4] + [Li_4(Ar)_4]$	10
Cu–Mg		
$[Cu_4MgPh_6\cdot Et_2O]$	$CuBr + 2MgPh_2$	66
$[Cu_4Mg_2(SAr)_4(Mes)_4]^d$	$[CuSAr]_3 + 1.5[Mg(Mes)_2(THF)_2]$	119,15

Table 2   (continued)

Isolated compound	Reagents used	Ref.
Cu–Ag		
[(Ph$_3$P)$_2$N]$^+$[Ag$_6$Cu$_7$(C≡CPh)$_{14}$]$^-$	[(Ph$_3$P)$_2$N]$^+$[Ag(C≡CPh)$_2$]$^-$ + 4[AgC≡CPh]$_n$ + 4[CuC≡CPh]$_n$	67
Organosilver compounds		
C$_{sp^3}$		
[{2-(TMS)$_2$C(Ag)C$_5$H$_4$N}$_2$]	AgBF$_4$ + Li[2-(TMS)$_2$CC$_5$H$_4$N]	28
[FeCp(η-C$_5$H$_3$CH$_2$NMe$_2$-2-Ag)]$_4$	AgI + Li[FeCp{η-C$_5$H$_3$CH$_2$NMe$_2$-2}]	68
[(CF$_3$)$_2$CFAg(CH$_3$CN)]	AgF + CF$_3$CF=CF$_2$ in MeCN	69
[Et$_4$N]$^+$[Ag(CF(CF$_3$)$_2$)$_2$]$^{-e}$	2[(CF$_3$)$_2$CFAg(CH$_3$CN)] + Et$_4$NBr	70
[Li(THF)$_4$]$^+$[Ag{C(TMS)$_3$}$_2$]$^-$	AgI + [Li(THF)$_4$]$^+$[Li{C(TMS)$_3$}$_2$]$^-$	146
K$^+$[Ag(DTT)$_2$]$^-$	AgNO$_3$ + 2DTT in KOH	71
C$_{sp^2}$		
[Ag$_4$(Mes)$_4$]	AgCl + MesMgBr	72,201
[Ag(2,6-Ph$_3$C$_6$H$_2$)]	AgCl + 2(2,4,6-Ph$_3$C$_6$H$_2$MgBr)	42,210
C$_{sp}$		
[Ag(PMe$_3$)(C≡CPh)]$_n$		73
Mixed		
[Ag(C$_6$F$_5$)(CH$_2$PPh$_3$)]	Ag(CF$_3$CO$_2$) + 3Li(C$_6$F$_5$) + [PPh$_3$Me][CF$_3$CO$_2$]	69
[Ag{CH(PPh$_3$)C(O)OEt}$_2$]$^+$[ClO$_4$]$^-$	AgClO$_4$ + 2CH(PPh$_3$)C(O)OEt	74
[Ag{CH(PPh$_3$)C(O)Ph}$_2$]$^+$[NO$_3$]$^-$	AgNO$_3$ + 2CH(PPh$_3$)C(O)Ph	74
[Ag$_2${μ-{{CH(COOEt)}$_2$PPh$_2$}}$_2$]	2Ag$_2$CO$_3$ + 2[Ph$_2$P(CH$_2$COOEt)$_2$]Cl	75
[Ag(CH$_2$PPh$_2$S)]$_2$	AgNO$_3$ + Li(CH$_2$PPh$_2$S)	76
Mixed-metal compounds		
Ag–Li		
[Li$_6$Br$_4$(Et$_2$O)$_{10}$]$^{2+}$[Li$_2$Ag$_3$Ph$_6$]$^-_2$	AgBr + 3LiPh	77

^{a} Vi = (4-MeC$_6$H$_4$)MeC=C(C$_6$H$_4$NMe$_2$-2). ^{b} L = macrocyclic ligand (see Table 5, note d). ^{c} Ar = C$_6$H$_4$CH$_2$NMe$_2$-2.
^{d} SAr = SC$_6$H$_4$[(R)-CH(Me)NMe$_2$]. ^{e} Owing to the high light and thermal instability of this compound, the crystal structure of the
tetraalkylammonium salt could not be obtained. Substitution of the cation for [Rh(dppe)$_2$]$^+$, however, made a structure determination possible.

coordinated DMS) and organocuprates from reactions of an organolithium reagent and CuBr in DMS, owing to the insolubility of LiBr in this solvent (Equation (5)). van Koten and co-workers[85] demonstrated that soluble copper(I) arenethiolates are also excellent starting materials which allow the synthesis of the uncomplexed compounds RCu in common organic solvents; the use of DMS as a solvent is not required when arenethiolate salts are used; see Equation (6) for a comparison with the results of the use of CuBr.[6a] The reaction of copper arenethiolates with lithium *t*-butylacetylide gave rise to the formation of a [Cu$_3$(C≡CBu^t)(SAr)$_2$]$_2$ aggregate (**16**)[51,52] and with mesitylcopper to a [Cu$_4$(Mes)$_2$(SAr)$_2$] species (**17**)[15] (Equations (7) and (8)).

$$\text{Bu}^t\text{—}\bigcirc\text{(Bu}^t)_2\text{—Li} + \text{CuBr} \xrightarrow[\text{DMS}]{\text{Et}_2\text{O}} \text{Bu}^t\text{—}\bigcirc\text{(Bu}^t)_2\text{—Cu(DMS)} + \text{LiBr} \quad (5)$$

$$[\text{Li}_4(\text{C}_6\text{H}_4\text{CH}_2\text{NMe}_2\text{-2})_4] \xrightarrow[\text{solvent}]{4\,\text{CuX},\,-4\,\text{LiX}} [\text{Cu}_4(\text{C}_6\text{H}_4\text{CH}_2\text{NMe}_2\text{-2})_4] \quad (6)$$
$$\textbf{(14)}$$

$$\begin{array}{lll} \text{solvent} = \text{Et}_2\text{O} & \text{X} = \text{Br} & 44\% \\ \text{solvent} = \text{Et}_2\text{O} & \text{X} = \text{SC}_6\text{H}_4\text{CH}_2\text{NMe}_2\text{-2} & 97\% \end{array}$$

$$4\,\text{CuSAr} + 2\,\text{CuC≡CBu}^t \longrightarrow [\text{Cu}_3(\text{SAr})_2(\text{C≡CBu}^t)]_2 \quad (7)$$
$$\textbf{(16)}$$

$$2\,\text{CuSAr} + 2\,\text{Cu(Mes)} \longrightarrow [\text{Cu}_4(\text{Mes})_2(\text{SAr})_2] \quad (8)$$
$$\textbf{(17)}$$

$$\text{SAr} = \text{---S—}\bigcirc\text{-CH}_2\text{NMe}_2$$

The use of CuOSO$_2$CF$_3$ (copper triflate, CuOTf) often leads to oxidative coupling of the organo group of the transmetallating reagent (note: a violent explosion occurred in an attempt to prepare a sodium cuprate from NaBu and CuOTf).[86] Further evidence for the destabilizing effect of the OTf^- anion is provided by the observation that pure arylcoppers undergo quantitative biaryl coupling with a catalytic amount of CuOTf[87] (see Reference 87 for a series of examples).

A complicating factor which is commonly encountered when copper halides are used as a precursor is the formation of mixed aggregates of the wanted organocopper with the starting copper salt. Strongly connected to this point is the importance of the order of addition of the reagents and the reaction temperatures for the synthesis of an organocopper compound of the type CuR. Ample evidence is now available which shows that these mixed organocopper copper halide aggregated species can be more stable than the desired organocopper, and even do not react further with excess organolithium or magnesium reagent to give the pure organocopper compound (see Section 2.2.2.2, e.g., Equations (21)–(25)).

Copper(II) halides can be used as starting materials, but in this case two equivalents of the organometallic reagent are required, one of which is consumed in the reduction of copper(II) to copper(I). For the synthesis of pure organocoppers this method is, however, not recommended, since (i) the organic products formed by oxidative coupling commonly hamper the isolation of a pure product (Equation (9)) and (ii) the formed organocopper can subsequently react with remaining copper(II) halide.[88,89] The presence of copper(II) impurities in copper(I) precursors, which is particularly evident in the cases of CuCl and CuBr, may be responsible for the formation of impure organocopper products. In the authors' laboratory, copper(I) bromide is prepared from $CuSO_4$ and NaBr in the presence of $NaHSO_3$ and NaOH.[90] The final workup procedure is carried out under a nitrogen atmosphere using Schlenk techniques, while the pure CuBr product is stored in the dark and further handling in organocopper synthesis is always carried out under nitrogen.

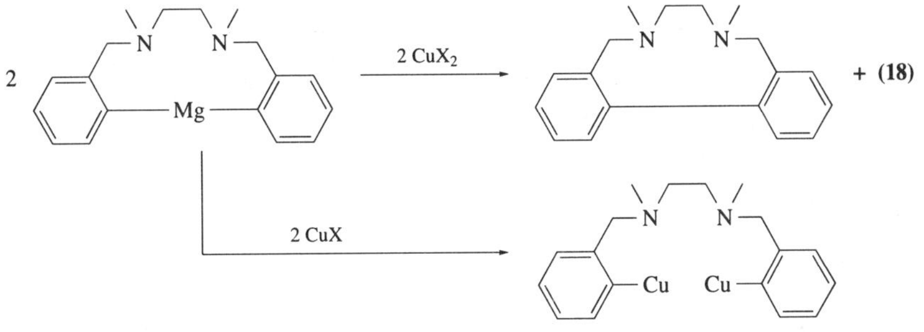

$$(9)$$

(18)

Copper(II) acetyl acetonate and $AlMe_2(OR)$ (R = Et^{91} or Pr^{i33}) react to give CuMe, which in the presence of a suitable triorganophosphine affords the corresponding pure $[MeCu(PR_3)_3]$ **(19)** complexes (Equation (10)).

$$AlMe_2(OR) + Cu^{II}(acac)_2 \xrightarrow{PPh_3} [CuMe(PPh_3)_3] + AlMe(OR)(acac) \qquad (10)$$

(19)

Other organometallic reagents derived from metals more electropositive than copper have been employed in transmetallation reactions with copper(I) salts. Organomagnesium halides have been most widely used.[20] Isolation of compounds of the type CuR from these reaction mixtures sometimes appeared impossible owing to complex formation between magnesium species and the organocopper compound itself (Equation (11)).[92] In these cases the corresponding organolithium reagent can be used. In other instances, however, the compound is only accessible via the organomagnesium route, for example, the synthesis of the dicopper compound **(18)** in Equation (9), while the synthesis of pure $[Cu_5Mes_5]$ **(20)** requires the use of MesMgBr and CuCl in THF[37] (Equation (12)).

$$3 MgPh_2 + 4 CuBr \xrightarrow{Et_2O} [Cu_4MgPh_6 \cdot Et_2O] + 2 MgBr_2 \qquad (11)$$

$$5 MgMesBr + 5 CuCl \xrightarrow{THF} [Cu_5(Mes)_5] + 5 MgBrCl \qquad (12)$$

(20)

The reaction of diarylzinc with copper(I) salts is an excellent route to pure arylcoppers.[93] It is of historic interest that the first attempted preparation of an organocopper compound, reported by Buckton in 1859, involved the reaction of CuCl with ZnEt$_2$.[94] Metathesis of trifluoromethylcadmium with a copper(I) salt afforded a solution of CuCF$_3$.[95] Interestingly, metathesis of CuBr with CdXCF$_3$ affords a stable copper(III) compound [Cu(CF$_3$)$_4$]$^-$.[96] CuCF$_3$ has been observed to insert CF$_2$ into copper–carbon bonds. A solution of CuCF$_3$ in DMF on standing at room temperature converts into CuCF$_2$CF$_3$.[95] In a related manner, CuCF$_3$ gives double insertion of CF$_2$ into the Cu–C bond of CuC$_6$F$_5$ to give CuCF$_2$CF$_2$C$_6$F$_5$ in 70–80% yield (Scheme 1).[30] The existence of a delicate but yet not understood balance between the nature of the transmetallating reagent and the reaction route chosen is demonstrated by the synthesis of another copper(III) compound [Cu(CF$_3$)$_2$(S$_2$CNMe$_2$)] (**21**), (i) via the oxidation reaction of *in situ* prepared [Cu(CF$_3$)$_2$]$^-$ with thiuram disulfide, and (ii) via transmetallation of [CuBr$_2$(S$_2$CNMe$_2$)] with Cd(CF$_3$)$_2$–Cd(CF$_3$)I (Equations (13) and (14)).[96] However, attempted metathesis of CuBr$_2$(S$_2$CNEt$_2$) with an organomagnesium or -lithium gave decomposition products (Equation (15)).

$$Cd(CF_3)X/Cd(CF_3)_2 \xrightarrow[-50\ °C\ to\ RT]{CuY} [CuCF_3] \xrightarrow[24\ h,\ RT]{DMF} CuCF_2CF_3$$

$$CuC_6F_5 + 2\ CuCF_3 \xrightarrow[70–80\ \%]{-30\ °C\ to\ RT} C_6F_5CF_2CF_2Cu$$

$$Y = Cl,\ Br,\ I,\ CN$$

Scheme 1

$$CdI^+[(CF_3)_2Cu^I]^- + [Et_2NC(S)S]_2 \xrightarrow{-30\ °C\ to\ RT} Cu^{III}(CF_3)_2(S_2CNEt_2)] + 1/2\ CdI_2 + 1/2\ Cd(edtc)_2 \qquad (13)$$
$$(\mathbf{21})$$

$$[Cu^{III}Br_2(edtc)] + 1/2\ Cd(CF_3) + CF_3CdI \xrightarrow{-30\ °C\ to\ RT} (\mathbf{21}) + 1/2\ CdI_2 + CdBr_2 \qquad (14)$$

edtc = *N,N*-diethylthiocarbamato

$$[Cu^{III}Br_2(edtc)] + RMgX(or\ RLi) \longrightarrow R–R + [Cu^I(S_2CNEt_2)] \qquad (15)$$

R = Me,Ph

The use of an organothallium is exemplified by the synthesis of a number of complexes [CuLCp] and [CuLCp*] (e.g., L = PR$_3$,[97] CO,[98] CNR,[98] THF[99]). Often the lithium, sodium or potassium analogues of these thallium compounds have been used with the same result. It has been reported that a cyclopentadienyl copper complex reacts with organolithiums to give organocopper compounds via loss of the Cp group, although the yields are low. Thus reaction of [Cu(PPh$_3$)Cp] with 2 equiv. of fluorenyllithium in THF, or equivalent amounts of LiC≡CBu^t and PCy$_3$ in THF, gave [Cu(fluorenyl)$_2$(PPh$_3$)]$^-$ (**22**) (Equation (16)) or [Cu$_2$(μ-C≡CBu^t)$_2$(PCy$_3$)(PPh$_3$)$_2$] (**23**) in yields of 17 and 21%, respectively.[100]

$$[Cu(PPh_3)Cp] + 2\ fluorenylLi \xrightarrow{THF} [Li(THF)_4][Cu(fluorenyl)_2] \qquad (16)$$

2.2.2.2 *Via transmetallation reactions: complex formation with organometallics and metal salts*

The tendency of organocopper compounds to form stable 'ate'-type complexes in the presence of excess organolithium reagent was first recognized by Gilman *et al.*[101] It appeared that ether-insoluble, yellow methylcopper dissolves on addition of 1 equiv. of methyllithium with the formation of a colourless complex with 1:1 stoichiometry. Complexes of this type, which are commonly named organocuprates, can generally be prepared by the reaction of 2 equiv. of an organolithium reagent with 1 equiv. of a copper(I) salt (Equation (17)) or by the reaction of 1 equiv. of an organocopper compound with 1 equiv. of an organolithium compound (Equation (18)). In cuprates with stoichiometry LiCuR$_2$, both metal cations and anionic ligands form one species. This class of compounds is commonly named 'neutral cuprates', of which [Li$_2$Cu$_2$(CH$_2$TMS)$_4$(DMS)$_2$] (**24**),[102] [Li$_2$Cu$_2$Ph$_4$(Et$_2$O)$_2$] (**25**),[65] [Li$_2$Cu$_2$Ph$_4$(DMS)$_3$] (**26**)[16] and [Li$_2$Cu$_2$(C$_6$H$_4$CH$_2$NMe$_2$-2)$_4$] (**3**)[8,10] constitute all compounds of which the structure in the solid state is known to date, and the two cuprates [Li$_2$Cu$_2$(4-Tol)$_4$(Et$_2$O)$_2$][103] and (**3**)[104] are those which have been fully characterized in solution. Weiss and co-workers reported that (**25**) is made preferably from [CuI(PBu$^n_3$)] and LiPh in hexane–diethyl ether, but this procedure still requires removal of LiI.[65] Alternatively, quantitative reaction to [Li$_2$Cu$_2$(4-Tol)$_4$] occurs on mixing pure Li(4-Tol) and pure Cu(4-Tol) in benzene, followed by titration of the suspension with L (e.g., Et$_2$O) to give a clear solution of pure [Li$_2$Cu$_2$(4-Tol)$_4$(L)$_2$].[103] Organocuprates with other stoichiometries have been arbitrarily divided into lower-order[105–7] (Li:Cu ratio <1, e.g., [Li$_2$Cu$_3$Ph$_5$] (**27**)) and higher-order[105–7] (Li:Cu ratio >1, e.g., [Li$_3$Cu$_2$Ph$_5$(DMS)$_4$] (**28**)) cuprates.[16] The nature of the cuprate, commonly prepared *in situ*, is of crucial importance when they are used in organic synthesis as only one of the organic groups is transferred to the substrate.

$$CuX + 2 LiR \longrightarrow LiCuR_2 + LiX \tag{17}$$

$$CuR + LiR \longrightarrow LiCuR_2 \tag{18}$$

In order to achieve economical use of the organic group, which is often valuable, the concept of cuprates with transferable R$_t$ and nontransferable R$_r$ groups was developed (Scheme 2).[108] Since the ease of transfer of a monoanionic ligand can be related to its ability to bridge between metal centres, various authors explored the choice of the best copper salt for the formation of LiCuR$_t$R$_r$.[1] For example, as R$_r$ the anions CN$^-$,[109] SCN$^-$,[101] $^-$SPh[110] or $^-$SC$_6$H$_4$CH$_2$NMe$_2$,[15,111,112] $^-$C≡CPh,[110] $^-$NR$_2$[113] and $^-$PR$_2$[114] have been used, and their effect on regioselectivity and chemoselectivity in C–C coupling reactions studied.[1] Discussion of the exact nature of these compounds concentrates on the question of whether copper and lithium, in addition to the transferable and nontransferable groups, constitute one aggregated species or whether they are a mixture of species. Earlier NMR studies[105] pointed to the possible existence of [LiCu$_2$Me$_3$], [Li$_2$CuMe$_3$] and [Li$_2$Cu$_3$Me$_5$] in methylcuprate solutions. The results of both Lipshutz *et al.*[106] and Bertz[115] and Bertz and Dabbagh,[116] who independently studied the composition of cuprate solutions by NMR, will be discussed in Section 2.3.4.2. For a discussion of the preparative aspects of cuprates, it is important to note that in solution higher-order cuprates which have discrete structures in the solid state do not exist as such in Et$_2$O or THF solution; for example, (**28**) in solution is best viewed as [Li$_3$CuPh$_3$]$^+$[CuPh$_2$]$^-$, whereas (**27**) does not exist at all in THF.[116] This can also explain the finding of Bau and co-workers,[14c] who isolated [Li(THF)$_4$][Cu$_5$Ph$_6$] (**10**) as well as the corresponding [Li(PMDTA)(THF)]$^+$ salt from a solution of CuBr–LiPh in Et$_2$O, with added PMDTA in the latter case. When crown ethers are applied in cuprate synthesis simple mononuclear homocuprate [CuR$_2$]$^-$ or heterocuprate [CuRBr]$^-$ species are formed, with the lithium cations stabilized as lithium crown ether cations (Equation (19)).[64] A similar approach to mononuclear cuprates is the reaction reported by Ghilardi and co-workers,[49] in which mesitylcopper is reacted with the bidentate phosphine dppe, which binds efficiently one copper cation as [Cu(dppe)$_2$]$^+$ and leaves [CuMes$_2$]$^-$ (Equation (20)).

R = R$_t$, Y = R$_r$ **Scheme 2**

$$2RLi + CuBr \xrightarrow{\text{12-crown-4}} [CuR_2][Li(12\text{-crown-4})_2] \tag{19}$$

R = Me, Ph, CH(TMS)$_2$

$$2 Cu(Mes) + 2 dppe \longrightarrow [Cu(dppe)_2][Cu(Mes)_2] \tag{20}$$

(**65**)

Organomagnesium reagents likewise form 'ate'-type complexes with organocopper compounds CuR and products containing both copper and magnesium such as [MgCu$_2$Ph$_4$(THF)$_n$] have been isolated from the reaction of CuBr with MgPh$_2$ in THF.[92,117] A further example is the isolation of [MgCu$_4$Ph$_6$(Et$_2$O), shown in Equation (11).[66,92]

Numerous reports dealing with attempts to prepare pure organocopper compounds describe the isolation of products contaminated with lithium or magnesium halides formed as a consequence of the transmetallation reaction.[19] In some cases it proved possible to remove these salts by appropriate washing procedures, an example being the isolation of pure [Cu$_4$(C$_6$F$_5$)$_4$], where the magnesium halide formed in the reaction of the copper halide with [MgBr(C$_6$F$_5$)] was removed as its insoluble dioxane complex.[118] It is has become clear, however, that lithium and magnesium halides present in isolated products are often not contained as strongly bound contaminants, but instead form an integral part of a mixed metal molecule. An interesting example is the quantitative reaction of a copper arenethiolate with MgMes$_2$ in benzene to give [{Mg(SC$_6$H$_4$CH$_2$NMe$_2$-2)$_2$}$_2${Cu$_4$Mes$_4$}] (**31**), a Mg$_2$Cu$_4$R$_4$X$_4$ 'ate'-type complex[15,119] (Section 2.3.2.5(i)), showing the high selectivity of these self-assembly reactions in organocuprate chemistry.

It is known that organocopper compounds are generally capable of forming stable, isolable complexes with metal salts.[7,88,120,121] $[Cu_6Br_2(C_6H_4NMe_2-2)_4]$ (2) was the first example of an organocopper–copper halide complex for which the structure in the solid state was obtained. This complex was isolated from the reaction of $Li(C_6H_4NMe_2-2)$ with CuBr in Et_2O in an attempt to synthesize pure $[Cu(C_6H_4NMe_2-2)]$.[7] In fact, bright-red (2) has been obtained in two ways: (i) by slow addition of $Li(C_6H_4NMe_2-2)$ in diethyl ether solution to a suspension of CuBr in diethyl ether (Equation (21)), or (ii) by reaction of pure, white $[Cu(C_6H_4NMe_2-2)]_n$ with CuBr in a slurry reaction in diethyl ether (Equation (22)). The direct reaction in a 1:1 molar ratio in diethyl ether first affords the red CuBr complex, further reaction to the pure organocopper requiring a change of solvent to benzene in order to solubilize the CuBr complex. The overall reaction is the formation of pure, insoluble $[Cu(C_6H_4NMe_2-2)]_n$ (Equations (23) and (24)). A number of cases are now known where this further reaction does not occur and the pure uncomplexed organocopper cannot be obtained from CuBr. Examples are the arylcopper complexes $[Cu_4\{C_6H_3(CH_2NMe_2)_2-2,6\}_2Br_2]$ (32) (see Equation (25)),[48] $[Cu_6Br_2(Oxl)_4]$ (33)[45] and the vinylcopper complexes $[Cu_4Br_2\{(Z)-(2-Me_2NC_6H_4)C=C(Me)(C_6H_4Me-4)\}_4]$ (34).[50]

$$4\,Li(C_6H_4NMe_2-2) + 6\,CuBr \longrightarrow [Cu_6Br_2(C_6H_4NMe_2-2)_4] + 4\,LiBr \qquad (21)$$
$$(2)$$

$$4/n\,[Cu(C_6H_4NMe_2-2)]_n + 2\,CuBr \xrightarrow[25\,°C]{Et_2O} (2) \qquad (22)$$

$$4\,Li(C_6H_4NMe_2-2) + 6\,CuBr \xrightarrow[-20\,°C]{Et_2O} (2) + 4\,LiBr \qquad (23)$$

$$(2) + 2\,(LiC_6H_4NMe_2-2) \xrightarrow[25\,°C]{C_6H_6} 6/n[CuC_6H_4NMe_2-2]_n + 2\,LiBr \qquad (24)$$

$$(25)$$

$$(32)$$

These results already indicated the importance of properly selecting the order of addition of the reactants in transmetallation reactions of organolithium compounds (and other organometallic reagents) with copper salts, which was further demonstrated with the synthesis of $[Cu(MeOxl)]$.[45,122] Addition of a solution of $Li(MeOxl)$ in diethyl ether to a suspension of CuBr in the same solvent (1:1 LiMeOxl:CuBr ratio) affords a bright-orange product which is a mixture of the organocopper compound $[Cu(MeOxl)]_2$ and its CuBr complex (33). However, reversed addition, that is, slow addition of solid CuBr to a solution of $Li(MeOxl)$ in diethyl ether, gave pure $[Cu(MeOxl)]$ (Scheme 3). Reaction of $[Cu(MeOxl)]_2$ with CuBr in a 2:1 ratio afforded the organocopper–copper bromide complex (33), which cannot be transformed into pure $[Cu(MeOxl)]_2$ by reaction with $Li(MeOxl)$, whereas a similar conversion was successful in the synthesis of $[Cu(C_6H_4NMe_2-2)]$ (Equation (24)). The effect of the order of addition on the outcome of the reaction is observed in the synthesis of $[Cu_4(C_6H_4CH_2NMe_2-2)_4]$ (14).[88] Addition of $Li(C_6H_4CH_2NMe_2-2)$ to a diethyl ether suspension of CuBr results in the formation of an inseparable mixture of copper bromide complexes of the type $(CuR)_x(CuBr)_y$. It appeared that only if CuBr is added to the solution of $Li(C_6H_4CH_2NMe_2-2)$ is the isolation of pure (14) possible. The organocopper species formed at reactive sites of the highly insoluble CuBr coordination polymer give rise initially to the formation of ether-insoluble complexes which in principle are susceptible to thermal decomposition via homolytic Cu–C bond cleavage. However, on addition of CuBr to $Li(C_6H_4CH_2NMe_2-2)$, further reaction of initially formed organocopper with the excess organolithium yields ether-soluble and thermally stable cuprate species $[Li_2Cu_2R_4]$ (3). The latter cuprate reacts smoothly with CuBr in the second half of the reaction to give the pure RCu (Equations (26) and (27)). The use of soluble copper salts, for example, $[CuBr(DMS)]$ or $[CuBr\{P(OMe)_3\}]$,[122] can circumvent these complications in preparing pure copper compounds, although the final removal of the solubilizing ligand can be unsuccessful or lead to lower yields. In various cases the use of CuO_2CPh,[39] $CuSR$[15,52,85,111]

(Equations (6) and (28)) or $[CuL(Cp)]^{100}$ and direct substitution with a suitable organometallic is the preferred route.

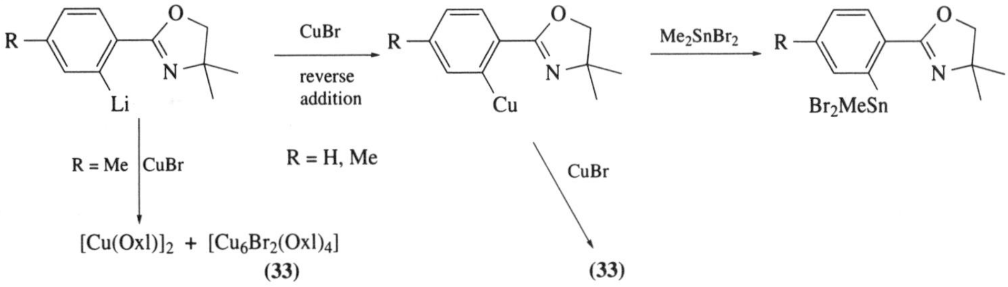

Scheme 3

$$[Li_4(C_6H_4CH_2NMe_2\text{-}2)_4] + 2\,CuBr \xrightarrow[-40\,°C]{Et_2O} [Cu_2Li_2(C_6H_4CH_2NMe_2\text{-}2)_4] + 2\,LiBr \qquad (26)$$
$$\textbf{(3)}$$

$$\textbf{(3)} + 2\,CuBr \xrightarrow[-40\,°C]{Et_2O} [Cu_4(C_6H_4CH_2NMe_2\text{-}2)_4] + 2\,LiBr \qquad (27)$$

$$CuY + n\,LiR \longrightarrow Li_nCuYR_n \qquad (28)$$
$$Y = CN, C\equiv CR, SAr$$
$$n = 1, 2$$

The copper–halide bond in organocopper–copper halide aggregates can react with organolithium reagents to form organocopper aggregates with different organic groups. These ligand-substitution reactions, which occur with retention of the aggregate structure of the molecule, have allowed the isolation of novel types of mixed organocopper cluster compounds (see Section 2.3.2.2(iii)). The reaction of $[Cu_6Br_2(C_6H_4NMe_2\text{-}2)_4]$ (**2**) with 2 equiv. of a lithium arylacetylide has made possible the isolation of mixed aryl–acetylide copper aggregated species of the type $[Cu_6(C_6H_4NMe_2\text{-}2)_4(C\equiv CC_6H_4Me\text{-}4)_2]$ (**35**) (Equation (29)).[58,123] Another example is the 1:1 reaction of $[Cu_4Br_2Vi_2]$ with $Li(C_6H_4NMe_2\text{-}2)$, which yielded the first representative of the class of mixed alkenyl–arylcopper aggregates, that is, (**6**) in Equation (30).[12,78] Mixed alkenyl–alkynylcopper compounds were also prepared, but these could not be obtained analytically pure.[78]

$$[Cu_6Br_2(C_6H_4NMe_2\text{-}2)_4] + 2\,Li(C\equiv CC_6H_4Me\text{-}4) \xrightarrow[25\,°C]{C_6H_6} [Cu_6(C_6H_4NMe_2\text{-}2)_4(C\equiv CC_6H_4Me\text{-}4)_2] + 2\,LiBr \qquad (29)$$
$$\textbf{(2)} \qquad\qquad\qquad\qquad\qquad\qquad \textbf{(35)}$$

$$[Cu_4Br_2Vi_2] + 2\,Li(C_6H_4NMe_2\text{-}2) \longrightarrow [Cu_4(C_6H_4NMe_2\text{-}2)_2Vi_2] + 2\,LiBr \qquad (30)$$

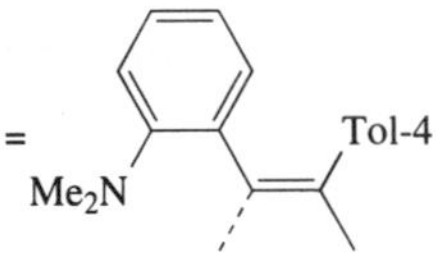

2.2.2.3 Via metallation reactions

Metallation is the usual method for the preparation of alkynylcopper compounds. Since the copper–carbon bond in compounds $CuC\equiv CR$ is not hydrolysed by water, alkynylcopper compounds can be simply prepared by the reaction of a terminal alkyne with an ammoniacal solution of copper(I) chloride (Equation (31)).[124–6] A copper–acetylide bond is also formed in the aryl–arylacetylide exchange reaction between polymeric $[Cu(C_6H_4NMe_2\text{-}2)]_n$ and 4-tolylacetylene, which gives the mixed aryl–alkynylcopper species (**35**) in 58% yield (Equation (32)).[59] Selective formation of a 1:1 complex of a copper acetylide with triphenylphosphine or 2-(diphenylphosphino)pyridine has been observed in the reaction of $[Cu(PR_3)_2(BH_4)]$ with phenylacetylene and KOH (Equation (33)).[53,55] In contrast to the tetrameric aggregates obtained in the previous reactions with monodentate phosphines, the use of dppm afforded a trimeric aggregate.[127] Reaction of the copper(II) complex $[Cu(II)_2L(OH)_2(ClO_4)_2]$ with phenylacetylene in a 2:1 MeCN–MeOH solvent mixture at reflux gave the tetranuclear compound $[Cu_4L_2(C\equiv CPh)(ClO_4)_3]$, in which L is a 20-membered macrocyclic ligand.[57] It should be noted that most of these compounds have also been obtained via direct transmetallation reactions. Various organocopper compounds have been prepared via metallation of C–H acidic compounds with alkyl- and arylcopper compounds. Cyanoalkanes including acetonitrile, propionitrile and malonitrile upon reaction with $[CuMe(PPh_3)_2]$ afforded the corresponding copper complexes having a direct copper–carbon bond (Equation (34)).[128] The reaction of arylcopper (phenyl, *o-*, *m-* and *p-*tolyl) with dppm afforded in quantitative yield $[Cu_3(Ph_2PCHPPh_2)_3]$ (**36**), although this compound is also accessible via the reaction of CuCl with $Li(Ph_2PCHPPh_2)$ (Equation (35)).[31] This selective metallation is remarkable as the corresponding reaction with excess of dppe results in selective P–C bond cleavage and formation of $[Cu_2(PPh_2)_2(dppe)_2]$ (Scheme 4).[129-32]

$$R\text{——}H + Cu(NH_3)_2^+ \longrightarrow [CuC\equiv CR] + NH_3 + NH_4^+ \tag{31}$$

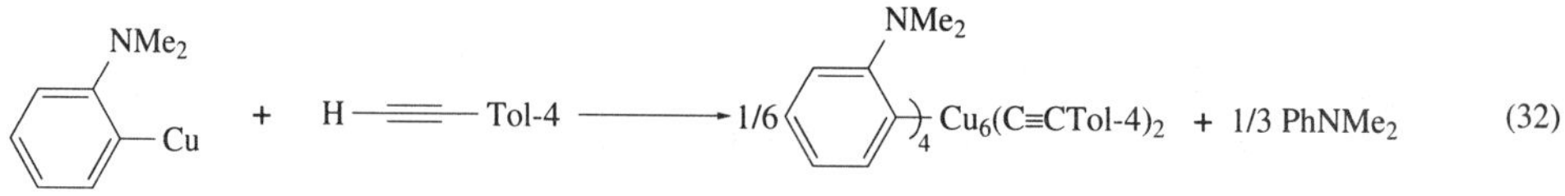

$$\text{(35)}$$

$$[CuL]_n^+ + n\ H\text{——}Ph \xrightarrow{KOH} n/x\ [Cu(C\equiv CPh)(L)]_x \tag{33}$$

$$L = PPh_3\ (n = 0,\ x = 4),\ PPh_2py\ (n = 2,\ x = 4),\ dppm\ (n = 2,\ x = 3)$$

$$[CuMe(PPh_3)_2] + MeCN \longrightarrow [Cu\{CH_2CN\}(PPh_3)_2] + CH_4 \tag{34}$$

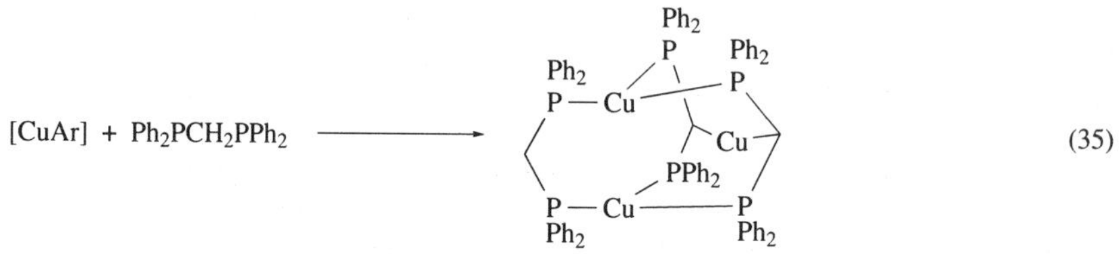

$$\text{(36)}$$

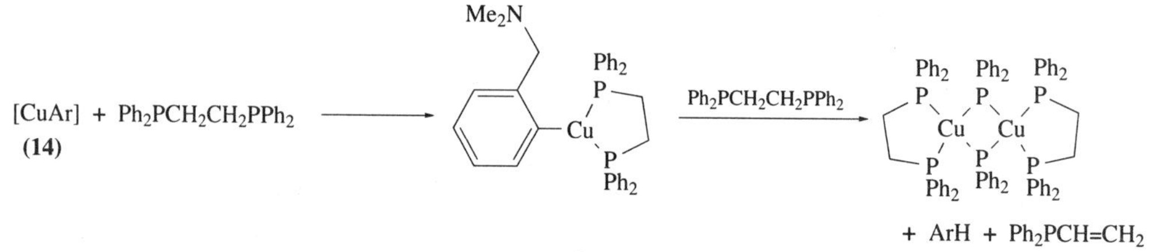

$$[CuAr] + Ph_2PCH_2CH_2PPh_2 \tag{35}$$

(14)

Scheme 4

A useful reagent for the metallation of compounds with an acidic hydrogen atom is copper *t*-butoxide, which has the advantage of being soluble in organic solvents.[133] This compound not only metallates terminal alkynes under mild conditions but also cyclopentadiene (at −78 °C), permitting the synthesis of [Cu(L)Cp] complexes (Equation (36)).[133,134] Copper(I) oxide may also be used for the synthesis of the latter type complexes (Equation (37))[135,136] as well as transmetallation reactions with lithium or thallium cyclopentadienyl and pentamethylcyclopentadienyl anions.[61,137,138] [CuLCp*] complexes have been used in metallation reactions, as is demonstrated by the synthesis of $[Cu(C\equiv CPh)(PPr^i_3)]$ from the reaction of phenylacetylene with $[Cu(PPr^i_3)Cp^*]$.[137]

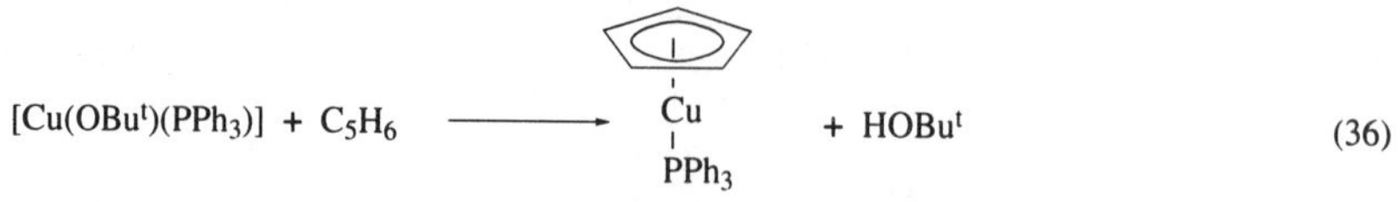

$$[Cu(OBu^t)(PPh_3)] + C_5H_6 \longrightarrow \quad + HOBu^t \tag{36}$$

(47)

$$Cu_2O + C_5H_6 + 2L \longrightarrow 2 \underset{\underset{PPh_3}{|}}{\overset{\overset{\text{(cyclopentadienyl)}}{|}}{Cu}} + H_2O \qquad (37)$$

$$(47)$$

Mesitylcopper, $[Cu_5Mes_5]$ (**20**), is a convenient starting material for the generation of organic copper complexes. An example is the reaction of (**20**) with benzoic acid, which afforded pure copper(I) benzoate in the 1:1 molar reaction but also provided access to unique organocopper–copper benzoate aggregates (**37**) in a 2:1 molar ratio (CuMes–benzoic acid) reaction (Scheme 5).[39,139]

$$RCO_2H + \underset{(20)}{CuMes} \longrightarrow 1/n\,[RCO_2Cu]_n + MesH \xrightarrow{CuMes} \underset{(37)}{[Cu_3(Mes)(O_2CR)_2]}$$

Scheme 5

2.2.2.4 Via interaggregate exchange/self-assembly

Apart from the synthesis of organocopper compounds via common transmetallation routes using a suitable copper salt and an organometallic reagent, routes are now known for the synthesis of well-defined organocopper aggregates of the type $[Cu_{n+m}R^1{}_nR^2{}_m]$ and $[Cu_nM_m\{(R^1)(R^2)\}_{n+m}]$. In these reactions the pure organometallic constituents of the $[Cu_{n+m}R^1{}_nR^2{}_m]$ or $[Cu_nM_m\{(R^1)(R^2)\}_{n+m}]$ aggregates are mixed in a solvent of low polarity. Most of the reactions discussed in Section 2.2.2.2 are of this type. They are mentioned here, however, in a different synthetic perspective following the view that the new aggregated species is assembled selectively from other aggregated species.[15,119,139] The first example was the reaction of $[Cu(C_6H_4NMe_2-2)]_n$ with copper arylacetylide affording quantitatively the hexanuclear copper aggregate (**35**) bearing two different organic groups (Equation (38)).[59,60,123] Even in the presence of excess of copper arylacetylide, this hexanuclear aggregate was formed selectively.[60] Further examples are (i) the formation of arylcuprates from the organocopper and the organolithium compounds[65,103,104] (Equation (39)), (ii) the formation of arylcopper–copper benzoate[39,139] (Equation (40)) and arenethiolate aggregates[15] (Scheme 6), (iii) the formation of a magnesium heterocuprate from the reaction of bisarylmagnesium with copper arenethiolates[15] (Equation (41)), and (iv) the formation of copper acetylide–copper arenethiolate aggregates[51,52] (Equation (42)). A series of interesting aggregates have been obtained from reactions of acetylide argentates and aurates with copper acetylides (Equations (43) and (44); PPN = $(Ph_3P)_2N$).[67,140]

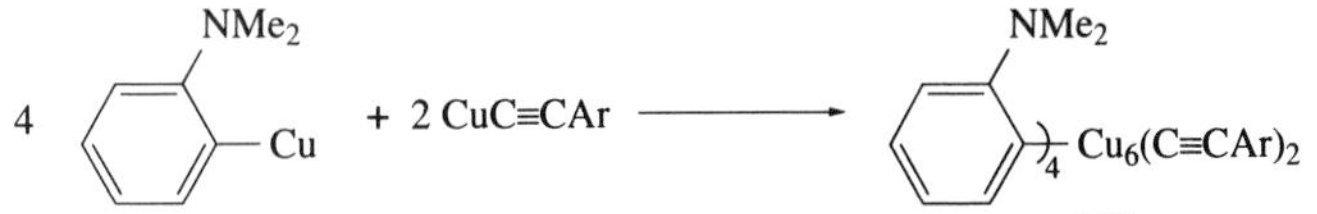

$$4\;\underset{}{\overset{NMe_2}{\underset{}{\bigcirc}}}\!\!\text{—}Cu + 2\,CuC{\equiv}CAr \longrightarrow \left[\overset{NMe_2}{\underset{}{\bigcirc}}\right]_4 Cu_6(C{\equiv}CAr)_2 \qquad (38)$$

$$(35)$$

$$2\;\underset{\text{tetramer}}{\overset{\text{—}NMe_2}{\underset{}{\bigcirc}}\!\!\text{—}Cu} + 2\;\underset{\text{tetramer}}{\overset{\text{—}NMe_2}{\underset{}{\bigcirc}}\!\!\text{—}Li} \longrightarrow \left[\overset{\text{—}NMe_2}{\underset{}{\bigcirc}}\right]_4 Cu_2Li_2 \qquad (39)$$

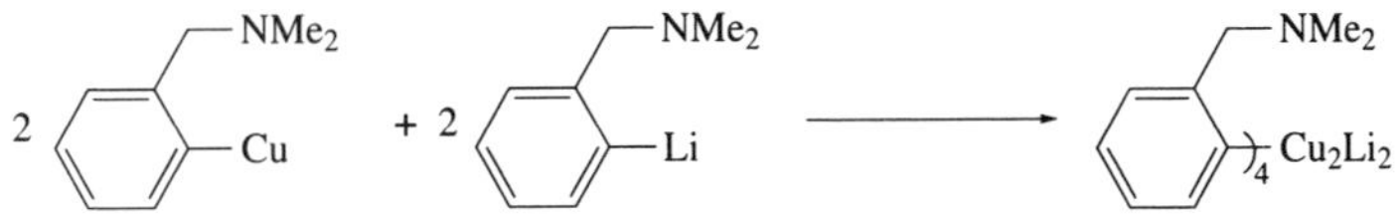

$$(3)$$

$$\underset{\text{pentamer}}{Cu(Mes)} + \underset{\text{tetramer}}{CuO_2CPh} \longrightarrow \underset{(37)}{[Cu_3(Mes)(O_2CPh)_2]} \qquad (40)$$

$$MgMes_2 + \underset{\text{trimer}}{CuSAr} \longrightarrow \underset{(31)}{[Cu_4(Mes)_4\{Mg(SAr)_2\}_2]} \qquad (41)$$

$$CuC{\equiv}CR + \underset{\text{trimer}}{CuSAr} \longrightarrow \underset{(16)}{[Cu_3(SAr)_2(C{\equiv}CR)]_2} \qquad (42)$$

$$[NBu^n{}_4][AuC{\equiv}CPh] + (AuC{\equiv}CPh)_n + Cu(C{\equiv}CPh)_n \longrightarrow [NBu^n{}_4][Au_3Cu_2(C{\equiv}CPh)_6] \qquad (43)$$

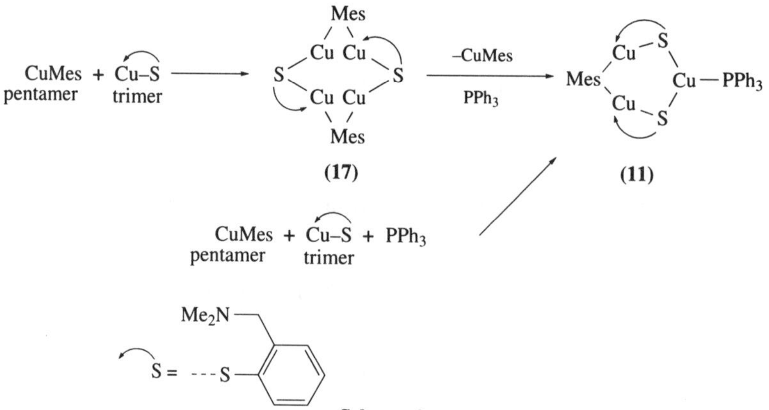

Scheme 6

$$[PPN][Ag(C{\equiv}CPh)_2] + Ag(C{\equiv}CPh)_n + Cu(C{\equiv}CPh)_n \longrightarrow [PPN][Ag_6Cu_7(C{\equiv}CPh)_{14}] \quad (44)$$

2.2.2.5 Miscellaneous preparative routes

Copper–halogen exchange reactions are of limited value in preparative organocopper synthesis, a notable exception being the preparation in 72% yield of perfluoro-*t*-butylcopper by this method (Equation (45)).[118] A preparatively useful decarboxylation of a copper(I) carboxylate has been reported, namely the synthesis of the quinoline complex of pentafluorophenylcopper (Equation (46)).[141] A second example is the decarboxylation of CuO_2CCH_2CN.[142] This reaction is completely reversible, that is, the alkylcopper $CuCH_2CN$ reacts readily with CO_2, whereas arylcopper compounds do not react with CO_2, except in the presence of triphenylphosphine in suitable solvents, when they undergo insertion to give the corresponding $[Cu(O_2CAr)(PPh_3)_2]$ complexes (Scheme 7).[143] Isocyanides insert into the copper–carbon bond of arylcopper compounds with the formation in excellent yield of a new type of organocopper compound (Equation (47)) which is dimeric in benzene.[144]

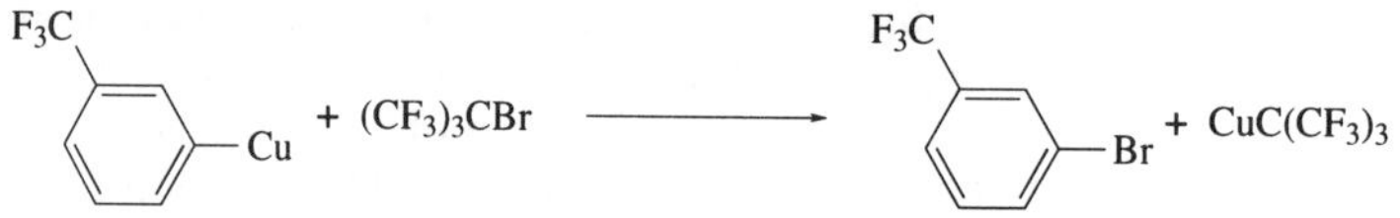

(45)

$$C_6F_5COOCu \xrightarrow[60\ ^\circ C]{\text{quinoline}} [Cu(C_6F_5)(quinoline)] + CO_2 \quad (46)$$

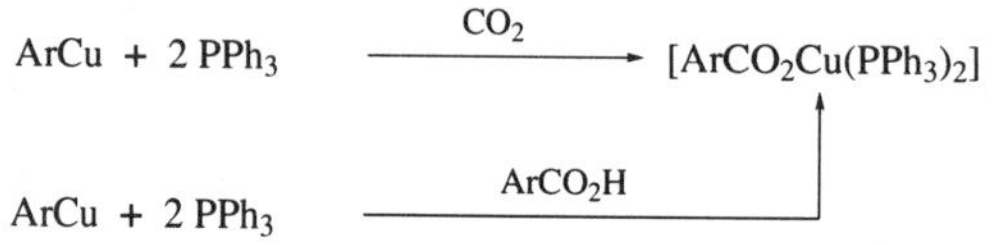

Scheme 7

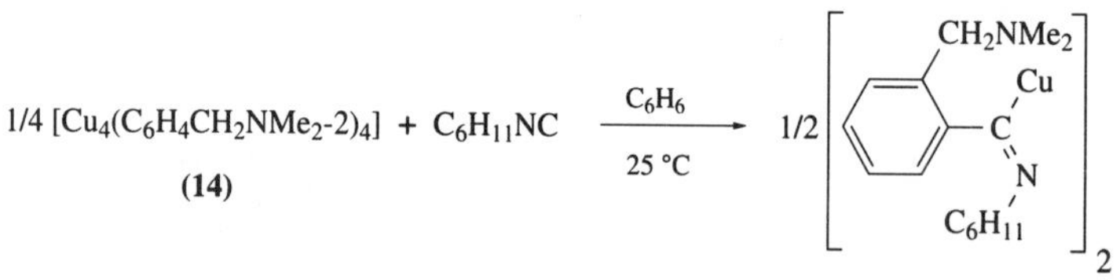

(47)

Perfluoroalkylcopper compounds have been prepared in a direct reaction from metallic copper and perfluoroalkyl halides in dipolar aprotic solvents such as DMSO at elevated temperatures.[145] Vinylcopper reagents have been derived from acyclic polyfluoroalkenes.[146] While alicyclic polyfluoroalkenes have been converted into the corresponding perfluoroalkenylcopper reagents in excellent yield, no attempts were reported to isolate the alkenylcopper derivatives. The syntheses involve the reaction of the iodo derivative with activated zinc and subsequent reaction of the organozinc derivative with copper(I) bromide (Equation (48)).[147] Wehmeyer and Rieke[148] reported the synthesis of alkylcopper reagents from alkyl halides using activated copper(0) prepared by the lithiumnaphthalide reduction of [CuI(PR$_3$)]. Interestingly, via this method alkylcopper reagents containing esters, nitrile and chloride functionalities can be obtained. No attempts to isolate the organocopper compounds were reported but subsequent reactions in conjugate additions indicated that the organocopper was formed in high yields.

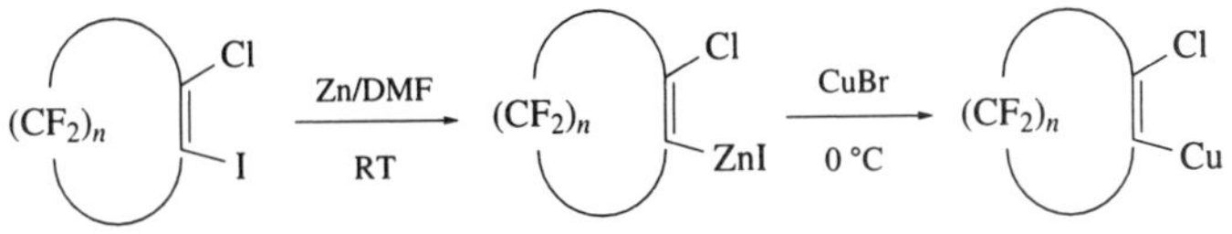

$$\text{(48)}$$

A particularly interesting pathway to very stable organocopper compounds involves the reaction of copper(I) chloride and a ylide (Equations (49) and (50)).[34,35,149-53]

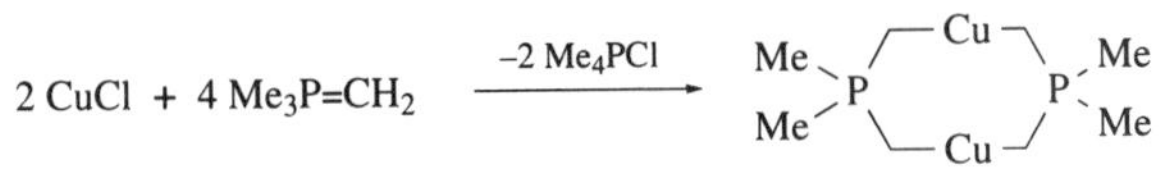

$$2\,\text{CuCl} + 4\,\text{Me}_3\text{P=CH}_2 \xrightarrow{-2\,\text{Me}_4\text{PCl}} \text{(49)}$$

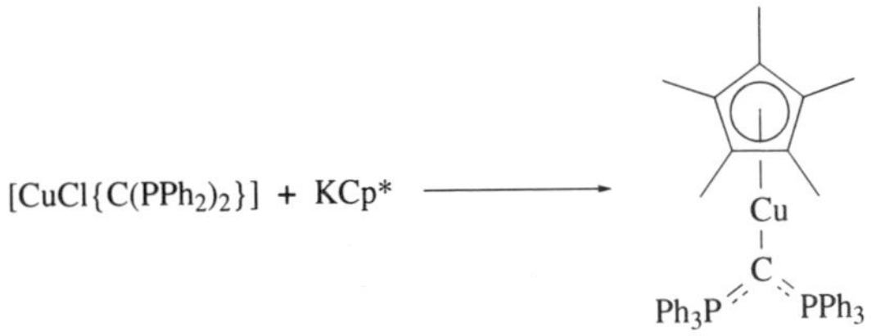

$$[\text{CuCl\{C(PPh}_2)_2\}] + \text{KCp*} \longrightarrow \text{(50)}$$

(81)

A single report exists on the electrochemical synthesis of an organocopper compound: phenylethynylcopper was prepared in high yield by the oxidation of a copper anode in a solution of phenylacetylene in acetone or acetonitrile (Equation (51)).[154]

$$\text{PhC≡CH} \xrightarrow{+\,e^-} \text{PhC≡C}^- \xrightarrow[\text{Cu}^0]{-\,e^-} \text{PhC≡CCu} \qquad \text{(51)}$$

2.2.3 Synthesis of Organosilver Compounds

With few exceptions, organosilver compounds (the organic group is a monoanionic C-ligand) have been prepared by the reaction of silver salts with an organometallic reagent derived from a metal more electropositive than silver.[21,22] Attempts to obtain ethylsilver by the reaction of silver chloride with diethylzinc were made as early as in 1859 by Buckton[94] and in 1861 by Wanklyn and Carius.[155] Until 1941 alkylsilver compounds were entirely unknown, merely because of their low stability (decomposition above −50 °C). The more stable arylsilver compounds (decomposition above 25 °C) were first prepared by Krause and Schmitz[156] in 1919. Alkynylsilver compounds (e.g., propenylsilver, decomposition above 150 °C) were already obtained in the second part of the nineteenth century, whereas the first alkenylsilver compound (e.g., isobutylensilver, decomposition above −20 °C) was prepared in 1955.[22]

The reaction of silver nitrate with tetraalkyllead compounds in methanol or ethanol at low temperatures (usually at −80 °C) yields alkylsilver compounds AgR as yellow to red-brown precipitates which are stable at this temperature for several hours (Equation (52)). However, upon warming to room temperature (note: alkylsilver compounds can explode on heating), decomposition takes place with the formation of metallic silver and a mixture of alkanes and alkenes. The stability of the alkylsilver compounds decreases with increasing chain length of R.[21,22] Alkyllithium and -magnesium reagents in combination with silver halides or complexes of the latter with phosphine ligands, for example, [AgI(PBu$_3$)], AgBF$_4$, AgNO$_3$ and AgO$_2$CCF$_3$, have also been employed in the transmetallation reactions (see Table 2).

$$\text{AgNO}_3 + \text{R}_4\text{Pb} \xrightarrow{-80\,°\text{C}} \text{AgR} + \text{R}_3\text{PbNO}_3 \qquad \text{(52)}$$
$$\text{R = alkyl or aryl}$$

To date, no examples of simple alkylsilver complexes have been isolated. In each case the composition of the initial AgR compounds formed has been derived from the analysis of the products formed in their thermal decomposition.[21] However, focusing again on the compounds that have been isolated and characterized by x-ray diffraction techniques (see Table 5) and NMR spectroscopy (see Section 2.3.4.2), a number of interesting alkylsilver compounds derived from mono-, di- and trisubstituted (sp^3)C$^-$ anions (see (38)–(42)), have been reported since the mid-1980s.

(38)   (39)

(40)   (41)   (42)

A representative monosubstituted methylsilver complex is the dimeric silver ylide (38), which has been obtained from the 1:1 reaction of AgNO$_3$ with the corresponding ylide.[76] Related to the latter compound is the disubstituted methylsilver compound (39), which has been obtained from a ligand-exchange reaction between [AgClO$_4$(PPh$_3$)] and the corresponding gold(III) ylide complex. It is interesting to note that this complex contains two chiral centres, but exists in solution almost exclusively as one enantiomeric pair.[157] A straightforward transmetallation reaction of AgBF$_4$ with a tertiary bis(silylpyridylmethyllithium) compound afforded the dinuclear silver species (40).[158]

The secondary β-disulfone carbanion forms the very stable, *alkali*-soluble argentate complex (41),[71] whereas the secondary substituted methyl ylide [(CH(PPh$_3$)C(O)R)$_2$] forms complexes of the type [Ag(CH(PPh$_3$)C(O)R)$_2$]X   (X = monoanion).[74] From the reaction of the phosphonium salts [Ph$_2$P(CH$_2$CO$_2$R)$_2$]Cl with Ag$_2$CO$_3$, stable dinuclear silver complexes [Ag$_2${(CHCO$_2$R)$_2$PPh$_2$}$_2$] (42) (R = Me, Et) were obtained.[75]

The silver–carbon bond in compounds containing a perfluoroalkyl group is considerably more stable than in nonfluorinated compounds. Perfluoroisopropylsilver, which has been obtained by addition of silver fluoride to perfluoropropene (Equation (53)), has been isolated as the 1:1 complex with acetonitrile and is stable at 60 °C.[159] The complex is heterolytically labile, existing in dynamic equilibrium with solvated Ag$^+$ and the bis(perfluoroisopropyl)silver anion (43) (Equation (53)).[70] This equilibrium provides a model for the transfer of groups between metal centres. The co-condensation of trifluoromethyl radicals with silver vapour at −196 °C to give AgCF$_3$ or [AgCF$_3$(PMe$_3$)] are the only examples of metal vapour syntheses.[160] The highly photosensitive trifluoromethylsilver complexes [AgCF$_3$(PR$_3$)] (R = Me, Et) have also been prepared by transmetallation with Cd(CF$_3$)$_2$·glyme (glyme = MeOCH$_2$CH$_2$OMe) and AgOAc in Et$_2$O, followed by addition of PR$_3$. The PMe$_3$ complex was more stable than the PEt$_3$ complex, which could not be isolated.[161] Exposure of these compounds to air gave the photolytically stable Ag^{III} ion [Ag(CF$_3$)$_4$]$^-$, previously identified by Dukat and Naumann.[162] Alkenylsilver compounds have been prepared via reaction of silver nitrate with tetraalkenyllead (e.g., tetravinyllead)[163] or trialkylalkenyllead (e.g., triethylstyrenyllead)[164] compounds. Alkenylsilver compounds are more thermally stable than alkylsilver compounds and this has allowed the isolation of a few representatives of this class of compounds, for example, isobutenylsilver, at low temperature (−20 °C). Styrenylsilver, [Ag(CH=CHPh)], shows this enhanced stability to an even greater extent: complete decomposition requires several days at room temperature, or several hours in boiling ethanol, and results in the formation of metallic silver and polymeric material.[164] The thermal stability of perfluoroalkenylsilver compounds is exemplified by perfluoroisopropenylsilver, [Ag{C(CF$_3$)=CF$_2$}], which has been prepared by addition of silver fluoride to perfluoroallene. This compound can be sublimed *in vacuo* at 160 °C.[165]

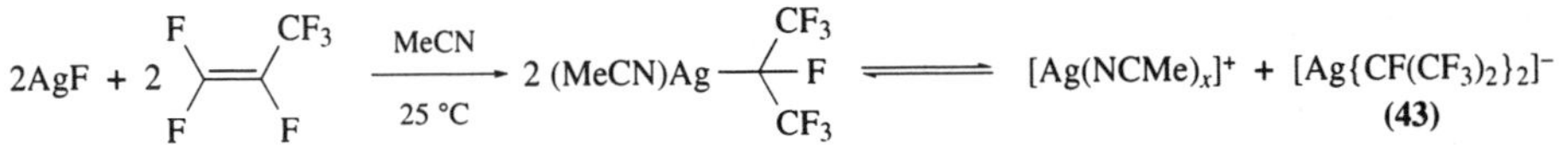

$$2AgF + 2 \underset{F}{\overset{F}{\rightthreetimes}} \underset{F}{\overset{CF_3}{=}} \xrightarrow[25\,°C]{MeCN} 2\,(MeCN)Ag \overset{CF_3}{\underset{CF_3}{\overset{|}{-C}}} F \rightleftharpoons [Ag(NCMe)_x]^+ + [Ag\{CF(CF_3)_2\}_2]^- \quad (53)$$
$$(\mathbf{43})$$

Arylsilver compounds are likewise air, moisture and light sensitive, but are sufficiently thermally stable in some cases to have been isolated and characterized. Following earlier preparations of arylsilver compounds starting from silver nitrate and aryltin or -lead compounds which gave impure products,[21] it was only in 1972 that pure, uncomplexed phenylsilver was isolated for the first time. Very slow addition of a solution of silver nitrate in ethanol to a solution of a large excess of trialkylphenyltin (at 15 °C) or trialkylphenyllead (at −10 °C) compounds in the same solvent resulted in the precipitation of phenylsilver as a colourless solid.[166] Subsequently, aryllithium[9,167] and arylzinc[168] compounds have been found to be more convenient arylating reagents. The reaction of finely divided solid silver nitrate with an ether solution of a diarylzinc reagent yielded the corresponding arylsilver compound almost quantitatively (Equation (54)).[168,169] A number of arylsilver compounds containing dimethylamino, dimethylaminomethyl or methoxy substituents in the aryl nucleus *ortho* to the silver–carbon bond have been isolated analytically pure via the 1:1 molar reaction of silver bromide and the corresponding aryllithium reagent (Equation (55)).[167] Floriani and co-workers[72,201] reported on the synthesis and x-ray structure of mesitylsilver, which was prepared by reacting silver chloride with mesitylmagnesium bromide in THF for 20 h at −20 °C (Equation (56)). In a similar reaction, 2,4,6-triisopropylphenylsilver was prepared.[170] The reaction of *C*-imidazoyllithium with [AgNO$_3$(DMS)] afforded the corresponding, fairly soluble trimeric (by molecular weight determinations) *C*-imidazoylsilver compound (Equation (57)).[171] In contrast to the air and moisture sensitivity of most of these arylsilver compounds, pentafluoro- and pentachlorophenylcopper are both white solids which are stable to water and oxygen. Their preparation involves the reaction of silver trifluoroacetate with an excess of C_6F_5Li or C_6Cl_5Li (Scheme 8).[172]

$$AgNO_3 + ZnAr_2 \longrightarrow AgAr + ArZnNO_3 \quad (54)$$

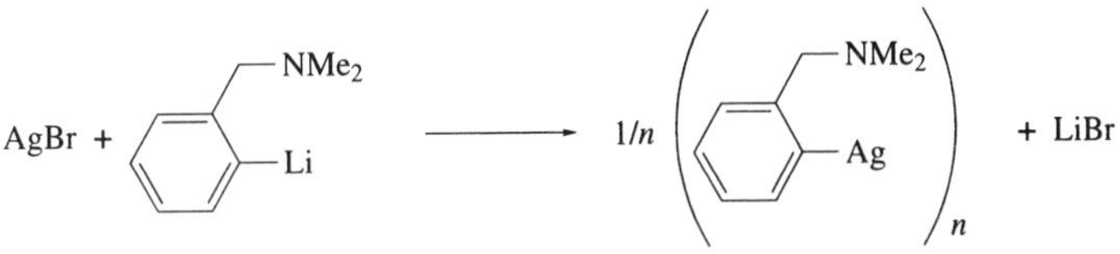

$$AgBr + \underset{Li}{\overset{NMe_2}{\text{[aryl]}}} \longrightarrow 1/n \left(\underset{Ag}{\overset{NMe_2}{\text{[aryl]}}} \right)_n + LiBr \quad (55)$$
$$n = 4 \text{ or } 6$$

$$AgCl + R\text{-}\underset{R}{\overset{R}{\text{[aryl]}}}\text{-}MgBr \longrightarrow R\text{-}\underset{R}{\overset{R}{\text{[aryl]}}}\text{-}Ag + MgBrCl \quad (56)$$
$$R = Me, Pr^i$$

$$AgNO_3(DMS) + \underset{R}{\overset{N}{\text{[imidazole]}}}\text{-}Li \longrightarrow \underset{R}{\overset{N}{\text{[imidazole]}}}\text{-}Ag + LiNO_3(DMS) \quad (57)$$

$$AgO_2CCF_3 + 2\,C_6F_5Li + MePPh_3^+ \xrightarrow{-LiO_2CCF_3;\,-C_6F_5H} \underset{F}{\overset{F\quad F}{\text{[}C_6F_5\text{]}}}\text{-}AgCH_2PPh_3$$
$$(\mathbf{44})$$
$$\xleftarrow{-[NBu_4]ClO_4}$$
$$[NBu_4][Ag(C_6F_5)_2] + Ag(CH_2PPh_3)(ClO_4)$$

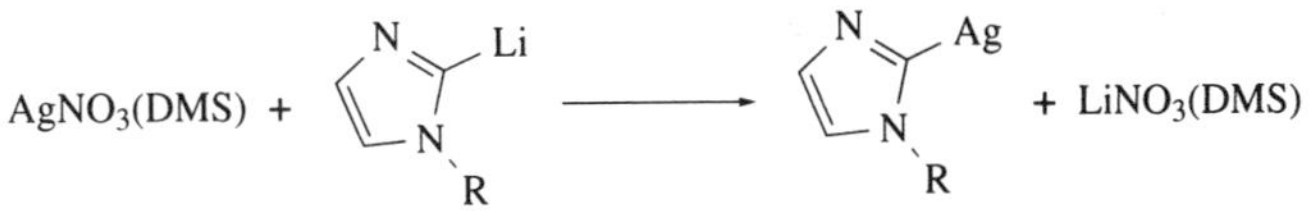

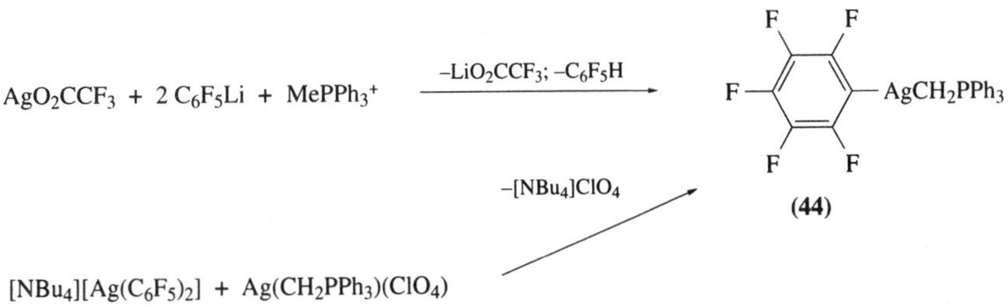

Scheme 8

The argentates [Bu$_4$N][AgR$_2$] (R = C$_6$F$_5$, C$_6$Cl$_5$) are accessible in this way, from which by reaction with a suitable silver salt a remarkable organosilver compound is formed; the reaction of RLi with a suitable silver salt afforded $[Bu_4N][AgR_2]$,[172] which can subsequently be converted into $[AgR]$[69] (Scheme 9). Reaction of $[AgR]$ with ylides (CH_2PPh_3, CH_2PPh_2Me, CH_2AsPh_3) gave rise to the formation of the 1:1 molar ratio complexes.[173]

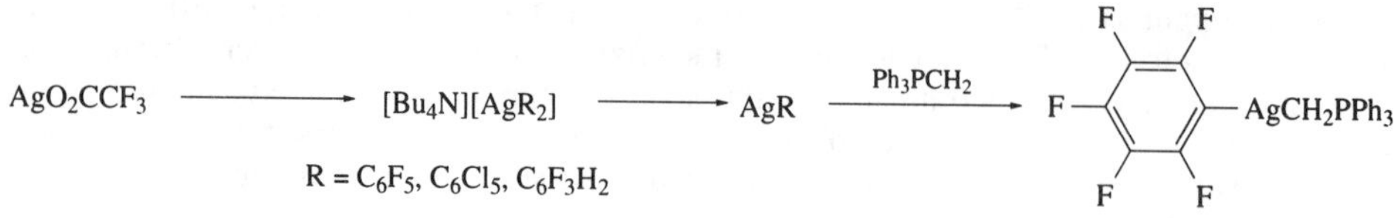

$$R = C_6F_5, C_6Cl_5, C_6F_3H_2$$

(44)

Scheme 9

Fully comparable to the situation in organocopper chemistry, the 1:2 reaction of silver halides with aryllithium compounds or the 1:1 reaction of an arylsilver compound with an aryllithium compound in an ether solvent[9,174] gives rise to the formation of diarylsilverlithium compounds (Scheme 10), which are sufficiently stable to allow their isolation and characterization. An early example dating back to 1973 is $[Li_2Ag_2(C_6H_4CH_2NMe_2-2)_4]$ (4) in Scheme 10, for which NMR spectroscopic studies[9] revealed spin–spin coupling of ^{107}Ag and ^{109}Ag nuclei with carbon and hydrogen nuclei in the organic group. This provided considerable insight into the structure and bonding in this arylsilverlithium compound (see Section 2.3.2.4). Subsequently the $[Li_2Ag_3Ph_6]^-$ mixed-metal aggregate was prepared by treating a cold suspension of finely divided silver bromide in diethyl ether with a solution of phenyllithium in the same solvent in a 1:3 molar ratio (Equation (58)).[77] An argentate with very bulky alkyl groups is $[Li(THF)_4][Ag\{C(TMS)_3\}_2]$ (45), which has been obtained from the reaction of $Li\{C(TMS)_3\}$ with silver iodide.[176]

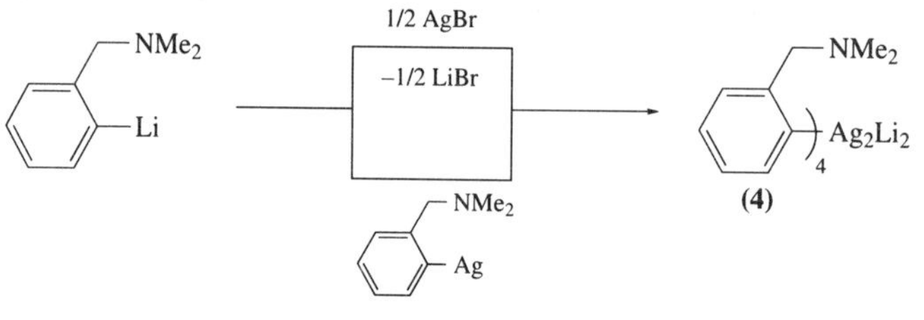

Scheme 10

$$PhLi + AgBr \xrightarrow[Et_2O]{<0\,°C} [Li_2Ag_3Ph_6]^- \qquad (58)$$

In the earlier preparative studies involving transmetallation reactions of silver nitrate, products were often isolated which contained silver nitrate starting material, for example, $AgMe·AgNO_3$, $(AgPh)_2·AgNO_3$ and $(AgPh)_5·(AgNO_3)_2$.[22] As in the case of organocopper reaction products containing the inorganic copper starting material (see Section 2.2.2.2), subsequent studies have revealed that in the mixed-ligand arylsilver compounds the inorganic anions form an integral part of a larger aggregate containing both the organic and the inorganic ligands bound to a central array of silver atoms. The intentional synthesis and isolation of a number of mixed-ligand arylsilver compounds obtained via interaction of arylsilver compounds with silver halides[167] is representative in this respect. Reaction of silver bromide with 2-[(dimethylamino)methyl]phenylsilver yields a rust-brown compound that analyses for the 1:1 complex virtually quantitatively (Equation (59)).[167] The hexanuclear compound $[Ag_6Br_2(C_6H_4NMe_2-2)_4]$ was isolated from the reaction of the aryllithium with silver bromide in a 2:3 molar ratio (Equation (60)).[82,167] Mixed hexanuclear aggregates have been isolated that contain anionic ligands (halogen) or groups ($O_3SCF_3^-$) and aryl groups bound to an array of silver and copper atoms,[177] silver and gold atoms[82] or copper and gold atoms[82] (Equations (61) and (62)). The quantitative synthesis of the copper–gold aggregate from the reaction of the aurate with 2 equiv. of copper bromide illustrates the ready substitution of lithium atoms in the lithium–gold aggregate.[178] More recently, a number of reactions have been reported in which the arylsilver compound functions as a transmetallating reagent, that is, the mixed aggregates are obviously not stable (Equations (63) and (64)).[179,180]

$$[Ag(C_6H_4CH_2NMe_2-2)] + AgBr \xrightarrow[1\,h]{Et_2O,\,<0\,°C} [Ag_2Br(C_6H_4CH_2NMe_2-2)] \qquad (59)$$

$$2[\text{Li}(C_6H_4NMe_2\text{-}2)] + AgBr \xrightarrow[-60\,°C;\ 0\,°C;\ 16\,h]{Et_2O,\ C_6H_6} \underset{\text{dimer}}{[Ag_3Br(C_6H_4NMe_2\text{-}2)_2]\cdot 1/3C_6H_6} + LiBr \qquad (60)$$

$$\underset{\textbf{(93)}}{[Au_2Li_2(C_6H_4NMe_2\text{-}2)_4]} + 4\,AgOTf \xrightarrow[25\,°C]{C_6H_6} [Ag_4Au_2(OTf)_2(C_6H_4NMe_2\text{-}2)_4] + 2\,LiOTf \qquad (61)$$

$$\underset{\textbf{(93)}}{[Au_2Li_2(C_6H_4NMe_2\text{-}2)_4]} + 4\,CuOTf \xrightarrow[25\,°C]{C_6H_6} [Au_2Cu_4(OTf)_2(C_6H_4NMe_2\text{-}2)_4] + 2\,LiOTf \qquad (62)$$

$$[AuCl(L)] + AgR \longrightarrow AgCl + [AuR(L)] \qquad (63)$$

$$L = \text{tetrahydrothiophene, } PPh_3$$
$$R = C_6F_5,\ C_6Cl_5$$

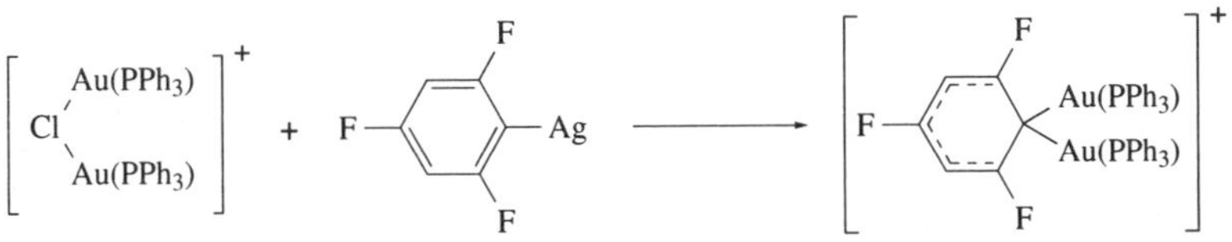

$$(64)$$

The number of well-defined alkynylsilver compounds is limited. Various cluster species have been synthesized and characterized, for example, $[Ag_6Cu_7(C{\equiv}CPh)_{14}]^-$ **(79)**,[67] $[Au_2Ag_2(C{\equiv}CPh)_4(PPh_3)_2]$,[181,182] $[Pt_2Ag_2(C{\equiv}CBu^t)_6]$[183] and $[Pt_2Ag_2(C_6F_5)(C{\equiv}CPh)_4]^{2-}$.[184]

2.3 STRUCTURE AND BONDING IN ORGANOCOPPER AND ORGANOSILVER COMPOUNDS

2.3.1 General Remarks: Thermal Stability

Since the mid-1980s, our knowledge of the structure of organocopper and -silver compounds has grown considerably (see Table 1). This is mainly due to the tremendous development of single-crystal x-ray structure determination techniques and NMR spectroscopy. Nowadays, NMR spectroscopy can routinely provide information not only about the organic entities in organocopper and -silver compounds, but also the metal centres (e.g., ^{6,7}Li in cuprates[10,15,19,185] and argentates,[76,185] ^{107,109}Ag in organosilver compounds and argentates[9,19,185]) from chemical shift and coupling data. However, equally important remains the determination of the molecular weight of the compounds in solution, mainly by cryoscopy and, when stability permits, also by osmometry or ebuliometry[19,20] (see Table 3). These colligative data remain crucial for the determination of the aggregation state of organocopper and -silver compounds, and also for the detection of interaggregate exchange processes in solution. They are also required to confirm whether or not the aggregated structure found in the solid state by x-ray diffraction is retained on dissolution. In a number of cases it has been possible to establish the aggregation state of the compound in the gas phase by mass spectrometry,[110,176] but this information is only of real value when accompanied by spectroscopic data for the compound in solution or as a solid.

Organocopper and -silver compounds are usually represented by simple formulae such as [CuR] or [AgR]. This is solely based on the fact that the compounds are empirically a combination of a monoanionic organic group and the monovalent copper or silver cation. It is now clear that the structures of organocopper and -silver compounds are much more complex than this representation suggests, and in fact comprise an array of fascinating structural elements. In view of the increased structural knowledge now available, it is surprising that only a few reports discuss attempts to relate the reactivity of *pure* organocopper and -silver species to their structural features.[60,87,111,112,188–90,194–6]

The main difficulties in obtaining structural information on organocopper reagents, and of organocopper and silver compounds in general, are (i) the intrinsic instability of the copper–carbon and the silver–carbon bonds, (ii) the unfavourable solubility properties of these compounds even in the 'new' solvents frequently used recently, such as DMS, and (iii) the fact that, as a result of their tendency to form mixed aggregates with metal salts and organometallic species, compounds with well-defined

Copper and Silver

Table 3 Thermal stability of some organocopper and -silver compounds.

Compound	Decomposition observed (°C)	Ref.	Degree of association	Ref.
[CuMe]	> −15	23		
[AgMe]	−80 to −50	21,186		
[CuCH$_2$TMS] (5)	78–79 (m.p.)	11	tetranuclear[a] hexanuclear[b]	11
[CuPh]	100	93	polymeric	190
[AgPh]	74[c]	168	polymeric	168
[Cu(C$_6$H$_4$Me-2)]	134[b] 110–120	93,187	tetranuclear	87,93
[Ag(C$_6$H$_4$Me-2)]	91[b]	168	trinuclear	168
[Cu(C$_6$H$_4$Me-3)]	>100	187		
[Cu(C$_6$H$_4$Me-4)]	100–120	187	tetranuclear	87,93
[Cu(C$_6$H$_4$CH$_2$NMe$_2$-2)] (14)	175–185	43	tetranuclear	176
[Ag(C$_6$H$_4$CH$_2$NMe$_2$-2)]	160–180	167	tetranuclear	167
[Cu(C$_6$H$_4$CH$_2$NMe$_2$-2-Cl-3)]	108–109	43	tetranuclear	176
[Cu(C$_6$H$_4$CH$_2$NMe$_2$-2-Cl-5)]	140–142	43	tetranuclear	176
[Cu(C$_6$H$_4$CH$_2$NMe$_2$-2-OMe-5)]	140–145	43	tetranuclear	176
[Cu(C$_6$H$_4$CH$_2$NMe$_2$-2-Me-5)]	170–210	43	tetranuclear	176
[Cu(PPh$_3$)Cp] (47)	sublimes at 70	188	mononuclear	81,188
[Ag(PPh$_3$)Cp]	75	189		
[Cu(PEt$_3$)Cp][b] (48)	122–124 (m.p.)	80		
[Cu(C$_6$H$_4$CF$_3$-3)]	158	118	octanuclear	118
[Cu(C$_6$F$_5$)]	210–220	5	tetranuclear	5
[Ag(C$_6$F$_5$)]	150	190		
[Cu{C(C$_6$H$_4$(CH$_2$)$_n$NMe$_2$-2)=C(Me)(C$_6$H$_4$Me-4)}] (1:1 CuBr complex) (34)				
$n = 0$	155	191	tetranuclear	191
$n = 1$	194	191	tetranuclear	191
[Ag(CH=CHPh)]	80	192		
[Ag(C(OEt)=NC$_6$H$_4$Me-4)]	>60	193	trinuclear	193
copper acetylides	>200	125,126	polymeric	125,126
silver acetylides	100–200[d]	21[e]	polymeric	125,126

[a] In benzene. [b] In cyclohexane. [c] Determined by TGA and DTA methods. [d] See Section 2.3.3. [e] Light sensitive.

stoichiometry are difficult to obtain. Before the structural aspects are discussed in more detail, the thermal stability of some organocopper and -silver compounds will be considered (see Table 3).

Both simple alkyl-copper and -silver compounds decompose below 0 °C, with the stability decreasing with increasing branching at C(α) due to increasing stability of the secondary and tertiary alkyl radicals. In the past, phenylcopper was also considered to be stable for only two or three days at room temperature.[197] However, it appears that this low stability is due to the presence of small impurities, for example, metal salts or colloidal copper. Extensively purified phenylcopper decomposes at about 100 °C. Table 3 also shows that the general order of stability is alkyl < aryl ~ alkenyl < alkynyl. Alkyl substitution at the 3- or 4-position of the phenyl ring in phenylcopper further influences the stability of the copper–carbon bond (decomposition at 110–120 °C).[187] A similar trend was found in the [Cu$_4$(C$_6$H$_4$R-2)$_4$] series.[43]

Earlier attempts to stabilize the copper–carbon and silver–carbon bonds in alkyl and aryl derivatives concentrated on substitution of hydrogen in the organo groups by fluorine, the substitution of β-hydrogen atoms in alkyl groups, and the use of external ligands such as phosphines or 2,2'-bipyridine. Introduction of fluorine into the alkyl and aryl groups results in considerably higher decomposition temperatures, for example, [CuPh] decomposes at 100 °C and [Cu(C$_6$F$_5$)] at 210 °C, and usually in higher solubility in common organic solvents.[5] Substitution of β-hydrogen atoms in alkylcopper and -silver compounds leads to improved stability (CuEt, decomposition below 0 °C[101]; [Cu$_4$(CH$_2$TMS)$_4$] (5), m.p. 78–79 °C;[11] [Me$_2$P(CH$_2$CuCH$_2$)$_2$PMe$_2$] (49), m.p. 136–138 °C[149]). Attempts to stabilize the copper–carbon and silver–carbon bonds by coordination with 2,2'-bipyridine,[131,132] Et$_3$N,[131] phosphines[129,132,198] or even bidentate phosphines[77,118,119] either failed or gave contrasting results (see Table 4). A surprising finding was the observation that the introduction of a heteroatom containing a coordinating group in the organic moiety bound to the copper or silver atom also influenced the decomposition temperature favourably (see Table 3).[43,120,121,176] Initially this increased stability was ascribed to the interaction of the heteroatom with the copper or silver atom, but later it appeared that introduction of mere steric hindrance near the metal–carbon bond had a similar stabilizing effect (see Section 2.3.4.1).[87,199] These observations apply to both aryl- and alkenylcopper and -silver compounds.[58]

Table 4 Thermal decomposition of complexes of some organocopper and -silver compounds with mono- and bidentate ligands.

Compound	Ligand	Complex (decomposition temperature in °C)	Ref.
[CuMe]	PPh_3	[CuMe(PPh$_3$)$_3$]toluene (70–75)	33b,132
	PPh_3	[CuMe(PPh$_3$)$_2$]1/2Et$_2$O (75–76)	33b,132
	$P(C_6H_{11})_3$	[CuMe(PCy$_3$)] (105–110)	33b,132
	dppe	[(CuMe)$_2$(dppe)$_3$] (149–150)	33b,132
[CuPh]	bipy	[(CuPh)$_n$(bipy)$_m$] (0)	131
[Cu$_4$(C$_6$H$_4$Me-2)$_4$]	dppe	[Cu$_2$(C$_6$H$_4$Me-2)$_2$(dppe)$_3$] (119–121)	131
[Cu$_4$(C$_6$H$_4$CH$_2$NMe$_2$-2)$_4$] (14)	PPh$_3$, MeCN or py	no breakdown of the tetranuclear cluster	130
	dppe[a]	[Cu(C$_6$H$_4$CH$_2$NMe$_2$-2)(dppe)] (160–165)	130
	dppe[b]	breakdown of cluster[c]	130
[Cu(PPh$_3$)Cp*]	PPh$_3$	[Cu(PPh$_3$)Cp*] (125)	98
[Cu(PPh$_3$)Cp] (47)	PPh$_3$	[Cu(PPh$_3$)Cp] (125)	81
[Cu$_4$(SAr)$_2$(Mes)$_2$] (17)	PPh$_3$	breakdown to [Cu$_3$(SAr)(Mes)(PPh$_3$)] (6) (125)	15
[Cu$_3$(SAr)$_2$(C≡CBu^t)]$_2$ (16)	PPh$_3$	breakdown to [Cu$_3$(SAr)$_2$(C≡CBu^t)(PPh$_3$)] (>150)	51,52
[Ag(C$_6$F$_5$)] (44)	CH$_2$PPh$_3$	[Ag(C$_6$F$_5$)(CH$_2$PPh$_3$)] (139)	69

[a] Cu:dppe = 1:1. [b] Cu:dppe = 1:2. [c] Products formed: [CuPPh$_2$(dppe)], H$_2$C=CHPPh$_2$ and PhCH$_2$NMe$_2$ in 1:1:1 molar ratio. A similar reaction has been observed by Camus and Marsich.[131]

In the following sections the structures of organocopper and -silver compounds will be discussed. Starting from these structures, a bonding picture has been developed which provides a rationale for the observed stability order and the reactivity of some of these copper and silver species in organic and organometallic synthesis.

2.3.2 Structures in the Solid State Solved by X-ray Analysis

Compounds for which the structure in the solid state has been solved are shown in Table 5. The compounds are listed according to the hybridization of the carbon atom of the organic group connected to the copper or silver atom, the number of metal atoms present and the overall charge of the organocopper or -silver entity. Also, the presence of other monoanionic ligands and additional neutral ligands is indicated in separate columns. The discussion of the organocopper and -silver structures is grouped according to the nature of the organic group. The variety of bonding in organocopper and -silver compounds will be discussed in Section 2.3.4. These x-ray studies have shown that organocopper and -silver compounds have structures ranging from species with a single metal atom to aggregates consisting of an array of copper or silver atoms to which the organo (and other) monoanions are bonded by either two-electron, two-centre (electron-precise) or two-electron, three-centre (electron-deficient) carbon–metal bonds. The architecture of the metal array is determined to a large extent by the nature of the groups bound to it[15,58,121,139,185,200] and by the preference for digonal or trigonal coordination geometry of the metals in the array.

2.3.2.1 Alkylcopper and -silver compounds

Structural information on alkylcopper and -silver compounds is still very limited. The reported structures of alkylcopper and -silver compounds are shown in Figure 1. The first alkylcopper for which the structure in the solid state was solved was [Cu$_4$(CH$_2$TMS)$_4$] (5). Also, the structures of some ylide complexes are now known. Compound (5) is a tetrameric species in which the four copper and the four CH$_2$ carbon atoms of the CH$_2$TMS groups are arranged in one plane.[11] The carbon atoms are five-coordinate, being surrounded by two hydrogen one silicon and two copper atoms. However, the geometry around each bridging carbon atom is clearly that of an electron-deficient two-electron, three-centre (2e–3c) bonded tetravalent carbon monoanion (see Section 2.3.4.1). Particularly diagnostic of this type of bonding in organocopper chemistry are acute Cu–C–Cu angles and short Cu···Cu distances. The copper atoms in (5) are linearly coordinated, making use of an empty *sp*- or *ds*-type orbital for bonding to the bridging carbon anion. Variations on this type of α-substituted methylcopper compound are [Cu$_4$(CH{2-py}{TMS})$_4$] (52)[28] and [Cu$_2$(C{2-py}{TMS}$_2$)$_2$][27,28,158] and the silver compound [Ag$_2$(C{2-py}{TMS}$_2$)$_2$] (40).[28] These compounds are formally derived from (5) by replacing the remaining α-H

Table 5 Organocopper and -silver compounds characterised by structural analysis.

		X^-	$L_n{}^d$	*Ref.*
Neutral organocopper compounds				
C_{sp^3}				
M_1	$[CuMe(PPh_3)_3]$ (**19**)		3 PR$_3$	33
M_2 homo M	$[Cu\{C(py)(TMS)_2\}]_2$		2 py	27,28
	$[Cu_2\{(CH_2)_2PMe_2\}_2]$ (**49**)			35
hetero M	$[Cu(Me)\{PBu^t_2Li(THF)_3\}]$ (**51**)		PR$_2$Li	63
M_3 homo M	$[Cu_3(Ph_2PCHPPh_2)_3]$ (**36**)		6 PR$_3$	31
M_4 homo M	$[Cu(CH_2TMS)]_4$ (**5**)			11
	$[Cu\{(CH(py)(TMS)\}]_4$ (**52**)		4 py	28
hetero M	$[Li_2Cu_2(CH_2TMS)_4(DMS)_2]_\infty$ (**24**)		SMe$_2$	102
C_{sp^2}				
M_1	$[CuPh\{(PPh_2CH_2)_3CMe\}]$ (**53**)		3 PR$_3$	36
	$[Cu(C_6H_2Bu^t_3\text{-}2,4,6)(DMS)]$ (**85**)		SR$_2$	208b
M_2 homo M	$[Cu_2(MeOxl)_2]^c$ (**54**)		2 R^{1}N=R^2	45
M_2	$[Cu_2(C_6H_2Ph_3\text{-}2,4,6)_2(DMS)_2]$ (**86**)		2 SR$_2$	208b
M_3 homo M	$[Cu_3Mes(O_2CPh)_2]^a$ (**37**)	2 BzO$^-$		39
	$[Cu_3Mes(SC_6H_4CH_2NMe_2\text{-}2\text{-}Cl\text{-}3)_2(PPh_3)]$ (**11**)	2 SAr$^-$	2 NR$_3$	15
M_4 homo M	$[Cu_4\{C_6H_2(CHMe_2)_3\text{-}2,4,6\}_4]$ (**55**)			41
	$[Cu_4(viph)_4]^b$			206
	$[Cu_4\{C_6H_3(CH_2NMe_2)_2\text{-}2,6\}_2Br_2]$ (**32**)	2 Br$^-$	4 NR$_3$	48
	$[Cu_4(C_6H_3CH_2NMe_2\text{-}2\text{-}Me\text{-}5)_4]$ (**1**)		4 NR$_3$	6
	$[Cu_4(C_6H_4CH_2NMe_2\text{-}2)_4]$ (**14**)		4 NR$_3$	85
	$[Cu_4(C_{10}H_6NMe_2\text{-}8)_4]$ (**56**)		4 NR$_3$	44
	$[Cu_4\{(C_5H_3CH_2NMe_2\text{-}2)FeCp\}_4]$ (**15**)		4 NR$_3$	29
	$[Cu_4Ph_4(DMS)_2]$ (**12**)		2 SMe$_2$	16
	$[Cu_4(C_6H_4Me\text{-}2)_4(DMS)_2]$ (**13**)		2 SMe$_2$	17
	$[Cu_4(Mes)_4(THT)_2]$ (**57**)		2 SR$_2$	201
	$[Cu_4(Vi)_2(C_6H_4NMe_2\text{-}2)_2]^e$ (**6**)		2 NR$_3$	12
	$[Cu_4(Vi)_2Br_2]^e$ (**34**)	2 Br$^-$	2 NR$_3$	50
M_4	$[Cu_4(C_6H_4Me\text{-}2)_4(DMS)_2]$ (**13**)		2 SR$_2$	17
	$[Cu_4(Mes)_2(SC_6H_4CH_2NMe_2\text{-}2)_2]$ (**17**)	SR$^-$	2 NR$_3$	13
hetero M	$[Li_2Cu_2(C_6H_4CH_2NMe_2\text{-}2)_4]$ (**3**)		4 NR$_3$	10
	$[Li_2Cu_2Ph_4(OEt_2)_2)]$ (**25**)		OEt$_2$	65
	$[Li_2Cu_2Ph_4(DMS)_3]$ (**26**)		3 SMe$_2$	16
M_5 homo M	$[Cu_5Mes_5]$ (**20**)			201
M_5	$[Cu_5Br_3\{C_6H_3(CH_2N(Me)CH_2CH_2NMe_2)_2\text{-}2,6\}_2]$	3 Br$^-$	8 NR$_3$	211
hetero M	$[Li_3Cu_2Ph_5(DMS)_4]$ (**28**)		4 SMe$_2$	16
	$[MgCu_4Ph_6(OEt_2)]$ (**58**)		OEt$_2$	66
M_6 homo M	$[Cu_6(C_6H_4NMe_2\text{-}2)_4Br_2]$ (**2**)	2 Br$^-$	4 NR$_3$	7
	$[Cu_6(MeOxl_4)Br_2]$ (**33**)	2 Br$^-$	4 R^{1}N=R^2	45,122
hetero M	$[Cu_4Mes_4(\mu\text{-}SAr)_2(MgSAr)_2]$ (**31**)	2 ArS$^-$		15,119
M_9	$[Li_5Cu_4Ph_9(DMS)_4]$ (**59**)		4 SMe$_2$	16
M_{10}	$[Cu_{10}(O_2)(Mes)_6]$ (**60**)	2 O^{2-}		202
mixed $C_{sp^2\text{-}sp}$				
M_6 homo M	$[Cu_6(C_6H_4NMe_2\text{-}2)_4(C{\equiv}CC_6H_4Me\text{-}4)_2]$ (**35**)		4 NR$_3$	58,123
C_{sp}				
M_2 homo M	$[Cu_2(\mu\text{-}C{\equiv}C^tBu)_2(PCy_3)(PPh_3)_2]$ (**23**)		2 PR$_3$	100
M_4 homo M	$[Cu_4(C{\equiv}CPh)_4(PPh_3)_4]$ (**61**)		4 PPh$_3$	55
	$[Cu_4(C{\equiv}CPh)_4(PMe_3)_4]$ (**7**)		4 PMe$_3$	13
	$[Cu_4(C{\equiv}CBu^t)_4(PPh_3)_4]$		4 PPh$_3$	203
	$[Cu_4(\mu_3\text{-}\eta\text{-}C{\equiv}CPh)_4(Ph_2Ppy\text{-}P)_4]$		4 PPh$_2$py	53
M_6 homo M	$[Cu_3(SC_6H_4CH_2NMe_2\text{-}2)_2(C{\equiv}CBu^t)]_2$ (**16**)	2 ArS$^-$	4 NR$_3$	51,52
M_{10} hetero M	$[Li_6Cu_4(C{\equiv}CPh)_{10}(Et_2O)_3]$ (**62**)		3 OEt$_2$	203
M_{24} homo M	$[Cu_{24}(C{\equiv}CBu^t)_{24}]\cdot2C_6H_{14}$			203
Charged organocopper compounds				
C_{sp^3}				
M_1	$[CuMe_2]^-$ (**30**)			64
	$[Cu\{C(TMS)_3\}_2]^-$ (**9**)			204
	$[Cu\{CH(TMS)_2\}Br]^-$ (**63**)	Br$^-$		64
	$[Cu(fluorenyl)_2PPh_3]^-$ (**22**)		PPh$_3$	100
C_{sp^2}				
M_1	$[CuPh_2]^-$ (**64**)			64
	$[CuMes_2]^-$ (**30**)			49
M_3 homo M	$[Cu_3Ph_2(PMDTA)_2]^+$ (**66**)	$[Cu_5Ph_6{}^-]$	6 NR$_3$	205
M_5 homo M	$[Cu_5Ph_6]^-$ (**10**)			14c
	$[Cu_5(viph)_2Br_4]^-$ (**67**)	4 Br$^-$	2 η^2-alkene	206
	$[Cu_5(viph)_4Br_2]^-$ (**68**)	2 Br$^-$	1 η^2-alkene	206
hetero M	$[LiCu_4Ph_6]^-$ (**69**)			66
	$[Li_2Cu_3Ph_6]^-$ (**8**)			14a

Table 5 (continued)

	X^-	L_n^d	*Ref.*
C_{sp}			
M$_3$ homo M [Cu$_3$(μ_3-η-C≡CPh)$_2$(dppm)$_3$]$^+$ (**70**)		6 PR$_3$	127
[Cu$_3$(μ_3-η-C≡CPh)(dppm)$_3$]$^{2+}$ (**71**)		6 PR$_3$	56
M$_4$ homo M [Cu$_4$(L)$_2$(C≡CPh]$^{3+}$		8 R^{1}N=R^2	57
M$_{13}$ hetero M [Cu$_7$Ag$_6$(C≡CPh)$_{14}$]$^-$ (**46**)			67
Neutral organosilver compounds			
C_{sp^3} [Ag(C$_3$F$_7$)(NCMe)] (**72**)		NCMe	70
[Ag$_2${Cpy(TMS)$_2$}$_2$] (**40**)		2 py	28
[Ag$_2${(CHCO$_2$Et)$_2$Ph$_2$P}$_2$] (**42**)			75
[Ag$_3${PPh$_2$CHPPh$_2$}$_3$] (**73**)		6 PR$_3$	201
[Ag$_2$(CH$_2$PPh$_2$S)$_2$] (**38**)		2 S=PR3	76
mixed $C_{sp^2-sp^3}$ [Ag(C$_6$F$_5$)(CH$_2$PPh$_3$)] (**44**)			69
C_{sp^2} [Ag$_4$Mes$_4$] (**74**)			201
[Ag$_4${C$_5$H$_3$(CH$_2$NMe$_2$)FeCp}$_4$] (**75**)			68
C_{sp} [Ag(C≡CPh)(PMe$_3$)]$_\infty$ (**76**)		PMe$_3$	73
Charged organosilver compounds			
C_{sp^3} [Ag{C(TMS)$_3$}$_2$]$^-$ (**45**)			14b
[Ag(C$_3$F$_7$)$_2$]$^-$ (**43**)			70
[Ag(CHPPh$_3$CO$_2$Et)$_2$]$^+$ (**77**)	ClO$_4^-$		74
[Ag(CHPPh$_3$CO$_2$Ph)$_2$]$^+$	NO$_3^-$		74
[Ag{$\overline{\text{CHSO}_2\text{(CH}_2)_3\text{SO}_2}$}$_2$]$^-$ (**41**)			71
C_{sp^2} [Li$_2$Ag$_3$Ph$_6$]$^-$ (**78**)			77
C_{sp} [Ag$_6$Cu$_7$(C≡CPh)$_{14}$]$^-$ (**46**)			67
Miscellaneous organocopper compounds			
[Cu(PEt$_3$)Cp] (**48**)		PEt$_3$	80
[Cu(PPh$_3$)Cp] (**47**)		PPh$_3$	81
[Cu(PPh$_3$)(C$_5$H$_4$Me)] (**80**)		PPh$_3$	236
[Cu{C(PPh$_3$)$_2$}Cp*] (**81**)		C(PR$_3$)$_2$	153
[Cu(CF$_3$)$_2$(S$_2$CNEt$_2$)] (**21**)	Me$_2$NCS$_2^-$		96
[CuPtW(μ-CC$_6$H$_4$Me-4)(CO)$_2$(PMe$_3$)$_2$Cp*] (**82**)			209

[a] Mes = C$_6$H$_2$Me$_3$-2,4,6; [b] viph = *o*-vinylphenyl;

[c] Oxl = ; [d] L = ; e, *p*-Tol

atoms with a 2-pyridyl substituent (cf. (**52**)), or a 2-pyridyl group and another TMS group (cf. (**40**) and [Cu$_2$(C{2-py}{TMS}$_2$)$_2$]). Note that none of these compounds contain β-hydrogens. In addition to the steric constraints imposed by the [CH$_{2-n}${2-py}{TMS}$_n$)$_4$]$^-$ anion, it can in principle function as a four-electron ligand by chelation or bridging. The molecular geometry arranges the carbon and nitrogen lone pairs in such a way that the C–Cu and N–Cu bonds are oriented almost parallel.

With the most sterically hindered anion [C{2-py}{TMS}$_2$]$^-$, stable compounds of the type [M$_2$R$_2$] (M = Cu or Ag) are formed. The N–M–C angles in these dimeric complexes are close to the ideal 180° for linear coordination of the copper and silver atoms, and as a consequence the M···M distances are short (Cu···Cu 0.241 2(1) nm vs. 0.284 3(3) nm in the dinuclear ylide copper compound [Cu$_2${(CH$_2$)$_2$PMe$_2$}$_2$] (**49**)). The M–C bonding in these compounds is of the simple, electron precise 2e–2c type. In the silver compound the Ag···Ag distance is likewise short, in particular when compared with the Ag···Ag distances in [Ag$_2${(CH$_2$)$_2$PMe$_2$}$_2$] (**83**) and [Ag$_2$(CH$_2$PPh$_2$S)$_2$] (**38**) which are 0.295 3(1) nm and 0.299 0(2) nm, respectively. An obvious difference between the [C(2-py)(TMS)$_2$]$^-$ anion and the ylide ligands is the larger bite angle of the latter, that is, with equal C–M and N–M distances different M···M distances are expected. When the sterically less bulky anion [CH{2-py}{TMS}]$^-$ is used in (**52**), a tetrameric instead of a dimeric aggregate is obtained. The structural features of (**52**) are fundamentally different from those of (**5**) as the [CH(2-py)(TMS)]$^-$ ligand bridges two copper centres by forming an electron-precise C–Cu bond with linear C–Cu–N coordination to one copper centre, and an N–Cu bond to a second copper atom. The ligands alternate on opposite sides of the Cu$_4$ plane, resulting in a structure that resembles those of many other Cu$_4$L$_4$ complexes in Cu^1 coordination chemistry.[201]

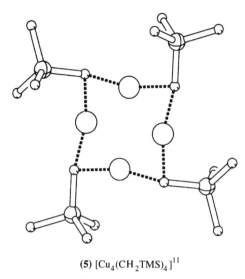

(5) $[Cu_4(CH_2TMS)_4]^{11}$

Cu-C	0.198–0.204 nm
Cu--Cu	0.2418 nm
Cu-C-Cu	73.4 and 73.8°
C-Cu-C	163.4 and 163.7°

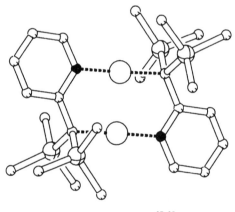

$[M_2\{C(TMS)_2py\}_2]^{27,\,28}$

	M = Cu	M = Ag (40)
M-C	0.1950(4) nm	0.2154(5) nm
M-N	0.1910(3) nm	0.2160(5) nm
M--M	0.2412(1) nm	0.2654(1) nm
N-M-C	178.0(5)°	174.5(1)°

Figure 1 Structures of alkyl-copper and -silver compounds. M = Cu for the actual compounds illustrated, except for (73) (M = Ag).

A special case is the structure of $[Cu_3\{CH(PPh_2)_2\}_3]$ (36).[31] Although it was initially proposed that this compound is a mixed-valence complex,[31] a more likely, alternative interpretation is that the trinuclear aggregate consists of a monoanionic cuprate unit comprising copper with an obtuse C–Cu–C angle, and a $Cu_2\{CH(PPh_2)_2\}^+$ cationic unit complexed by two additional phosphine ligands of the cuprate unit. The structures of both (36) and its silver analogue $[Ag_3\{CH(PPh_2)_2\}_3]$ (73) have been redetermined.[201] The silver atoms in (73) are arranged in an almost regular isoceles triangle with trigonal AgP_3 units. The methyne carbon atoms bonded to silver make an angle of 149.8(4)° at silver; the geometry of the carbon atoms is close to tetrahedral, suggesting sp^3 hybridization. The third, uncoordinated carbon atom has sp^2 hybridization, and the P–C bond distances to this carbon suggest considerable double-bond character.

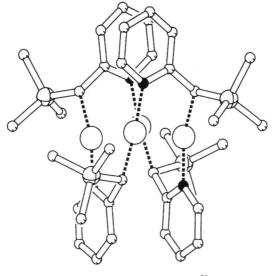

(52) [Cu$_4${CH(TMS)py}$_4$] [28]

Cu-C	0.1934(5) nm
Cu-N	0.1935(4) nm
Cu--Cu	0.2668(2) nm
C-Cu-N	165.5(2)°

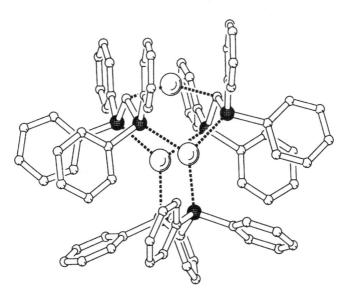

[M(Ph$_2$PCHPPh$_2$)]$_3$ [31, 200]

	M = Cu **(36)**	M = Ag **(73)**
M-C	0.196(2) and 0.200(2) nm	0.219(1) and 0.220(1) nm
M-P$_{av}$	0.2317 nm	0.2516 nm
M--M	0.284 – 0.315 nm	0.2933(2) – 0.3456(1) nm

Figure 1 (continued)

2.3.2.2 Arylcopper and -silver compounds

(i) With noncoordinating substituents

The structures of only a few unligated arylcopper compounds are known, namely [Cu$_5$Mes$_5$] (**20**),[37,201] [Cu$_4$(C$_6$H$_2$Pr$^i_3$-2,4,6)$_4$] (**55**)[41] and [Cu(C$_6$H$_2$Ph$_3$-2,4,6)] (**84**),[42,210] and two arylsilver compounds, namely [Ag$_4$Mes$_4$] (**74**)[72,201] and [Ag(C$_6$H$_2$Ph$_3$-2,4,6)][42,210] (see Figure 2).

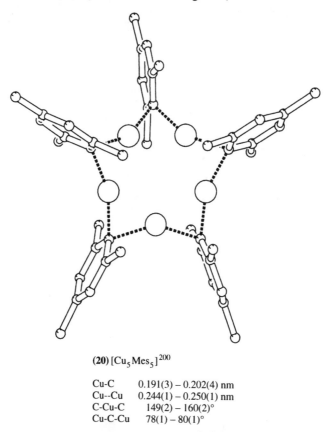

(**20**) [Cu$_5$Mes$_5$][200]

Cu–C	0.191(3) – 0.202(4) nm
Cu--Cu	0.244(1) – 0.250(1) nm
C–Cu–C	149(2) – 160(2)°
Cu–C–Cu	78(1) – 80(1)°

Figure 2 Unsolvated arylcopper and -silver compounds.

Of these structures, only (**84**) and the analogous silver compound [Ag(C$_6$H$_2$Ph$_3$-2,4,6)] seem to be monomeric.[42] However, these results have been contested, as the crystals used for the x-ray structure determination seem to have contained co-crystallized [BrC$_6$H$_2$Ph$_3$-2,4,6].[210] The other structures in Figure 2 show that these compounds are aggregated species comprising rings of either four or five metal–mesityl units. The mesityl groups are 2e–3c bonded with the characteristic acute M–C–M angle (comprising a four-coordinate, tetravalent monoanionic carbon atom), and short M···M distances (the distance in metallic copper is 0.256 nm). The Ag$_4$C$_4$ ring in (**74**) is planar, whereas the Cu$_5$C$_5$ ring in (**20**) is folded.[201] It should be noted that the angle made by the three atoms involved in this bonding is more acute than the angles between the MOs used, that is, when the copper atom is involved in 2e–3c bonding, the C–Cu–C angle will deviate from linearity but the MO involved is linear. Following a simple geometrical approach, assuming that a linear *sp* copper hybrid orbital is involved in bonding, a typical estimate for the C–Cu–C angle is 164°,[58] which is near the C–M–C angles found in (**20**) and (**74**).[201] The mesityl groups are oriented almost perpendicular to the M$_n$C$_n$ ring, as this arrangement gives the least steric interaction between the *ortho*-positioned alkyl groups[19,104,139,200] (see Section 2.3.4).

The influence of steric interactions on the nature of the bridge bonding can be inferred from the geometric differences between (**20**) and (**55**).[41] In the latter structure the Cu$_4$ frame is folded with the bridging carbon atoms alternating above and below the Cu$_4$ mean plane. The steric interactions of the bulky *o*-Pr^i groups in (**55**) result in a fairly asymmetric Cu–C–Cu bridge. The bonding in this bridge can be described as 2e–2c bonding to one copper atom, with additional π-type bonding to the second copper atom (see Section 2.3.4.1, Figure 23).[41] Earlier discussions[139,208] have also included the presumed[210] monomeric [Cu(C$_6$H$_2$Ph$_3$-2,4,6)] (**84**), which has even more bulky *ortho*-substituents, but this must now await the remeasurement of its structure in the solid state.

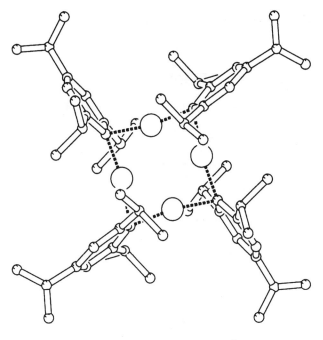

(55) $[Cu_4(C_6H_2Pr^i_3\text{-}2, 4, 6)_4]^{41}$

Cu-C   0.1958(7) and 0.2018(7) nm
Cu--Cu   0.2445(1) nm
C-Cu-C 169.8(3)°

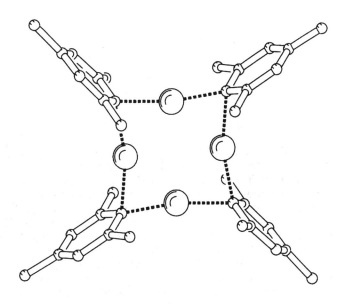

(74) $[Ag_4Mes_4]^{200}$

Ag-C   0.216(3) – 0.222(3) nm
Ag--Ag 0.2733(3) and 0.2755(3) nm
C-Ag-C 165.2(8) and 169.0(9)°
Ag-C-Ag 76.1(7) and 78.1(7)°

Figure 2   (continued)

    The use of DMS as either a reaction medium or as a solvent for recrystallization of organocopper and -silver compounds has resulted in the isolation and characterization of a number of DMS-solvated/coordinated organocopper compounds (see Figure 3).

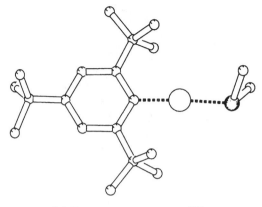

(85) [Cu(C$_6$H$_2$Bu$^t_3$-2,4,6)(DMS)]^{207b}

Cu-C	0.1916(3) nm
Cu-S	0.2185(1) nm
C-Cu-S	175.7(1)°

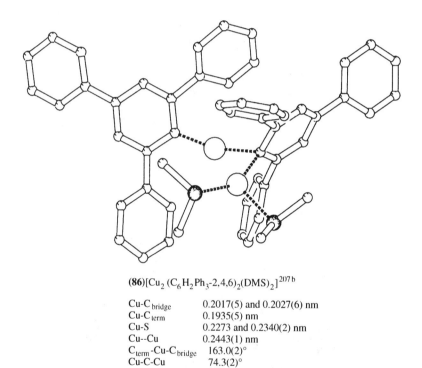

(86) [Cu$_2$(C$_6$H$_2$Ph$_3$-2,4,6)$_2$(DMS)$_2$]^{207b}

Cu-C$_{bridge}$	0.2017(5) and 0.2027(6) nm
Cu-C$_{term}$	0.1935(5) nm
Cu-S	0.2273 and 0.2340(2) nm
Cu--Cu	0.2443(1) nm
C$_{term}$-Cu-C$_{bridge}$	163.0(2)°
Cu-C-Cu	74.3(2)°

Figure 3 Structures of DMS-solvated/coordinated arylcopper compounds.

Both the mononuclear [Cu(C$_6$H$_2$Bu$^t_3$-2,4,6)(DMS)] (**85**) and the dinuclear [Cu$_2$(C$_6$H$_2$Ph$_3$-2,4,6)$_2$(DMS)$_2$] (**86**) have very bulky *ortho*-substituents; the formation of higher aggregates for these di-*ortho*-substituted aryl ligands does not occur, and most probably is precluded by the steric constraints of these groups. The coordination at copper is completed by DMS.^{208b} That sulfur-donor solvents can have sufficient coordinating power to alter the aggregation state of arylcopper compounds is shown by the isolation of [Cu$_4$Mes$_4$(THT)$_2$] (**57**) (THT = tetrahydrothiophene) when [Cu$_5$Mes$_5$] (**20**) is crystallized in the presence of THT.^{37,20f} [Cu$_4$Mes$_4$(THT)$_2$] (**57**), [Cu$_4$Ph$_4$(DMS)$_2$] (**12**)^{16} (not shown) and [Cu$_4$(C$_6$H$_4$Me-2)$_4$(DMS)$_2$] (**13**)^{17} have mutually similar structural features. The *ipso* carbons are 2e–3c bonded to two copper atoms, the aryl rings are oriented perpendicular to the Cu$\cdots$Cu vector with a short Cu$\cdots$Cu distance (0.245 nm (mean)) and the Cu$_4$ array has a flattened butterfly geometry with two two-coordinate and two three-coordinate copper centres arranged in transannular fashion. Interestingly, [Cu$_4$(C$_6$H$_4$Me-2)$_4$(DMS)$_2$] (**13**) exists as two rotamers in the solid state, one having all *o*-tolyl methyl groups on the same side of the Cu$_4$ plane (75% abundance: Figure 3), and another in which these methyl

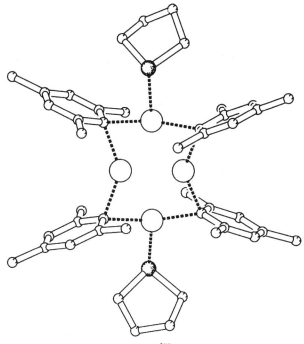

(57) $[Cu_4(Mes)_4(THT)_2]^{200}$

Cu-C	0.2023(10) – 0.2068(9) nm
Cu-S	0.2369(4) and 0.2401(4) nm
Cu--Cu	0.2439(2) – 0.2451(2) nm
Cu---Cu (across)	0.2600(3) and 0.4138(3) nm
C-Cu$_{digonal}$-C	140.6(4) and 141.5(4)°
Cu-C-Cu	73.1(4) – 73.6(4)°

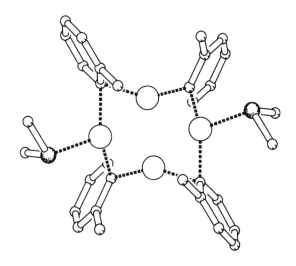

(13) $[Cu_4(C_6H_4Me-2)_4(DMS)]^{17}$

Cu-C	0.1967(3) – 0.2087(3) nm
Cu-S	0.23160(9) nm
Cu--Cu (bridged)	0.24126(5) and 0.24626(5) nm
Cu---Cu (across)	0.26874(6) and 0.40105(7) nm
Cu-C-Cu	72.7(2) – 74.8(1)°
C-Cu-C	145.4(2) – 157.0 (3)°

Figure 3   (continued)

groups are alternately oriented above and below this plane (25% abundance: see Section 2.3.4.2(ii)).[17] Finally, the different Cu–C distances within these compounds suggest that they may also be seen as two cuprate species [CuR$_2$]$^-$ (with the shorter Cu–C bonds) linked together by [Cu(DMS)]$^+$ cations. In this light, these [Cu$_4$R$_4$(DMS)$_2$] structures are isovalent with those of the neutral cuprates [Cu$_2$Li$_2$R$_4$], that is, in the latter structures [CuL$_x$]$^+$ is replaced by [LiL$_y$]$^+$.

(ii) With functionalized heteroatom-containing substituents

Another, larger series of *ortho*-functionalized arylcopper compounds is known. These compounds emerged from successful attempts to improve the stability of the copper–carbon bond in alkyl- and arylcopper and -silver compounds by the introduction of substituents containing heteroatoms (N-, O-, C(NMe$_2$)=O)[46,120,139,167] instead of fluorine or trialkylsilyl substituents. Following the view that a phosphorus-containing ligand would coordinate too strongly to copper(I), in most cases nitrogen-based substituents were used. These *ortho*-substituents have two possible functions. When coordinated to the metal centre of the (aryl)C–M bond they can act as a well positioned intramolecular ligand, that is, the copper–ligand interaction does not dominate the type of aggregate structure formed, but can function as a stabilizing interaction. When not coordinated to the metal, they will act merely as very bulky *ortho*-substituents.[139,185]

Figure 4 shows the structures of a series of arylcopper compounds with heteroatom-containing substituents. As in the case of the [Cu$_4$R$_4$(DMS)$_2$] structures (Figure 3), the CH$_2$NMe$_2$-substituted compounds are tetranuclear aggregates with 2e–3c bonded aryl groupings.

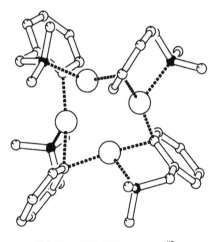

(14) [Cu$_4$(C$_6$H$_4$CH$_2$NMe$_2$-2)$_4$][85]

Cu–C	0.197 – 0.216 nm
Cu–N	0.219 and 0.224 nm
Cu--Cu	0.2389 and 0.2377 nm
Cu---Cu	0.3075 and 0.3393 nm
Cu–C–Cu	70.74 and 71.35°

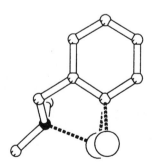

Figure 4 Structures of arylcopper compounds with heteroatom-containing substituents and the puckering in the five-membered chelate ring.

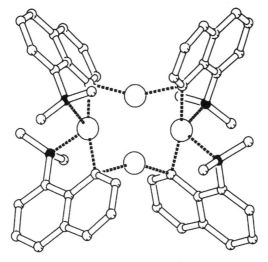

(56) $[Cu_4(1\text{-}C_{10}H_6NMe_2\text{-}8)_4]^{44}$

Cu-C	0.2019(9) – 0.2076(9) nm
Cu-N	0.2243(8) and 0.2269(7) nm
Cu--Cu	0.2407(2) and 0.2430(2) nm
Cu---Cu	0.2708(2) and 0.3998(2) nm
Cu-C-Cu	68.1(1) – 72.8(5)°
C-Cu-C	154.9(5) and 168.5(4)°

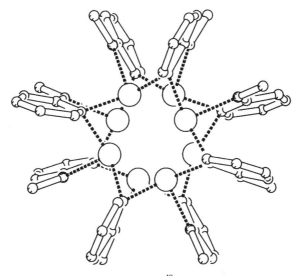

(87) $[Cu_8(C_6H_4OMe\text{-}2)_8]^{40}$

Cu-C	0.199(2) – 0.208(3) nm
Cu-O	0.231(2) – 0.243(2) nm
Cu--Cu (C-bridged)	0.2436(5) – 0.2499(5) nm
Cu--Cu (nonbridged)	0.2680(5) – 0.2779(4) nm
C-Cu-O	88.5(8) – 95.7(8)°
Cu-C-Cu	73.3(8) – 76.7(8)°

Figure 4 (continued)

An obvious difference between *o*-CH$_2$NMe$_2$-substituted compounds [Cu$_4$(C$_6$H$_3$CH$_2$NMe$_2$-2-Me-5)$_4$] (1) (not shown)[6] and [Cu$_4$(C$_6$H$_4$CH$_2$NMe$_2$)$_4$] (14) and the solvated compounds [Cu$_4$R$_4$(DMS)$_2$] (e.g., (13)) is that in (1) and (14) each of the copper atoms is three-coordinate.

Also, the variation in Cu–C distances is not so pronounced as in [Cu$_4$R$_4$(DMS)$_2$]. Accordingly, the 2 [CuR$_2$]$^-$2 [CuL]$^+$ arrangement is not so obvious in these structures. Only one rotamer of (14) is present in the solid state, with the *ortho*-substituents oriented alternately above and below the Cu$_4$ mean plane.[85]

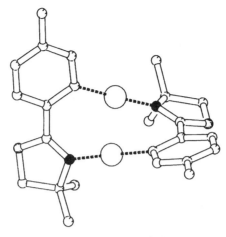

(54) $[Cu_2(MeOxl)_2]^{45, 122}$

Cu–C	0.1899(5) nm
Cu–N	0.1902(4) nm
Cu--Cu	0.24708(9) nm
N–Cu–C	177.8(2)°

Figure 4   (continued)

The additional Cu–N coordination involves one of the two copper atoms bridged by the aryl group to which the *o*-CH$_2$NMe$_2$ group is connected. This gives rise to a five-membered $\overline{CuCCCH_2N}Me_2$ chelate ring with pronounced puckering (see Figure 4). In the case of the chiral compound [Cu$_4$(C$_6$H$_4$CH(Me)NMe$_2$-2)$_4$], this ring puckering is detectable in solution.[208a] The presence of a benzylic Me substituent in this compound results in two possible ring conformers for each [Cu$_2$Ar] unit, one with the Me substituent coplanar with the aryl nucleus and the other with the Me perpendicular to the aryl ring. The latter arrangement is the most favourable, and it is this diastereomer which is most abundant in solution.[208]

The importance of the flexibility of the chain connecting the carbon and nitrogen donor atoms in the *C,N*-chelate ring in influencing the bonding in the aggregate is demonstrated in the structure of [Cu$_4$(1-C$_{10}$H$_6$NMe$_2$-8)$_4$] (**56**). This 1-naphthyl compound has a rigid C,N arrangement and contains two four-coordinate and two two-coordinate copper atoms,[44] a structure which is similar to that of [Cu$_4$R$_4$(DMS)$_2$] (**12**). The structure of (**56**) has been discussed in terms of two [CuR$_2$]$^-$ anions which are held together by two [CuL$_2$]$^+$ cations, by analogy with the corresponding neutral cuprate structures [Li$_2$Cu$_2$R$_4$] (see below and Figure 10).

C,N-Chelate binding, as observed in (**56**), requires that the two electron pairs involved in the coordination can be directed toward the same metal centre. This is clearly impossible in the case of the *o*-oxazoline-substituted arylcopper compound [Cu$_2$(MeOxl)$_2$] (**54**), which adopts a dinuclear structure as a result of parallel (2e–2c) Cu–C and Cu–N bonds.[45,122] The coordination geometry of the copper atoms in this dimeric unit is close to linear (cf. the dimeric structures of [Cu$_2$(C{2-py}{TMS}$_2$)$_2$] (see above)). Since chelate coordination is not possible for the *o*-oxazoline-substituted aryl anion, [C$_6$H$_4$C=NC(Me$_2$)CH$_2$O-2]$^-$, it makes a very suitable ligand for binding to three copper centres (see below for a discussion of [Cu$_3$Br(MeOxl)$_2$]$_2$ (**33**)).[45,122] Likewise, [Cu(C$_6$H$_4$NMe$_2$-2)]$_n$ (**88**) and [Cu$_8$(C$_6$H$_4$OMe-2)$_8$] (**87**) have structures which reflect the directional properties of the [C$_6$H$_4$NMe$_2$-2]$^-$ and [C$_6$H$_4$OMe-2]$^-$ anions, having carbon and nitrogen (or oxygen) lone pairs oriented in a parallel fashion, suited to binding to three copper atoms. The structure of (**87**) comprises eight CuAr units in an octanuclear aggregate. The main features are the 2e–3c bonding mode of C$_{ipso}$ of the aryl group to two copper atoms (0.247 nm) and the Cu–O bonding to a third copper atom. The copper atoms in this octanuclear aggregate are arranged square antiprismatically, with the triangular Cu$_3$ faces bridged by C$_6$H$_4$OMe-2 anions.[40] For [Cu(C$_6$H$_4$NMe$_2$-2)]$_n$ (**88**), a polymeric network structure has been proposed (see Section 2.3.3).[46] Molecular models show that the increased number of methyl groups at the heteroatom (NMe$_2$ vs. OMe) would indeed destabilize a similar octanuclear aggregate for (**88**), owing to steric hindrance. The more open structure of (**88**) is further corroborated by the fact that it reacts readily with a variety of metal salts, and even other organocopper compounds (see below).

(iii) Heteroleptic organocopper compounds

The monoanionic bidentate ligands $[C_6H_4NMe_2-2]^-$, $[C_6H_4OMe-2]^-$ and $[MeOxl]^-$ are all suitable for binding three copper atoms in one face of a polyhedron. This inherent ability to stabilize aggregated species means that arylcopper compounds bearing such ligands often incorporate copper halide or organocopper units (without heteroatom-containing substituents) into their structures. Examples are $[Cu_6Br_2(C_6H_4NMe_2-2)_4]$ (2), $[Cu_6(C{\equiv}CC_6H_4Me-4)_2(C_6H_4NMe_2-2)_4]$ (35) and $[Cu_6Br_2(MeOxl)_4]$ (33) (see Figure 5).

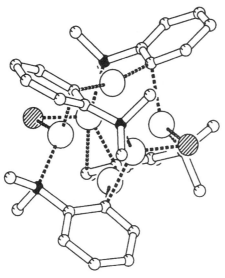

(2) $[Cu_6Br_2(C_6H_4NMe_2-2)_4]^7$

Cu-C$_{av}$	0.197 and 0.208 nm
Cu-N$_{av}$	0.211 nm
Cu--Cu$_{av}$ (C-bridged)	0.248 nm
Cu--Cu (Br-bridged)	0.270 nm
Cu--Cu	0.264 nm
Cu-C-Cu$_{av}$	70.5°
Cu-Br-Cu	66.7°

schematic structure shown

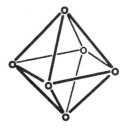

Figure 5 Heteroleptic organo- and organohalide-copper aggregates.

Aggregates (2)[7,121] and (35)[59,123] both contain a distorted octahedral Cu_6 array, to which the four $C_6H_4NMe_2-4$ groups and the bromide or $C{\equiv}CC_6H_4Me-2$ anions are bonded. The $[Cu_6(C_6H_4NMe_2-2)_4]^{2+}$ dication can be seen as the basic structural element, that is, six copper atoms are held together in an array by only four $C_6H_4NMe_2-2$ groups, each capping three copper atoms. The distortion of the octahedral Cu_6 core arises from differences between C^--bridged (average 0.243 nm) and unbridged (average 0.280 nm) Cu–Cu distances. The two apical copper atoms in these structures are two-coordinate with a C–Cu–C angle of 164°. This is consistant with linear (*ds* or *sp*) hybridization at these apical copper atoms (see Figure 6(a)). The equatorial copper atoms have trigonal coordination geometries, because the two remaining ligands, that is, Br in (2) or $C{\equiv}CC_6H_4Me-4$ in (35), each bridge two equatorial copper atoms. Alternatively, these structures can be considered to be comprised of anionic ($[CuR_2]^-$) and cationic ($[Cu_4X_2]^{2+}$) building blocks, in which X = Br or $C{\equiv}CC_6H_4Me-4$. The structure of (35) shows how mixed organocopper aggregates can exist in which two different organic

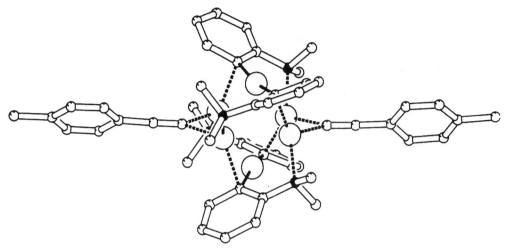

(35) $[Cu_6(C_6H_4NMe_2\text{-}2)_4(C{\equiv}CC_6H_4\text{-}Me\text{-}4)_2]$ [58, 59, 123]

Cu-C$_{Ar, av}$	0.1984(7) – 0.2130(6) nm
Cu-C$_{Ac}$	0.2015 – 0.2054 nm
Cu--Cu$_{av}$ (Ar-bridge)	0.2521(5) nm
Cu--Cu (Ac-bridge)	0.2466 and 0.2474 nm
Cu-C$_{Ar}$-Cu$_{av}$	75.5(3)°
Cu-C$_{Ac}$-Cu	75.2 and 74.6°
C$_{Ar}$-Cu-C$_{Ar}$	168.3°

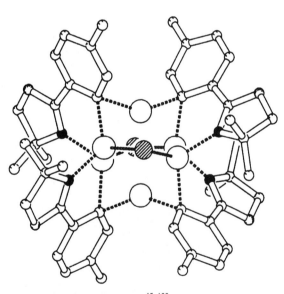

(33) $[Cu_6Br_2(MeOxl)_4]$ [45, 122]

Cu-C$_{av}$	0.2031(3) nm
Cu-N$_{av}$	0.1968(7) nm
Cu--Cu (C-bridged)	0.2426(5) – 0.2451(5) nm
Cu--Cu (Br-bridged)	0.3012(6) and 0.3033(5) nm
Cu--Cu	0.2600(5) and 0.2634(5) nm
C-Cu-C	143.9(4) and 147.1(4)°
Cu-C-Cu	72.9(5) – 75.1(5)°
Cu-Br-Cu	73.0(3) and 74.04(20)°

Figure 5   (continued)

groups are bound to the same central copper array. This is important for the discussion of selective cross-coupling reactions of (**35**) to 4-MeC$_6$H$_4$C≡CC$_6$H$_4$NMe$_2$-2 (see *COMC-I*,[19] Volume 2, Chapters 14.4.2.3 and 14.4.4.1).

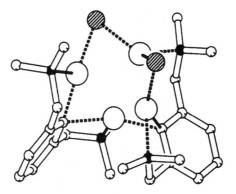

(32) $[Cu_4Br_2(C_6H_3(CH_2NMe_2)_2\text{-}2,6)_2]^{48}$

Cu-C	0.2003(8) – 0.2172(9) nm
Cu-N	0.2124(7) – 0.2247(8) nm
Cu--Cu (C-bridged)	0.2412(2) and 0.2406(2) nm
Cu--Cu (Br-bridged)	0.2737(2) and 0.2429(2) nm
Cu--Cu	0.2894(2) and 0.3475(4) nm
C-Cu-C	155.8(4)°
Cu-C-Cu	70.4(5) and 71.5(5)°
Cu-Br-Cu	59.85(8) and 70.66(10)°

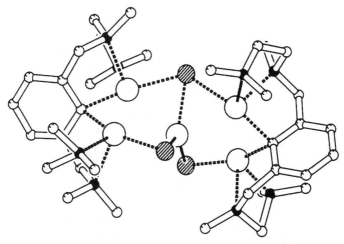

(89) $[Cu_5Br_3\{C_6H_3(CH_2N(Me)CH_2CH_2NMe_2)_2\text{-}2,6\}_2]^{210}$

Cu-C	0.2020(6) – 0.2076(6) nm
Cu-N	0.2203(6) – 0.2334(6) nm
Cu--Cu	0.2454(1) and 0.2458(1) nm
Cu-Cu$_{center, av}$	0.277(8) nm
Cu-C-Cu	73.6(2) and 73.8(2)°

Figure 5 (continued)

The structure of $[Cu_6Br_2(MeOxl)_4]$ (33) comprises such a highly distorted central Cu_6 array that it can be thought of as a dimer of two trinuclear copper species '$[Cu_3Br(MeOxl)_2]$'.[45,122] The C_{ipso} of each ligand bridges two copper atoms with a 2e–3c bond, while the oxazoline nitrogen atom bonds to the third copper of a trigonal face (see Figure 5). In Figure 6(b) the preferred bonding mode of the various heteroatom-substituted aryl groups is shown.

The influence of the number and mutual orientation of heteroatoms in the *ortho*-substituents is clearly demonstrated by the structures of $[Cu_4Br_2(C_6H_3\{CH_2NMe_2\}_2\text{-}2,6)_2]$ (32)[48] and $[Cu_5Br_3(C_6H_3\{CH_2N(Me)CH_2CH_2NMe_2\}_2\text{-}2,6)_2]$ (89).[211] Figure 6(b) shows that bis-*ortho*-chelation leads to stabilization of a $[CuR_2]^-$ building block which combines with $[Cu_3Br_2]^+$ in the case of $[Cu_4Br_2(C_6H_3\{CH_2NMe_2\}_2\text{-}2,6)_2]$, and of two $[Cu_2R]^+$ building blocks combining with the known (planar)[212] $[CuBr_3]^{2-}$ dianion in the case of (89). These structures presumably arise because it is only in these aggregates that all donor atoms can participate in bonding to copper. Interesting features are the

Copper and Silver

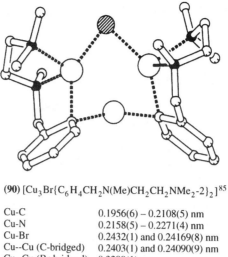

(90) $[Cu_3Br\{C_6H_4CH_2N(Me)CH_2CH_2NMe_2-2\}_2]^{85}$

Cu–C	0.1956(6) – 0.2108(5) nm
Cu–N	0.2158(5) – 0.2271(4) nm
Cu–Br	0.2432(1) and 0.24169(8) nm
Cu--Cu (C-bridged)	0.2403(1) and 0.24090(9) nm
Cu--Cu (Br-bridged)	0.3299(1) nm
Cu–C–Cu	72.6(2)°
Cu–Br–Cu	85.75(3)°

Figure 5 (continued)

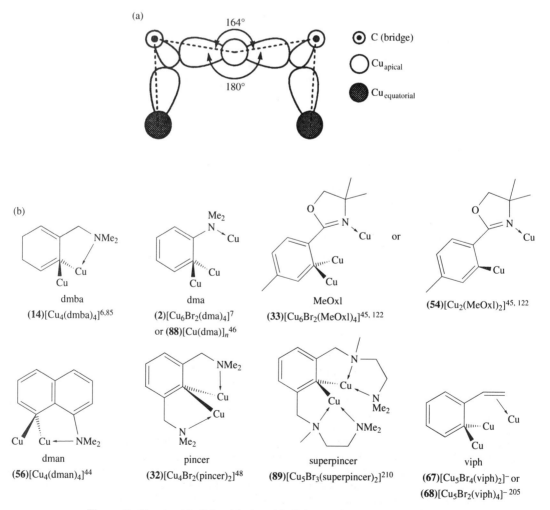

Figure 6 Structural building blocks with their associated chelate orientation.

asymmetry of the $[Cu_4Br_2R_2]$ structure, the acute angles not only at C_{ipso}, which is the expected feature of a 2e–3c-bonded C^- anion, but also at the halide atoms, for example, 71.5° in (32), and the $[Cu_2R]^+$ cationic building block in which all four nitrogen-donor atoms are well oriented for Cu–N coordination. Note that in this structure the two N(Me) centres become stereogenic when coordinated; only the *R,R*-diastereoisomer is present in the solid state.

The structure of $[Cu_3Br(C_6H_4\{CH_2N(Me)CH_2CH_2NMe_2\}-2)]$ (90) in the solid state has been obtained.[85] This structure is of interest for a number of reasons. First, it possesses an anionic cuprate $[CuR_2]^-$ and a cationic $[Cu_2Br]^+$ building block. In this latter unit the copper atoms are four coordinate due to intramolecular Cu–N coordination and Cu–Br and Cu–C interactions. The C–Cu–C angle of the two-coordinate copper centre is 156.3(1)°. Second, it reveals the flexibilty of arylcopper and copper halide units to assemble into new, stable aggregates. Third, this compound may be a model for trinuclear species which have been proposed to explain the reactivity of higher-order cuprates, $[Li_2CuR_2X]$.[104,106,212] Note that on heating quantitative biaryl formation occurs.[85]

The structure of $[Cu_3Mes(O_2CPh)_2]$ (37) (see Figure 7) is a good example of selective heteroorganic aggregate formation by self-assembly from the homoleptic aggregates $[Cu_4(O_2CPh)_4]$ and $[Cu_5Mes_5]$ (20) (Equation (40)).[39] This structure was the first example of a trinuclear copper species. Each of the copper atoms achieves an almost linear coordination by taking part either in 2e–3c bonding to C_{ipso} of the mesityl group and 2e–2c bonding with a carboxylate oxygen atom, or in two 2e–2c bonds with the carboxylate oxygen atoms. This simple structure shows (i) the expected 76.4(3)° angle made by the mesityl ring with the Cu⋯Cu vector, (ii) the bridge bonding mode of the carboxylato group using orbitals on oxygen which are oriented mutually parallel, and (iii) the short (C_{ipso}-bridged, 0.242 1(2) nm) and longer (carboxylato-bridged, 0.288 8(2) nm) Cu⋯Cu distances.

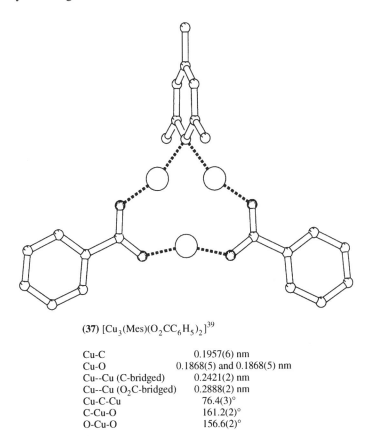

(37) $[Cu_3(Mes)(O_2CC_6H_5)_2]$[39]

Cu-C	0.1957(6) nm
Cu-O	0.1868(5) and 0.1868(5) nm
Cu--Cu (C-bridged)	0.2421(2) nm
Cu--Cu (O_2C-bridged)	0.2888(2) nm
Cu-C-Cu	76.4(3)°
C-Cu-O	161.2(2)°
O-Cu-O	156.6(2)°

Figure 7 Heteroleptic organocopper aggregates.

Aggregates $[Cu_3Mes(SC_6H_4CH_2NMe_2-2)_2(PPh_3)]$ (11) and $[Cu_4Mes_2(SC_6H_4CH_2NMe_2-2)_2]$ (17) are stable species that self-assemble when solutions of $[Cu_5Mes_5]$ (20) and $[Cu_3(SC_6H_4CH_2NMe_2-2)_3]$ are mixed in toluene, in the presence or absence of PPh$_3$, respectively (Equation (1)).[15] The structure of (17) has a $Cu_4S_2C_2$ ring in a twist-boat conformation, with adjacent two- and three-coordinate copper atoms, and aryl groups bonded to the sulfur atoms in equatorial positions. Both the mesityl and the arenethiolate groups bridge the copper atoms symmetrically via 2e–3c Cu_2C or Cu_2S interactions, respectively.

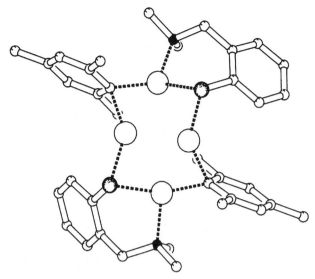

(17) $[Cu_4(Mes)_2(SC_6H_4CH_2NMe_2-2)_2]$[15]

Cu-C	0.2005(2) – 0.2022(3) nm
Cu-N	0.2130(2) and 0.2143(2) nm
Cu-S	0.21812(9) – 0.22590(9) nm
Cu--Cu (C-bridged)	0.23805(7) and 0.24075(7) nm
Cu--Cu (S-bridged)	0.26983(7) and 0.27952(7) nm
Cu---Cu	0.26563(7) and 0.4346(1) nm
Cu-C-Cu	72.54(8) and 73.22(9)°
Cu-S-Cu	74.82(2) and 78.09(2)°

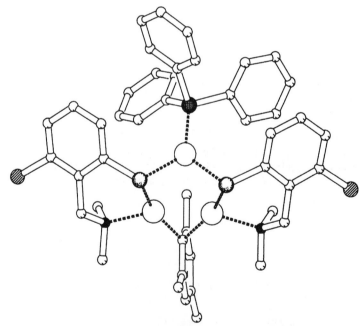

(11) $[Cu_3(Mes)(SC_6H_4CH_2NMe_2-2-Cl-3)_2(PPh_3)]$[15]

Cu-C	0.2021(4) and 0.2040(4) nm
Cu-S	0.2234(1) – 0.2280(1) nm
Cu-N	0.2150(4) and 0.2187(4) nm
Cu-P	0.2214(1) nm
Cu--Cu (C-bridged)	0.24361(8) nm
Cu--Cu (S-bridged)	0.27525(8) and 0.28056(9) nm
Cu-C-Cu	73.7(1)°
Cu-S-Cu	75.13(4) and 76.87(4)°

Figure 7 (continued)

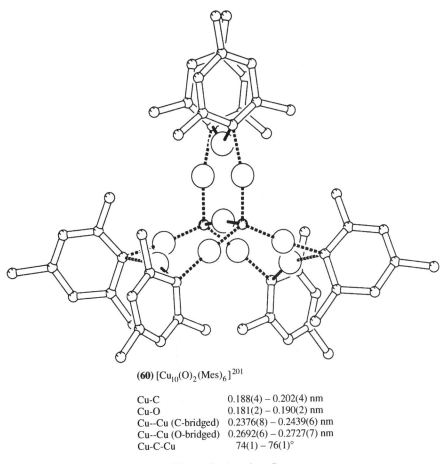

(60) [Cu$_{10}$(O)$_2$(Mes)$_6$]^{201}

Cu–C	0.188(4) – 0.202(4) nm
Cu–O	0.181(2) – 0.190(2) nm
Cu--Cu (C-bridged)	0.2376(8) – 0.2439(6) nm
Cu--Cu (O-bridged)	0.2692(6) – 0.2727(7) nm
Cu-C-Cu	74(1) – 76(1)°

Figure 7 (continued)

Addition of PPh$_3$ to a solution of (17) results in the quantitative formation of [Cu$_3$Mes(SC$_6$H$_4$CH$_2$NMe$_2$-2)(PPh$_3$)]. The 3-chloro derivative (11) of this compound is based on a six-membered Cu$_3$S$_2$C ring, in a boatlike conformation, with pairs of copper atoms bridged by either the sulfur of the arenethiolate anions or C$_{ipso}$ of the mesityl group.

A unique compound is [Cu$_{10}$Mes$_6$(O)$_2$] (60), having two oxide ligands and a central copper(I) atom situated on an approximate threefold axis. The remaining nine copper(I) centres form three wings, each with three copper atoms participating in 2e–3c bonds with two mesityl groups (Figure 7).[202] The oxide ligands are trigonal pyramidally coordinated by four Cu^I atoms.

2.3.2.3 Alkenylcopper compounds

Alkenylcopper compounds have been frequently used as *in situ*-generated organometallic reagents in organic synthesis.[1] Only for two compounds has the structure in the solid state been elucidated (see Figure 8).[12,50] These heteroleptic organocopper compounds have a structure consisting of a rhombus-type Cu$_4$ core, of which each of the two propenyl and the two other anions (C$_6$H$_4$NMe$_2$-2 in (6) or Br in (34)) occupy adjoining edges. The dimethylamino group of the propenyl ligand is coordinated to copper, whereas those of the bridging aryl ligand are uncoordinated. In this way, the copper atoms alternate between being two- and three-coordinate. Analogous to the 2e–3c bonding observed for the alkyl and aryl bridges, the two *cis*-positioned propenyl groups each bridge two copper atoms via a 2e–3c-bonded carbon atom. Interesting structural features are the planarity of the 2e–3c bridge-bonded C$_6$H$_4$NMe$_2$-2 ligand (i.e., this NMe$_2$ group does not coordinate) in (6), the extremely acute Cu–Br–Cu angle of 63.31(9)° in (34) and the planarity of the propenyl ligand of which the position with respect to the Cu$\cdots$Cu vector is essentially perpendicular. The structural features of the 2e–3c-bonded propenyl groups are very similar to those reported for the alkenyl group in [Al$_2$(CH=CHBu^t)$_2$(Bu^i)$_4$].[213]

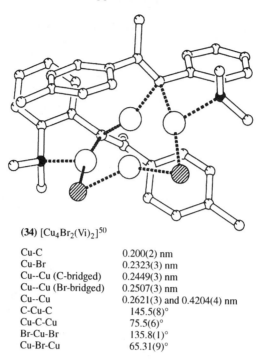

(34) [Cu$_4$Br$_2$(Vi)$_2$][50]

Cu-C	0.200(2) nm
Cu-Br	0.2323(3) nm
Cu--Cu (C-bridged)	0.2449(3) nm
Cu--Cu (Br-bridged)	0.2507(3) nm
Cu--Cu	0.2621(3) and 0.4204(4) nm
C-Cu-C	145.5(8)°
Cu-C-Cu	75.5(6)°
Br-Cu-Br	135.8(1)°
Cu-Br-Cu	65.31(9)°

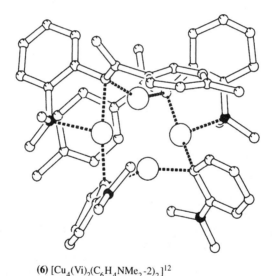

(6) [Cu$_4$(Vi)$_2$(C$_6$H$_4$NMe$_2$-2)$_2$][12]

Cu-C	0.198(3) – 0.216(3) nm
Cu-N	0.226(4) and 0.233(4) nm
Cu-C$_{Ar}$	0.198(4) – 0.207(4) nm
Cu--Cu (C$_{alk}$-bridged)	0.2426(6) and 0.2451(6) nm
Cu--Cu (C$_{ar}$-bridged)	0.2474(6) and 0.2475(6) nm
C$_{alk}$-Cu-C$_{alk}$	149(1)°
C$_{ar}$-Cu-C$_{ar}$	157(2)°
Cu-C$_{alk}$-Cu	70(1) and 75(1)°
Cu-C$_{ar}$-Cu	74(2) and 77(1)°

Figure 8   Structure of two heteroleptic alkenylcopper compounds.

2.3.2.4 Homo- and heteroleptic alkynylcopper and -silver compounds

In Figure 9, the structures of some neutral homo- and heteroleptic alkynylcopper compounds are shown. In Sections 2.3.2.5(ii) and 2.3.2.6, anionic alkynylcuprates and -argentates and cationic alkynyl complexes are discussed. Alkynylcopper and -silver compounds are generally almost insoluble in noncoordinating solvents. This suggests that they have polymeric structures. X-ray analysis of [CuC≡CPh]$_n$ reveals infinite zig-zag chains of copper atoms lying roughly in one plane. The phenylethynyl group seems to be σ-bonded to one copper atom, and symmetrically π-bonded to a second and a third copper atom.[54] However, the structure may be explained equally well in terms of the alkynyl group being 2e–3c bonded with Cu-1 and Cu-2 and π-bonded with Cu-3 (see encircled part in Figure 9). The Cu···Cu distances of 0.242 nm and 0.247 nm seem to be in accord with this view (see Section 2.3.4.1).

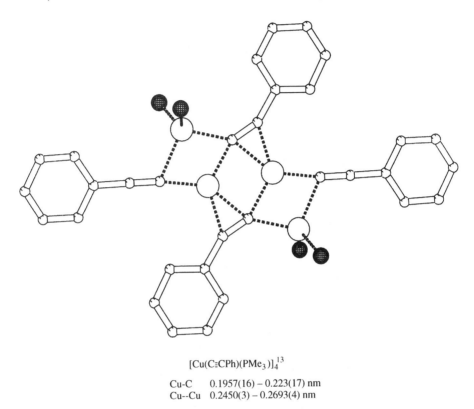

[Cu(C≡CPh)(PMe$_3$)]$_4$[13]

Cu–C   0.1957(16) – 0.223(17) nm
Cu--Cu   0.2450(3) – 0.2693(4) nm

Figure 9 Structures of alkynylcopper compounds (Me groups of PMe$_3$ in (**61**) omitted).

Three alkynylcopper structures are known with [Cu(C≡CPh)L] stoichiometry. They are all tetranuclear (i.e., [Cu$_4$(C≡CPh)$_4$L$_4$]). Depending on the nature of L, they adopt structures in which the copper atoms are arranged either in a zig-zag chain (L = PMe$_3$)[13] (see Figure 9), or in a tetrahedral core (L = PPh$_3$,[55] and PPh$_2$(2-pyridyl)[53]). Two distinct Cu···Cu distances of 0.2450 nm and 0.2693 nm are present in [Cu$_4$(C≡CPh)$_4$(PMe$_3$)$_4$], of which the shorter distance appears to be bridged by a 2e–3c-bonded phenylethynyl group and the longer one belongs to an unbridged copper atom pair. Two of the four ethynyl groups are both 2e–3c bonded and π-bonded.

The structure of [Cu$_4$(C≡CPh)$_4$(PPh$_3$)$_4$] (**61**) consists of an essentially tetrahedral Cu$_4$ skeleton bearing four terminally bonded PPh$_3$ molecules and four μ_3-η^1-bonded phenylacetylide ligands.[55] The copper acetylide bridges are markedly asymmetric, with Cu–C contacts in the range 0.2072(4)–0.2380(4) nm. The C≡C bond lengths (0.115 4(6) nm and 0.119 3(6) nm) are typical for triple bonds, and are consistent with a purely σ-type acetylide–metal interaction. Accordingly, the bonding in the electron-deficient Cu$_3$C≡CPh moiety can be described as typically 2e–4c (see Section 2.3.4.1). Similar structural features are found in [Cu$_4$(C≡CPh)$_4$(PPh$_2$(2-pyridyl))$_4$].[53]

A distinctly different structure[19] has been found for [Ag(C≡CPh)(PMe$_3$)]$_n$ (**76**).[73] In the solid state this compound consists of almost straight chains of silver atoms. The silver atoms in the chain alternate between being bonded to two σ-phenylethynyl groups, or two phosphine ligands with two additional weak π-ethynyl interactions.

An example of a heteroleptic organocopper compound, containing both an aryl and an acetylide ligand bonded to the same copper array, is [Cu$_6$(C≡CC$_6$H$_4$Me-4)$_2$(C$_6$H$_4$NMe$_2$-2)$_4$] (**35**). The structure of this compound was discussed in Section 2.3.2.2(iii), but it is of interest here to recall that in this compound the ethynyl–Cu$_2$ interaction is purely of the μ_2-η^1 type, that is, it is a typical 2e–3c bond.[58,59,123]

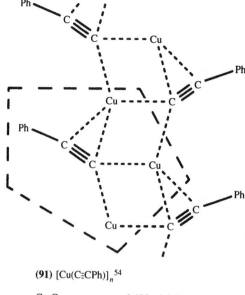

(91) $[Cu(C{\equiv}CPh)]_n{}^{54}$

Cu–C	0.198 – 0.213 nm
Cu--Cu (C-bridged)	0.242 and 0.247 nm

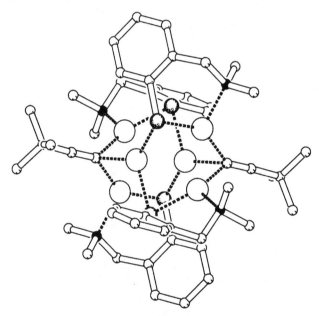

(16) $[Cu_3(C{\equiv}CBu^t)(SC_6H_4CH_2NMe_2\text{-}2)_2]_2{}^{51,\,52}$

Cu–C	0.196(1) nm
Cu'–C	0.213(1) nm
Cu--Cu (C-bridged)	0.2458(2) nm
Cu--Cu' (C-bridged)	0.2555(2) nm
Cu--Cu (S-bridged)	0.2841(2) and 0.3026(2) nm
Cu–C–Cu	77.7(5)°
Cu–C–Cu'	77.4(5) and 78.4(5)°
Cu–S–Cu'	77.1(1) and 84.4(1)°

Figure 9   (continued)

Two further examples of heteroleptic alkynylcopper compounds are $[Cu_3(C{\equiv}CBu^t)$-$\{SC_6H_4(CH(R)NMe_2)\text{-}2\}_2]_2$, in which R = H (**16**) or Me (**92**).[51,52] These compounds have been obtained by assembly from the trimeric copper arenethiolate $[Cu_3\{SC_6H_4(CH(R)NMe_2)\text{-}2\}_3]$ and $[CuC{\equiv}CBu^t]$. Each structure contains two trinuclear $Cu_3(C{\equiv}CBu^t)\{SC_6H_4(CH(R)NMe_2)\text{-}2\}_2$ units that have Cu_3S_2C

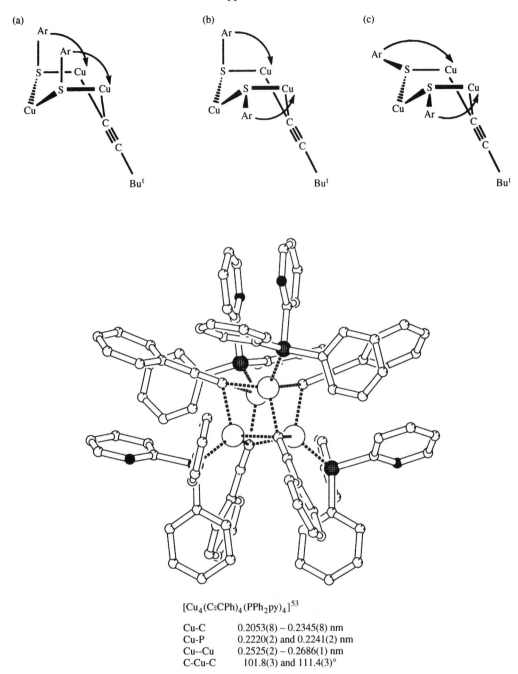

(a)   (b)   (c)

$[Cu_4(C{\equiv}CPh)_4(PPh_2py)_4]$[53]

Cu–C	$0.2053(8) - 0.2345(8)$ nm
Cu–P	$0.2220(2)$ and $0.2241(2)$ nm
Cu--Cu	$0.2525(2) - 0.2686(1)$ nm
C–Cu–C	$101.8(3)$ and $111.4(3)°$

Figure 9 (continued)

six-membered rings, in a boatlike conformation. The three possible conformers, (a)–(c), which are shown in Figure 9, are based on axial or equatorial positioning of the aryl groups on the sulfur atoms with respect to the Cu_3S_2C ring. Note that the acetylide-bridged copper atoms are three-coordinate, while the copper atom bonded to the bridging sulfur atoms is coordinatively unsaturated (see below). In both trinuclear units of $[Cu_3(C{\equiv}CBu^t)\{SC_6H_4(CH_2NMe_2)\text{-}2\}_2]_2$ (**16**), one of the aryl groups is equatorial and the other axial (conformer (b)). In the chiral compound $[Cu_3(C{\equiv}CBu^t)\{SC_6H_4(CH(Me)NMe_2)\text{-}2\}_2]_2$ (**92**), one trinuclear unit is like that in the achiral compound (**16**), but in the other unit all aryl groups are axially positioned (conformer (a)). The third possibility of two equatorially positioned aryl groups, conformer (c), is not found in these heteroleptic acetylide compounds, but has been observed in $[Cu_3Mes\{SC_6H_3(CH_2NMe_2)\text{-}2\text{-}Cl\text{-}3\}_2(PPh_3)]$ (**11**) (see Section 2.3.2.2(iii), Figure 7).[15] In this case, through side-on coordination of the bridging alkynyl group of one unit to a coordinatively unsaturated copper atom of a second Cu_3S_2C unit, hexanuclear structures are formed. Consequently, the alkynyl group is asymmetrically bridging, and its bonding mode is of the $(\mu_3\text{-}C_\alpha, \mu_1\text{-}C_\beta)\text{-}\eta^2$-type.

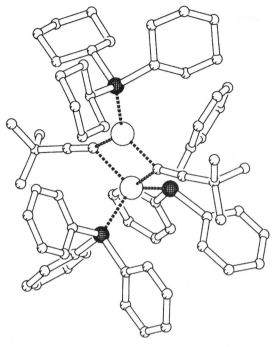

(23) $[Cu_2(C\equiv CBu^t)_2(PCy_3)(PPh_3)_2]^{100}$

(P)Cu-C	0.2001(9) and 0.2053(9) nm
(P_2)Cu-C	0.2142(9) and 0.2203(9) nm
Cu--Cu	0.23892(14) nm
Cu-P(Cy)	0.2224(2) nm
Cu-P(Ph)	0.2272(2) and 0.2273(3) nm
C-Cu(P)-C	116.4(3)°
C-Cu(P_2)-C	105.0(3)°
Cu(P)-C-Cu(P_2)	69.1(3) and 69.4(3)°

Figure 9 (continued)

The planar structure of $[Cu_2(C\equiv CBu^t)_2(PCy_3)(PPh_3)_2]$ (**23**) consists of a dinuclear Cu_2C_2 core.[100] The $C\equiv CBu^t$ groups bridge the copper atoms in a 2e–3c manner (the Cu–C–Cu angle is 69.2°), which results in a very short Cu···Cu distance of 0.239 8(1) nm. One of the copper atoms is three-coordinate due to coordination of the most bulky phosphine (PCy_3), while the second copper atom is four-coordinate due to coordination of two less bulky phosphines (PPh_3). In fact, (**23**) can be seen as a combination of a three-coordinate cuprate species $[Cu(C\equiv CBu^t)_2(PCy_3)]^-$ complexed to a $[Cu(PPh_3)_2]^+$ cation. Both alkynyl groups asymmetrically bridge the copper atoms, and are tilted in the direction of the $[Cu(PPh_3)_2]$ unit.

2.3.2.5 *Organocuprates and -argentates*

Organocopper and -silver compounds have a high tendency to form complexes with other organometallic species. These complexes can be either neutral, having the general formula $M^1_xM^2_yR_{x+y}$, or monoanionic species comprising mononuclear anions $[M^2R_2]^-$, polynuclear anions $[M^2_yR_{y+1}]^-$ and $[M^1_xM^2_yR_{x+y+1}]^-$. This distinction is not only useful from a structural point of view, but also with respect to the mechanistic discussions of the reactivity of cuprates in organic synthesis. It is further of interest to note that the first structure of a cuprate dates back to 1982[14c] and that of an argentate to 1983.[71] Before that time only the solution-state structures of the neutral cuprate and argentate $[Li_2M_2(C_6H_4CH_2NMe_2-2)_4]$ (M = Cu (**3**), Ag (**4**)) were known in detail (Section 2.3.4.2(i)). Some aspects of the structures of heterocuprates and organocuprates have been reviewed.[214]

(i) Neutral homo- and heteroleptic cuprates

Only the structures of a limited number of organocuprates have been reported. The fact that similar characterization of neutral argentates is still lacking is a reflection of the synthetic problems involved. However, a further reason may be that in a number of cases the structure of neutral argentates could be determined in solution from ^{107,109}Ag, ^{13}C and ^{1}H NMR data.

For the most part, the structurally characterized neutral organocuprates consist of a combination of neutral units of an organocopper with an organolithium or a diorganomagnesium species. Two further examples are known in which the organocopper species is found in combination with a lithium ($[Li_4Cl_2(Et_2O)_{10}]^{2+}[Li_2Cu_3Ph_6]^-{}_2$ (**8**)[14d]) or a magnesium salt ($[Cu_4Mes_4][Mg_2(SC_6H_4CH_2NMe_2-2)_4]$ (**31**)[15]). It should be noted that the above formulations used to describe the composition of these neutral cuprates is not necessarily related to the way they have been synthesized. In Figure 10 the structures of three arylcuprates and one alkylcuprate are shown. A common feature is the *trans*-metal Li_2Cu_2 arrangement, which may be planar or slightly folded at the copper atoms. The C_{ipso} atoms of the organic groups each bridge in an asymmetric fashion between one copper and one lithium atom. In all four cases the coordination at the copper centres is close to linear with C–Cu–C angles of about 170°. In $[Li_2Cu_2(C_6H_4CH_2NMe_2-2)_4]$ (**3**),[10] tetrahedral geometry at each lithium centre is achieved with two intramolecular Li–N coordinative bonds. In $[Li_2Cu_2Ph_4(DMS)_3]$, (**26**),[16] the coordination of the three DMS molecules results in one trigonal and one tetrahedral lithium, while in $[Li_2Cu_2Ph_4(Et_2O)_2]$ (**25**),[65] coordination at each lithium atom by diethyl ether gives rise to two trigonal lithium centres. Note that $[Li_2Cu_2(CH_2TMS)_4(DMS)_2]$ (**24**) has a polymeric structure in the solid state, as the dimeric units are connected by DMS molecules which bridge lithium atoms of neighbouring dimers.[102] In (**3**) the *ortho*-substituents are alternately oriented above and below the Cu···Li axes of the Li_2Cu_2 butterfly arrangement. When the *ortho*-substituent bears a chiral benzylic carbon atom, as is the case in $[Li_2Cu_2(C_6H_4CH(Me)NMe_2-2)_4]$, this leads to diastereoisomers which have been studied by NMR spectroscopy (see Section 2.3.4.2(ii)).

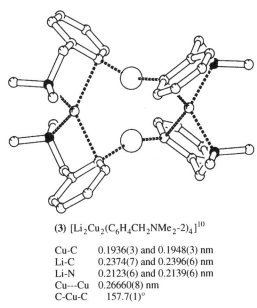

(**3**) $[Li_2Cu_2(C_6H_4CH_2NMe_2-2)_4]^{10}$

Cu–C	0.1936(3) and 0.1948(3) nm
Li–C	0.2374(7) and 0.2396(6) nm
Li–N	0.2123(6) and 0.2139(6) nm
Cu---Cu	0.26660(8) nm
C–Cu–C	157.7(1)°

Figure 10 Structures of neutral homoleptic cuprates with $[Li_2Cu_2R_4]$ stoichiometry.

The asymmetric bridging of the aryl groups between lithium and copper will be discussed in Section 2.3.4.1. The comparison of the above-mentioned cuprate structures reveals close relationships; all skeletal frameworks can be derived from the combination of $[CuR_2]^-$ monoanions and $[LiL_x]^+$ cations. In the structure of $[Li_2Au_2(C_6H_4CH_2NMe_2-2)_4]$ (**93**), which is related to that of (**3**), the coordination at the gold centres is strictly linear, indicating 2e–2c Au–C bonding, and the $[AuR_2]^-$ and $[LiL_x]^+$ anionic–cationic units are arranged in a square fashion with additional π-coordination of C_{ipso} at the lithium centres.[234]

So-called neutral higher-order cuprates have been obtained by reaction of phenyllithium with various under-stoichiometric amounts of copper bromide in DMS. The characterized species have $[Li_3Cu_2Ph_5(DMS)_4]$ (**28**) and $[Li_5Cu_4Ph_9(DMS)_4]$ (**59**) overall stoichiometry. However, representation of the compounds as $[Li_3(CuPh_2)(CuPh_3)(DMS)_4]$ and $[Li_5(CuPh_2)_3(CuPh_3)(DMS)_4]$, respectively, reveals in more detail the intriguing composition of these neutral aggregates (see Figure 11).[16]

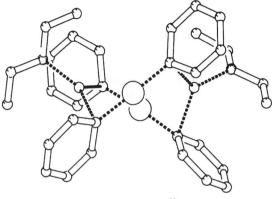

(25) [Li$_2$Cu$_2$Ph$_4$(Et$_2$O)$_2$][65]

Cu-C	0.1906(5) – 0.1928(7) nm
Li-C	0.2213(12) – 0.2264(10) nm
Cu---Cu	0.2865(2) nm
C-Cu-C	167.7(3) and 168.0(2)°

twist

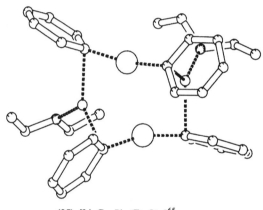

(25) [Li$_2$Cu$_2$Ph$_4$(Et$_2$O)$_2$][65]

Cu-C	0.1912(6) – 0.928(7) nm
Li-C	0.2211(12) – 0.2282(13) nm
Cu---Cu	0.3075(3) nm
C-Cu-C	163.3(3) and 165.2(3)°

Figure 10   (continued)

The structure of [Li$_3$(CuPh$_2$)(CuPh$_3$)(DMS)$_4$] (**28**) can be considered as a cation–anion association of [Li$_3$(CuPh$_3$)(DMS)$_4$]$^+$ and [CuPh$_2$]$^-$. The [CuPh$_3$]$^{2-}$ moiety has an average Cu-C distance of 0.202 nm and the sum of the angles at the copper atom is 357.1° with the copper atom 0.0202 nm above the CuC$_3$ plane. The phenyl rings are almost perpendicular to this plane. Two of the lithium cations bridge two *ipso* carbon atoms of the [CuPh$_3$]$^{2-}$ moiety and an *ipso* carbon atom of the [CuPh$_2$]$^-$ moiety, while the third lithium ion bridges exclusively two *ipso* carbon atoms of the [CuPh$_3$]$^{2-}$ moiety. The structural details of the [CuPh$_2$]$^-$ unit are similar to those encountered in the single species.

The aggregate [Li$_5$(CuPh$_2$)$_3$(CuPh$_3$)(DMS)$_4$] (**53**) consists of three linear [CuPh$_2$]$^-$ ions, which are triply bridged by two lithium cations, and a trigonal [CuPh$_3$]$^{2-}$ ion, which is associated with three lithium cations and coordinated by four DMS ligands. The two resulting units, [Li$_2$(CuPh$_2$)$_3$]$^-$ and [Li$_3$(CuPh$_3$)(DMS)$_4$]$^+$, are linked to each other primarily via a bridging phenyl *ipso* carbon atom. Each of the three lithium atoms in the [Li$_3$(CuPh$_3$)(DMS)$_4$]$^+$ unit approaches the bridging lithium atoms in [Li$_2$(CuPh$_2$)$_3$]$^-$ from the side. Consequently, the planes of the phenyl rings are approximately perpendicular to the Li$_3$Cu plane.

The structure of [MgCu$_4$Ph$_6$(Et$_2$O)] (**58**) is of interest as it represents one of the few cuprate aggregates containing copper and magnesium.[66] Compound (**58**) has much in common with the

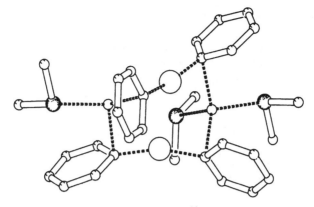

(26) $[Li_2Cu_2Ph_4(DMS)_3]$[16]

Cu-C	0.1934(3) – 0.1940(2) nm
Li-C	0.2267(5) – 0.2302(5) nm
Li-S	0.2524(5) – 0.2552(5) nm
Cu---Cu	0.2869(1) nm
C-Cu-C	162.2(1) and 163.5°
Cu-C-Li	79.6(2) – 84.8(1)°

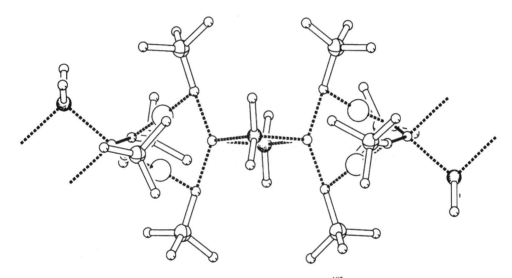

(24) $[Li_2Cu_2(CH_2TMS)_4(DMS)_2]_n$[102]

Cu-C	0.1949(5) – 0.1961(6) nm
Li-C	0.2198(12) – 0.2218(13) nm
Li-S	0.2654(9) – 0.2709(9) nm
Cu--Cu	0.2984(1) nm
Cu--Li	0.2618(9) – 0.2648(9) nm
C-Cu-C	171.5(3) and 173.8(3)°

Figure 10   (continued)

structural features of monoanionic $[M_5Ph_6]^-$ aggregates; for example, $[LiCu_4Ph_6]^-$ **(69)** and $[MgCu_4Ph_6(Et_2O)]$ **(58)** have the same basic trigonal-bipyramidal core geometry. The six phenyl groups bridge the M(axial)–M(equatorial) bonds in a perpendicular fashion. Complex **(58)** is the only member of this family of $[M_5Ph_6]$ aggregates to contain a coordinated solvent molecule.

Neutral heterocuprates LiCuRX (X = heteroatom ligand) play an important role in organic synthesis because they avoid waste of the sometimes valuable R groups, which is inherent in the use of homocuprates LiCuR₂. The structure of the phosphidocuprate $[LiCu(PBu^t_2)Me(THF)_3]$ **(51)** is shown in Figure 12. This species is monomeric, probably because of the bulky Bu^t groups. The copper atom exhibits linear geometry, and the Cu–C distance is similar to those found in $[CuMe_2]^-$. In contrast to other cuprates, in which lithium is involved in 2e–3c bonding with the organic group, the lithium atom in **(51)** is bonded to the heteroatom (P) and three THF molecules.[63]

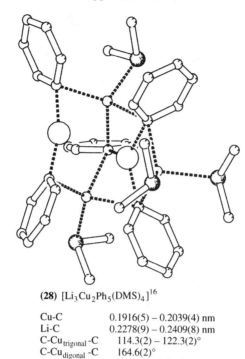

(28) [Li$_3$Cu$_2$Ph$_5$(DMS)$_4$][16]

Cu-C	0.1916(5) – 0.2039(4) nm
Li-C	0.2278(9) – 0.2409(8) nm
C-Cu$_{trigonal}$-C	114.3(2) – 122.3(2)°
C-Cu$_{digonal}$-C	164.6(2)°

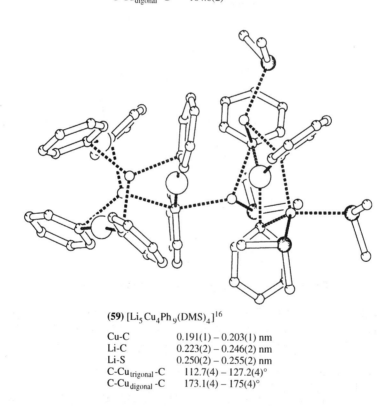

(59) [Li$_5$Cu$_4$Ph$_9$(DMS)$_4$][16]

Cu-C	0.191(1) – 0.203(1) nm
Li-C	0.223(2) – 0.246(2) nm
Li-S	0.250(2) – 0.255(2) nm
C-Cu$_{trigonal}$-C	112.7(4) – 127.2(4)°
C-Cu$_{digonal}$-C	173.1(4) – 175(4)°

Figure 11 Structures of neutral, homoleptic cuprates.

Another neutral heterocuprate is the arylcopper magnesium arenethiolate [Cu$_4$Mes$_4$][Mg(SC$_6$H$_4$CH-(Me)NMe$_2$-2)$_2$]$_2$ (**31**), which self-assembles in solutions of [Cu$_3$(SC$_6$H$_4$CH(Me)NMe$_2$-2)$_3$] with MgMes$_2$. The structure in the solid state shows that during its formation a remarkable ligand exchange occurs, forming a tetranuclear central organocopper unit, to which [Mg(SC$_6$H$_4$CH(Me)-NMe$_2$-2)$_2$] moieties are bonded at copper via sulfur-donor atoms. The overall structure of the central tetranuclear arylcopper unit is comparable to that found in [Cu$_4$(Mes)$_4$(THT)$_2$] (**57**) (THT = tetrahydrothiophene) and the *o*-Tol

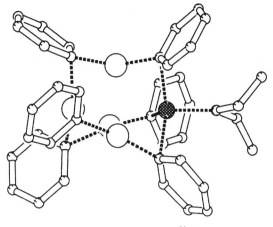

(58) [MgCu$_4$Ph$_6$(OEt$_2$)][66]

Cu$_{ax}$-C$_{av}$	0.209(1) nm
Cu$_{eq}$-C$_{av}$	0.195(1) nm
Cu$_{eq}$--Cu$_{eq,av}$	0.3019(3) nm
Cu$_{eq}$--Cu$_{eq,av}$	0.2427(2) nm
C-Cu$_{eq}$-C$_{av}$	160.9(4)°
C-Cu$_{ax}$-C$_{av}$	119.7(4)°
C-Mg-C$_{av}$	117.1(4)°

Figure 11 (continued)

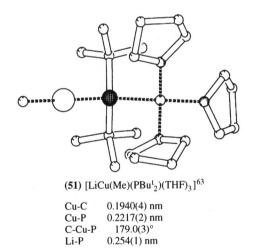

(51) [LiCu(Me)(PBu$^t_2$)(THF)$_3$][63]

Cu-C	0.1940(4) nm
Cu-P	0.2217(2) nm
C-Cu-P	179.0(3)°
Li-P	0.254(1) nm

Figure 12 Structures of neutral heteroleptic cuprates.

compound [Cu$_4$(*o*-Tol)$_4$(DMS)$_2$] **(13)**[17,201] with two two- and two three-coordinate copper atoms. The arenethiolate ligands are *N,S*-chelate-bonded to the magnesium atoms. For the sulfur atom bridging between copper and magnesium, there are two extreme overall bonding descriptions possible. In one description (i) there is a tetranuclear organocopper unit, [Cu$_4$Mes$_4$], to which two [Mg{SC$_6$H$_4$CH(Me)-NMe$_2$-2}$_2$] units are connected through thiolate S–Cu dative coordination (cf. the THT and DMS coordination in the homoleptic arylcopper compounds). The second description (ii) is based on a [Cu$_4$(Mes)$_4$(SC$_6$H$_4$CH(Me)NMe$_2$-2)$_2$]$^{2-}$ anion and two [Mg{SC$_6$H$_4$CH(Me)NMe$_2$-2}]$^+$ cations, the latter being bonded to a thiolato sulfur atom of the cuprate by a S–Mg bond.

(ii) Anionic, mononuclear homo- and heteroleptic cuprates and argentates

The first structural determination of a mononuclear diorganocuprate was that of [Cu(Mes)$_2$][Cu(dppe)$_2$] **(30)** (see Figure 13). This compound resulted from a disproportionation reaction of [Cu$_5$Mes$_5$] with dppe. The [Cu(Mes)$_2$]$^-$ anion exhibits linear coordination at copper and two electron-precise 2e–2c Cu–C bonds.[49]

Copper and Silver

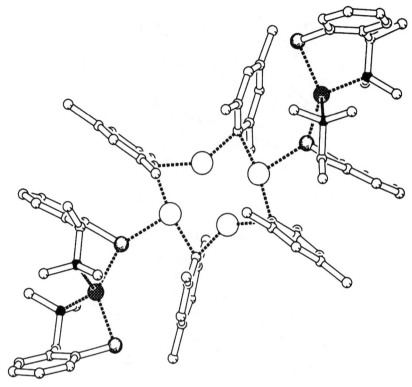

(31) [Cu$_4$(Mes)$_4$][Mg(SC$_6$H$_4$CH(Me)NMe$_2$-2)$_2$]^{15}

Cu-C	0.1981(8) – 0.2064(8) nm
Cu-S	0.2389(2) nm
Mg-S (bridge)	0.2427(4) nm
Cu--Cu	0.2431(1) and 0.2464(1) nm
Cu---Cu	0.2700(1) and 0.4079(1) nm
C-Cu-C	142.8(3) and 168.1°
Cu-C-Cu	73.4(3) and 75.0(3)°
Mg-S-Cu	136.4(1)°

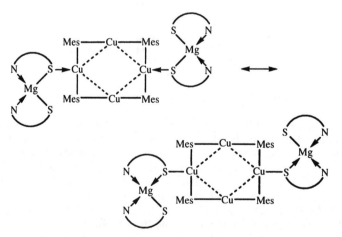

Figure 12 (continued)

Further examples have since been obtained by using hindered substituents as in [Cu{C(TMS)$_3$}$_2$][Li(THF)$_4$] (**9**), which was the first example of an alkylcuprate,[204] or by using 12-crown-4 ethers to stabilize a lithium counter-cation. In this way, well-separated cuprate species are formed. Examples of the latter approach are [CuMe$_2$]$^-$, [CuPh$_2$]$^-$ (**64**) and most interestingly the first example of a heterocuprate, [CuBr(CH{TMS}$_2$)]$^-$ (**63**).[64] The use of hindered alkyl anions also led to the dialkylargentate species, [Ag(C{TMS}$_3$)$_2$][Li(THF)$_4$].[14b] The argentate anion is linear and the C(TMS)$_3$

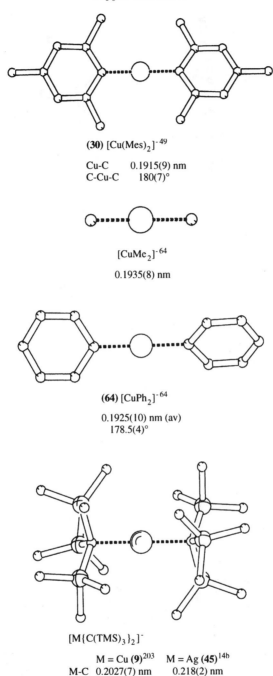

(**30**) [Cu(Mes)$_2$]$^-$ [49]

Cu-C	0.1915(9) nm
C-Cu-C	180(7)°

[CuMe$_2$]$^-$ [64]

0.1935(8) nm

(**64**) [CuPh$_2$]$^-$ [64]

0.1925(10) nm (av)
178.5(4)°

[M{C(TMS)$_3$}$_2$]$^-$

M = Cu (**9**)[203]   M = Ag (**45**)[14b]
M-C 0.2027(7) nm   0.218(2) nm

Figure 13 Structures of anionic mononuclear homo- and heteroleptic cuprates and argentates.

groups are staggered about the C–M–C axis. The perfluoroalkylargentate anion [Ag(CF{CF$_3$}$_2$)$_2$]$^-$, with a [Rh(dppe)$_2$]$^+$ cation, was sufficiently light stable to permit an x-ray structure determination. The overall geometry exhibits a linear C–Ag–C arrangement with the two isopropyl groups oriented about the C–Ag–C axis such that the C–F bonds are nearly perpendicular.[70] A special case is the structure of the bis(β-disulfone carbanion)argentate complex (**41**), which was the first example of an sp^3-hybridized carbon bonded to silver in a largely covalent, linear bond.[71]

The above examples suggest that all mononuclear cuprate and argentate compounds have [MR$_2$]$^-$ stoichiometry and are linear. However, for copper salt species a variety of anions with [CuX$_2$]$^-$, [CuX$_3$]$^{2-}$ and [CuX$_4$]$^{3-}$ stoichiometry (X = halide) are known. Also, in polynuclear cuprates and arylcopper copper halide aggregates, anions of this type have been suggested as important building blocks. Species of the

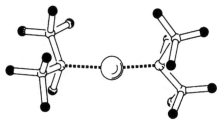

(43) $[Ag\{CF(CF_3)_2\}_2]^{-}$[70]

Ag-C   0.2015(35) and 0.2191(17) nm
C-Ag-C   170(1)°

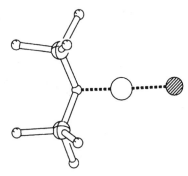

(63) $[CuBr\{CH(TMS)_2\}]^{-}$[64]

Cu-C   0.1920(6) nm
Cu-Br   0.2267(2) nm
Br-Cu-C   178.7(2)°

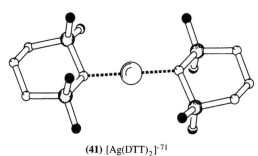

(41) $[Ag(DTT)_2]^{-}$[71]

Ag-C   0.2147 nm
C-Ag-C   180°

DTT =

Figure 13   (continued)

type $[CuR_2L]^{-}$ were considered unlikely because of the anionic charge of the $[CuR_2]^{-}$ group. However, by using the appropriate alkyl fragment a $[CuR_2L]^{-}$ anion has been structurally characterized, namely $[Cu(fluorenyl)_2(PPh_3)]^{-}$ **(22)**.[100]

(iii) Anionic, polynuclear homo- and heteroleptic cuprates

The greater part of the compounds that have been structurally characterized are monoanionic species and have $[M_5R_6]^{-}$ stoichiometry. In Figure 14 their molecular geometries are shown. $[Cu_5Ph_6]^{-}$ **(10)** has

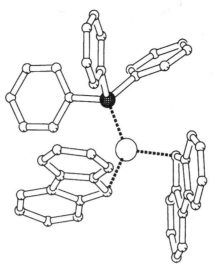

(22) [Cu(fluorenyl)$_2$(PPh$_3$)]$^-$ [100]

Cu–C	0.2079(9) and 0.2170(9) nm
Cu–P	0.2223(3) nm
C–Cu–P	112.6(3) and 134.7(3)°
C–Cu–C	112.7(4)°

Figure 13 (continued)

a pentanuclear copper skeleton that can be described as a trigonal bipyramid.[14c] The Cu(ax)–Cu(eq) distances (mean 0.244 9(9) nm) are much shorter than the Cu(eq)–Cu(eq) distances (mean 0.315 1(6) nm). The six phenyl groups bridge the Cu(ax)–Cu(eq) edges of the trigonal bipyramid. The three equatorial copper atoms each participate in two 2e–3c bonds, corroborated by the C–Cu–C angles of around 167°. The two axial copper atoms are roughly trigonal planar, suggesting the presence of [CuPh$_3$]$^{2-}$ units linked together by the equatorial copper atoms via C$_{ipso}$–Cu interactions. An alternative way of visualizing the bonding in this species is to consider it as an aggregate of three linear Ph–Cu–Ph$^-$ units held together by two copper(I) cations. This latter view seems to be supported by the bonding in the neutral cuprate and aurate, [Li$_2$M$_2$(C$_6$H$_4$CH$_2$NMe$_2$-2)$_4$] (M = Cu (**3**), Au (**93**)), which consist of two [MAryl$_2$]$^-$ anions held together by [LiL$_x$]$^+$ cations.[10,215]

Subsequently, the other structures can be rationalized by formal substitution of copper for lithium atoms. Comparison of (**10**), [LiCu$_4$Ph$_6$]$^-$ (**69**),[66] [Li$_2$Cu$_3$Ph$_6$]$^-$ (**8**)[14a] and [Li$_2$Ag$_3$Ph$_6$]$^-$ (**78**)[77] reveals that this substitution is highly specific, in each step one of the axial copper atoms being replaced by lithium. This may result from the preference of copper(I) for linear coordination, which is the geometry at the equatorial copper sites. Consequently, these structures consist of [CuPh$_2$]$^-$ anions held together by either copper or lithium cations.

Similar structures can be expected for alkenyl and alkynylcuprates and argentates. For example, [Au$_3$Cu$_2$(C≡CPh)$_6$]$^-$ is known[67,140] and [Ag$_6$Cu$_7$(C≡CPh)$_{14}$]$^-$ (**79**) has been structurally characterized. The latter constitutes a central linearly coordinated copper atom linked through Cu–Ag interactions to three tetranuclear [Ag$_2$Cu$_2$(C≡CPh)$_4$] subclusters. Within each unit of the subclusters, copper and silver atoms are bonded via bridging phenylethynyl ligands.[67,140]

The presence of coordinating substituents in the *ortho*-position of arylcuprates can influence the relative stability of different aggregates and favour the formation of a specific type of aggregate. An example is the [Cu$_5$(C$_6$H$_4$CH=CH$_2$-2)$_2$Br$_4$]$^-$ (**67**) anion.[206] The structure consists of a square pyramidal copper core, in which the four basal copper atoms are bridged by four bromine atoms, thus forming an eight-membered ring (see Figure 15). The apical copper atom is bridged via 2e–3c bonds of the C$_6$H$_4$CH=CH$_2$-2 ligands to two of the basal copper atoms, while the remaining two basal copper atoms are π-coordinated to the *o*-vinyl groups. The basic structural motif is the binding of the C$_6$H$_4$CH=CH$_2$-2 ligand to a face of three copper atoms (see (**67**) in Figure 6(b)). Related to this cuprate aggregate is the [Cu$_5$(C$_6$H$_4$CH=CH$_2$-2)$_4$Br$_2$]$^-$ (**68**) anion, the bonding in which is complex and difficult to rationalize.[206] One C$_6$H$_4$CH=CH$_2$-2 ligand binds to a trigonal face of three copper atoms, and in addition there are two-, three- and four-coordinate copper atoms and both terminal and bridging bromine atoms.

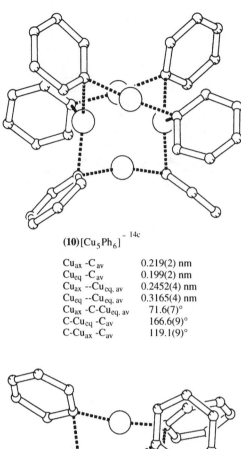

(10) $[Cu_5Ph_6]^-$ [14c]

Cu_{ax} -C_{av}	0.219(2) nm
Cu_{eq} -C_{av}	0.199(2) nm
Cu_{ax} --$Cu_{eq,\,av}$	0.2452(4) nm
Cu_{eq} --$Cu_{eq,\,av}$	0.3165(4) nm
Cu_{ax} -C-$Cu_{eq,\,av}$	71.6(7)°
C-Cu_{eq} -C_{av}	166.6(9)°
C-Cu_{ax} -C_{av}	119.1(9)°

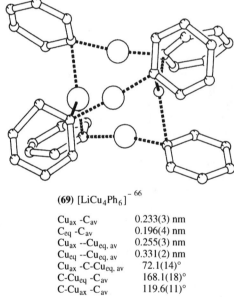

(69) $[LiCu_4Ph_6]^-$ [66]

Cu_{ax} -C_{av}	0.233(3) nm
C_{eq} -C_{av}	0.196(4) nm
Cu_{ax} --$Cu_{eq,\,av}$	0.255(3) nm
Cu_{eq} --$Cu_{eq,\,av}$	0.331(2) nm
Cu_{ax} -C-$Cu_{eq,\,av}$	72.1(14)°
C-Cu_{eq} -C_{av}	168.1(18)°
C-Cu_{ax} -C_{av}	119.6(11)°

Figure 14 Structures of polynuclear homoleptic cuprates and argentates with $[M_5R_6]^-$, $[MR_4]^{3-}$ and $[MR_2]^-$ stoichiometry.

2.3.2.6 *Ligand stabilized neutral and cationic organocopper and -silver compounds*

The aggregated structures of organocopper compounds such as methylcopper,[117] phenylcopper[77] and 2-, 3- and 4-tolylcopper[49] break down on reaction with amines or phosphines. Two neutral mononuclear copper complexes have been structurally characterized (see Figure 16). In [CuMe(PPh$_3$)$_3$] **(19)**[33] the copper atom is surrounded by the methyl and the three phosphine ligands with P–Cu–P and C–Cu–P angles which are remarkably close to the theoretical value for a tetrahedral angle. With triphos a mononuclear phenylcopper complex [CuPh(MeC{CH$_2$PPh$_2$}$_3$)] **(35)** was isolated.[36]

In this structure, the geometry around copper is severely distorted with large deviations from ideal tetrahedral bond angles (<P–Cu–P = 90.0(1)–93.3(1)°, <P–Cu–C = 119.4(1)–126.9(1)°). The complex

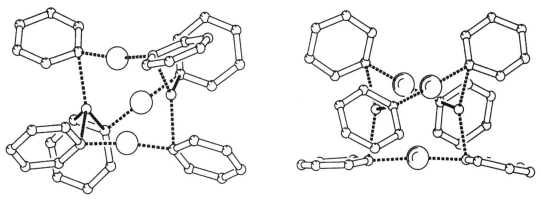

$[Li_2M_3Ph_6]^-$

	M = Cu (8) [14a]	M = Ag (78) [77]
M-C	0.1923(6) – 0.1942(6) nm	0.213(1) nm (av)
Li-C	0.2240(14) – 0.2257(13) nm	0.227(2) nm (av)
Li---Li	0.363(3) nm	0.383(4) nm
M---M	0.3223(13) – 0.3382(13) nm	0.3427(2) nm (av)

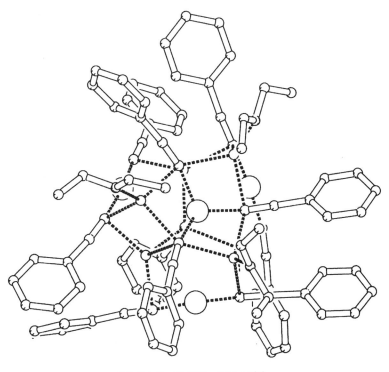

(62) $[Li_6Cu_4(C{\equiv}CPh)_{10}(OEt_2)_3]$ [202]

Figure 14 (continued)

$[Ag(CF(CF_3)_2)(NCMe)]$ (72) is the only alkylsilver–ligand complex of which the structure in the solid state has been elucidated (Figure 16).[70,162] This two-coordinate, neutral alkylsilver complex has a linear geometry at silver. The pentafluorophenylsilver compound $[Ag(C_6F_5)(CH_2PPh_3)]$ (44) is the only arylsilver compound of which the structure in the solid state is known (two forms exist that differ in the orientation of one phenyl ring).[69] The linear coordination at the silver centre is as expected, but there is severe distortion of the angles about C_{ipso} (see Figure 16). Similar distortions have also been observed in other M–C_6F_5 fragments.

It is frequently found, on addition of a ligand, that an aggregated structure breaks up into a ligand-stabilized cationic part and an anionic part, as opposed to the formation of neutral complexes of the type

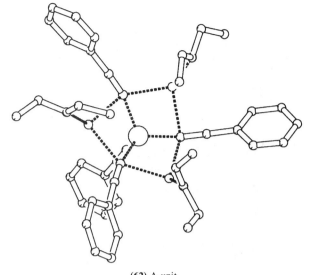

(62) A-unit

$[Li(OEt_2)]_3[Cu(C{\equiv}CPh)_4]$

Cu-C	0.197 – 0.202 nm
Li-C	0.230 – 0.236 nm
C-Cu-C	108 – 111°

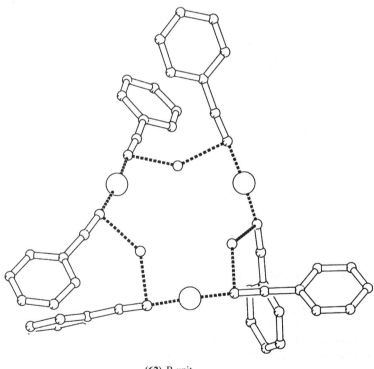

(62) B-unit

$Li_3[Cu(C{\equiv}CPh)_2]_3$

Cu-C	0.183 – 0.189 nm
C-Cu-C	170 – 173°

Figure 14   (continued)

$[MR(L)_x]$. Either one of these ionic species can be organometallic. In the mononuclear cuprate and argentate structures discussed in Section 2.3.2.5(ii) and shown in Figure 13, it is the anion that contains the organometallic part. The main driving force for the formation of these cuprates and argentates is

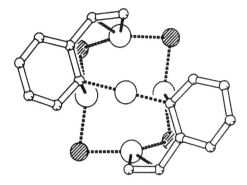

(67) $[Cu_5Br_4(C_6H_4CH=CH_2-2)_2]^-$ [205]

Cu-C$_{Ar}$	0.195(4) – 0.212(4) nm
Cu-Cα	0.212(4) and 0.217(4) nm
Cu-Cβ	0.204(4) and 0.208(4) nm
Cu-Cu (Br-bridged)	0.2690(7) – 0.2926(7) nm
Cu-Cu$_{ap}$	0.2458(7) – 0.2889(7) nm
Cu-C-Cu	74(1) and 75(1)°
C-Cu-C	170(1)°

Figure 15 Molecular structure of $[Cu_5(C_6H_4CH=CH_2-2)_2Br_4]^-$ **(67)**.

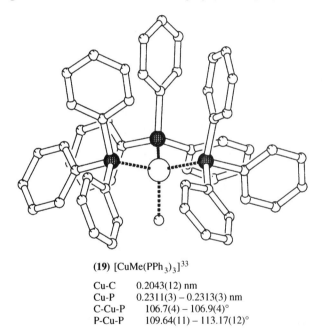

(19) $[CuMe(PPh_3)_3]$ [33]

Cu-C	0.2043(12) nm
Cu-P	0.2311(3) – 0.2313(3) nm
C-Cu-P	106.7(4) – 106.9(4)°
P-Cu-P	109.64(11) – 113.17(12)°

Figure 16 Structures of neutral organocopper and silver ligand complexes.

ligand stabilization of the cation. An example is **(72)**, which exists in solution in equilibrium with a solvated silver cation $[Ag(NCMe)_2]^+$ and the argentate anion $[Ag(CF(CF_3)_2)_2]^-$ **(43)**.[70] Likewise, the addition of PMe$_3$ to pure methylcopper gives the stable cation $[Cu(PMe_3)_4]^+$ and the cuprate anion $[CuMe_2]^-$.[201] In a similar way, by using a ligand that has a strong preference for the binding of lithium cations neutral cuprates have been converted into ionic species (cf. the formation of [Li(12-crown-4)$_2$][CuMe$_2$]).[102]

In a few cases, attempts to prepare mononuclear organocopper complexes have resulted in the isolation of ionic compounds in which the cationic part is the organometallic species. The first example of a cationic trinuclear organocopper complex is $[Cu_3(\mu_3-\eta^1-C\equiv CPh)_2(\mu-Ph_2PCH_2PPh_2)_3][BF_4]$ **(70)**.[127] The structure comprises a triangle of copper atoms which is asymmetrically bicapped by two phenylacetylide groups in a linear $\mu_3-\eta^1$-bonding mode. Most probably it is the presence of the Ph$_2$PCH$_2$PPh$_2$ ligands, each of which bridges a Cu$_2$ edge of the triangle, that stabilizes the $[Cu_3(\mu_3-\eta^1-C\equiv CPh)_2(\mu-Ph_2PCH_2PPh_2)_3]^+$ cation. The dicationic species $[Cu_3(\mu_3-\eta^1-C\equiv CPh)(\mu-Ph_2PCH_2PPh_2)_3]^{2+}$ **(71)** likewise has a triangular Cu$_3$ arrangement stabilized by three Ph$_2$PCH$_2$PPh$_2$ ligands, but here only one $\mu_3-\eta^1$-C$\equiv$CPh acetylide anion is bonded to the $[Cu_3L_3]^{3+}$ cation.[56]

Copper and Silver

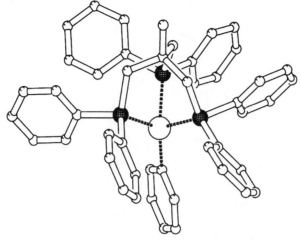

(53) [CuPh{(Ph$_2$PCH$_2$)$_3$CMe}][36]

Cu-C	0.2020(4) nm
Cu-P	0.2276(2) – 0.2342(2) nm
C-Cu-P	119.4 – 126.9(1)°
P-Cu-P	90.0(1) – 93.3(1)°

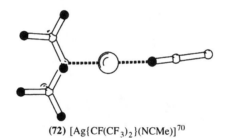

(72) [Ag{CF(CF$_3$)$_2$}(NCMe)][70]

Ag-C	0.2104(11) nm
Ag-N	0.2083(7) nm
C-Ag-N	176.32°

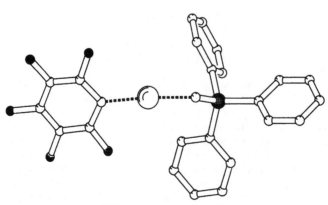

(44) [Ag(C$_6$F$_5$)(CH$_2$PPh$_3$)][69]

Ag-C$_{Ar}$	0.2105(6) nm
Ag-C	0.2144(5) nm
C$_{Ar}$-Ag-C	178.2(2)°

Figure 16   (continued)

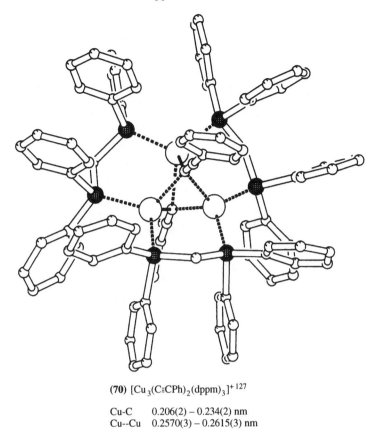

(**70**) $[Cu_3(C{\equiv}CPh)_2(dppm)_3]^{+\,127}$

Cu—C   0.206(2) – 0.234(2) nm
Cu--Cu   0.2570(3) – 0.2615(3) nm

Figure 17   Structures of cationic organocopper compounds and $[Ag\{CH(PPh_3)(COPh)\}_2]^+$.

The use of PMDTA for the complexation of phenylcopper has resulted in the formation of not only a cationic phenylcopper species $[Cu_3Ph_2(PMDTA)_2]^+$ but also a phenylcuprate anion $[Cu_5Ph_6]^-[Li(PMDTA)(THF)]^+$ (**29**) (see Section 2.2.2.2). The cation $[Cu_3Ph_2(PMDTA)_2]^+$ consists of a linear arrangement of three copper atoms of which the outer copper atoms each bind to one of the terdentate amine ligands, and the two phenyl groups each bridge two copper atoms via 2e–3c bonds.[205]

In organosilver chemistry two cationic complexes are known: $[Ag_2\{CH(PPh_3)COR\}_2]^+$ (OR = OPh or OEt (**77**)).[74] The structures contain linearly coordinated silver atoms (C–Ag–C = 174° (**77**)).

2.3.2.7 Miscellaneous organocopper and -silver compounds

In the mid-1970s it was discovered that phosphorylide compounds form stable σ-organometallic complexes. The first structurally characterized copper complex was prepared from Me_3PCH_2, which is isoelectronic with the $[TMS\text{-}CH_2]^-$ anion. The dimeric compound obtained, $[Cu(CH_2PMe_3)]_2$ (**49**), consists of two linear C–Cu–C arrangements (175.8(8)°) and normal, almost equal Cu–C bond lengths[137] (see Figure 18).

Also $[Ag(C_6F_5)(CH_2PPh_3)]$ (**44**)[69] and $[Cu(C\{PPh_3\}_2)Cp^*]$ (**81**)[61] are examples of ylide complexes. Introduction of carbonyl groupings next to the ylidic carbon atom results in stabilization of the ylide. Ylides of this type have been used to prepare stable cationic silver complexes (see, for example, Figure 17). The structure of $[Ag_2(\mu\text{-}CH_2PPh_2S)]_2$ (**38**) (Figure 18) is comparable to that of (**49**), in which the silver atoms are linearly coordinated by the ylide-carbon and the sulfur atom of a second unit. Likewise, $[Ag(\mu\text{-}\{CH(COOEt)\}_2PPh_2)]_2$ is a dimer with the same molecular geometry as the previous structures.

A number of Cp_nL_n copper complexes are known, all of which have the linear arrangement of the C(ring centroid)–Cu–L axis as a common structural feature (see Figure 19). The structures of $[CuCpL]$ (L = PEt_3 (**48**)[80] and PPh_3 (**47**)[81]) were published in 1970. They provided unequivocal proof of the η-bonding mode of the cyclopentadienyl group. The study of $[Cu(\eta\text{-}C_5H_4Me)(PPh_3)]$ (**80**)[236] revealed a distinct correlation between the C(ring-centroid)–Cu–L angle and the bulk of the phosphine and cyclopentadienyl ligands. Deviation from linearity increases steadily from the PEt_3 complex (**48**) (179.9°), via PPh_3 (**47**) (175.2°) to 173.1° in (**80**), with the addition of the methyl group to the

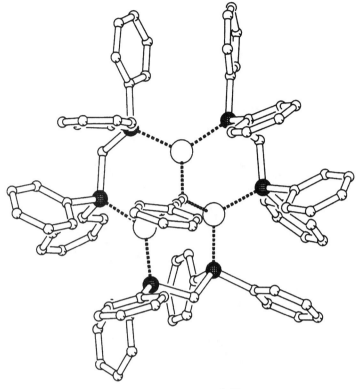

(71) $[Cu_3(C{\equiv}CPh)(dppm)_3]^{2+}$ [56]

Cu-C	$0.1964(14) - 0.2081(13)$ nm
Cu-P	$0.2257(4) - 0.2279(4)$ nm
Cu--Cu	$0.2813(3) - 0.3274(3)$ nm

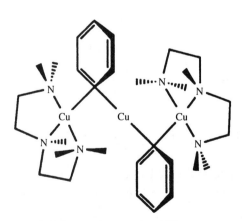

(66) $[Cu_3(C_6H_5)_2(PMDTA)]^+$ [204]

Cu-C	$0.1989(5)$ and $0.2006(7)$ nm
Cu-N	$0.2087(3) - 0.467(5)$ nm
Cu--Cu	$0.2444(1)$ nm
Cu-C-Cu	$75.4(2)°$

Figure 17 (continued)

cyclopentadienyl ligand. Complex **(81)**[153] has instead of a phosphine ligand an η^1-C bonded hexaphenylcarbodiphosphorane ligand. This ligand has a perfectly planar carbon-donor centre while the C–Cu distance is comparable to that of 2e–2c Cu–C bonds.

The observation that the Cp*Cu fragment is isolobal with methylene led to the study of the 'carbene-like' chemistry of this fragment. In the structure of $[CuPtW(\mu\text{-}CC_6H_4Me\text{-}4)(CO)_2CpCp^*]$ **(82)** (see Figure 19), the copper atom binds to the WC moiety and there is no direct Cu–Pt interaction.[209]

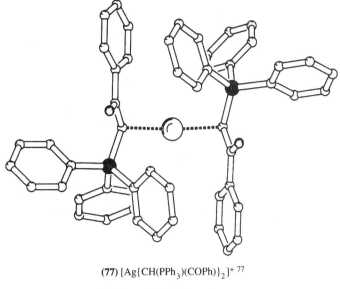

(77) [Ag{CH(PPh$_3$)(COPh)}$_2$]$^{+}$ [77]

Ag–C	0.2219(9) nm
Ag–O	0.2751(10) nm
C–Ag–C	175.6(4)°

Figure 17 (continued)

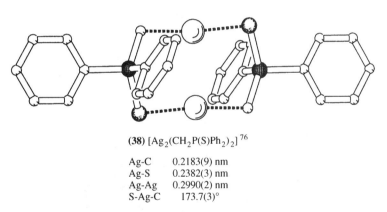

(38) [Ag$_2$(CH$_2$P(S)Ph$_2$)$_2$] [76]

Ag–C	0.2183(9) nm
Ag–S	0.2382(3) nm
Ag–Ag	0.2990(2) nm
S–Ag–C	173.7(3)°

Figure 18 Structure of [Ag$_2$(CH$_2$P(S)Ph$_2$)$_2$] (**38**).

One ferrocenylcopper compound (**15**) and its silver analogue (**75**) have been reported[29,68] (see Figure 19). These compounds, which have a σ-metal–Cp bond, have structures that are comparable to that of the arylcopper [Cu$_4$(C$_6$H$_3$CH$_2$NMe$_2$-2-Me-5)$_4$] (**1**),[6] that is, they have an almost planar Cu$_4$ or Ag$_4$ array. The ferrocenyl groups are 2e–3c bonded to this metal core. The bridging carbon atoms are slightly bent out of the M$_4$ plane, alternating above and below. The CH$_2$NMe$_2$ ligands are only weakly coordinated to the copper centres, as deduced from the long Cu–N distance of 0.309 nm. The coordination geometry in (**15**) can therefore best be described as digonal. The observed C–Cu–C angle of 165.8° and the almost planar Cu$_4$ array are in accord with this view. The Ag–N distance in (**75**) is likewise too long (0.294 nm) for significant Ag–N bonding to be taking place. This is supported by the almost planar Ag$_4$ core and the C–Ag–C angle of 170.7° which suggest digonal coordination of the silver atom. A final point of interest is the short Ag···Ag distance of 0.276 nm.

One structure of a diorgano(*N,N*-diorganodithiocarbamato)copper compound, [Cu(CF$_3$)$_2$(S$_2$CNEt$_2$)] (**21**), has been reported (see Figure 20), which is the first stable organocopper compound containing copper in its trivalent state.[96] The structure shows a distorted planar geometry at the copper atom, similar to [CuBr$_2$(S$_2$CNBu$_2$)].[237] The two Cu–C bond lengths are markedly different, one being 0.186 3(5) nm and the other 0.2026(5) nm. The former Cu–C distance is the shortest reported in organocopper chemistry.

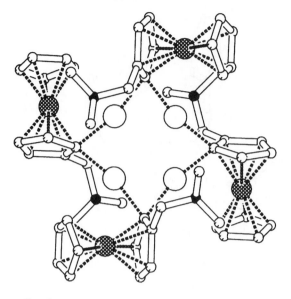

σ-Complexes:

[M(1-ferrocenylCH$_2$NMe$_2$-2)]

M = Cu **(15)**[29]   M = Ag **(75)**[29]

M-C	0.205(4) nm	0.214(2) and 0.220(2) nm
M---M	0.2443(4) nm	0.2740(2) nm
M-N	0.309(2) nm	0.2935(17) nm
M-C-M	73.2(1)°	78.3(6)°
C-M-C	165.8(4)°	170.7(6)°

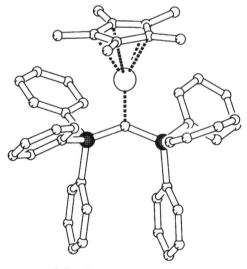

η^5-Complexes:

(81) [Cu{C(PPh$_2$)$_2$}Cp*][153]

Cu-C $_{(centroid)}$	0.194 nm
Cu-C $_{(ligand)}$	0.1922(6) nm
C$_{(centroid)}$-Cu-C$_{(ligand)}$	177.9°

Figure 19 σ- and η-Cyclopentadienylcopper and -silver complexes.

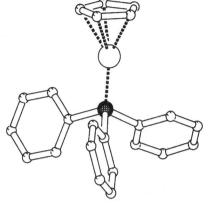

(47) [Cu(PPh₃)Cp][81]

Cu-C (centroid)	0.1864 nm
Cu-P	0.2135(1) nm
C(centroid) -Cu-P	175°

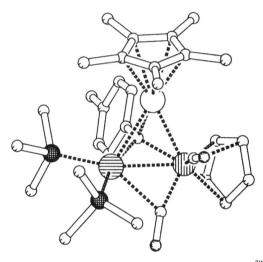

(82) [CuPtW(μ₃-CC₆H₄Me-4)(CO)₂(PMe₃)₂Cp Cp*][208]

Cu-C(centroid)	0.192 nm
Cu-C	0.196(2) nm
Cu-W	0.2648(3) nm
Cu-Pt	0.2807(3) nm

Figure 19 (continued)

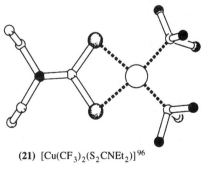

(21) [Cu(CF₃)₂(S₂CNEt₂)][96]

Cu-C	0.1863(5) and 0.2026(5) nm
Cu-S	0.2224(1) and 0.2186(1) nm
C-Cu-C	92.2(2)°

Figure 20 Structure of the first diorganocopper(III) compound.

2.3.3 Polymeric Structures (Including ^{63,65}Cu NQR Spectroscopy)

The structures initially proposed for alkyl-, aryl- and alkynylcopper and -silver compounds were polymeric. This notion was mainly based on the insolubility of these compounds in common organic solvents such as diethyl ether, THF, benzene and toluene. X-ray structure determinations (see Section 2.3.2) demonstrated the polymeric nature of some organocopper and -silver compounds in the solid state, for example, the infinite zig-zag chains of metal atoms in $[CuC\equiv CPh]_n$ (91)[73] and $[Ag(C\equiv CPh)(PMe_3)]_n$ (76),[173] the intriguing oligomeric structure of $[Cu_{24}(C\equiv CBu^t)_{24}]$ (an aggregate of 24 $[Cu(C\equiv CBu^t)]$ units)[203] and the polymeric structure of $[Li_2Cu_2(CH_2TMS)_4(DMS)_2]$ (24) in which the dimeric units are connected by DMS molecules bridging lithium atoms of different dimers.[102] However, care has to be taken in drawing conclusions on the basis of solubility properties, as demonstrated by the discrete aggregate structure of $[Cu_4(1-C_{10}H_6NMe_2-8)_4]$ (56).[44] This arylcopper compound is extremely insoluble in common organic solvents, so much so that a ^{1}H NMR spectrum could not be measured. Suitable crystals for x-ray analysis were only obtained by carrying out the reaction of $[Li(1-C_{10}H_6NMe_2-8)]$ with $[CuBr\{P\{OMe_3\}_3\}]$ very slowly in frozen benzene. The determined structure showed unambiguously that this arylcopper species is not polymeric but consists of discrete $[Cu_4(1-C_{10}H_6NMe_2-8)_4]$ aggregates. However, in many cases it has not been possible to obtain x-ray quality crystals of insoluble organocopper compounds, and structural conclusions are then based exclusively on IR and Raman spectroscopic data[19] and photoconductivity properties.[19]

The insolubility of phenylcopper has likewise been cited by several authors as an argument for a polymeric structure arising from intermolecular interaction of filled copper d-orbitals with antibonding π-orbitals of the phenyl group.[198] However, this proposal was made at a time when x-ray structural information concerning organocopper compounds was scarce. Note that insoluble arylcoppers have been converted into soluble complexes by using the coordinating solvent DMS (see Section 2.3.2.2(i)). For example, in the case of phenylcopper, a structure in the solid state comprising discrete tetrameric $[Cu_4Ph_4(DMS)_2]$ aggregates (12) (cf. (13 in Figure 3) was obtained.[16] Of course, this finding does not exclude the possibility that pure, uncomplexed phenylcopper is polymeric. A plausible structure could consist of a chain of copper atoms with each phenyl group bridging two copper atoms by 2e–3c bonds (see Figure 21).[46] A similar proposal was made for the structure of uncomplexed phenylsilver.[168]

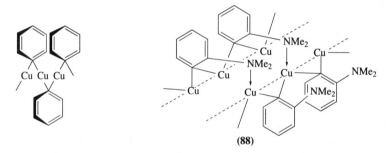

(88)

Figure 21 Proposed structures of uncomplexed phenylcopper and 2-(dimethylamino)phenylcopper (88).[46]

In these proposed arrangements, the metal atoms (copper or silver) are assumed to have digonal coordination geometries with C–M–C angles of about 160° (see the two-coordinate copper atoms in the structures of (57) and (13) in Figure 3). It must be noted, however, that the geometrical rearrangement necessary for the conversion of a 2e–3c bond into a 2e–2c bond is very small. A 2e–2c bond structure for phenylcopper could consist of alternating $[Ph_2Cu]^-$ units and C_{ipso} π-coordinated Cu^+ cations. This proposal seems to be corroborated by elements present in a number of recent structures of arylcoppers in the solid state (see Section 2.3.2).

The structure proposed for phenylcopper (see Figure 21) may serve as a basis for the structures of insoluble methoxy- and dimethylamino-substituted phenylcopper compounds.[46] For example, in $[Cu(C_6H_4NMe_2-2)]_n$ (88), possible interchain Cu–N coordination could result in a three-dimensional network structure as is shown in Figure 21 (cf. the two-coordinate copper sites in the structures of (2) and $[Cu_6(C_6H_4NMe_2-2)_4(C\equiv CC_6H_4Me-4)_2]$ (35) in Figure 5, and the structural unit of the $C_6H_4NMe_2$ anion in Figure 6). Attempts have been made to find support for this proposal by ^{63,65}Cu NQR spectroscopy.[19,239] In some instances resonances were recorded, for which the data are collected in Table 6. Without going into the detail which is needed for a full interpretation of these results, some interesting conclusions can be drawn. In $[Cu_6Br_2(C_6H_4NMe_2-2)_4]$ (2) and (35), the presence (in 1:2 atomic ratio) of apical (two-coordinate) and equatorial (three-coordinate) copper atoms identified by x-ray crystallography was confirmed by the ^{63}Cu NQR data. The resonance at about 30–35 MHz is indicative of a copper atom with trigonal geometry (cf. Figure 5 and the data for (**1**) in Figure 4), and the resonance at 21–23 MHz indicates digonally coordinated copper. The relative intensity of the resonances is 2:1. These data also confirm the absence of direct Cu–Cu interactions in these clusters (see Section 2.3.4.1). It is of interest that the ^{63}Cu NQR data observed for insoluble (**88**) suggest a structure containing mainly trigonal copper atoms (cf. the resonance at 35.892 MHz, in agreement with the structure proposed in Figure 20). The weak resonance at 21.794 MHz may indicate the presence of, for example, two-coordinate copper (end) groups, defects in the bulk material or distortion of two-coordinate copper atoms from a linear coordination geometry.

Table 6 ^{63}Cu NQR spectroscopic data for some arylcopper compounds.

Compound	^{63}Cu at ca. 290 K[a] (MHz)	
[Cu$_4$(C$_6$H$_4$CH$_2$NMe$_2$-2-Me-5)$_4$] (**1**)	30.114	28.045
	30.470	28.636
[Cu$_n$(C$_6$H$_4$NMe$_2$-2)$_n$] (**88**)	35.892 (strong)	21.794 (weak)
[Cu$_6$Br$_2$(C$_6$H$_4$NMe$_2$-2)$_4$] (**2**)	35.906[b,c]	21.762[b]
[Cu$_6$(C$_6$H$_4$NMe$_2$-2)$_4$(C≡CC$_6$H$_4$Me-4)$_2$] (**35**)	35.994	22.330
	37.062	22.852

[a] Correct isotope frequency ratio observed. [b] The intensity of the resonance at 35.906 MHz is larger (about 2:1) than that at 21.762 MHz. [c] ^{79}Br NQR resonance at 76.378 MHz (pointing to a calculated charge on bromine of −6.0 e).

2.3.4 Bonding and Stereochemical Aspects (Including Theoretical Studies)

2.3.4.1 The nature of the copper– and silver–carbon bonds and of the metal–metal interaction

In Section 2.3, a variety of structures were presented. Together these provide a much clearer view of the bonding between organic groups and metals in organocopper and -silver compounds than was possible at the time when the copper and silver chapter in *COMC-I*[19] was prepared. To date, still relatively few mononuclear copper and silver compounds have been reported. Most of the structures are polynuclear, aggregated species possessing short metal–metal distances, and bridging organic groups between two or more metals with acute M–C–M bond angles as characteristic structural features. These features are comparable to those observed for aggregated organolithium compounds. In fact, the remarkable ability of copper and lithium cations to interchange in a series of related aggregates is demonstrated in the compounds [Cu$_2$Ph$_4$(DMS)$_2$] (**12**), [Li$_2$Cu$_2$Ph$_4$(DMS)$_3$] (**26**) and [Li$_4$Ph$_4$(DMS)$_4$][16,102] and also [Cu$_4$(C$_6$H$_4$CH$_2$NMe$_2$-2)$_4$] (**14**), [Li$_2$Cu$_2$(C$_6$H$_4$CH$_2$NMe$_2$-2)$_4$] (**3**) and [Li$_4$(C$_6$H$_4$CH$_2$NMe$_2$-2)$_4$][215] (cf. References 139, 184 and 207a). The aggregates in these series have different Li:Cu ratios but all are tetranuclear. Comparison of the two series also corroborates the view that chelating *ortho*-substituents and weakly coordinating ligands (solvent molecules such as DMS) have much the same function in stabilizing the aggregate.[139,185] The complexes discussed here are not isostructural because the Cu$^+$ ion has a strong preference for linear, two-coordination, whereas Li$^+$ has a more flexible coordination behaviour. As a consequence of this preference for linear coordination, in the hypothetical series of aggregates beginning with CuR and progressing to Cu$_2$R$_2$, Cu$_3$R$_3$ and Cu$_4$R$_4$ (see Figure 22 and Table 5), it is only Cu$_4$R$_4$ which is known to exist as such. This is because it can be formed from four digonal R–Cu–R units without distortion of bond lengths or angles. All other species, owing to the acute Cu–C–Cu angles of the bridging groups, require that additional ligands complete the copper(I) coordination sphere (see examples in Figure 22). Also, the incorporation of metal salts in the aggregated structure can sometimes be explained by the release of strain when an anion other than a carbon anion is included in the final aggregated structure (cf. the structure of [Cu$_3$Br(C$_6$H$_4$CH$_2$N(Me)CH$_2$CH$_2$NMe$_2$-2)$_2$] (**90**)).[85] Carbon anions can sometimes also be exchanged for heteroatom anions, while preserving the overall structure of the aggregates: compare [Cu$_4$Mes$_4$(THT)$_2$] (**57**) and [Cu$_4$Mes$_2$(SAr)$_2$] (**17**), in which two mesityl anions are replaced by the covalent SAr$^-$ anion, and the THT ligands by intramolecular Cu–S coordination. By the same token [Cu$_4$Mes$_4$(THT)$_2$] (**57**) and [Cu$_4$Mes$_4$][Mg(SC$_6$H$_4$CH(Me)NMe$_2$-2)$_2$]$_2$ (**31**) (Z = Me) are comparable, as the Cu–S coordination by THT ligand in (**57**) is replaced by Cu–S coordination by the Mg(SC$_6$H$_4$CH(Me)NMe$_2$-2)$_2$ fragment in (**31**).

Copper and Silver

(a) 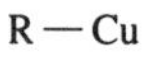   R — Cu

(85) [Cu(C$_6$H$_2$Bu$^t_3$-2,4,6)(DMS)]

(b) 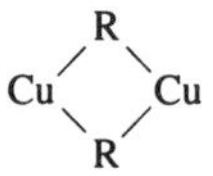
$$Cu \diamondsuit Cu$$
with R at top and R at bottom

(23) [Cu$_2$(C≡CBu^t)$_2$(PCy$_3$)(PPh$_3$)$_2$] and
(86) [Cu$_2$(C$_6$H$_2$Ph$_3$-2,4,6)$_2$(DMS)$_2$]

(c) 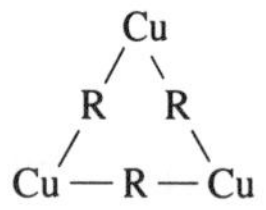
```
        Cu
       /  \
      R    R
     /      \
   Cu — R — Cu
```

(90) [Cu$_3$Br(C$_6$H$_4$CH$_2$N(Me)CH$_2$CH$_2$NMe$_2$-2)$_2$]
(one R replaced by Br)

(d) 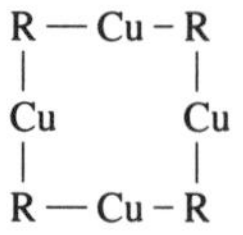
```
R — Cu – R
|        |
Cu       Cu
|        |
R — Cu – R
```

(12) [Cu$_4$Ph$_4$(DMS)$_2$] and
(14) [Cu$_4$(C$_6$H$_4$CH$_2$NMe$_2$-2)$_4$]

Figure 22 Aggregate formation based on assembly of CuR fragments and linear coordination at the copper(I) centres.

Apart from the bridge-type bonding of the organic anions in these aggregates, the presence of very short copper–copper and silver–silver distances in the various structures of organocopper and -silver compounds (cf. the figures and legends in Section 2.3) is an intriguing structural feature. There is continuing interest in the possibility that direct copper–copper and silver–silver bonding occurs in these compounds. These distances are even shorter than the corresponding distances in the pure metals (0.256 nm and 0.289 nm, respectively).[216] Since, with few exceptions, the organocopper and -silver compounds are derived from the monovalent metals, the presence of a direct metal–metal bond would imply attraction between two closed-shell (d^{10}) metal centres. Various calculations have been carried out in order to shed more light on this point.[217] These calculations have shown that in the absence of metal s and p functions, the expected closed-shell repulsion between the d^{10} centres is indeed operative, but that mixing of metal s and p orbitals converts this repulsion into a slight attraction in copper(I) aggregates of any size.

It must be noted that no spectroscopic evidence for the presence of metal–metal bonds in polynuclear organocopper or silver compounds has been reported. Raman spectra of [Cu$_4$(C$_6$H$_3$CH$_2$NMe$_2$-2-Me-5)$_4$] (1) and of related compounds lack strong absorptions in the region where metal–metal vibrations are expected.[175] Further, ^{63,65}Cu NQR data (see Table 6) provide no evidence for the presence of metal–metal interactions in these compounds. Such interactions would, of course, alter the formal digonal or trigonal geometry around the metal atoms. Likewise, ^{196}Au Mössbauer spectroscopy, a technique particularly suited for the study of the geometry around gold atoms, of [Au$_2$Cu$_4$Br$_2$(C$_6$H$_4$NMe$_2$-2)$_4$] (94) and [Au$_2$M$_2$(C$_6$H$_4$CH$_2$NMe$_2$-2)$_4$] (M = Li (3) or Cu (95)) provides no evidence for metal–metal interactions. The structure of (94) is analogous to that of the corresponding [Cu$_6$Br$_2$(C$_6$H$_4$NMe$_2$-2)$_4$] (2), with the gold atoms occupying apical, two-coordinate sites (see Section 2.3.2.1(iii), Figure 5).[178]

If there is at best only a weak attractive metal–metal interaction, then we must begin elsewhere for a discussion of the range of metal–metal distances observed, which depends largely on the type of group bridging the metals. A quantitative discussion is not possible at present, owing to the large perturbation caused by the introduction of bridging ligands in calculations. The best qualitative approach[218] to an interpretation of the M–C and M···M bonding in organocopper and -silver aggregates is still the localized molecular orbital (MO) approach illustrated in Figure 23.

In aggregated structures, for example, [Cu$_4$R$_4$], the bonding is no longer electron precise (2e–2c) but is electron deficient with metal–carbon bonds in which the carbon anion interacts with two or more metal centres (2e–3c or 2e–4c). The bridging organic anions fall into three groups: (i) alkyl anions which act as two-electron donors by overlap of a filled sp^3 orbital; (ii) aryl and alkenyl anions, which can act as either two- or four-electron ligands using sp^2 and $p\pi$ electron pairs; and (iii) alkynyl anions, which can participate in a maximum of three molecular orbitals (one sp and two $p\pi$ overlaps), thus acting as two-, four- or six-electron donors. A variety of these possible metal–carbon bonding interactions are summarized in Table 7.[58,175] The aryl– and alkynyl–copper(silver) bonding will now be discussed in greater detail.

The MOs describing aryl–copper bonding are depicted in Figure 23. The lowest MO is a bonding combination of the filled sp^2-C$_{ipso}$ orbital with mutually bonding orbitals on the two copper atoms (Figure 23(a)). A second, higher-energy MO is due to combination of a π-C$_{ipso}$ orbital with a mutually antibonding combination of copper orbitals (Figure 23(b)). Back donation from the copper atoms to the aryl bridging ligand occurs via overlap of filled mutually antibonding orbitals of the copper atoms to a π^*-C$_{ipso}$ orbital (Figure 23(c)). The shortest Cu···Cu distance is expected when only the first MO is

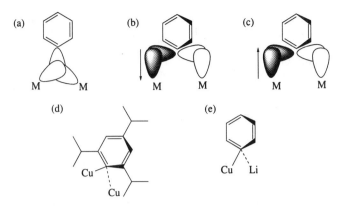

Figure 23 Electron deficient bonding schemes in C-M$^1_2$ and C-M^1(M^2) structural units. (a), (b) and (c) between equal M centres, (d) unsymmetrical bonding caused by bulky *ortho* groups and (e) between unlike M centres.

Table 7 Bonding of alkyl, aryl, alkenyl and alkynyl anions to one or more copper or silver atoms.

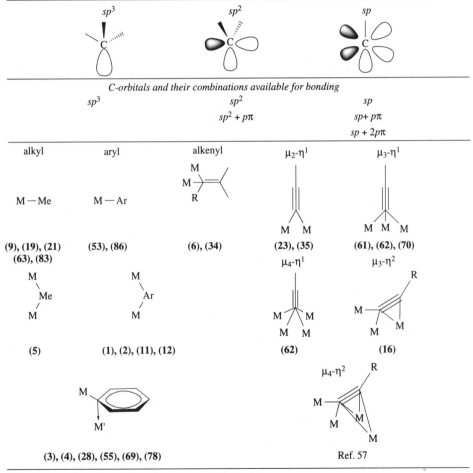

^{a}The compound numbers refer to the structures in Section 3.

occupied. Closed-shell repulsion between the two d^{10} centres is in this case overcome. The contribution of the second MO to Cu–C bonding increases the electron density at C$_{ipso}$, and thus the kinetic stability of the bond. This helps to explain why bulky *ortho*-substituents stabilize organocopper aggregates. Since these substituents, by their steric influence, tend to orient the aryl ring perpendicular to the metal–metal axis, orbital overlap (Figure 23(b)) is maximized. In the following series of R groups, the stability of the corresponding organocopper compounds decreases: 2-Me$_2$NCH(Z)C$_6$H$_4$ (Z = H or Me) ~ 2-Me$_2$CCH$_2$C$_6$H$_4$ ~ 8-phenyl-1-naphthyl ~ 2-Me$_2$NC$_6$H$_4$ ~ 2,6-(MeO)$_2$C$_6$H$_3$ ~ mesityl > 2-MeOC$_6$H$_4$ >

2-MeC$_6$H$_4$ > 4-MeC$_6$H$_4$ ~ phenylcopper.[87,104] The steric effects of *ortho*-substituents can be very pronounced, as shown by the unsymmetrical bridge observed in [Cu$_4$(C$_6$H$_2$Pr$^i_3$-2,4,6)$_4$] (**55**). The bonding in this unit can alternatively be interpreted as 2e–2c Cu–C$_{ipso}$ bonding, which bridges the copper atom of another arylcopper unit via a π-type Cu'–C$_{ipso}$ interaction (see Figure 23(d)). This kind of unsymmetrical metal–C$_{ipso}$ bonding is commonly encountered in cuprate structures (see Figures 10 and 23(e)). As a final word, the additional stabilisation provided by bulky *ortho*-substituents, due to their mere steric protection of the metal–carbon bond, should not be discounted.

In agreement with the 2e–3c bonding description is the acute <Cu–C–Cu' bond angle (around 75°) at the bridging carbon atoms. An interesting example is [Cu$_6$(C$_6$H$_4$NMe$_2$-2)$_4$(C≡CC$_6$H$_4$R-4)$_2$] (**35**), in which two different organic groups, four aryl and two alkynyl, are bound to the same Cu$_6$ core.[58,59] Both the aryl- and the alkynyl-bridging carbon atoms show acute <Cu–C–Cu' angles of 75.5° and 74.6°, respectively, clearly indicating that both groups are 2e–3c bonded. This is again reflected by the short Cu···Cu distances of 0.242 nm and 0.247 nm, respectively. The Cu···Cu distance increases to 0.270 nm when the alkynyl group is replaced by bromine, that is, in [Cu$_6$Br$_2$(C$_6$H$_4$NMe$_2$-2)$_4$] (**2**).

Alkynyl–copper bonding is even more versatile than the aryl–copper interaction. In addition to 2e–3c bonding as encountered in a number of alkynylcopper aggregates (see Figures 9 and 17), 2e–4c bonding has also been observed (see Figure 17). In these compounds, the C–Cu interaction involves exclusively the alkynyl *sp*-orbital, and not the π-orbitals. This can be concluded from the unchanged C≡C bond length. Other bonding situations have been encountered as in, for example, [Cu$_3$(C≡CBu^t)(SAr)$_2$]$_2$ (**16**) (see Figure 9). In this aggregate two trinuclear Cu$_3$(C≡CBu^t)(SAr)$_2$ units are combined via a (μ$_3$-C$_α$, μ$_1$-C$_β$)–η^2 bridging C≡CBu^t anion.[52]

The bonding picture discussed above concerns organic anions bridging like metal atoms. In this case, symmetrical overlap of the organic fragment and the metal atom orbitals can be expected. However, a different situation is encountered when the organic fragment bridges two unlike metals, as in the aggregated cuprate and argentate compounds (or mixed metal species such as [Au$_2$Cu$_4$Br$_2$(C$_6$H$_4$NMe$_2$-2)$_4$] (**94**) and [Au$_2$Cu$_2$(C$_6$H$_4$CH$_2$NMe$_2$-2)$_4$] (**95**)). Examples are [Li$_2$Cu$_2$(Ph)$_4$(Et$_2$O)$_2$] (**25**), [Li$_2$Cu$_2$(Ph)$_4$(DMS)$_3$] (**26**), [Li$_2$Cu$_2$(C$_6$H$_4$CH$_2$NMe$_2$-2)$_4$] (**3**) and [Li$_2$Ag$_2$(C$_6$H$_4$CH$_2$NMe$_2$-2)$_4$] (**4**). So far, unsymmetrical bridges have been observed for *sp*^{2}- and *sp*-hybridized carbon anions, that is, aryl and alkynyl. In Figure 23(e) the situation for the binding of an aryl group in an arylcuprate or -argentate is shown. The observation in the solid state is that the plane of the aryl group is perpendicular to the M···Li axis, as in the homometallic case described above, but that the C$_{ipso}$–C-4 axis is tilted towards the lithium centre. This asymmetry suggests that heterometallic bridging is better described as involving 2e–2c bonding of the aryl anion to copper or silver, and interaction of the lithium cation with the filled *p*π-arene orbital. Alternatively, this latter interaction may be purely electrostatic in nature. It is interesting to recall here a comparison of the bonding in the series [Li$_2$M$_2$(C$_6$H$_4$CH$_2$NMe$_2$-2)$_4$] (M = Cu (**3**), Ag (**4**), Au (**93**)). In the case of the cuprate (**3**), the ^{13}C NMR spectrum showed a small ^{6,7}Li–^{13}C coupling, larger than this coupling in the corresponding argentate (**14**), whereas in the aurate (**93**) this coupling is absent (see Section 2.3.4.2). The fact that monovalent gold has the strongest preference for linear coordination stabilizes the [R–Au–R]$^-$ anionic grouping and these anions are held together by the lithium cations by electrostatic forces. In the [R–Au–R]$^-$ anions the negative charge will be concentrated at the C$_{ipso}$ centres (cf. the extremely large IS and QS values of (**93**) in its ^{197}Au Mössbauer spectra)[178] and it is this C$_{ipso}$ centre that interacts with the lithium cation. This interpretation is corroborated by the observation that the structure of (**93**) in the solid state consists of a square Au$_2$Li$_2$ arrangement.[234] The fact that in argentate (**4**) and cuprate (**3**) a ^{6,7}Li–^{13}C coupling is observed indicates that in these compounds greater orbital contribution exists in the bonding, in addition to electrostatic interactions. The nature of the bonding of the organic anion in ate complexes of copper and silver implies that the asymmetric bonding discussed above will be less likely to occur in the case of alkylcuprates, because of the steric and electronic constraints of an *sp*^{3}-hybridized C-anion. Accordingly, alkylcuprates are expected to have structures consisting of [CuR$_2$]$^-$ anions held together by solvated lithium cations by electrostatic forces. This conclusion, which is based on the extrapolation of the results from structural and spectroscopic studies of the above series of aryl ate complexes, is supported by the results of various MO calculations on organocuprates.[219] The structures of higher-order cuprates of the type [Li$_2$CuR$_3$] have been explored theoretically.[220] A trinuclear species of type (**96**) rather than of type (**97**) was predicted as the most likely structure (see Figure 24). Structure (**96**) contains a near-linear R–Cu–R fragment and monovalent, two-coordinate copper aggregated with a methyl anion and bridged by a pair of lithium cations, that is, an aggregate of MeLi and LiCuMe$_2$.[220] Structure (**97**) consists of a CuMe$_3$ dianion complexed to two lithium cations. With respect to the structural features of (**96**), it is interesting to reconsider the trinuclear structures in Table 5. In this respect the similarity of the structures of (**96**) and (**90**) is striking. Could it be that these structures are in fact a combination of an R–Cu–R anion and an M–X–M cation, in which M has a preference for more extended coordination spheres?[85] Further work is needed to clarify this important point.

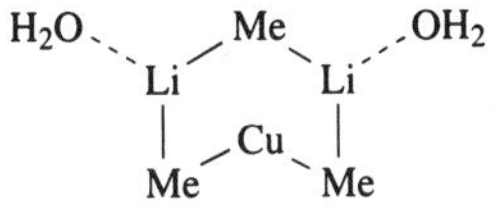

$$H_2O \cdots Li \overset{Me}{\underset{Me}{\diagdown}} Li - OH_2$$

(96)

$$H_2O \sim Li \overset{\overset{Me}{|}}{\underset{Me \quad Me}{Cu}} Li - OH_2$$

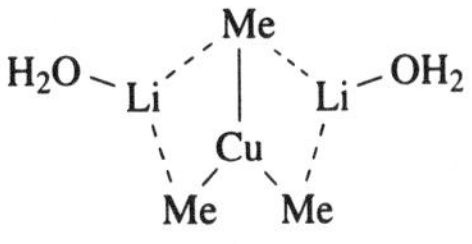

(97)

Figure 24 Calculated structures for [Li$_2$CuR$_3$] cuprates.

2.3.4.2 Structural aspects of organocopper and -silver compounds in solution (including NMR studies and EXAFS spectroscopic studies)

(i) Structure and bonding in solution

Colligative studies provide valuable information about the aggregation state of species in solution (see Table 3). NMR spectroscopic studies have become increasingly important for the study of the structural aspects of organocopper and -silver compounds in solution. The fact that these compounds undergo inter- and intra-aggregate exchange processes, in which the solvent can also play a role, complicates these studies. Since organocopper compounds are among the most versatile reagents for bringing about C–C bond formation,[1] extensive NMR studies on cuprate[221,222] reagents and reaction mixtures[223,224] have also been conducted. These studies have revealed the presence of complex equilibria, and pointed to the possible composition of the species involved. However, a basis for the understanding of the fundamental processes and bonding aspects of homo- and heteroleptic organocopper (and -silver) compounds, in solution, is still lacking. Several reports have, however, concentrated on the solution behaviour of *pure* compounds and reagents.

In a number of cases (see Table 3), colligative measurements (commonly in benzene) indicated that the aggregation state in the solid is retained in solution. Examples are [Cu$_4$(C$_6$H$_3$CH$_2$NMe$_2$-2-Me-5)$_4$] (1), [Cu$_6$Br$_2$(C$_6$H$_4$NMe$_2$-2)$_4$] (2) and [Li$_2$M$_2$(C$_6$H$_4$CH$_2$NMe$_2$-2)$_4$] (M = Cu (3) or Ag (4)). For (1) and (2) it is assumed that the main structural features of these compounds in the solid state (see Figures 4 and 5, respectively) are also retained in solution, since direct proof is not available from ^{1}H or ^{13}C NMR measurements. In a variety of compounds a range of aggregation numbers has been observed, for example, two, three, four, and six to eight. In some cases concentration dependence of the aggregation state is observed. For [Cu{C$_6$H$_2$(OMe)$_3$-2,4,6}]$_n$, n ranges from 5 to 7,[46] and [Ag(C$_6$H$_4$CH$_2$NMe$_2$-2)] seems to be a mixture of tetra- and hexanuclear species.[167] Sometimes the compounds have different aggregation states in different solvents. This is the case for [Cu$_4$(CH$_2$TMS)$_4$] (5), which is a tetramer in benzene (see Figure 1 for the tetrameric structure in the solid state), but is a hexamer in cyclohexane.[11] Moreover, organocopper aggregates can undergo interaggregate exchange processes, as was established by a mass spectrometric study of a solution of two comparable tetranuclear aggregates [Cu(C$_6$H$_3$CH$_2$NMe$_2$-2-Me-5)]$_4$ ((1), [Cu$_4$Ar$^1_4$]) and [Cu(C$_6$H$_4$CH$_2$NMe$_2$-2)]$_4$ ((14), [Cu$_4$Ar$^2_4$]) in benzene solution, which showed the formation of a statistical mixture of the homoleptic aggregates (1) (*m/z* 864), (14) (*m/z* 790) and the heteroleptic aggregates [Cu$_4$Ar^{1}Ar$^2_3$] (*m/z* 804), [Cu$_4$Ar$^1_2$Ar$^2_2$] (*m/z* 818) and [Cu$_4$Ar$^1_3$Ar^2] (*m/z* 832).[176]

For some compounds, the connectivity in the aggregate has been established unambiguously by NMR spectroscopy. This is feasible when direct spin–spin coupling information involving one of the metal nuclei is present in the NMR spectra of organocopper and -silver compounds. Illustrative of this is the ^{1}H, ^{13}C, ^{109}INEPT Ag and ^{6}Li NMR study of [Li$_2$Ag$_2$(C$_6$H$_4$CH$_2$NMe$_2$-2)$_4$] (4), of which the spectra are shown in Figures 25 and 26.

The schematic structure of the ate complexes with an *o*-Me$_2$NCH$_2$ substituent, (3), (4) and (93), as deduced from NMR studies and colligative measurements, is shown in Figure 27. Both ^{7}Li ($I = 3/2$) and ^{107}Ag and ^{109}Ag (natural abundance 51.8 and 48.2%, respectively, $I = 1/2$) can give rise to spin–spin coupling with C$_{ipso}$ provided that interaggregate exchange is slow on the NMR timescale. In argentate (4) this is clearly the case, as concluded from the multiplicity of the C$_{ipso}$ resonance (see Figure 25). Each C$_{ipso}$ nucleus couples with one silver and one lithium nucleus, providing proof of a multicentre interaction of each C$_{ipso}$ centre with one Ag–Li pair. This observation also points to a transannular arrangement of metal atoms, shown schematically in Figure 27. The amine donor atoms coordinate to lithium, giving four-coordinate lithium centres and two-coordinate silver centres. Also, for [Li$_2$Ag$_2$Ph$_4$(OEt$_2$)$_2$], proof of the presence of an Ag–C interaction through observation of the $J(^{109}$Ag, ^{13}C) (132 Hz[174]) was obtained. Moreover, in the case of argentate (4) direct silver–lithium coupling (see Figure 26) was observed, which supports the view (see Section 2.3.4.1) that in the unsymmetrical bridge

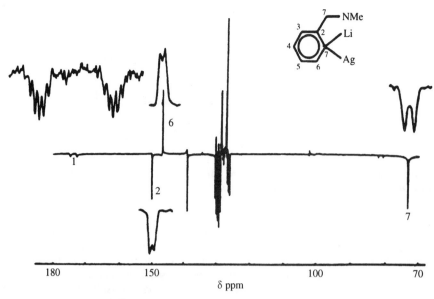

Figure 25 *J*-modulated ECHO ^{13}C NMR spectrum of $[Li_2Ag_2(C_6H_4CH_2NMe_2-2)_4]$ (**4**) at 293 K. $J(^{109}Ag, ^{13}C) = 136.0 \pm 0.81$ Hz; $J(^7Li, ^{13}C) = 7.2 \pm 0.2$ Hz.[9,184]

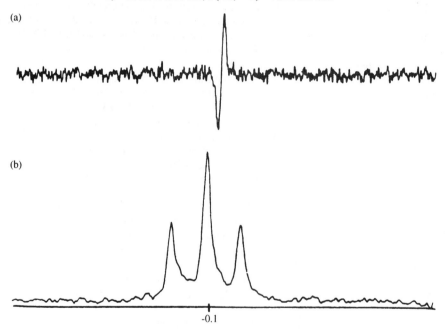

Figure 26 (a) ^{1}H coupled ^{109}Ag INEPT NMR spectrum of (**4**) in toluene-d_8 at 293 K. δ^{109}Ag = 897.7 ppm (1 M AgNO$_3$/D$_2$O external reference). (b) ^{6}Li NMR spectrum of (**4**) in toluene-d_8 at 2931 K. δ^6Li = −0.10 ppm (1 M LiCl/D$_2$O external reference). $J(^{107,109}$Ag, ^{6}Li) = 1.461 Hz; $J(^{107,109}$Ag, ^{7}Li) = 3.91 Hz.[9,184]

bonding of C_{ipso} with the Ag–Li pair in the argentate (and the cuprate (**3**)) a substantial orbital interaction between C_{ipso} and the lithium cation exists. Also, in the cuprate (**3**) only $^{13}C_{ipso}-^7$Li coupling is observed (cf. Figure 9 in Reference 207a), and ^{63,65}Cu–^{7}Li coupling is not observed. The structure of (**3**) and related neutral cuprates with $[Li_2M_2R_4]$ stoichiometry in the solid state is shown in Figure 10.

The exclusive formation of the $[Li_2M_2R_4]$ aggregates, both in solution and in the solid state, having the lithium atoms arranged in a *trans* fashion shows that these aggregates are more stable than the other possible aggregates with $[LiM_3R_4]$ or $[Li_3MR_4]$ or *cis*-arranged $[Li_2M_2R_4]$ cores. The basic building block is obviously the linear $[R-M-R]^-$ unit held together by solvated–intramolecularly coordinated lithium cations.

In solution, the synthetically important cuprate $[Li_2Cu_2Me_4]$ likewise has an open *trans*-Li_2Cu_2 core with the methyl groups bridging CuLi edges, rather than a tetrahedral structure like that observed in $[Li_4Me_4]$.[225] Data obtained from solution x-ray scattering methods gave Cu⋯Cu distances in

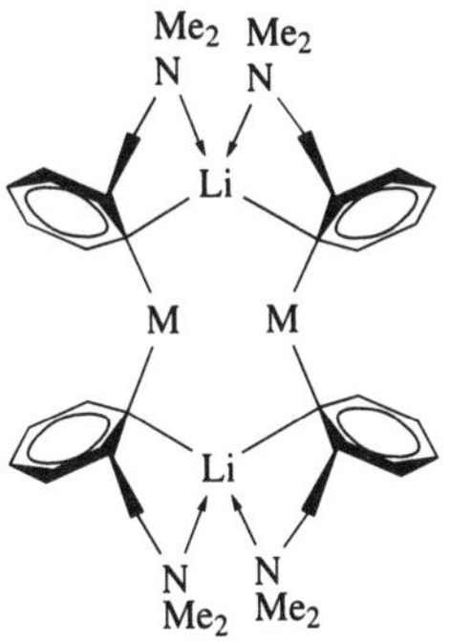

Figure 27 The schematic structure of [Li$_2$M$_2$(C$_6$H$_4$CH$_2$NMe$_2$-2)$_4$] (M = Cu (**3**), Ag (**4**) or Au (**93**)).

[Li$_2$Cu$_2$Me$_4$] of 0.44 ± 0.07 nm. However, the precision of these data was low owing to interference from the ether solvent.[194] This long distance becomes even more doubtful when compared with the Cu···Cu distances for a number of neutral cuprates, which were ca. 0.27 nm (see Figure 10).

The structures of simple arylcopper and cuprate species in various solvents have been the subject of NMR studies. It was known that the reaction of M(C$_6$H$_4$Me-4) with LiC$_6$H$_4$Me-4 results in the formation of an insoluble species [LiM(C$_6$H$_4$Me-4)$_2$] (M = Cu,[103] Ag,[174] or Au[103]), which in the case of the cuprate upon careful addition of Et$_2$O affords benzene-soluble [Li$_2$Cu$_2$(C$_6$H$_4$Me-4)$_4$(OEt$_2$)$_2$]. This was concluded on the basis of molecular weight determinations and ^{1}H NMR spectroscopic data (large shift of the Et$_2$O protons).[103] Extensive ^{13}C NMR studies of phenylcopper and diphenylcuprate dissolved in DMS indicated that phenylcopper is an equilibrium mixture of a 'tetramer' and a 'trimer' in DMS.[226] The chemical shift of the C$_{ipso}$ atom characterizes the aggregation state of CuPh and LiCuPh$_2$ (see Figure 1 in Reference 225). The data show that [Li$_3$Cu$_2$Ph$_5$(DMS)$_4$] (**28**) is in fact [Li$_3$CuPh$_3$]$^+$[CuPh$_2$]$^-$. The results of this study also indicate that 'higher-order cyanocuprates'[106,107] are better represented as [LiCuR$_2$][LiCN] (R = Ph).[227] The latter cuprate has also been subjected to EXAFS spectroscopy[228] and computational studies.[229] The EXAFS spectra of various combinations of CuCN and BuLi show that in a 1:2 molar mixture of CuCN with BuLi, there is no coordinated cyanide, consistent with NMR studies.[227] The results of the computational study on species of composition [Li$_2$CuMe$_2$(CN)(H$_2$O)$_2$] suggest that the compound is best described as either a [LiCuMe$_2$] species complexed by LiCN, or a [LiCuMe(CN)] species interacting with MeLi. A range of trinuclear structures is proposed to exist (cf. (**98**) in Figure 28), which are characterized by anionic units NC–Cu–Me or Me–Cu–Me with two-coordinate copper, rather than species with three-coordinate copper dianionic units [CuR$_3$].[229] Intermediates such as (**98**) have much in common with the trinuclear, heteroleptic copper compound (**90**) (see Figure 28), which consists of a [diarylcopper]$^-$ anion and a [Cu$_2$Br]$^+$ cation. An interesting observation is that (**90**) on heating to 60 °C produces quantitatively the corresponding biaryl coupling product.[85] Care is needed, however, when this kind of information is used for the interpretation of the kinetic reactivity of cuprates in organic synthesis, as there is no guarantee that the species observed *in situ*, or isolated, lie on the kinetic route to products.

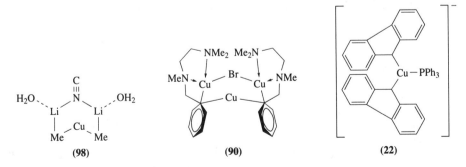

(**98**)　　(**90**)　　(**22**)

Figure 28 Schematic structures of a 'higher order' cuprate, a trinuclear heteroleptic copper compound and a three-coordinate cuprate anion.

Cuprate reagents are frequently used in conjugate addition reactions to α,β-unsaturated ketones and esters, reactions that occur with high regio- and stereoselectivity.[1] In order to gain further insight into the reaction mechanism, various ^{13}C NMR studies on model reaction solutions have been carried out. The first indication for an interaction of the substrate with dimethyl cuprate came from the reaction mixture of [LiCuMe$_2$][LiI] with methyl cinnamate. The conclusion of this study was that upon mixing two complexes are formed, (**99**) and (**100**) in Figure 29, which are in equilibrium with each other.[230] The alkene–copper π-complex (**99**) is the major component at low temperature and with excess [LiCuMe$_2$]. The lithium carbonyl complex (**100**) is the major component at the end of the reaction, when the concentration of [LiCuMe$_2$]–alkene π-complex is reduced by formation of the product enolate.

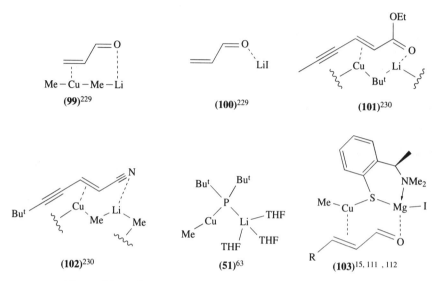

Figure 29   π-Complexation of copper to unsaturated substrates.

Detailed studies of the 1,6- and 1,8-addition reactions of alkylcuprates to 2-en-4-ynoates and 2,4-dien-6-ynoates gave some conclusive data. When selectively ^{13}C-labelled 2-en-4-ynoate substrates (labelled at C-2, C-3 and C-5, respectively) were used in a reaction with [Li$_2$CuBu$^t_2$(CN)] and a 2-en-4-ynenitrile with [LiCuMe$_2$][LiI],[231] comparison of the ^{13}C–^{13}C coupling constants of the alkene moiety of the π-complex (**101**) with those of the uncomplexed substrate unambiguously proved that the cuprate is coordinated to the C–C double bond of the enynoates, whereas no interaction of the π-system of the triple bond is taking place. Furthermore, it could be concluded that the hybridization of the double bond in the π-complex is similar to that of a single bond connecting two carbon atoms with sp^3 hybridization. In other words, one of the alkyl groups functions as a bridge to one copper and one lithium atom, while the second alkyl group is transferred to the substrate. The copper–alkyl–lithium unit then forms the matrix on which the substrate is anchored in a bidentate fashion via the alkene and the carbonyl oxygen, activated by the Li–O interaction, and is subsequently converted into the product enolate.[231] As a consequence, species (**101**) or (**102**) is formed from [Li$_2$CuBu$^t_2$(CN)] or [LiCuMe$_2$][LiI] only in the presence of the substrate. The stable heteroleptic cuprate (**51**) in which the bridging alkyl group is replaced by a nontransferable phosphido group[63] is a stable analogue of the intermediates (**101**) and (**102**). In the enantioselective 1,4-addition reactions of MeMgI to benzylideneacetone catalysed by the chiral, enantiopure copper arenethiolate [Cu(SC$_6$H$_4$CH(Me)NMe$_2$-2)]$_3$, it is the intermediate (**103**) that is formed in a self-assembly process between MeMgI, the substrate (via alkene–carbonyl oxygen chelate bonding) and a single unit of the chiral catalyst.[15,111,112,139,232] A similar species is also formed as a key intermediate in the selective substitution (α or γ) of allylic esters with alkyl Grignard reagents.[112] In (**103**) it is the arenethiolate sulfur atom that acts as a nontransferable bridging group and generates the appropriate Cu–S–Mg matrix, together with the substrate, for selective product formation. Similar bridging units are present in enantiopure [Cu$_4$Mes$_4$][Mg(SC$_6$H$_4$CH(Me)-NMe$_2$-2)$_2$]$_2$ (**31**), of which the structure in the solid state is known (see Figure 12). Complex (**31**) can be considered as a model for the resting state of the catalyst from which intermediate (**103**) is generated by interaction with the substrate.[232] This point of view is supported by theoretical studies of nucleophilic additions of organocoppers and -cuprates to acrolein by Morokuma and co-workers.[217]

(ii) Bonding and stereochemistry

The dynamic behaviour of arylcopper and -silver compounds and of the corresponding cuprates and argentates has been the subject of extensive study. Before elaborating on the structural features of the aggregates as a whole, it is essential to consider the unique properties of the 2e–3c copper(or silver)–carbon bond and its stereochemical consequences. As presented in Figure 23, each aryl group is almost perpendicularly oriented to the Cu···Cu vector; this is the rotamer (a) representing a minimum energy situation in a rotation process of the aryl group around the C_{ipso}–C-4 axis. In the maximum-energy rotamer (b), the metal–metal axis is coplanar with the aryl ring, and there is maximum steric encumberence between the *ortho*-hydrogens (and/or -substituents) and the metals. For *ortho*-substituted phenyl compounds this leads to the existence of different conformations in the aggregate. An example is $[Cu_4(C_6H_4Me-2)_4(DMS)_2]$ (**13**), which in the solid state was found as a mixture of two possible conformations, one with the four *o*-methyl groups on the same side of the $Cu_4(C_{ipso})_4$ plane and the other with the methyl groups alternating above and below this plane (see Figure 3).[17] Figure 30 illustrates that rotamer (a) C_{ipso} in (**13**) has tetrahedral geometry, and can therefore have either *R*- or *S*-configuration. Rotation (180°) around the C_{ipso}–C-4 axis gives the opposite configuration. Combination of four of these units to give the tetranuclear aggregate structure of (**13**) then leads to a number of different stereoisomers, of which the diastereoisomeric ones can be differentiated by NMR spectroscopy, provided that interaggregate exchange and rotation processes are slow on the NMR timescale. In the case of (**13**) this slow exchange limit can be reached, and a number of methyl ^1H resonances can be observed in the ^1H NMR spectrum at −30 °C in toluene.[17] This proves (in combination with the result of molecular weight determinations) that the Cu_4R_4 aggregate is not only present in the solid state but also in solution.

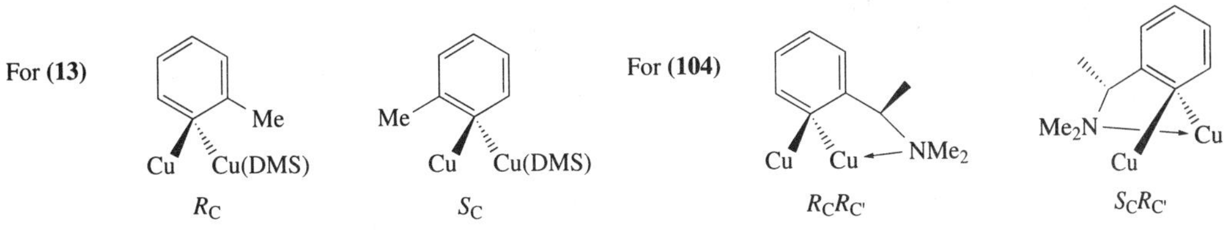

Figure 30 Diastereomeric Cu_2-aryl building blocks.

In the case of the *o*-CH(Z)NMe$_2$-substituted phenylcopper compounds (**14**) (Z = H) and enantiopure (**104**) (Z = Me), the situation is again more spectacular. For (**14**) a single resonance pattern is observed, which is expected for the situation when all copper atoms are three-coordinate, that is, chelate ring formation occurs either clockwise or counter-clockwise without generating chirality (cf. Figures in Reference 207a). However, on introduction of a stereogenic centre in the *ortho*-substituent, that is, a methyl group, we have a different situation. First, this chirality can unambiguously establish the presence of Cu–N coordination in solution, and second, in the case of stable Cu–N coordination, it can provide evidence for stereochemistry at the C_{ipso} centres in the tetranuclear aggregate. The latter supposition can be made because the stable configuration at the C_{ipso} centres is now accompanied by a given configuration at the benzylic centres. Consequently, the Cu$_2$–aryl units in (**104**) (see Figure 30) are either $R_C R_{C'}$ or $S_C R_{C'}$ diastereoisomers, that is, they are NMR distinguishable. Assembly of the diastereomeric units under the condition that each copper atom is three-coordinate leads to a structure with alternating $R_C R_{C'}$ and $S_C R_{C'}$ diastereoisomeric units. This occurrence in solution is confirmed by the resonance patterns of the CH(Me)NMe$_2$ protons and carbon nuclei and of C_{ipso} in the ^{13}C NMR spectra of enantiopure (**104**). For racemic (**104**) seven lines are observed for C_{ipso} originating from the seven possible diastereoisomers that can be formed (see Reference 208a for data and spectra). A similar approach has been followed to elucidate the stereochemistry and dynamic behaviour in solution of the neutral cuprate, argentate and aurate $[Li_2M_2(C_6H_4CH_2NMe_2-2)_4]$ (M = Cu (**3**), Ag (**4**), Au (**93**)) and of the racemic and enantiopure derivatives $[Li_2M_2(C_6H_4CH(Me)NMe_2-2)_4]$ (M = Cu or Ag) (see also the figures in *COMC-I*).[19,104,185,200,208,233] In particular, the experiments with enantiopure cuprates showed that formation of the tetranuclear $[Li_2Cu_2Ar_4]$ aggregate is diastereoselective and occurs under complete control of the chosen configuration of the stereogenic benzylic carbon centre.

ACKNOWLEDGEMENTS

The preparation of the figures by Mr Dennis Kruis is greatly appreciated.

2.4 REFERENCES

1. (a) B. H. Lipshutz and S. Sengupta, *Org. Reacts.*, 1992, **41**, 135; (b) T. Ibuka and Y. Yamamoto, *Synlett.*, 1992, 769; (c) R. J. K. Taylor, *Synthesis*, 1985, 364; (d) B. E. Rossiter and N. M. Swingle, *Chem. Rev.*, 1992, **92**, 771; (e) M. J. Chapdelaine and M. Hulce, *Org. Reacts.*, 1990, **38**, 225; (f) Y. Yamamoto, *Angew. Chem., Int. Ed. Engl.*, 1986, **25**, 947; E. Erdik, *Tetrahedron*, 1984, **40**, 641; (g) J. F. Normant and A. Alexakis, *Synthesis*, 1981, 841; (h) G. H. Posner, 'An Introduction to Synthesis Using Organocopper Reagents', Wiley, New York, 1980; (i) G. H. Posner, *Org. React.*, 1975, **22**, 253; 1972, **19**, 1.
2. B. H. Lipshutz, in 'Organometallics in Synthesis', ed. M. Schlosser, Wiley, New York, 1994, chap. 4, p. 283.
3. P. Wipf, *Synthesis*, 1993, 537.
4. R. Reich, *C. R. Hebd. Seances Acad. Sci.*, 1923, **177**, 322.
5. A. Cairncross and W. A. Sheppard, *J. Am. Chem. Soc.*, 1968, **90**, 2186; A. Cairncross, H. Omura and W. A. Sheppard, *ibid.*, 1971, **93**, 248.
6. (a) G. van Koten, A. J. Leusink and J. G. Noltes, *J. Chem. Soc., Chem. Commun.*, 1971, 1107; (b) J. M. Guss, R. Mason, I. Søtofte, G. van Koten and J. G. Noltes, *ibid.*, 1972, 446.
7. J. M. Guss, R. Mason, K. M. Thomas, G. van Koten and J. G. Noltes, *J. Organomet. Chem.*, 1972, **40**, C79.
8. G. van Koten and J. G. Noltes, *J. Chem. Soc., Chem. Commun.*, 1972, 940.
9. A. J. Leusink, G. van Koten, J. W. Marsman and J. G. Noltes, *J. Organomet. Chem.*, 1973, **55**, 419.
10. G. van Koten, J. T. B. H. Jastrzebski, F. Muller and C. H. Stam, *J. Am. Chem. Soc.*, 1985, **107**, 697.
11. (a) and R. Pearce, *J. Chem. Soc., Chem. Commun.* 1973, 24; J. A. J. Jarvis, B. T. Kilbourn, R. Pearce and M. F. Lappert, *J. Chem. Soc., Chem. Commun.*, 1973, 475; (b) J. A. J. Jarvis, R. Pearce and M. Lappert, *J. Chem. Soc., Dalton Trans.*, 1977, 999.
12. J. G. Noltes, R. W. M. ten Hoedt, G. van Koten, A. L. Spek and J. C. Schoone, *J. Organomet. Chem.*, 1982, **225**, 365.
13. P. W. R. Corfield and H. M. M. Shearer, *Acta Crystallogr.*, 1966, **21**, 957.
14. (a) H. Hope, D. Oram and P. P. Power, *J. Am. Chem. Soc.*, 1984, **106**, 1149; (b) C. Eaborn, P. B. Hitchcock, J. D. Smith and A. C. Sullivan, *J. Chem. Soc., Chem. Commun.*, 1984, 870; (c) P. G. Edwards, R. W. Gellert, M. W. Marks and R. Bau, *J. Am. Chem. Soc.*, 1982, **104**, 2072.
15. D. M. Knotter, D. M. Grove, W. J. J. Smeets, A. L. Spek and G. van Koten, *J. Am. Chem. Soc.*, 1992, **114**, 3400.
16. M. M. Olmstead and P. P. Power, *J. Am. Chem. Soc.*, 1990, **112**, 8008.
17. B. Lenders, D. M. Grove, W. J. J. Smeets, P. van der Sluis, A. L. Spek and G. van Koten, *Organometallics*, 1991, **10**, 786.
18. H. Lang, K. Köhler and S. Blau, *Coord. Chem. Rev.*, in press; B. Lenders and W. Kläui, *Chem. Ber.*, 1990, **123**, 2233 and references therein.
19. G. van Koten and J. G. Noltes, in 'COMC-I', vol. 2, p. 709.
20. A. E. Jukes, *Adv. Organomet. Chem.*, 1974, **12**, 215.
21. C. D. M. Beverwijk, G. J. M. van der Kerk, A. J. Leusink and J. G. Noltes, *Organomet. Chem. Rev. (A)*, 1970, **5**, 215.
22. 'Gmelin Handbuch der anorganischen Chemie', Springer, Berlin, 1975, vol. 61, part B5.
23. G. Bahr and G. Burba, in 'Methoden Der Organischen Chemie (Houben-Weyl)', Thieme, Stuttgart, 1970, vol. 13/1, p. 763.
24. A. Camus, N. Marsich, G. Nardin and L. Randaccio, *Inorg. Chim. Acta*, 1977, **23**, 131.
25. G. Bahr and G. Burba, in 'Methoden Der Organischen Chemie (Houben-Weyl)', Thieme, Stuttgart, 1970, vol. 13/1, p. 731.
26. J. F. Normant, in 'New Applications of Organometallic Reagents in Organic Synthesis', ed. D. Seyferth, Elsevier, Amsterdam, 1976, p. 219.
27. R. I. Papasergio, C. L. Raston and A. H. White, *J. Chem. Soc., Chem. Commun.*, 1983, 1419.
28. R. I. Papasergio, C. L. Raston and A. H. White, *J. Chem. Soc., Dalton. Trans.*, 1987, 3085.
29. A. N. Nesmeyanov, Yu. T. Struchkov, N. N. Sedova, V. G. Andrianov, Yu. V. Volgin and V. A. Sazonova, *J. Organomet. Chem.*, 1977, **137**, 217.
30. Z.-Y. Yang, D. M. Wiemers and D. J. Burton, *J. Am. Chem. Soc.*, 1992, **114**, 4402.
31. A. Camus, N. Marsich, G. Nardin and L. Randaccio, *J. Organomet. Chem.*, 1973, **21**, C39.
32. D. F. Dempsey and G. S. Girolami, *Organometallics*, 1988, **7**, 1208.
33. (a) P. S. Coan, K. Folting, J. C. Huffman and K. G. Caulton, *Organometallics*, 1989, **8**, 2724; (b) A. Miyashita and A. Yamamoto, *Bull. Chem. Soc. Jpn.*, 1977, **50**, 1102.
34. H. Schmidbaur and W. Richter, *Chem. Ber.*, 1975, **108**, 2656.
35. G. Nardin, L. Randaccio and E. Zangrando, *J. Organomet. Chem.*, 1974, **74**, C23.
36. S. Gambarotta, S. Strologo, C. Floriani, A. Chiesi-Villa and C. Guastini, *Organometallics*, 1984, **3**, 1444.
37. S. Gambarotta, C. Floriani, A. Chiesi-Villa and C. Guastini, *J. Chem. Soc., Chem. Commun.*, 1983, 1156.
38. E. M. Meyer, S. Gambarotta, C. Floriani, A. Chiesi-Villa and C. Guastini, *Organometallics*, 1989, **8**, 1067.
39. H. L. Aalten, G. van Koten, K. Goubitz and C. H. Stam, *J. Chem. Soc., Chem. Commun.*, 1985, 1252; H. L. Aalten, G. van Koten, K. Goubitz and C. H. Stam, *Organometallics*, 1989, **8**, 2293.
40. · A. Camus, N. Marsich, G. Nardin and L. Randaccio, *J. Organomet. Chem.*, 1979, **174**, 121.
41. D. Nobel, G. van Koten and A. L. Spek, *Angew. Chem., Int. Ed. Engl.*, 1989, **28**, 208.
42. R. Lingnau and J. Strähle, *Angew. Chem., Int. Ed. Engl.*, 1988, **27**, 436.
43. G. van Koten, A. J. Leusink and J. G. Noltes, *J. Organomet. Chem.*, 1975, **84**, 117.
44. E. Wehman *et al.*, *J. Organomet. Chem.*, 1987, **325**, 293.
45. E. Wehman, G. van Koten, J. T. B. H. Jastrzebski, M. A. Rotteveel and C. H. Stam, *Organometallics*, 1988, **7**, 1477.
46. G. van Koten, A. J. Leusink and J. G. Noltes, *J. Organomet. Chem.*, 1975, **85**, 105.
47. R. W. M. Ten Hoedt, G. van Koten and J. G. Noltes, *J. Organomet. Chem.*, 1980, **201**, 327.
48. E. Wehman, G. van Koten, C. J. M. Erkamp, D. M. Knotter, J. T. B. H. Jastrzebski and C. H. Stam, *Organometallics*, 1989, **8**, 94.
49. P. Leoni, M. Pasquali and C. A. Ghilardi, *J. Chem. Soc., Chem. Commun.*, 1983, 240.
50. W. J. J. Smeets and A. L. Spek, *Acta Crystallogr., Sect. C*, 1987, **43**, 870.
51. D. M. Knotter, A. L. Spek, D. M. Grove and G. van Koten, *J. Chem. Soc., Chem. Commun.*, 1989, 1738.
52. D. M. Knotter, A. L. Spek and G. van Koten, *Organometallics*, 1992, **11**, 4083.
53. M. P. Gamasa, J. Gimeno, E. Lastra and X. Solans, *J. Organomet. Chem.*, 1988, **346**, 277.

54. P. W. R. Corfield and H. M. M. Shearer, cited in G. E. Coates, M. L. H. Green and K. Wade, 'Organometallic Compounds', Methuen, London, 1968, vol. 2, p. 274; 'Abstracts of the American Crystallographic Association, Bozenman, Montana', 1964, p. 96.
55. L. Naldini, F. Demartin, M. Manassero, M. Sansoni, G. Rassu and M. A. Zoroddu, *J. Organomet. Chem.*, 1985, **279**, C42.
56. M. P. Gamasa, J. Gimeno, E. Lastra, A. Aguirre and S. García-Granda, *J. Organomet. Chem.*, 1989, **378**, C11.
57. M. G. B. Drew, F. S. Esho and S. M. Nelson, *J. Chem. Soc., Chem. Commun.*, 1982, 1347.
58. R. W. M. ten Hoedt, J. G. Noltes, G. van Koten and A. L. Spek, *J. Chem. Soc., Dalton. Trans.*, 1978, 1800.
59. R. W. M. ten Hoedt, G. van Koten and J. G. Noltes, *J. Organomet. Chem.*, 1977, **133**, 113.
60. G. van Koten, R. W. M. ten Hoedt and J. G. Noltes, *J. Org. Chem.*, 1977, **42**, 2705.
61. C. Zybill and G. Müller, *Organometallics*, 1987, **6**, 2489.
62. M. M. Olmstead and P. P. Power, *Organometallics*, 1990, **9**, 1720.
63. S. F. Martin *et al.*, *J. Am. Chem. Soc.*, 1988, **110**, 7226.
64. H. Hope, M. M. Olmstead, P. P. Power, J. Sandell and X. Xu, *J. Am. Chem. Soc.*, 1985, **107**, 4337.
65. N. P. Lorenzen and E. Weiss, *Angew. Chem., Int. Ed. Engl.*, 1990, **29**, 300.
66. S. I. Khan, P. G. Edwards, H. S. H. Yuan and R. Bau, *J. Am. Chem. Soc.*, 1985, **107**, 1682.
67. O. M. Abu-Salah, M. Sakhawat Hussein and E. O. Schlemper, *J. Chem. Soc., Chem. Commun.*, 1988, 212.
68. A. N. Nesmeyanov, N. N. Sedova, Yu. T. Struchkov, V. G. Andrianov, E. N. Stakheeva and V. A. Sazonova, *J. Organomet. Chem.*, 1978, **153**, 115.
69. R. Usón, A. Laguna, A. Usón, P. G. Jones and K. Meyer-Bäse, *J. Chem. Soc., Dalton Trans.*, 1988, 341.
70. R. R. Burch and J. C. Calabrese, *J. Am. Chem. Soc.*, 1986, **108**, 5359.
71. J. R. DeMember, H. F. Evans, F. A. Wallace and P. A. Tariverdian, *J. Am. Chem. Soc.*, 1983, **105**, 5647.
72. S. Gambarotta, C. Floriani, A. Chiesi-Villa and C. Guastini, *J. Chem. Soc., Chem. Commun.*, 1983, 1087.
73. P. W. R. Corfield and H. M. M. Shearer, *Acta Crystallogr.*, 1966, **20**, 502.
74. J. Vicente *et al.*, *J. Organomet. Chem.*, 1987, **331**, 409.
75. J. Vicente, T. M. Chicote, I. Saura-Llamas and P. G. Jones, *Organometallics*, 1989, **8**, 767.
76. S. Wang, J. P. Fackler, Jr., and T. F. Carlson, *Organometallics*, 1990, **9**, 1973.
77. M. Y. Chiang, E. Böhlen and R. Bau, *J. Am. Chem. Soc.*, 1985, **107**, 1679.
78. R. W. M. ten Hoedt, G. van Koten and J. G. Noltes, *J. Organomet. Chem.*, 1979, **179**, 227.
79. A. N. Nesmeyanov, N. N. Sedova, V. A. Sazonova and S. K. Moiseev, *J. Organomet. Chem.*, 1980, **185**, C6.
80. L. T. J. Delbaere, D. W. McBride and R. B. Ferguson, *Acta Crystallogr., Sect. B*, 1970, **26**, 515.
81. F. A. Cotton and J. Takats, *J. Am. Chem. Soc.*, 1970, **92**, 2353.
82. G. van Koten, J. T. B. H. Jastrzebski and J. G. Noltes, *Inorg. Chem.*, 1977, **16**, 1782.
83. S. H. Bertz and G. Dabbagh, *Tetrahedron*, 1989, **45**, 425.
84. H. O. House, C.-Y. Chu, J. M. Wilkins and M. J. Umen, *J. Org. Chem.*, 1975, **40**, 1460.
85. M. D. Janssen, D. M. Grove, A. L. Spek and G. van Koten, in 'XVIth International Conference on Organometallic Chemistry, 1994', Royal Society of Chemistry (FECS event No. 196), Book of Abstracts, OC.9.
86. S. H. Bertz, C. P. Gibson and G. Dabbagh, *Organometallics*, 1988, **7**, 227.
87. G. van Koten, J. T. B. H. Jastrzebski and J. G. Noltes, *J. Org. Chem.*, 1977, **42**, 2047.
88. G. van Koten and J. G. Noltes, *J. Organomet. Chem.*, 1975, **84**, 419.
89. G. van Koten and J. G. Noltes, *J. Organomet. Chem.*, 1976, **104**, 127.
90. W. C. Fernelius (ed.), 'Inorganic Syntheses', McGraw-Hill, New York, 1946, vol. 2, p. 3.
91. S. Pasynkiewicz, S. Pikul and J. Poplawska, *J. Organomet. Chem.*, 1985, **293**, 125.
92. L. M. Seitz and R. Madl, *J. Organomet. Chem.*, 1972, **34**, 415.
93. H. K. Hofstee, J. Boersma and G. J. M. van der Kerk, *J. Organomet. Chem.*, 1978, **144**, 255.
94. G. Buckton, *Justus Liebigs Ann. Chem.*, 1859, **109**, 225.
95. D. M. Wiemers and D. Burton, *J. Am. Chem. Soc.*, 1986, **108**, 832.
96. M. A. Willert-Porada, D. J. Burton and N. C. Baenziger, *J. Chem. Soc., Chem. Commun.*, 1989, 1633.
97. H. Werner, H. Otto, T. Ngo-Khac and Ch. Burschka, *J. Organomet. Chem.*, 1984, **262**, 123.
98. D. W. Macomber and M. D. Rausch, *J. Am. Chem. Soc.*, 1983, **105**, 5325.
99. G. A. Carriedo, J. A. K. Howard and F. G. A. Stone, *J. Chem. Soc., Dalton Trans*, 1984, 1555.
100. A. J. Edwards, M. A. Paver, P. R. Raithby, M. A. Rennie, C. A. Russel and D. S. Wright, *Organometallics*, 1995, in press.
101. H. Gilman, R. G. Jones and L. A. Woods, *J. Org. Chem.*, 1952, **17**, 1630.
102. M. M. Olmstead and P. P. Power, *Organometallics*, 1990, **9**, 1720.
103. G. van Koten, J. T. B. H. Jastrzebski and J. G. Noltes, *J. Organomet. Chem.*, 1977, **140**, C23.
104. G. van Koten and J. G. Noltes, *J. Am. Chem. Soc.*, 1979, **101**, 6593.
105. E. C. Ashby and J. J. Lin, *J. Org. Chem.*, 1977, **42**, 2805.
106. B. H. Lipshutz, S. Sharma and E. L. Ellsworth, *J. Am. Chem. Soc.*, 1990, **112**, 4032.
107. B. H. Lipshutz, J. A. Kozlowski and R. S. Wilhelm, *J. Org. Chem.*, 1983, **48**, 546.
108. E. J. Corey and D. J. Beames, *J. Am. Chem. Soc.*, 1972, **94**, 7210.
109. B. H. Lipshutz, R. S. Wilhelm and D. M. Floyd, *J. Am. Chem. Soc.*, 1981, **103**, 7672; B. H. Lipshutz, R. S. Wilhelm and J. M. Kozlowski, *Tetrahedron*, 1984, **40**, 5005.
110. G. H. Posner, C. E. Whitten and J. J. Sterling, *J. Am. Chem. Soc.*, 1973, **95**, 7788; S. H. Bertz, G. Dabbagh and G. M. Villacorta, *ibid.*, 1982, **104**, 5824.
111. F. Lambert, D. M. Knotter, M. D. Janssen, M. van Klaveren, J. Boersma and G. van Koten, *Tetrahedron Asymmetry*, 1991, **2**, 1097; A. Haubrich, M. van Klaveren, G. van Koten, G. Handke and N. Krause, *J. Org. Chem.*, 1993, **58**, 5849.
112. M. van Klaveren, E. S. M. Persson, D. M. Grove, J.-E. Bäckvall and G. van Koten, *Tetrahedron Lett.*, 1994, **35**, 5931; M. van Klaveren, F. Lambert, D. J. F. M. Eijkelkamp, D. M. Grove and G. van Koten, *ibid.*, 1994, **35**, 6135.
113. B. E. Rossiter and N. M. Swingle, *Chem. Rev.*, 1992, **92**, 771; S. H. Bertz, G. Dabbagh and G. Sundararajan, *J. Org. Chem.*, 1986, **51**, 4953.
114. S. H. Bertz and G. Dabbagh, *J. Am. Chem. Soc.*, 1984, **94**, 1119.
115. S. H. Bertz, *J. Am. Chem. Soc.*, 1990, **112**, 4031.
116. S. H. Bertz and G. Dabbagh, *J. Am. Chem. Soc.*, 1988, **110**, 3668.
117. G. Costa, A. Camus, L. Gatti and N. Marsich, *J. Organomet. Chem.*, 1966, **5**, 568.

118. A. Cairncross and W. A. Sheppard, *J. Am. Chem. Soc.*, 1968, **90**, 2186; A. Cairncross and W. A. Sheppard, *ibid.*, 1971, **93**, 247.
119. D. M. Knotter, W. J. J. Smeets, A. L. Spek and G. van Koten, *J. Am. Chem. Soc.*, 1990, **112**, 5895.
120. G. van Koten, J. G. Leusink and J. G. Noltes, *Inorg. Nucl. Chem. Lett.*, 1971, **7**, 227.
121. G. van Koten and J. G. Noltes, *J. Organomet. Chem.*, 1975, **102**, 551.
122. E. Wehman, G. van Koten and J. T. B. H. Jastrzebski, *J. Organomet. Chem.*, 1986, **302**, C35.
123. G. van Koten and J. G. Noltes, *J. Chem. Soc., Chem. Commun.*, 1974, 575.
124. B. Bähr and G. Burba, in 'Methoden der organischen Chemie (Houben-Weyl)', Thieme, Stuttgart, 1970, vol. 13/1, p. 731.
125. A. M. Sladkov and I. R. Gol'ding, *Russ. Chem. Rev. (Engl. Transl.)*, 1979, **48**, 868.
126. A. M. Sladkov and L. Yu. Ukhin, *Russ. Chem. Rev.(Engl. Transl.)*, 1968, **37**, 748.
127. J. Diéz, M. P. Gamasa, J. Gimeno, A. Aguirre and S. García-Granda, *Organometallics*, 1991, **10**, 380.
128. T. Yamamoto, M. Kubota, A. Miyashita and A. Yamamoto, *Bull. Chem. Soc. Jpn.*, 1978, **51**, 1835.
129. G. van Koten and J. G. Noltes, *J. Chem. Soc., Chem. Commun.*, 1972, 452.
130. G. van Koten, J. G. Noltes and A. L. Spek, *J. Organomet. Chem.*, 1978, **159**, 441.
131. A. Camus and N. Marsich, *J. Organomet. Chem.*, 1970, **21**, 249.
132. T. Ikariya and A. Yamamoto, *J. Organomet. Chem.*, 1974, **72**, 145; A. Miyashita, T. Yamamoto and A. Yamamoto, *Bull. Chem. Soc. Jpn.*, 1977, **50**, 1109.
133. T. Tsuda, T. Hashimoto and T. Saegusa, *J. Am. Chem. Soc.*, 1972, **94**, 658.
134. T. Tsuda, H. Habu, S. Horiguchi and T. Saegusa, *J. Am. Chem. Soc.*, 1974, **96**, 5930.
135. G. Wilkinson and T. S. Piper, *J. Inorg. Nucl. Chem.*, 1956, **2**, 32.
136. T. Saegusa, Y. Ito and S. Tomita, *J. Am. Chem. Soc.*, 1971, **93**, 5656.
137. H. Werner, H. Otto, T. Ngo-Khac and Ch. Burschka, *J. Organomet. Chem.*, 1984, **262**, 123.
138. D. W. Macomber and M. D. Rausch, *J. Am. Chem. Soc.*, 1983, **105**, 5325.
139. G. van Koten, *J. Organomet. Chem.*, 1990, **400**, 283.
140. O. M. Abu-Salah, A. R. A. Al-Ohaly and C. B. Knobler, *J. Chem. Soc., Chem. Commun.*, 1985, 1502.
141. A. Cairncross, J. R. Roland, R. M. Henderson and W. A. Sheppard, *J. Am. Chem. Soc.*, 1970, **92**, 3187.
142. T. Tsuda, T. Nakatsuka, T. Hirayama and T. Saegusa, *J. Chem. Soc., Chem. Commun.*, 1974, 557.
143. N. Marsich, A. Camus and G. Nardin, *J. Organomet. Chem.*, 1982, **239**, 429.
144. G. van Koten and J. G. Noltes, *J. Chem. Soc., Chem. Commun.*, 1972, 59.
145. V. C. R. McLoughlin, *Tetrahedron*, 1969, **25**, 5921.
146. D. J. Burton and S. W. Hansen, *J. Am. Chem. Soc.*, 1986, **108**, 4229.
147. Y.-T. Jeong, J.-H. Jung, S.-K. Shin, Y.-G. Kim, I.-H. Jeong and S.-K. Choi, *J. Chem. Soc., Perkin Trans. 1*, 1991, 1601.
148. R. M. Wehmeyer and R. D. Rieke, *J. Org. Chem.*, 1987, **52**, 5056.
149. H. Schmidbaur, J. Adlkofer and W. Buchner, *Angew. Chem., Int. Ed. Engl.*, 1973, **12**, 415.
150. Y. Yamamoto and H. Schmidbaur, *J. Organomet. Chem.*, 1975, **96**, 133.
151. Y. Yamamoto and H. Schmidbaur, *J. Organomet. Chem.*, 1975, **97**, 479.
152. H. Schmidbaur, J. Adlkofer and M. Heimann, *Chem. Ber.*, 1974, **107**, 3697.
153. C. E. Zybill and G. Müller, *Organometallics*, 1987, **6**, 2489.
154. R. Kumar and D. G. Tuck, *J. Organomet. Chem.*, 1985, **281**, C47.
155. J. A. Wanklyn and L. Carius, *Justus Liebigs Ann. Chem.*, 1861, **120**, 70.
156. E. Krause and M. Schmitz, *Ber. Dtsch. Chem. Ges.*, 1919, **52**, 2159.
157. R. Usón *et al.*, *J. Chem. Soc., Dalton Trans.*, 1990, 333.
158. R. I. Papasergio, C. L. Raston and A. H. White, *J. Chem. Soc., Chem. Commun.*, 1984, 612.
159. W. T. Miller and R. J. Burnard, *J. Am. Chem. Soc.*, 1968, **90**, 7367.
160. M. A. Guerra, T. R. Bierschenk and R. J. Lagow, *J. Organomet. Chem.*, 1986, **307**, C58.
161. H. K. Nair and J. A. Morrison, *J. Organomet. Chem.*, 1989, **376**, 149.
162. W. Dukat and D. Naumann, *Rev. Chim. Miner.*, 1986, **23**, 589.
163. A. K. Holliday and R. E. Pendlebury, *J. Organomet. Chem.*, 1967, **7**, 281.
164. F. Glockling and D. Kingston, *J. Chem. Soc.*, 1959, 3001.
165. R. E. Banks, R. N. Haszeldine, D. R. Taylor and G. Webb, *Tetrahedron Lett.*, 1970, 5215.
166. C. D. M. Beverwijk and G. J. M. van der Kerk, *J. Organomet. Chem.*, 1972, **43**, C11.
167. A. J. Leusink, G. van Koten and J. G. Noltes, *J. Organomet. Chem.*, 1973, **56**, 379.
168. H. K. Hofstee, J. Boersma and G. J. M. van der Kerk, *J. Organomet. Chem.*, 1978, **168**, 241.
169. J. Boersma, F. J. A. des Tombe, F. Weijers and G. J. M. van der Kerk, *J. Organomet. Chem.*, 1977, **124**, 229.
170. D. Nobel, G. van Koten and A. L. Spek, cf. Ref. 41.
171. F. Bonati, A. Burini, B. R. Pietroni and B. Bovio, *J. Organomet. Chem.*, 1989, **375**, 147.
172. R. Usón, A. Laguna and J. A. Abad, *J. Organomet. Chem.*, 1983, **246**, 341.
173. R. Usón, A. Laguna, M. Laguna, A. Usón, P. G. Jones and M. C. Gimeno, *J. Chem. Soc., Dalton. Trans.*, 1988, 701.
174. J. Blenkers, H. K. Hofstee, J. Boersma and G. J. M. van der Kerk, *J. Organomet. Chem.*, 1979, **168**, 251.
175. C. Eaborn, P. B. Hitchcock, J. D. Smith and A. C. Sullivan, *J. Chem. Soc., Chem. Commun.*, 1984, 870.
176. G. van Koten and J. G. Noltes, *J. Organomet. Chem.*, 1975, **84**, 129.
177. G. van Koten and J. G. Noltes, *J. Organomet. Chem.*, 1975, **102**, 551.
178. G. van Koten and J. G. Noltes, *J. Organomet. Chem.*, 1974, **82**, C53; G. van Koten, C. A. Schaap, J. T. B. H. Jastrzebski and J. G. Noltes, *ibid.* 1980, **186**, 427.
179. A. Laguna, M. Laguna, J. Jiménez and A. J. Fumanal, *J. Organomet. Chem.*, 1990, **396**, 121.
180. R. Usón, A. Laguna, E. J. Fernandez, A. Mendia and P. G. Jones, *J. Organomet. Chem.*, 1988, **350**, 129.
181. O. M. Abu-Salah and C. B. Knobler, *J. Organomet. Chem.*, 1986, **302**, C10.
182. O. M. Abu-Salah, *J. Organomet. Chem.*, 1990, **387**, 123.
183. P. Espinet, J. Forniés, F. Martínez, M. Tomás, E. Lalinde, M. T. Moreno, A. Ruiz and A. J. Welch, *J. Chem. Soc., Dalton Trans.*, 1990, 791.
184. P. Espinet, J. Forniés, F. Martínez, M. Sotes, E. Lalinde, M. T. Moreno, A. Ruiz and A. J. Welch, *J. Organomet. Chem.*, 1991, **403**, 253.

185. G. van Koten, J. T. B. H. Jastrzebski, C. H. Stam and C. Brevard, in 'Biological and Inorganic Copper Chemistry', eds. K. D. Karlin and J. J. Zubieta, Adenine Press, Guilderland, 1985, p. 267.
186. G. Semerano and L. Riccoboni, *Chem. Ber.*, 1941, **74**, 1089.
187. A. Camus and N. Marsich, *J. Organomet. Chem.*, 1968, **14**, 441.
188. F. A. Cotton and T. J. Marks, *J. Am. Chem. Soc.*, 1969, **91**, 7281.
189. H. K. Hofstee, J. Boersma and G. J. M. van der Kerk, *J. Organomet. Chem.*, 1976, **120**, 313.
190. K. K. Sun and W. T. Miller, *J. Am. Chem. Soc.*, 1970, **92**, 6985.
191. R. W. M. ten Hoedt, G. van Koten and J. G. Noltes, *Organomet. Chem.*, 1979, **179**, 227.
192. F. Glockling and D. Kingston, *J. Chem. Soc.*, 1959, 3001.
193. G. Minghetti, F. Bonati and M. Massobrio, *Inorg. Chem.*, 1975, **14**, 1974.
194. R. G. Pearson and C. D. Gregory, *J. Am. Chem. Soc.*, 1976, **98**, 4098.
195. H. O. House, *Acc. Chem. Res.*, 1976, **9**, 59.
196. P. W. R. Corfield and H. M. M. Shearer, *Acta. Crystallogr.*, 1966, **20**, 502.
197. G. Costa, A. Camus, L. Gatti and N. Marsich, *J. Organomet. Chem.*, 1966, **5**, 568.
198. G. Costa, A. Camus, N. Marsich and L. Gatti, *J. Organomet. Chem.*, 1967, **8**, 339.
199. G. van Koten and J. G. Noltes, *J. Am. Chem. Soc.*, 1979, **101**, 6593.
200. G. van Koten and J. G. Noltes, in 'Fundamental Research in Homogeneous Catalysis', eds. M. Tsutsui and R. Ugo, Plenum, New York, 1979, vol. 3, p. 953.
201. E. M. Meyer, S. Gambarotta, C. Floriani, A. Chiesi-Villa and C. Guastini, *Organometallics*, 1989, **8**, 1067.
202. M. Håkansson, M. Örtendahl, J. Jagner, M. P. Sigalas and O. Eisenstein, *Inorg. Chem.*, 1993, **32**, 2018.
203. F. Olbrich, J. Kopf and E. Weiss, *Angew. Chem., Int. Ed. Engl.*, 1993, **32**, 1077.
204. C. Eaborn, P. B. Hitchcock, J. D. Smith and A. C. Sullivan, *J. Organomet. Chem.*, 1984, **263**, C23.
205. X. He, K. Ruhlandt-Senge, P. P. Power and S. H. Bertz, *J. Am. Chem. Soc.*, 1994, **116**, 6963.
206. H. Eriksson, M. Örtendahl and M. Håkansson, *Organometallics*, 1995, in press.
207. T. Tsuda, K. Watanabe, K. Miyata, H. Yamamoto and T. Saegusa, *Inorg. Chem.*, 1981, **20**, 2728; T. Tsuda, T. Yazawa, K. Watanabe, T. Fujii and T. Saegusa, *J. Org. Chem.*, 1981, **46**, 192.
208. (a) G. van Koten and J. T. B. H. Jastrzebski, *Tetrahedron*, 1989, **45**, 569; (b) X. He, M. M. Olmstead and P. P. Power, *J. Am. Chem. Soc.*, 1992, **114**, 9668.
209. G. A. Carriedo, J. A. K. Howard and F. G. A. Stone, *J. Organomet. Chem.*, 1983, **250**, C28.
210. A. Haaland, K. Rypdal, H. P. Verne, W. Scherer and W. R. Thiel, *Angew. Chem., Int. Ed. Engl.*, 1994, **33**, 2443.
211. G. M. Kapteijn, I. C. M. Wehman-Ooyevaar, D. M. Grove, W. J. J. Smeets, A. L. Spek and G. van Koten, *Angew. Chem., Int. Ed. Engl.*, 1993, **32**, 72.
212. S. Jagner and G. Helgesson, *Adv. Inorg. Chem.*, 1991, **37**, 10.
213. M. J. Albright, W. M. Butler, T. J. Anderson M. D. Glick and J. P. Oliver, *J. Am. Chem. Soc.*, 1976, **98**, 3995.
214. P. P. Power, *Prog. Inorg. Chem.*, 1991, **39**, 75.
215. J. T. B. H. Jastrzebski, G. van Koten, M. Konijn and C. H. Stam, *J. Am. Chem. Soc.*, 1982, **104**, 5490.
216. A. F. Wells, 'Structural Inorganic Chemistry', 5th edn., Oxford University Press, Oxford, 1984, p. 1104.
217. A. E. Dorigo and K. Morokuma, *J. Am. Chem. Soc.*, 1989, **111**, 6524; A. E. Dorigo and K. Morokuma, *ibid.*, 1989, **111**, 4653; J. P. Fackler, Jr., *Prog. Inorg. Chem.*, 1976, **21**, 55; A. Avdeef and J. P. Fackler, Jr., *Inorg. Chem.*, 1978, **17**, 2182; F. J. Hollander, D. Coucouvanis, J. P. Fackler, Jr. and K. Knox, *J. Am. Chem. Soc.*, 1974, **96**, 5646; P. K. Mehrotra and R. Hoffmann, *Inorg. Chem.*, 1978, **17**, 2187; D. M. P. Mingos, *J. Chem. Soc. Dalton Trans.*, 1976, 1163.
218. R. Mason and D. M. P. Mingos, *J. Organomet. Chem.*, 1973, **50**, 53.
219. K. R. Stewart, J. R. Lever and M.-H. Whangbo, *J. Org. Chem.*, 1982, **47**, 1472.
220. J. P. Snyder, G. H. Tipsword and D. P. Spangler, *J. Am. Chem. Soc.*, 1992, **114**, 1507.
221. E. C. Ashby and J. J. Watkins, *J. Chem. Soc. Chem. Commun.*, 1976, 784.
222. E. C. Ashby and J. J. Watkins, *J. Am. Chem. Soc.*, 1977, **99**, 5312.
223. D. J. Clive, V. Frina and P. L. Beaulieu, *J. Org. Chem.*, 1982, **25**, 2572.
224. E. C. Ashby, J. J. Lin and J. J. Watkins, *J. Org. Chem.*, 1977, **42**, 1099.
225. T. Brown, *Adv. Organomet. Chem.*, 1966, **3**, 365.
226. S. H. Bertz, G. Dabbagh, X. He and P. P. Power, *J. Am. Chem. Soc.*, 1993, **115**, 1640.
227. S. H. Bertz, *J. Am. Chem. Soc.*, 1990, **112**, 4031; S. H. Bertz, *ibid.*, 1991, **113**, 5471.
228. T. Stemmler, J. E. Penner-Hahn and P. Knockel, *J. Am. Chem. Soc.*, 1993, **115**, 348.
229. J. P. Snyder, D. P. Spangler, J. R. Behling and B. E. Rossiter, *J. Org. Chem.*, 1994, **59**, 2665.
230. G. Hallnemo, T. Olsson and C. Ullenius, *J. Organomet. Chem.*, 1985, **282**, 133.
231. N. Krause, R. Wagner and A. Gerold, *J. Am. Chem. Soc.*, 1994, **116**, 381.
232. G. van Koten, *Pure Appl. Chem.*, 1994, **66**, 1455.
233. G. van Koten and J. G. Noltes, *J. Organomet. Chem.*, 1979, **174**, 367; G. van Koten and J. G. Noltes, *ibid.*, 1979, **171**, C39.
234. G. van Koten, J. T. B. H. Jastrzebski, C. H. Stam and N. C. Niemann, *J. Am. Chem. Soc.*, 1984, **106**, 1880.
235. U. Schumann and E. Weiss, *Angew. Chem., Int. Ed. Engl.*, 1988, **27**, 584.
236. T. P. Hanusa, T. A. Ulibarri and W. J. Evans, *Acta Crystallogr., Sect. C*, 1985, **41**, 1036.
237. P. T. Beurskens, J. A. Cras and J. J. Steggerda, *Inorg. Chem.*, 1968, **7**, 810.
238. M. R. Churchill, B. G. de Boer, F. J. Rotella, O. M. Abu Salah and M. I. Bruce, *Inorg. Chem.*, 1975, **14**, 2051.
239. H. van Dam and G. van Koten, 1982, cf. Ref. 19.
240. D. E. Bergbreiter, T. J. Lynch and S. Shimazu, *Organometallics*, 1983, **2**, 1354.

3
Mercury

ALWYN G. DAVIES
University College London, UK

and

JAMES L. WARDELL
University of Aberdeen, UK

3.1 INTRODUCTION   135

3.2 FORMATION OF THE MERCURY–CARBON BOND   136

   3.2.1 Transmetallation   136
   3.2.2 Aliphatic Mercury–Hydrogen Exchange   139
   3.2.3 Aromatic Mercury–Hydrogen Exchange   141
   3.2.4 Decarboxylation   142
   3.2.5 Solvomercuration   144
      3.2.5.1 Alkenes   144
      3.2.5.2 Alkynes   148
      3.2.5.3 Cyclopropanes   151
   3.2.6 Carbene Insertion   153
   3.2.7 Miscellaneous Methods   154

3.3 STRUCTURES AND PROPERTIES OF ORGANOMERCURY COMPOUNDS   155

   3.3.1 Methylmercury(II) Compounds   155
   3.3.2 α-Mercuriated Carbonyl Compounds   157
   3.3.3 Arylmercury(II) Compounds   159
   3.3.4 Cyclopentadienylmercury Compounds   160
   3.3.5 Polydentate Lewis Acids   162
   3.3.6 Organomercury Hydrides   165
   3.3.7 Radicals and Radical Reactions   166

3.4 REFERENCES   171

3.1 INTRODUCTION

The period since the publication of *COMC-I*[1] has been largely one of development rather than innovation in organomercury chemistry.

Structural investigations have played a major part. Mercury(II) salts react as electrophiles with arenes and activated alkanes (e.g., carbonyl compounds), to replace hydrogen by mercury. Frequently, polymercuration occurs and, as the mercury is divalent, this can lead to the formation of linear, network or three-dimensional polymers. These compounds are often insoluble and intractable, and although they have been known for many years, it is only recently that their structures have been elucidated by single-crystal x-ray diffraction. High-resolution solid-state ^{13}C NMR spectroscopy can help in determining the structure, but solid-state ^{199}Hg NMR spectroscopy is rendered difficult by the large anisotropy of the chemical shift, and as yet has proved of little value.[2]

In the environment, inorganic mercury can undergo biological methylation, and the methylmercury compounds which are formed are lipophilic and can enter into the food chain. They are highly toxic, and this has been shown to result from their interaction with the base components of nucleic acids. A lot of effort has therefore been put into determining the nature of these complexes, again largely by x-ray crystallography.

A comprehensive list of the structures which have been determined is given in Volume 13. They confirm that in most compounds RHgX and RHgR, the geometry about the mercury is essentially linear; if secondary interactions occur, they usually do not cause a deviation by more than 10° from linearity. Cyclic compounds containing a number of mercury atoms can behave as 'anticrown' polydentate Lewis acids in binding anions and other Lewis bases.

The principal use of organomercury compounds in organic synthesis continues to be in the solvomercuriation of alkenes and alkynes, and this reaction has been surveyed in a book by Larock.[3] In the field of reaction mechanisms, there is an increasing awareness of the importance of processes involving electron-transfer and free-radical intermediates, and some of these reactions have been developed into useful preparative procedures. Precautions against environmental contamination have restricted the use of organomercury compounds outside the laboratory, but the search for better methods of trace analysis has contributed to the identification of the organomercury hydrides, RHgH, which were previously recognized only as reactive intermediates.

Annual reviews of organomercury chemistry have appeared in the Royal Society of Chemistry's Specialist Periodical Report on *Organometallic Chemistry*,[4] which, up to 1990, also listed structures that had been determined by diffraction methods. The second edition of the *Dictionary of Organometallic Compounds* indexes 1250 important organomercury compounds, with brief details of preparations and properties and leading references.[5]

3.2 FORMATION OF THE MERCURY–CARBON BOND

3.2.1 Transmetallation

Transmetallation reactions, as illustrated in Equations (1) and (2), provide the most general route to organomercury compounds.[1] The metal M is usually magnesium or lithium, although organoboron compounds, which are readily available through hydroboration reactions, are also frequently used.

$$RM + HgX_2 \longrightarrow RHgX + MX \tag{1}$$

$$R^1M + R^2HgX \longrightarrow R^1HgR^2 + MX \tag{2}$$

Some examples of recent applications of these reactions are shown in Scheme 1, Equations (3)–(7), Scheme 2 and Equations (8) and (9).[6–13]

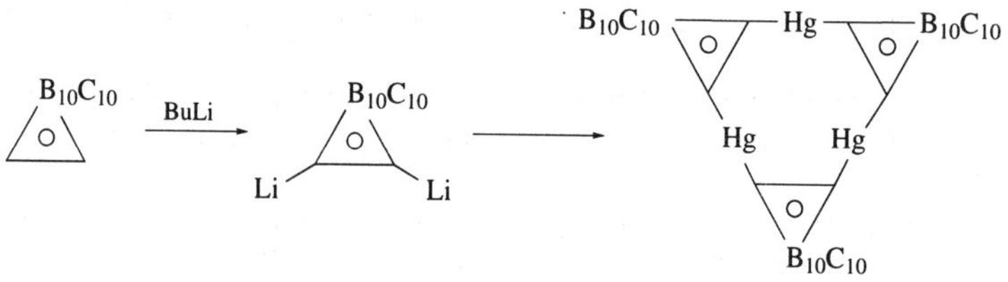

(1)

Scheme 1

$$MeOSiMe_2(TMS)_2CLi + HgBr_2 \xrightarrow[\substack{-110\,°C\ to\ RT \\ 30\%}]{THF} \left[MeOSiMe_2 - \overset{TMS}{\underset{TMS}{\mid}} - Hg \right]_2 \tag{3}$$

$$(Me_2PhSi)_3CLi + ClHgCH_2Ph \longrightarrow (Me_2PhSi)_3CHgCH_2Ph \tag{4}$$

(2)

$$Me_nHgX_{2-n} + NaBEt_4 \xrightarrow{\text{H}_2\text{O}} Me_nHgEt_{2-n} + NaBEt_3X \tag{5}$$

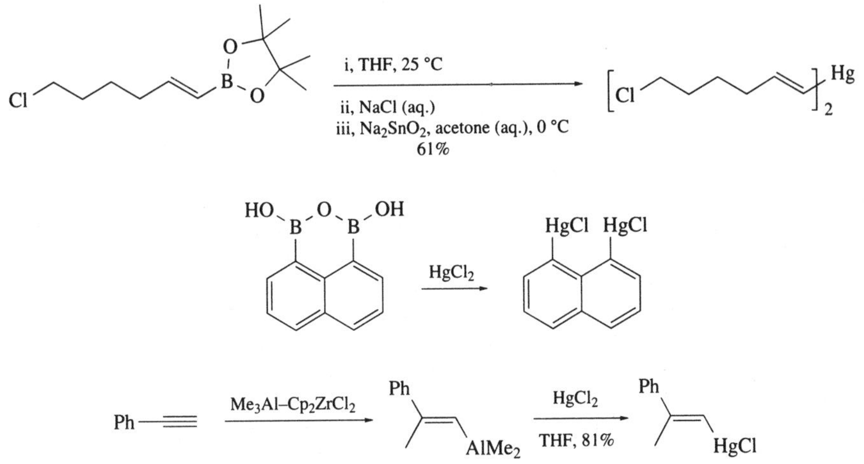

(6)

i, THF, 25 °C
ii, NaCl (aq.)
iii, Na$_2$SnO$_2$, acetone (aq.), 0 °C
61%

(7)

Scheme 2

$$EtO_2C(CH_2)_4ZnI + Hg_2Cl_2 \xrightarrow[\substack{RT, 2 h \\ 87\%}]{\text{THF}} [EtO_2C(CH_2)_4]_2Hg \tag{8}$$

$$NC(CH_2)_5ZnI + Hg_2Cl_2 \xrightarrow[\substack{RT, 2 h \\ 82\%}]{\text{THF}} [NC(CH_2)_5]_2Hg \tag{9}$$

A number of cyclic tri- and tetrameric mercury(II) compounds (e.g., (**1**)) derived from 1,2-dicarbadodecaboranes have been prepared via the lithium derivatives as shown in Equation (1). These act as polydentate Lewis acids ('anticrowns') towards anions and other Lewis acids, and they are discussed in Section 3.3.5.[14]

The sterically hindered benzylmercury(II) compound (**2**) shows a remarkable thermal stability compared with dibenzylmercury, and it has been suggested that this may be because, within the collisionally activated molecule, there is only slow transmission through the metal atom.[8]

Alkylation with a tetraalkylborate, BR_4^-, uses only one of the four alkyl groups, and the reagent is expensive, but it can be used in aqueous solution; the reaction is quantitative and is used for the analysis, by spectroscopically coupled gas–liquid chromatography, of organomercury pollutants in the environment. The absolute determination limit for MeHgCl in aqueous solution with BEt_4^- is 167 pg. It was the use of BH_4^- under similar conditions that resulted in the identification of the hydrides RHgH (see Section 3.3.6).

The alkylalumination reaction in Scheme 2 provides a method for the *cis*-alkylmercuriation of alkynes. Hydroalumination, and thence hydromercuriation, can be achieved with Me$_2$AlH as the reagent.[12]

Methylcobalamine has been used for the synthesis of Me^{203}HgCl from ^{203}HgCl$_2$,[15] and it has been shown that mercury(II) can be methylated in sea water by methyltin or methyllead compounds.[16]

Transmetallation of alkylzinc halides with mercury(II) chloride gives a very incomplete reaction, but the reaction with mercury(I) chloride (e.g., Equations (8) and (9)) is complete in a few hours at room temperature. Presumably the alkylmercury(I) compound RHgHgR is first formed, but this rapidly extrudes mercury.[13] This principle might be extended to the reaction with other organometallic reagents.

Another promising variant of the usual protocol involves generating the organolithium compound in the presence of HgCl$_2$. Thus, in a benzamide, PhCONR$_2$, the amide group activates the *ortho* positions, and treatment with lithium 2,2,5,5-tetramethylpiperidine (LiTMP) gives a small equilibrium concentration of the *o*-lithiobenzamide. This can be trapped by HgCl$_2$, driving the reaction to completion and giving the 2-mercuri- and 2,6-dimercuri-benzamide. These are not easy to isolate in the pure state, but the crude material reacts with bromine to give the 2-bromo- and 2,6-dibromo-benzamides, free from any 4-bromo isomer, or with butyllithium to give the 2-lithio- and 2,6-dilithio-benzamides, which cannot be prepared directly (Scheme 3).[17]

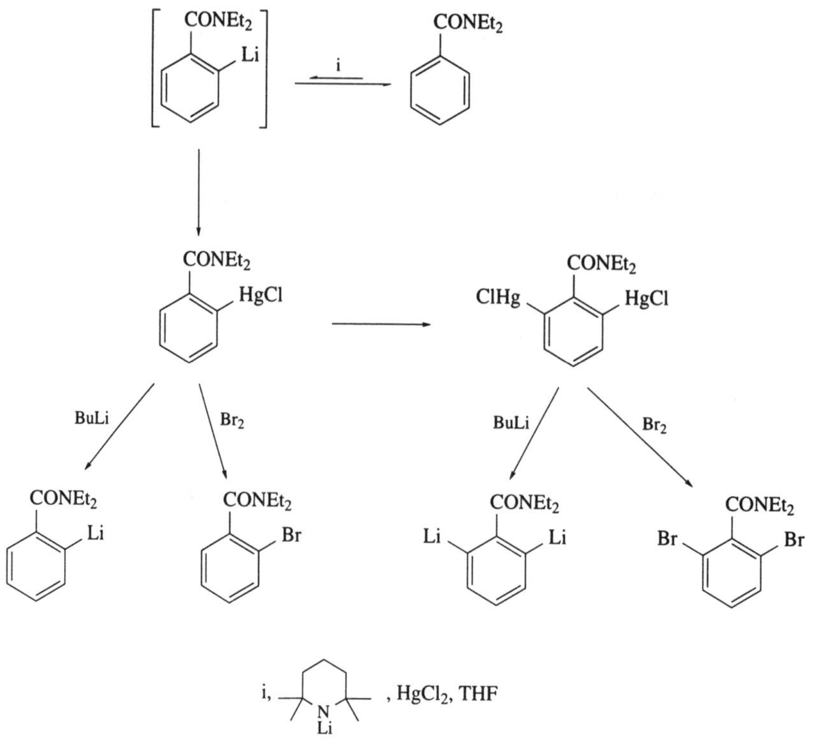

Scheme 3

The amide group will also activate a 'saturated' system. Thus the amidocubane (**3**) reacts with LiTMP to give only 3% of the 'ortho'-lithiated compound, but in the presence of mercury(II) chloride, the conversion into the chloromercury derivative (**4**) is almost complete. If this is now treated with methyllithium, the lithiated cubane is obtained in high yield in what has been called a reverse transmetallation reaction (Scheme 4). Similarly, the dicarboxamide (**5**) gives the dimercury derivative, from which the cubyllithium or cubylmagnesium compounds can be prepared (Scheme 5).[18] In a similar way, the cyclopropylcarboxamide (**6**) can be mercuriated and then converted into the corresponding Grignard reagent (Scheme 6).[18]

It is convenient to include the *o*-phenylene- and the biphenylenemercurials in this section, although some of the reactions involved are not transmetallations. *o*-Phenylenemercury, $(o\text{-}C_6H_4Hg)_n$, can be prepared from 1,2-dibromobenzene and sodium amalgam. It was originally thought to be a tetramer, but a full x-ray structure analysis showed it to be in fact a trimer, $n = 3$ (as is the corresponding perfluorophenylene compound). However, molecular models suggest that the hexamer and the tetramer are both structurally feasible, and a search has been made for these compounds carrying various ring substituents (Me, MeO, F, Cl), using mass spectrometric analysis.

2,3-Dibromotoluene, 3,4-dibromotoluene, 1,2-dibromo-4,5-dimethylbenzene and 1,2-dibromo-4,5-dimethoxybenzene react with sodium amalgam to give the corresponding *o*-phenylenemercurials (**7**)–(**9**) in 1–2% yield; clearly, with one substituent in the ring, other isomers are possible than those shown here, but the products have sharp melting points, and they are too insoluble to allow purification by chromatography. The mass spectra established the trimeric structure, and gave no indication of the presence of tetramers.[19]

Perfluorophenylenemercury is best prepared by decarboxylating mercury tetrafluorophthalate, and is also a trimer. Perchlorophenylenemercury, again a trimer, cannot be obtained by an analogous route, but it can be prepared in almost quantitative yield by heating 1,2-diiodotetrachlorobenzene with mercury in a sealed evacuated tube at 260–300 °C, and the mixed trimers $(C_6Cl_4)(C_6F_4)_2Hg_3$ and $(C_6Cl_4)_2(C_6F_4)Hg_3$ can be obtained from a mixture of $(C_6F_4Hg)_3$ and $(C_6Cl_4Hg)_3$.[19]

2,2'-Biphenylenemercury can be prepared from the reaction of 2,2'-dilithiobiphenyl with mercury(II) chloride (Equation (10)). Again, this compound, which had been thought to be a tetramer, has now been shown by mass spectrometry to be a trimer. Perfluoro-2,2'-biphenylenemercury, prepared similarly, is again a trimer.[19]

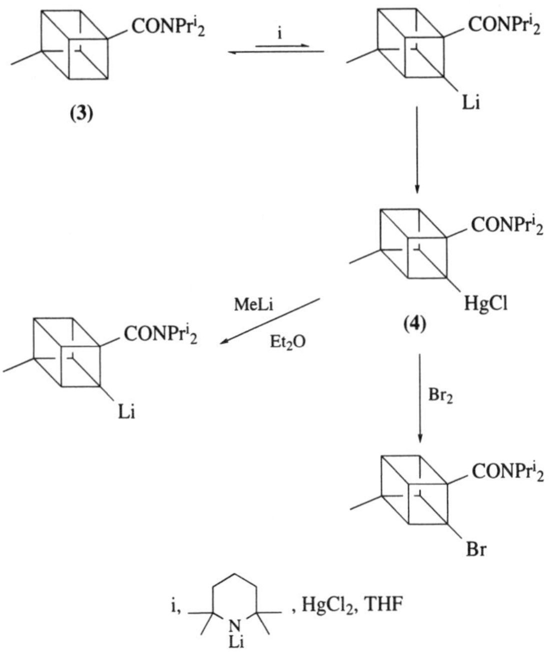

Scheme 4

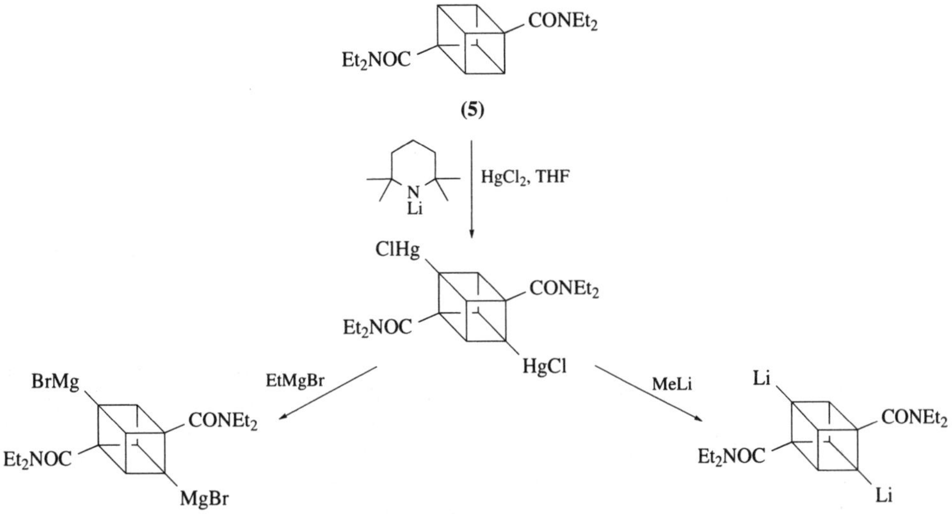

Scheme 5

3.2.2 Aliphatic Mercury–Hydrogen Exchange

The replacement of hydrogen by mercury is an electrophilic substitution (Equation (11)), and it therefore occurs with arenes, cyclopentadienes and terminal alkynes, and also with aliphatic compounds such as carbonyl compounds or nitriles in which the hydrogen has enhanced acidic character.

When there are a number of equivalent hydrogen atoms, polymercuriation often occurs. Many of these reactions have been known for almost a century, although the detailed molecular structures of the products were often obscure, and are only now becoming clear with the application of x-ray crystallography. These crystallographic studies are dealt with in a later section; we are concerned here only with new preparative aspects.

Acetone slowly undergoes progressive mercuriation by mercury(II) salts in aqueous solution, and compounds such as $MeCOCH_2HgI$ and $(MeCOCH_2)_2Hg$ can be isolated. These products have not yet been examined by x-ray crystallography, but the progressive replacement of hydrogen by mercury has

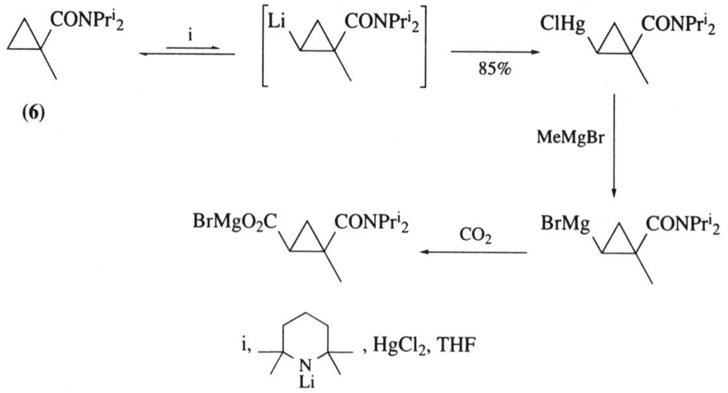

Scheme 6

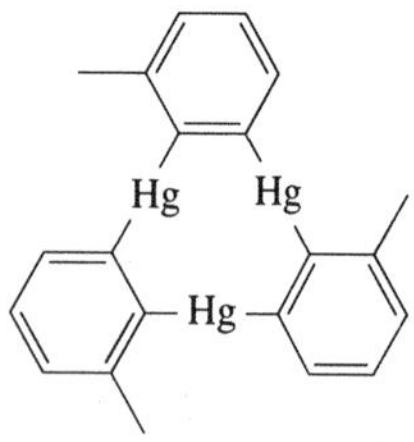

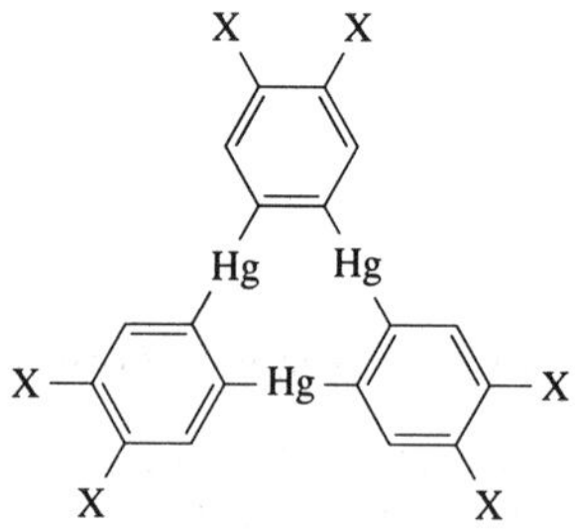

X = Me, OMe, F or Cl

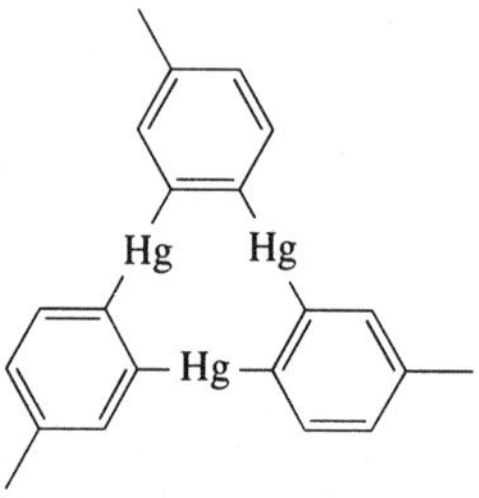

(7)   (8)   (9)

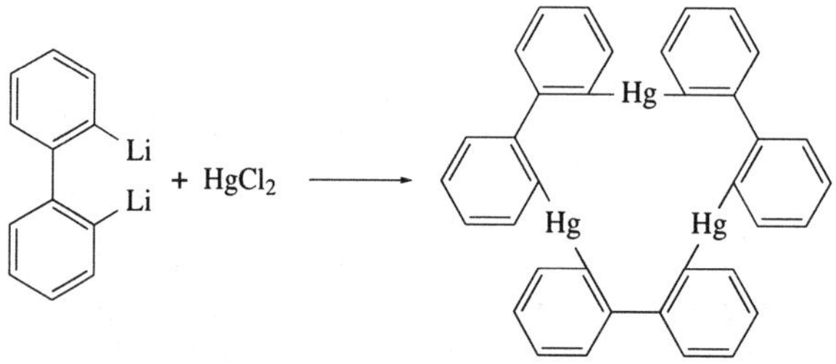

(10)

$$HgX_2 + H-\overset{|}{\underset{|}{C}}-A=B \longrightarrow XHg-\overset{|}{\underset{|}{C}}-A=B + HX \qquad (11)$$

been followed by ^1H and ^{199}Hg NMR spectroscopy. Over a period of about 50 h, the formation and further reaction of compounds containing the structural elements $MeCOCH_2HgX$, $XHgCH_2COCH_2HgX$, $MeCOCH(HgX)_2$, $MeCOC(HgX)_3$, $XHgCH_2COCH(HgX)_2$, $(XHg)_2CHCOCH(HgX)_2$, $XHgCH_2COC(HgX)_3$, $(XHg)_2CHCOC(HgX)_3$ and $(XHg)_3CCOC(HgX)_3$ can be followed.[20]

Ethyl *t*-butyl ketone gives $Bu^tCOCHMeHgBr$ and $(Bu^tCOCHMe)_2Hg$, which have been identified by cryoscopy, NMR spectroscopy and mass spectrometry.[21]

The nature of the products from the mercuriation of β-diketones, $MeCOCHRCOMe$, depends on the nature of R. When R is Me, Et or Ph, reaction with mercury(II) occurs at C-3 to give a 2:1 compound. but when R is more bulky, reaction occurs at the less acidic C-1 to give a 1:1 copolymer (Scheme 7).[22]

Similarly, decahydronaphthalene-1,8-dione does not give an OHg-bonded product, but again gives a polymer bonded through the α-methylene group (Equation (12)).[22]

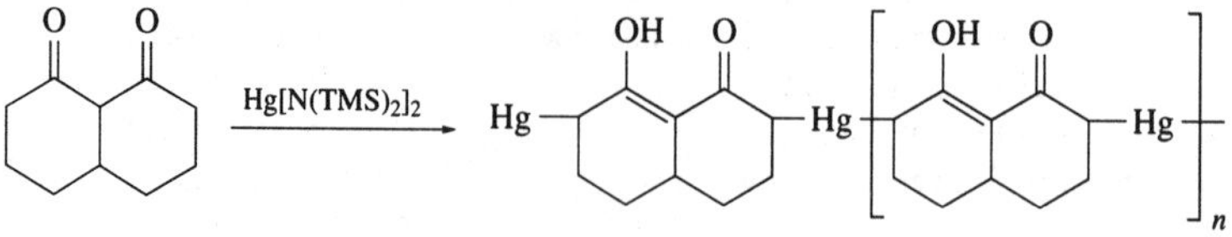

(12)

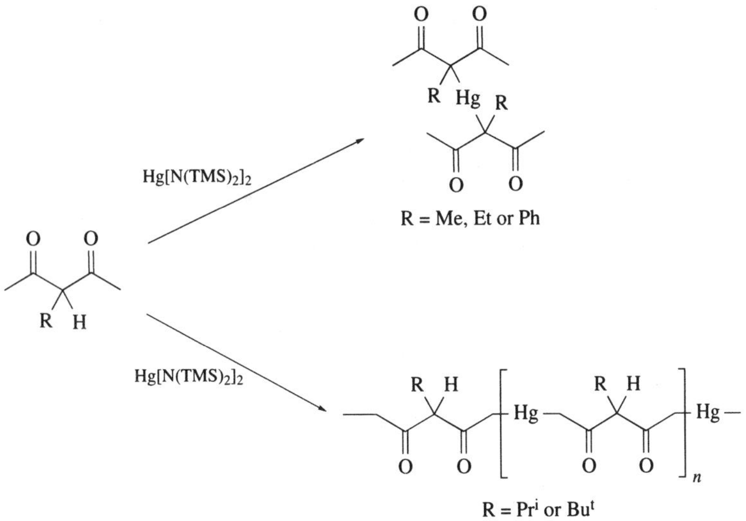

Scheme 7

3.2.3 Aromatic Mercury–Hydrogen Exchange

The mercuriation of an arene by a reagent HgX_2 involves the formation of a charge-transfer complex, then a Wheland σ-bonded intermediate, and is completed by loss of HX (Scheme 8).[23]

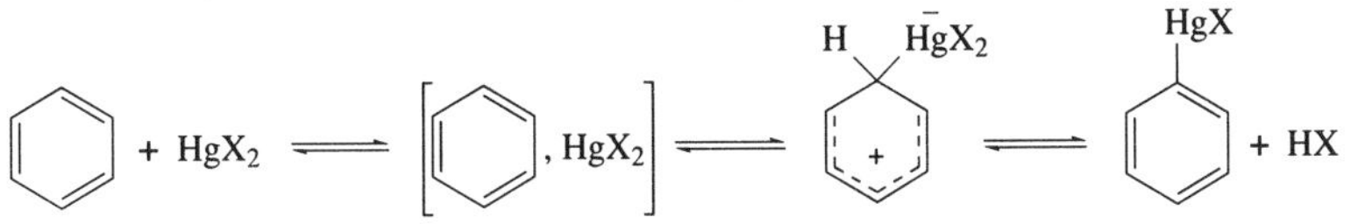

Scheme 8

The directive effect of substituents is usually weak, and frequently polymercuriation occurs. This is illustrated by the mercuriation of ruthenocenes (see Section 3.3.4), where pentamercuriation of the cyclopentadienyl ring is observed. Dimercuriation occurs when 2-nitromethoxybenzene, 3-methyl-4-hydroxybenzaldehyde or 4-methoxybenzoic acid is treated with mercury(II) trifluoroacetate, and the products provided the first examples of mercury–mercury NMR coupling in aromatic mercury compounds (e.g., Equation (13); $^4J(^{199}Hg\,^{199}Hg) = 2163$ Hz).[24]

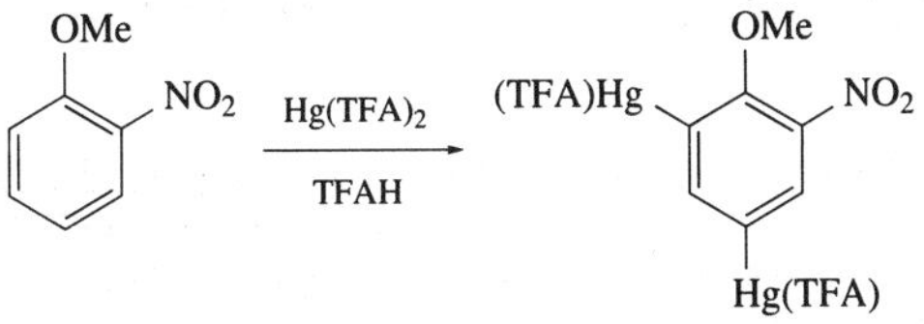

(13)

The presence of the charge-transfer complex is made obvious by its colour, and one such complex has now been isolated and examined crystallographically. Removal of the solvent from a mixture of hexamethylbenzene and mercury(II) trifluoroacetate in trifluoroacetic acid gave yellow crystals of the 1:1 $Me_6C_6 \cdot Hg(TFA)_2$ complex, which has the structure (**10**). Two molecules of each component are involved, the benzene rings being η^2-bonded to the mercury, and the trifluoroacetate groups acting as bidentate ligands.[25]

Kinetic analysis of the reaction confirms the mechanism shown in Scheme 8, but under photolytic conditions a second mechanism involving electron transfer is available (Scheme 9). If a solution of an arene and mercury(II) trifluoroacetate is irradiated with light in the absorption band of the charge-transfer complex, photoinduced electron transfer occurs, and the EPR spectrum of the corresponding arene radical cation can often be observed.[25] It is thought that energy-wasting back-transfer of the electron is avoided by dissociation of the mercury radical anion. This is one of the most common methods for generating radical cations in solution for study by EPR spectroscopy.[26]

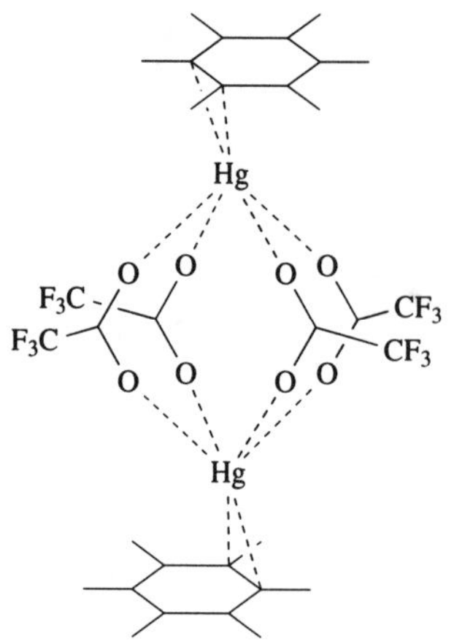

(10)

$$[ArH, Hg(TFA)_2] \xrightleftharpoons{h\nu} ArH^{\bullet+} Hg(TFA)_2^{\bullet-} \longrightarrow ArH^{\bullet+} \cdot Hg(TFA) \, TFA^-$$

Scheme 9

The hyperfine coupling pattern in the spectrum identifies the nature and location of the groups bound to the ring, and as photolysis is continued, the spectrum shows that, with certain arenes, hydrogen bonded to the ring becomes replaced by mercury.[27] A good example is provided by biphenylene, where progressive replacement of the four β-protons can be followed (Scheme 10).[28]

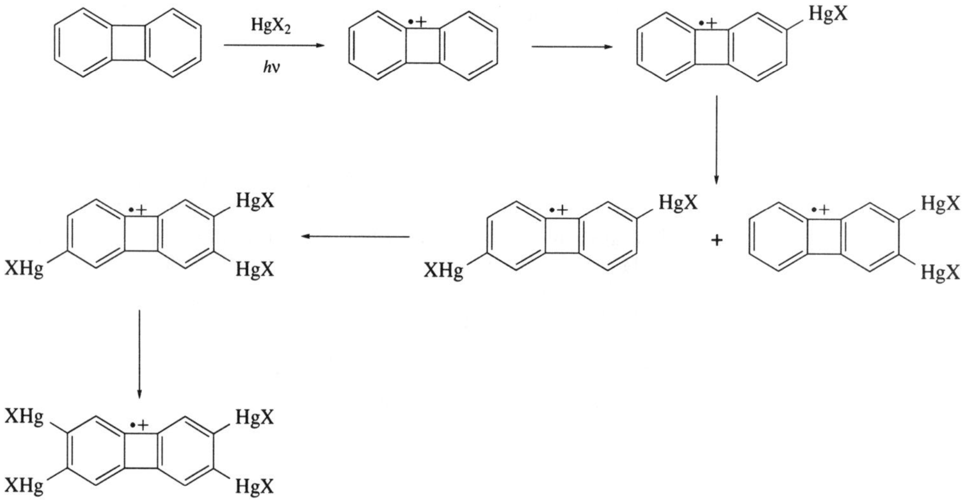

Scheme 10

This type of behaviour has been observed for about 20 arenes. The reactivity follows the sequence which is observed in conventional electrophilic mercuration, and occurs at the position where the local coefficient of the SOMO is largest, that is, where the proton hyperfine coupling constant is greatest. There therefore appears to be an alternative radical ion mechanism forming the Wheland intermediate from the charge-transfer complex (Scheme 11), which can bypass the heterolytic mechanism of Scheme 8.

3.2.4 Decarboxylation

The decarboxylation of mercury(II) carboxylates to give organomercury compounds (Equation (14)) can be brought about thermally or with free radical initiation.

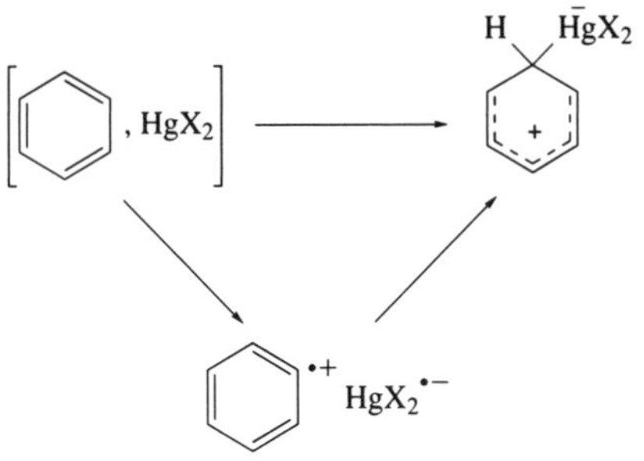

Scheme 11

$$RCO_2HgX \longrightarrow RHgX + CO_2 \tag{14}$$

The thermal reaction occurs when R, which can be an alkyl or alkenyl, or aryl group, carries electron-withdrawing substituents. Details of the synthesis of bis(trifluoromethyl)mercury by the thermal decarboxylation of mercury(II) trifluoroacetate (Equation (15)) have been given in *Inorganic Syntheses*; on a small scale, yields of up to 90% can be obtained, and the product can then be used for making the trifluoromethyl derivatives of other metals.[29] Some examples of thermal decarboxylation are shown in Equations (16)–(18),[30-2] Schemes 12[33] and 13,[34] and Equation 19.

$$Hg(OCOCF_3)_2 \xrightarrow[120-180\ °C]{K_2CO_3} Hg(CF_3)_2 \tag{15}$$

$$(MeO)_nC_6H_{5-n}CO_2H + Hg(OAc)_2 \longrightarrow (MeO)_nC_6H_{5-n}HgO_2CC_6H_{5-n}(OMe)_n \tag{16}$$

$$[(CF_3)_2CFCF=CFCO_2]_2Hg \xrightarrow[220-240\ °C]{K_2CO_3} [(CF_3)_2CFCF=CF]_2Hg \tag{17}$$

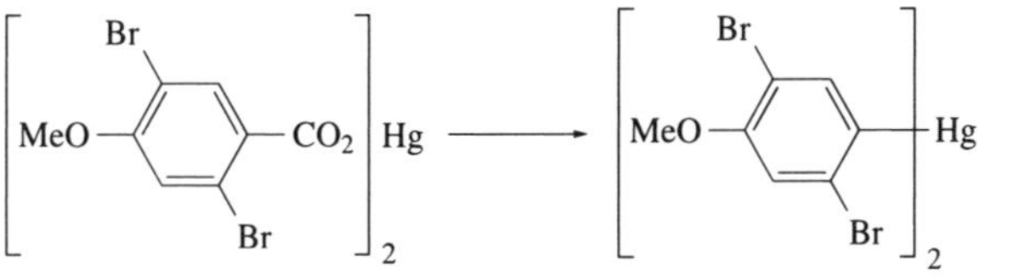

$$\tag{18}$$

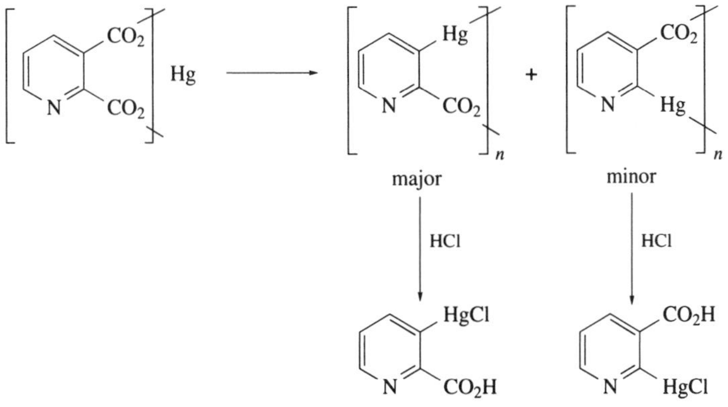

Scheme 12

It is surprising that the decarboxylation of 2,6-dimethoxy-, 2,3,4-trimethoxy- and 2,4,6-trimethoxyphenylmercury compounds (Equation (16)) occurs at room temperature, since the methoxy groups are electron releasing rather than electron withdrawing, and might be expected to induce mercurideprotonation rather than decarboxylation. The mercury carboxylates are not precipitated, as is usually observed, when the corresponding acids are treated with mercury(II) acetate in aqueous methanol, but carbon dioxide is evolved and the polymethoxymercury(II) polymethoxybenzoate is

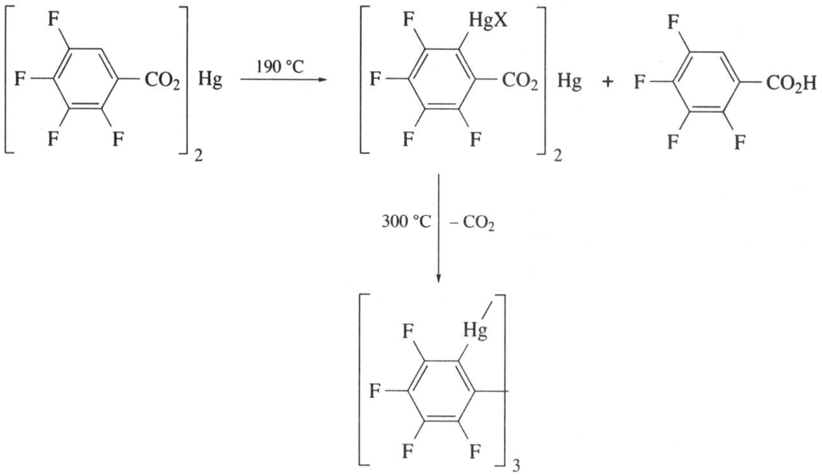

Scheme 13

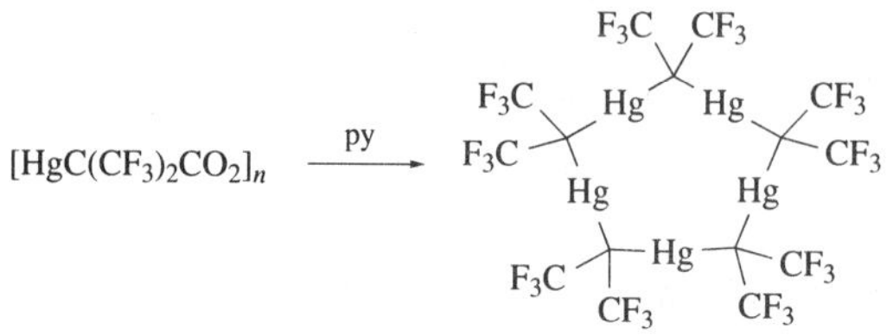

$$[HgC(CF_3)_2CO_2]_n \xrightarrow{\text{py}} \text{(structure)} \qquad (19)$$

deposited in 55–90% yield. A free-radical mechanism appears to be ruled out, and the substituent effect suggests that CO_2 elimination occurs as a result of a classical electrophilic substitution.[30]

In Scheme 13, mercurideprotonation of the ring precedes decarboxylation, resulting in an *o*-dimercuri compound.[34] The α-mercurialkyl carboxylate in Equation (19) decarboxylates to give an oligomeric *gem*-dialkylmercury compound, which has been shown by x-ray crystallography to be a near-planar pentamer as shown.[35]

3.2.5 Solvomercuriation

3.2.5.1 Alkenes

The solvomercuriation reaction (Equation (20)) occurs with unactivated alkenes and a variety of nucleophiles HY under mild conditions; it takes place with high stereo- and regioselectivity, and tolerates a wide range of functional groups.[1] Similar reactions occur with alkynes and cyclopropanes, as discussed below.

$$R^1R^2C=CR^3R^4 + HgX_2 + HY \longrightarrow \underset{\overset{|}{Y} \ \overset{|}{HgX}}{R^1R^2C-CR^3R^4} \qquad (20)$$

$X = OCOR^5, ONO_2, Cl, \text{etc.}$

$Y = OH, OR^5, O_2H, O_2R^5, NR^5_2, OCONR^5_2, OCOR^5, N_3, \text{etc.}$

The mercury can then be removed by reduction with sodium tetrahydroborate or by reaction with bromine (Scheme 14). The reduction involves an intermediate organomercury(II) hydride, R^1HgH (see Section 3.3.6), which decomposes to give the radical $R^1\bullet$, and if the reduction is carried out in the presence of a suitable alkene, this radical can be diverted by the alkene with the formation of a new carbon–carbon bond. These reactions are considered in Section 3.3.7. If the mercury compound is to be isolated, it is often treated with aqueous sodium chloride to convert it into the organomercury(II) halide, which is easy to handle.

Larock's book[3] comprehensively covers the work up to mid-1983, and over half of the book is devoted to tabulations of examples.

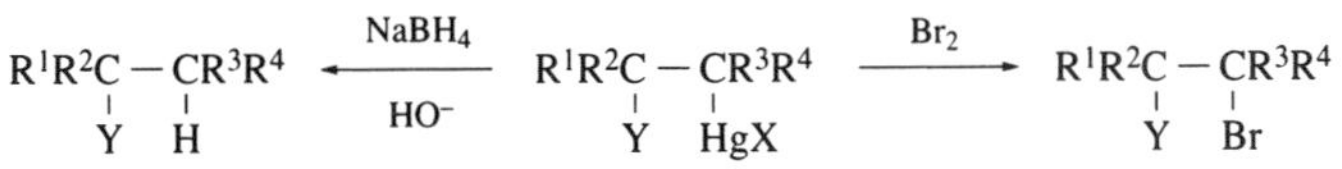

$$R^1R^2C - CR^3R^4 \xleftarrow[\text{HO}^-]{\text{NaBH}_4} R^1R^2C - CR^3R^4 \xrightarrow{\text{Br}_2} R^1R^2C - CR^3R^4$$
$$\;\;\;\; | \quad\;\; | \qquad\qquad\qquad\qquad | \qquad | \qquad\qquad\qquad\qquad | \qquad\;\; |$$
$$\;\;\;\; Y \quad H \qquad\qquad\qquad\qquad\;\; Y \quad HgX \qquad\qquad\qquad\qquad\; Y \quad Br$$

Scheme 14

The reaction is accepted to proceed by electrophilic attack by mercury(II) on the alkene to give an intermediate mercurinium ion, followed by attack of the nucleophile, with the normal result of *trans* addition of HgX^+ and Y^- in the Markownikov direction (Equations (21) and (22)). The precise structure of the cationic intermediate may vary from a symmetrically bridged species to an open β-mercurialkyl cation, depending on the number and nature of the substituents on the alkene.

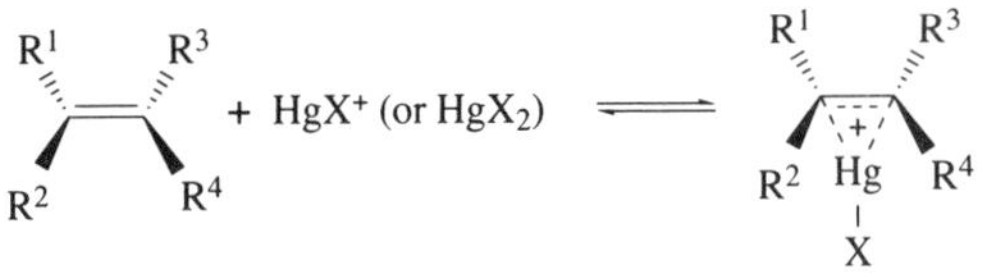

(21)

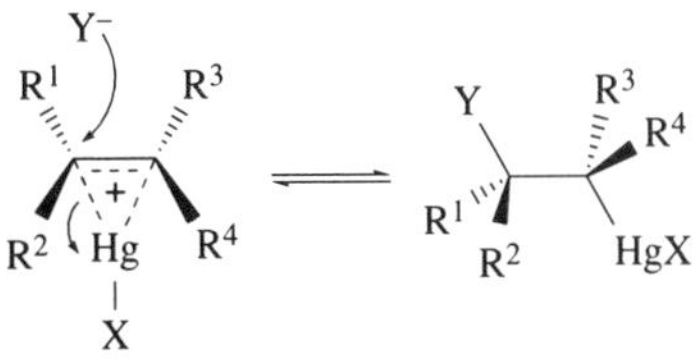

(22)

Some recent examples of solvomercuration reactions of simple alkenes are shown in Table 1.

To establish the best conditions for the oxymercuration, experiments have been carried out with various alkenes using the mercury salts $Hg(OAc)_2$, $Hg(TFA)_2$, $Hg(NO_3)_2$ and $Hg(OSO_2Me)_2$ and the alcohols MeOH, EtOH, Pr^iOH and Bu^tOH. Mercury(II) trifluoroacetate was found to be best reagent, giving high yields of the ether even for tri- and tetra-substituted alkenes (Scheme 15).[49]

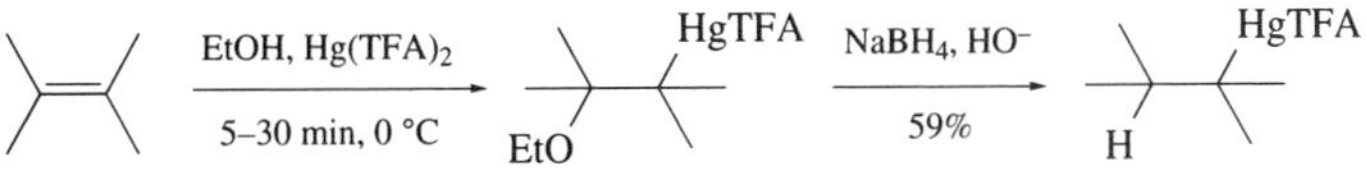

Scheme 15

Alkylaminomercuriations can be carried out with primary amines, and at 40 atm pressure ethene reacts with aromatic primary amines to give the bis-adduct (Table 1, entry 1).

Ammonia cannot usually be used for aminomercuration because it complexes too strongly to the mercury salt, but this problem can be avoided by using amides, urethanes or ureas (Table 1, entries 2 and 3); treatment of the amidomercuration product with sodium tetrahydroborate gives the amide in good yield, then the protecting group can be removed by hydrolysis. Entry 4 (Table 1) shows a simple example of intramolecular amidomercuration in the synthesis of *trans*-dialkylpiperidines, and entry 5 (Table 1) gives an example of tandem sulfonamidomercuration in the synthesis of *cis*-2,5-dimethylpyrrolidine. Amidomercuration can also be achieved with mercury(II) nitrate and acetonitrile (Table 1, entry 6; 70% yield).

Entries 7, 8 and 9 (Table 1) give examples of sulfinatomercuration, nitratomercuration and azidomercuration, respectively. Solvomercuration of alkenes usually proceeds by *anti* addition, but strained cycloalkenes, such as norbornene, may show *syn* addition. The azidomercuration of cyclobutene in methanol gives a 2:3 mixture of the *trans* and *cis* adducts (Table 1, entry 9), but the reaction with cyclohexene gives only the *trans* product. Methyl 1,2-diphenylcyclopropane-3-carboxylate (Table 1, entry 10) reacts with $Hg(OAc)_2$ in MeOH to give the *trans* adduct, but the corresponding dimethyl compound (Table 1, entry 11) with HgN_3 gives the *cis* adduct, and with $Hg(OAc)_2$ gives a mixture of *cis* and *trans* isomers.

Hydroperoxymercuriation has been achieved previously with concentrated hydrogen peroxide, which is not readily available and carries an explosion hazard. It can be replaced by 30% aqueous hydrogen peroxide if this is used in eightfold excess in the presence of mercury(II) acetate (Table 1, entry 12). Some hydroxymercuration sometimes occurs in parallel, but the two products can be separated chromatographically.

Table 1 Solvomercuriation of simple alkenes.

Entry	Alkene	Conditions	Product	Ref.
1	$H_2C=CH_2$	$Hg(OAc)_2$, $PhNH_2$	$PhN(CH_2CH_2HgOAc)_2$	36
2	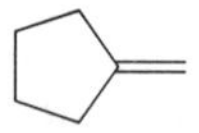 (methylenecyclopentane)	$Hg(NO_3)_2$, H_2NCOMe	 (1-methyl-cyclopentyl)–NHCOMe	37
3	$PhHC=CH_2$	$Hg(NO_3)_2$, H_2NCONH_2	$(PhCHMeNH)_2CO$	37
4	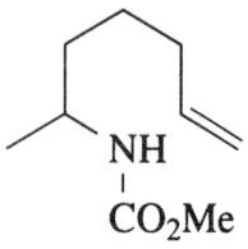 (structure) NH, CO_2Me	$Hg(OAc)_2$	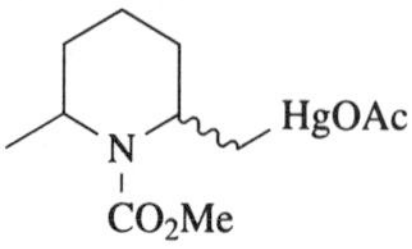 (piperidine) HgOAc, N, CO_2Me	38
5	 (diene)	$Hg(NO_3)_2$, $TsNH_2$	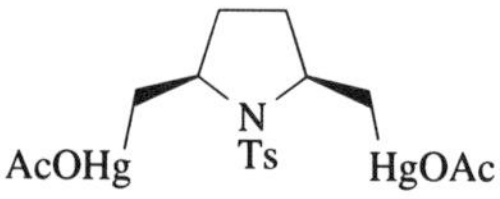 (pyrrolidine) $AcOHg$, N–Ts, $HgOAc$	39
6	$H_2C=CH(CH_2)_8CO_2Me$	$Hg(NO_3)_2$, $MeCN$	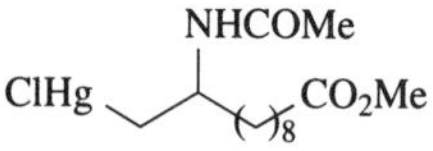 $ClHg$, NHCOMe, CO_2Me, ${}_8$	40
7	$PrCH=CH_2$	$HgCl_2$, $TsNa$	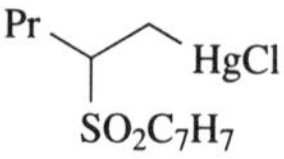 Pr, HgCl, $SO_2C_7H_7$	41
8	$R^1CH=CHR^2$	$Hg(NO_3)_2$	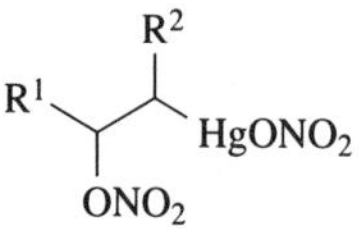 R^1, R^2, $HgONO_2$, ONO_2	42
9	 (cyclobutene)	HgN_3	 N_3, HgCl + N_3, HgCl 2:3	43
10	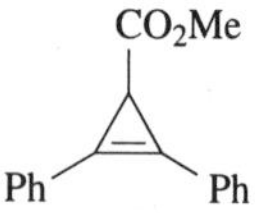 CO_2Me, Ph, Ph (cyclopropene)	$Hg(OAc)_2$, MeOH	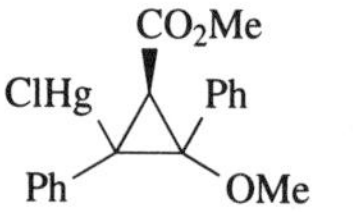 CO_2Me, ClHg, Ph, Ph, OMe	44
11	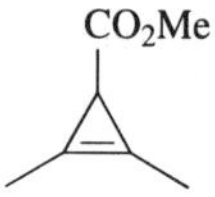 CO_2Me (cyclopropene)	HgN_3 or $Hg(OAc)_2$	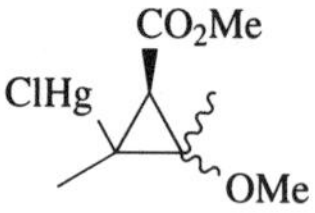 CO_2Me, ClHg, OMe	45
12	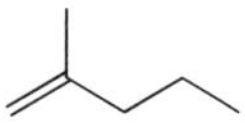 (2-methyl-1-pentene)	$Hg(OAc)_2$, 30% H_2O_2	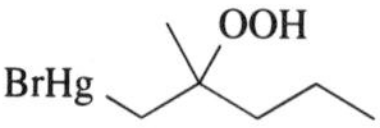 OOH, BrHg	46
13	 (cyclohexene)	$Hg(SO_3Me)_2$, Br_2	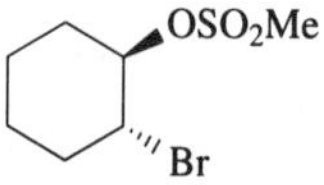 OSO_2Me, Br	47
14	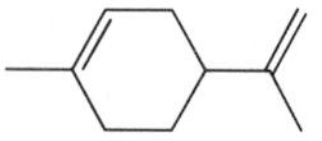 (terpene)	HgO, $ArSO_3H$, THF–H_2O, sound	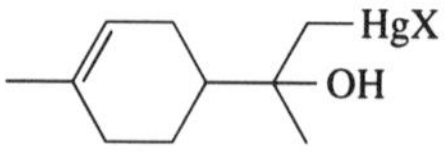 HgX, OH	48

Hydrodemercuriation of the hydroperoxide products cannot be carried out directly because sodium tetrahydroborate will reduce the hydroperoxide group, but reduction to the alkyl hydroperoxide can be achieved if the O_2H group is first protected as its 2-methoxypropyl ether[50] (e.g., Scheme 16). This gives a useful route to secondary alkyl hydroperoxides, which are difficult to prepare by alternative methods.

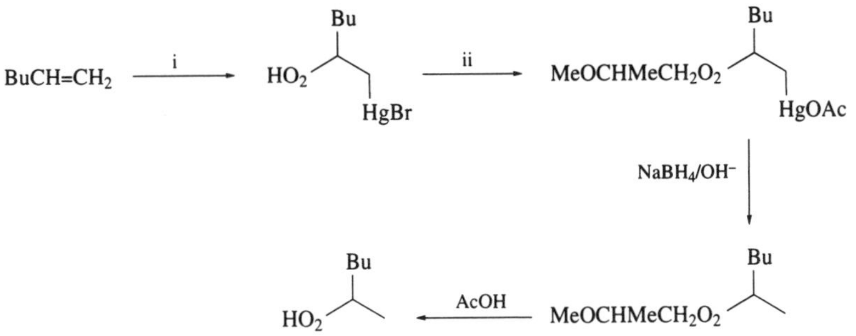

i, 30% H_2O_2, Hg(OAc)$_2$ then KBr; ii, MeOCMe=CH$_2$, pyH$^+$OTs$^-$ catalyst

Scheme 16

Intramolecular peroxymercuriation with unsaturated hydroperoxides has been used for preparing a variety of cyclic peroxides.[51] With Hg(OAc)$_2$, the reaction appears to be kinetically controlled, but with Hg(TFA)$_2$ it is thermodynamically controlled. Two examples are given in Equations (23) and (24).

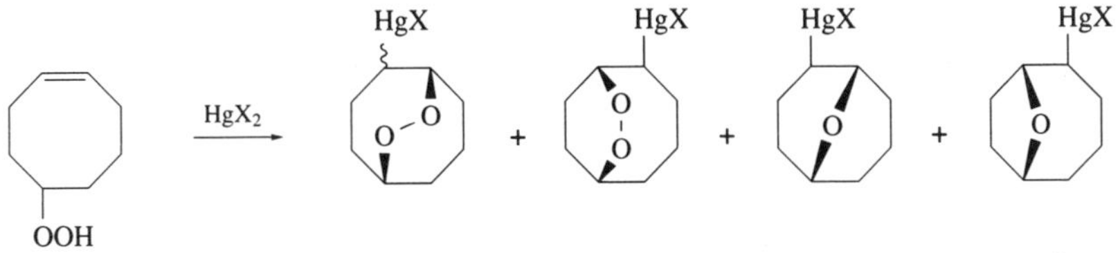

Hg(O$_2$CMe)$_2$:	74	: 18	: 5	: 3
Hg(O$_2$CCF$_3$)$_2$:	40	: 14	: 36	: 10

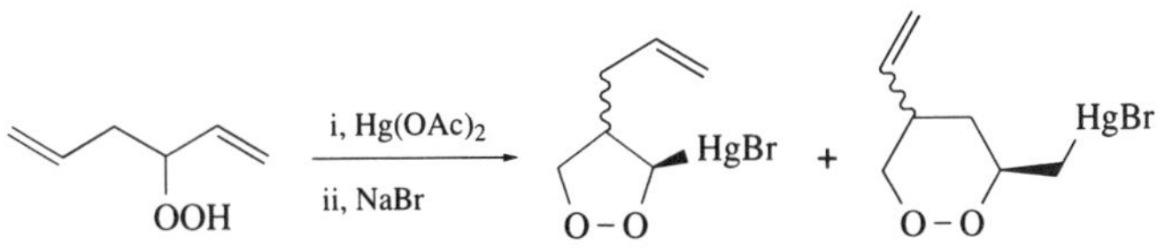

(23)

(24)

The mercury salt–halogen, HgX$_2$–hal$_2$, combination provides a versatile reagent for the stereoselective addition of halX to alkenes, where X can be F, Cl, Br, N$_3$, NCO, NO$_3$, MeSO$_3$, ArSO$_2$ or ArSO$_3$. The example shown in entry 13 in Table 1 gives a 90% yield of adduct.

Mercury salts, Hg(OCOR)$_2$ and Hg(OSO$_2$R)$_2$, which cannot be purchased, can be prepared by sonication of a suspension of HgO in a solution of the acid in THF. Oxymercuriation can be carried out by making the mercury salt *in situ*. Thus, if a suspension of HgO in aqueous THF containing toluene-*p*-sulfonic acid and limonene is sonicated until the colour is lost (Table 1, entry 14), the tertiary alcohol is selectively obtained after reduction in 63% yield. With 1-vinylcyclohexene, the regioselectivity for the exocyclic double bond was best (100% selectivity, 67% yield of alcohol) when perfluorobutyric acid was used, and under micellar conditions.

An example of a double intramolecular oxymercuriation resulting in the formation of spiroacetals is shown in Equation (25), the carbonyl group behaving as a *gem*-dihydroxide.[52]

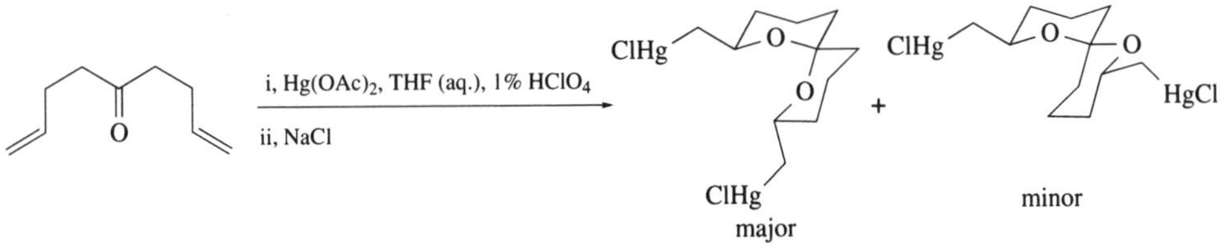

(25)

Solvomercuriation of alkenes usually proceeds more readily that of cyclopropanes, and acetoxymercuriation of the methylenecyclopropane in Equation (26) or of cyclopropylidene-cyclopropane (Equation 27) occurs principally at the double bond.[53]

Bloodworth and Tallant[54] established the route to the 1,2,4-trioxane antimalarials shown in Scheme 17, in which the hemiperacetal formed between an allylic hydroperoxide and an aldehyde is caused to undergo an intramolecular ring-forming oxymercuriation.

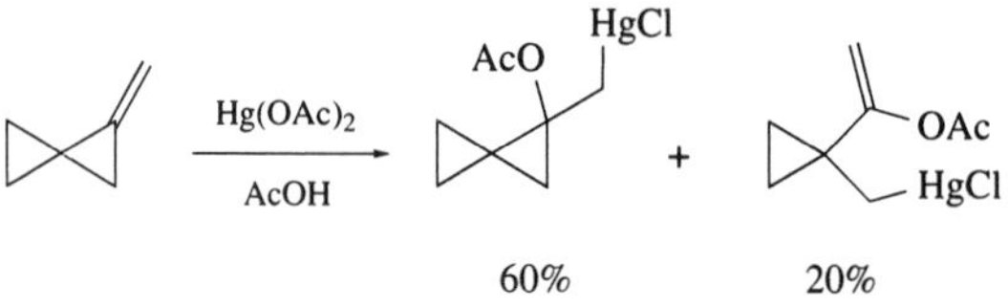

$$(26)$$

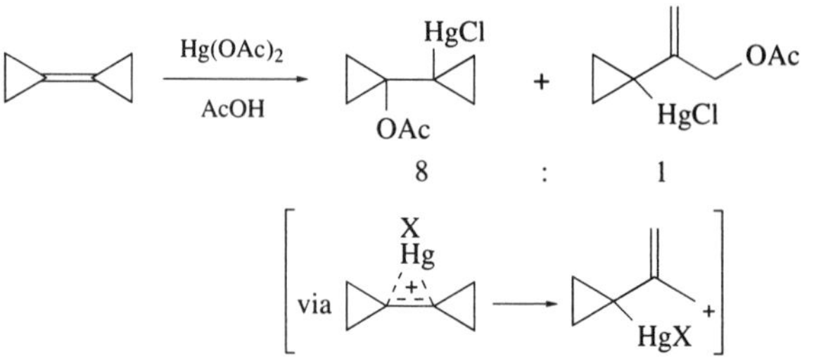

60%   20%

$$(27)$$

8 : 1

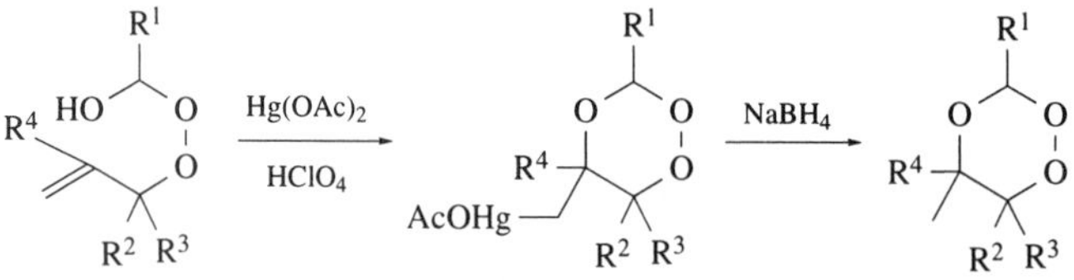

Scheme 17

The synthesis can be improved by exploiting the principle that metal alkoxides act as the synthetic equivalents of alcohols, and oxymercuriations can be carried out with trialkyltin alkoxides, Bu_3SnOR, and peroxymercuriations with trialkyltin peroxides, Bu_3SnO_2R, in place of the usual alcohols or alkyl hydroperoxides. Allylperoxytin compounds can readily be prepared from the reaction of singlet oxygen with allylstannanes. They add to carbonyl compounds to give the organotin derivatives of hemiacetals, which then undergo intramolecular oxymercuriation in the absence of any catalyst to give the precursors of the final trioxanes (Scheme 18).[55]

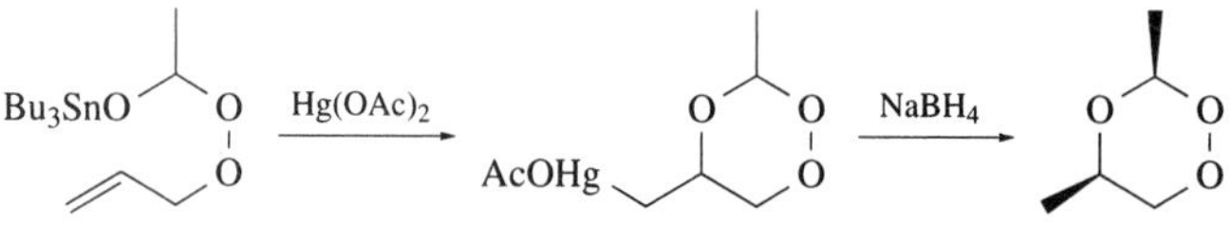

Scheme 18

3.2.5.2 Alkynes

Solvomercuriation of an alkyne to give a vinyl mercurial is illustrated in Equation (28). Sometimes further mercuriation may occur to give a *gem*-dimercurialkane (Equation (29)). The enol resulting from hydroxymercuriation rearranges to the ketone (Equation (30)), and the reaction of terminal alkynes may be complicated by initial mercury–hydrogen exchange (Equation (31)).[13]

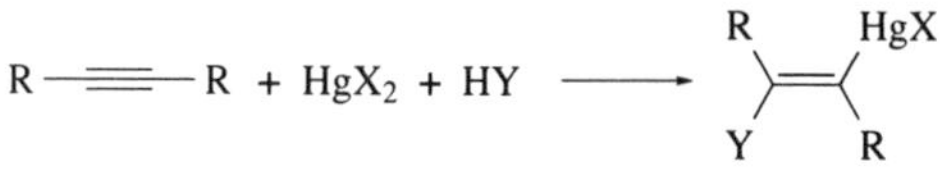

$$(28)$$

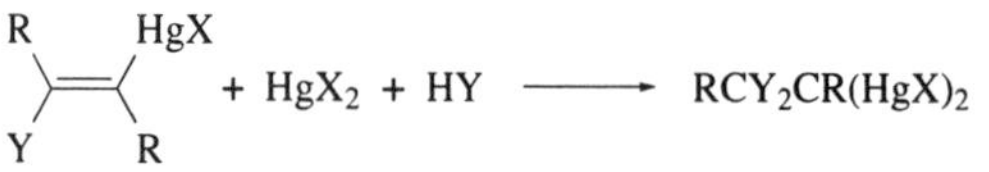

$$(29)$$

$$(30)$$

$$R \!-\!\!\!\equiv\!\!\!- + HgX_2 \longrightarrow R\!-\!\!\!\equiv\!\!\!-HgX \qquad (31)$$

Much less work has been carried out on the mechanism of the solvomercuriation of alkynes than that of alkenes, but again it is believed that an intermediate mercurinium ion is involved (Equations (32) and (33)).

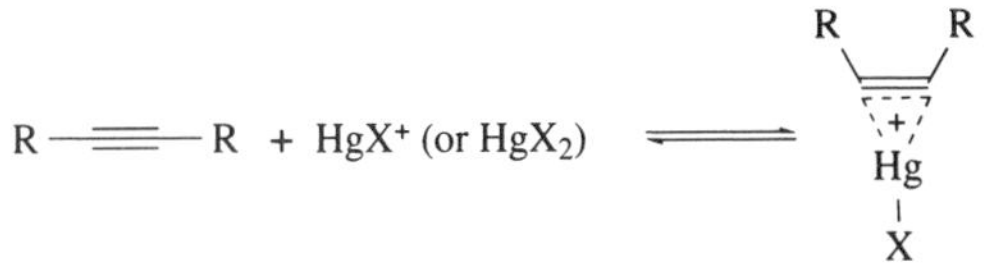

$$R\!-\!\!\!\equiv\!\!\!-R + HgX^+ \text{ (or } HgX_2) \rightleftharpoons \qquad (32)$$

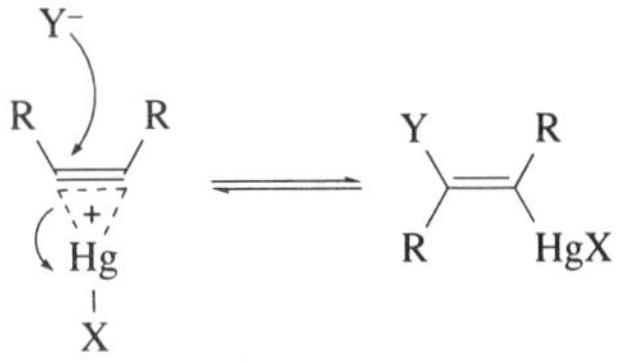

$$(33)$$

Examples of the solvomercuriation of simple alkynes are given in Table 2.

Table 2 Solvomercuriation of simple alkynes.

Entry	Alkyne	Conditions	Product	Ref.
1	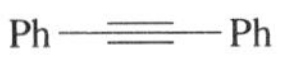 Ph—≡—Ph	Hg(TFA)$_2$, ROH	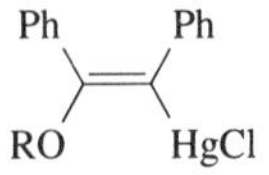 Ph, Ph / RO, HgCl	56
2	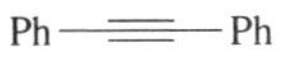 Ph—≡—Ph	Hg(OAc)$_2$	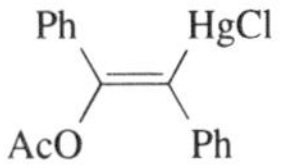 Ph, HgCl / AcO, Ph	57
3	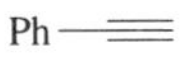 Ph—≡	i, Hg(OAc)$_2$, PrNH$_2$; ii, NaBH$_4$	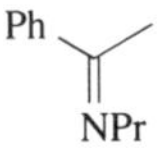 Ph / NPr	58
4	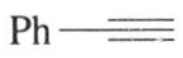 Ph—≡	HgCl$_2$, Bu$_2$NH	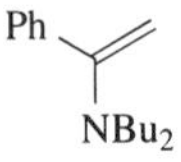 Ph / NBu$_2$	58
5	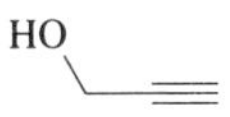 HO—≡	Hg(OAc)$_2$, RNH$_2$	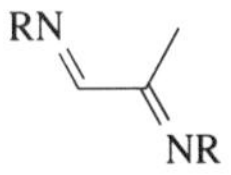 RN / NR	59
6	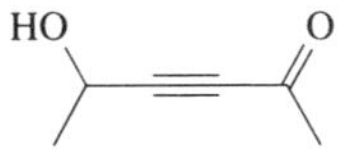 HO—≡—O	HgCl$_2$, NaCl	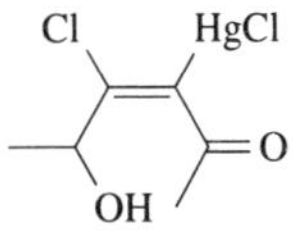 Cl, HgCl / OH, O	60
7	 Me$_2$N—≡—	HgCl$_2$, HCl	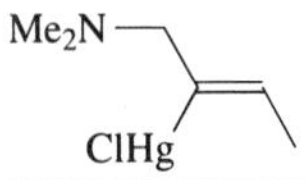 Me$_2$N / ClHg	61

The above simple model of the mechanism of the reaction implies that *anti* addition of HgX and Y should occur, but a variety of stereochemistries have been reported, depending on the nature of X and Y and on the reaction conditions. For example, diphenylethyne reacts with Hg(OAc)$_2$ in methanol or ethanol to give a mixture of the *Z*- and *E*-adducts, but with Hg(TFA)$_2$ in ethanol or propan-2-ol, only the *Z*-adduct is formed (Table 2, entry 1).[56] Again, the acetoxymercuriation of diphenylethyne was initially reported to give a mixture of *Z*- and *E*-products, and then only the *Z*-adduct, but since then the *E*-adduct has been identified by x-ray crystallography (Table 2, entry 2).[57]

This confused situation has been clarified by the demonstration that the Z-adduct is a tertiary rather than a primary product, and that the nature of the products depends on the time of the reaction. The initial product of the methoxymercuration of diphenylethyne undergoes a second methoxymercuration, and with a 1:1 ratio of alkyne and Hg(OAc)$_2$, after 2 h the secondary adduct can be isolated in 50–55% yield. If the reaction is continued, demethoxymercuration occurs, to give the Z-adduct, and if a 1:3 ratio of reagents is used, the Z-adduct can be isolated in 96% yield (Scheme 19).[62]

Scheme 19

The products of the aminomercuration of terminal alkynes with primary amines depends of the ratio of the reagents. In CH$_2$Cl$_2$ solvent, and with alkyne:Hg(OAc)$_2$ >2, the principal product is the mercury acetylide, (RC≡C)$_2$Hg, but with a ratio <2, the imine is formed; for example, the reaction of entry 3 in Table 2 gives a 78% yield of imine after 14 h at room temperature. The reaction proceeds through the mercury acetylide, and, for example, hexylamine reacts with phenylethynylmercury chloride in boiling THF to give the corresponding imine (Equation (34)).[58]

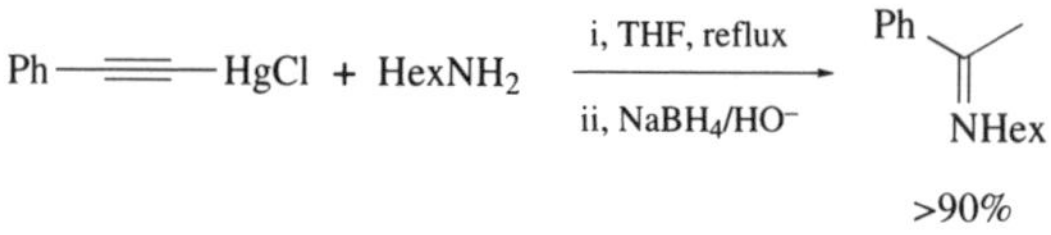

$$Ph-\!\!\!≡\!\!\!-HgCl + HexNH_2 \xrightarrow[\text{ii, NaBH}_4/\text{HO}^-]{\text{i, THF, reflux}} \qquad (34)$$

>90%

Below about 100 °C, a secondary amine reacts under the same conditions to give only the acetylide (RC≡C)$_2$Hg, but in boiling dioxane (100 °C), phenylethyne reacts with dibutylamine to give the enamine in 72 h in 71% yield (Table 2, entry 4); metallic mercury is deposited during the reaction, and the reaction is complete without the usual tetrahydroborate reduction.[58]

Under suitable conditions, the reaction of an amine with prop-2-ynol (Table 2, entry 5) can give mainly an α-imino ketone (**11**), an α-diimine (**12**) or an α-aminopropionamidine (**13**).[59] The probable course of the reaction is shown in Scheme 20.

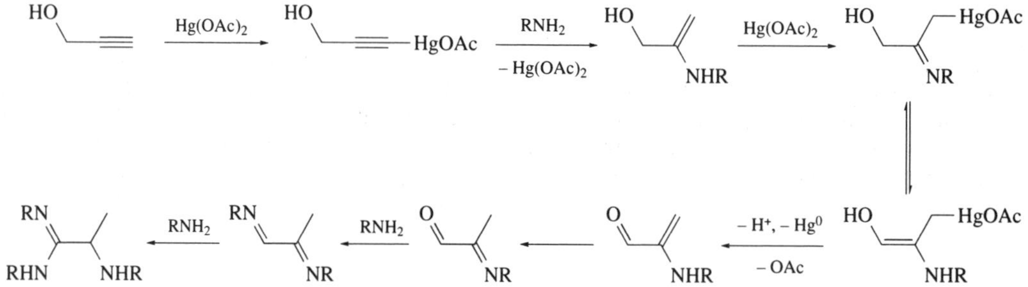

Scheme 20

Mercury(II) chloride adds readily to the triple bond of certain 4-hydroxyalk-2-yn-1-ones to give vinylmercurials which appear to be the first established examples of genuine *syn* addition; cyclization of the adduct can then give a furan (Scheme 21). The stereochemistry of the addition can be rationalized in terms of the interaction of the hydroxyl and carbonyl groups.[60]

Propargylamines react with HgCl$_2$ in the presence of HCl (to prevent aminomercuration) to give the product of chloromercuration in high yield (Table 2, entry 7).[61]

Homo- and heterocyclizations can be achieved by provision of an intramolecular nucleophile to react with the mercurinium ion; two examples are given in Scheme 22 and Equation (35).[63,64]

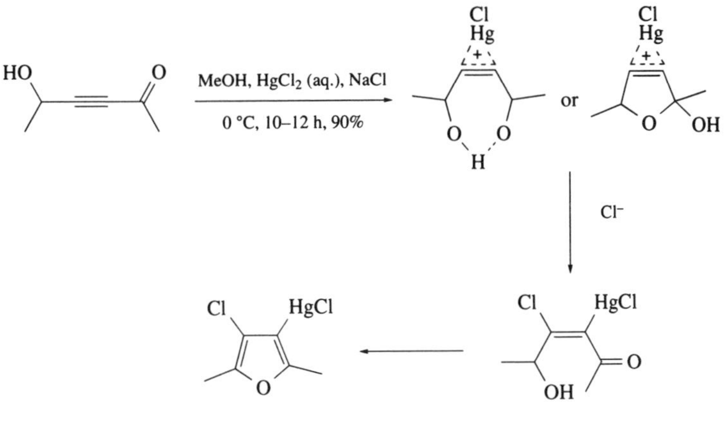

Scheme 21

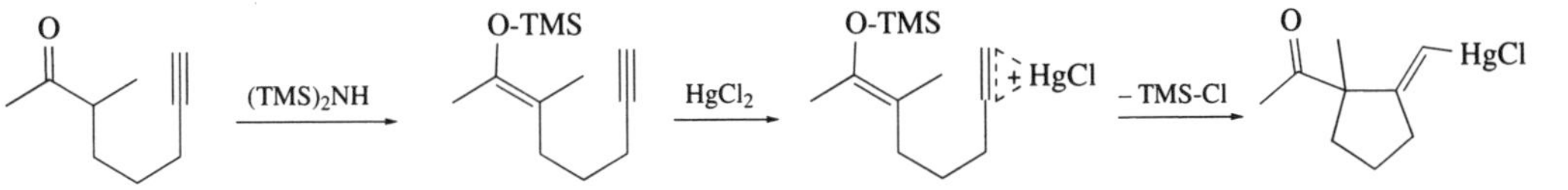

Scheme 22

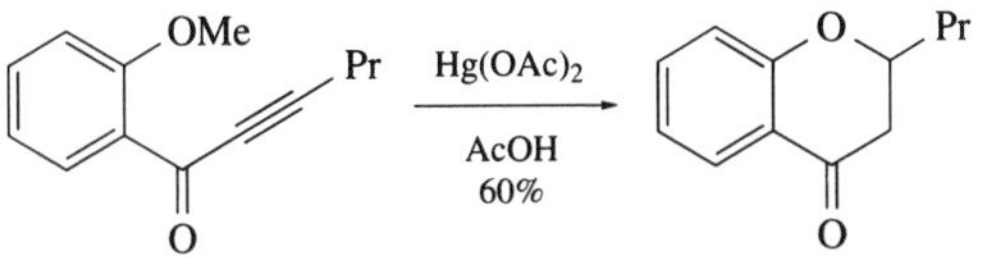

(35)

3.2.5.3 Cyclopropanes

Ring-opening solvomercuriation of cyclopropanes occurs less readily than addition to alkenes (e.g., Equations (26) and (27)); the mercury attacks at the least alkyl-substituted carbon and may give either inversion or retention of configuration, and the nucleophile attacks at the carbon which is then best able to support a positive charge, and usually gives substantial inversion of configuration.[1]

Some examples of the solvomercuriation of cyclopropanes are given in Table 3.

The first example of the solvomercuriation of cyclopropane itself, free of any substituents, is shown in entry 1 (Table 3), and gives a mixture of the compounds $XCH_2CH_2CH_2HgBr$ where $X = Bu^tO_2$, AcO and $OCH_2CH_2CH_2HgBr$. Acetoxymercuriation of *cis,cis*-1,2,3-d_3-cyclopropane gives AcOCHD-CHDCHDHgOAc, in which the groups about the terminal carbon atoms are in an *erythro* relationship, implying that inversion occurs in both the electrophilic attack of $AcOHg^+$ and nucleophilic attack of AcO^-.

Arylcyclopropanes containing an *o*-nitro or *p*-acetyl substituent react with mercury(II) nitrate in alcohol solvent at 20 °C in 24 h to give the 3-aryl-3-alkoxypropylmercury chloride (Table 3, entry 3). The reactivity is enhanced in formic acid solvent, when mercury(II) formate is the reactive species.

trans-1,2-Diphenylcyclopropane reacts with mercury(II) acetate in the presence of *t*-butyl hydroperoxide by scission of the C-1–C-3 bond to give a mixture of the *threo* and *erythro* adducts in up to 68% yield (Table 3, entry 6) (along with products resulting from the shift of a phenyl group), but the *cis*-isomer (Table 3, entry 7) reacts mainly by scission of the C-1–C-2 bond, and gives a mixture of the isomeric 1,3-diperoxides and the 1,2-dioxolanes via oxidative mercuriation (Scheme 23).

This conversion of cyclopropanes into dioxolanes has been developed into a preparatively useful reaction (Scheme 24).[69]

Bicyclo[*n*.1.0]alkanes undergo cleavage of the [0] and [1] bridges (Equation (36)) in a ratio which depends on the value of *n*. Two examples are given in entries 8 and 9 in Table 3, and the results are summarized in Table 4.

Table 3 Solvomercuriation of cyclopropanes.

Entry	Cyclopropane	Conditions	Product	Ref.
1	△	$Hg(OAc)_2$, Bu^tOOH, $HClO_4$	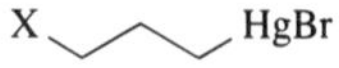 X⌒⌒⌒HgBr X = $OOBu^t$ or OAc	65
2	 D D D triangle	i, $Hg(OAc)_2$; ii, Br^-	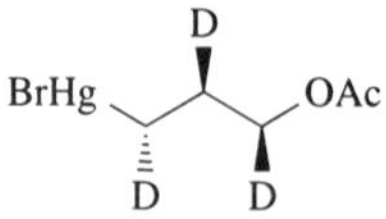 BrHg⟋⟍D OAc D D	66
3	 Ar⌒△	$Hg(NO_3)_2$, ROH	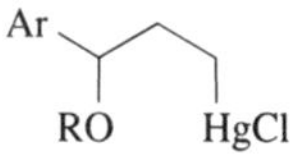 Ar⌒⌒ RO HgCl	67
4	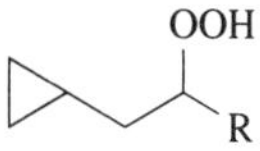 △⌒ OOH R	$Hg(OAc)_2$	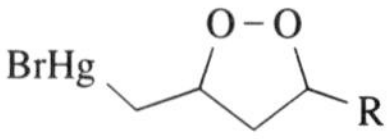 BrHg⌒ O–O R	68
5	 ⋎△	$Hg(OAc)_2$, Bu^tOOH, $HClO_4$	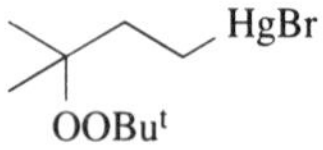 ⌒⌒HgBr $OOBu^t$	69
6	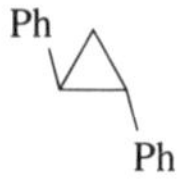 Ph △ Ph	$Hg(OAc)_2$, Bu^tOOH	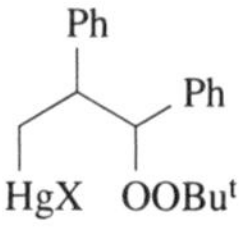 Ph Ph HgX $OOBu^t$	70
7	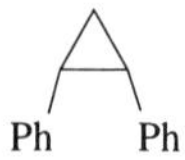 △ Ph Ph	$Hg(OAc)_2$, Bu^tOOH, $HClO_4$	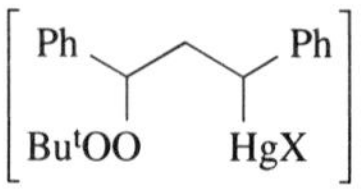 [Ph Ph Bu^tOO HgX]	70
8	 ◇△	$HgCl_2$	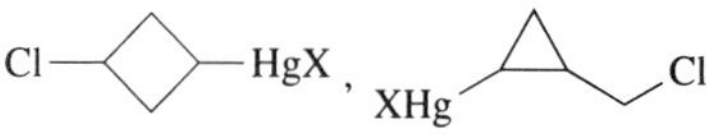 Cl—□—HgX , XHg△⌒Cl	71
9	 ⬠	$Hg(OAc)_2$, Bu^tOOH, $HClO_4$	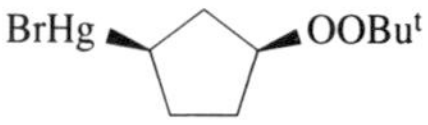 BrHg⟋⌒$OOBu^t$	72
10	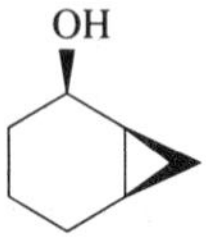 OH bicycle	$Hg(OAc)_2$	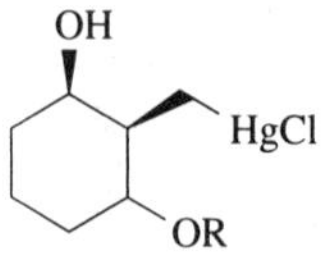 OH HgCl OR	73
11	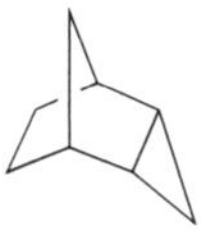 bicyclooctane	$Hg(OAc)_2$, MeOH	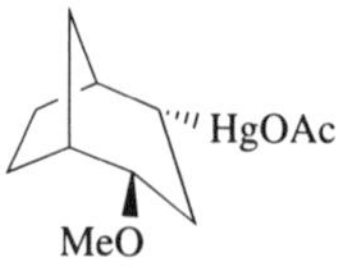 HgOAc MeO	74
12	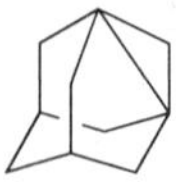 adamantane	$Hg(OAc)_2$, EtOH	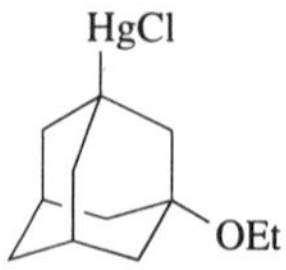 HgCl OEt	75

With a *cis*-OH group in the bicycloheptane or -octane, hydroxymercuriation occurs with retention in the electrophilic attack of mercury, and inversion in the nucleophilic attack of hydroxyl ion (e.g., Table 3, entry 10), but inversion occurs at both centres in entries 11 and 12 (Table 3).

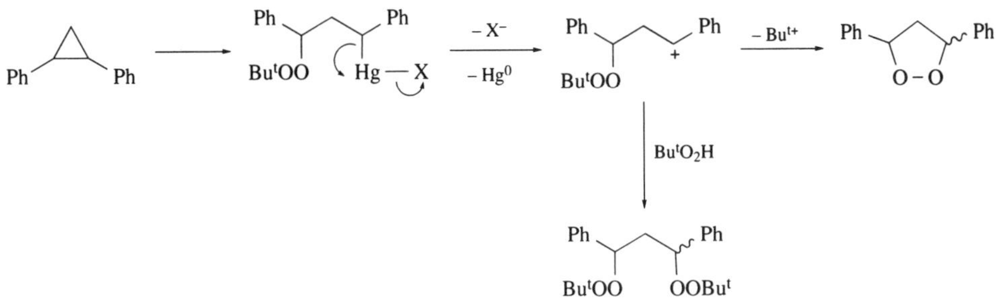

Scheme 23

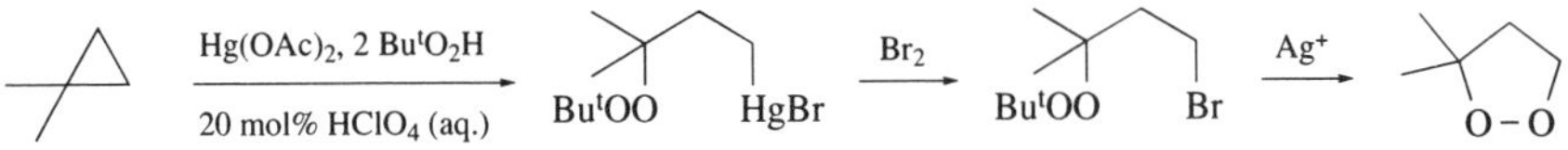

Scheme 24

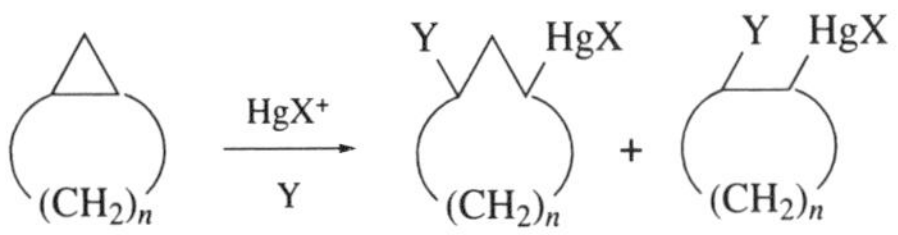

(36)

Table 4 Solvomercuration of bicyclo[n.1.0]alkanes.

Bicyclo[n.1.0.]alkane	Reagent	Ratio [0]:[1] cleavage
[1.1.0]Butane	$HgCl_2$	33:67
[2.1.0]Pentane	$Hg(OAc)_2/Bu^tOOH$	100:0
[3.1.0]Hexane	$Hg(OAc)_2/Bu^tOOH$	Both
[4.1.0]Heptane	$Hg(OAc)_2/Bu^tOOH$	0:100

3.2.6 Carbene Insertion

Carbenes derived from diazo compounds will insert into a mercury–halogen bond to give α-halogenoalkyl mercurials.[1] Improved conditions have been recommended for the preparation of chloromethylmercury chloride[76] and bromomethylmercury bromide,[77] in which ethereal diazomethane is distilled into a suspension of a slight excess of the mercury halide in diethyl ether (with precautions against explosion!) (Equation 37). The halogenomethylmercury halide separates from solution as it is formed.

$$CH_2N_2 + HgX_2 \longrightarrow XCH_2HgX + N_2 \qquad (37)$$

$$X = Cl \text{ or } Br$$

A similar mechanism appears to be involved in the reaction of a hydrazone, $R^1R^2C=NNH_2$, with mercury(II) oxide in the presence of mercury(II) acetate, the hydrazone first being oxidized to the diazo compound, $R^1R^2CN_2$. If the α-acetoxymercurial $R^1R^2C(OAc)HgOAc$ which is formed is treated with sodium tetrahydroborate, it is reduced to the mercury hydride $R^1R^2C(OAc)HgH$, which decomposes to give the radical $R^1R^2C(OAc)\bullet$, and this can be trapped by an activated alkene $CH_2=CH-X=Y$ to give the adduct $R^1R^2C(OAc)CH_2CH_2-X=Y$. A simple example is shown in Scheme 25.

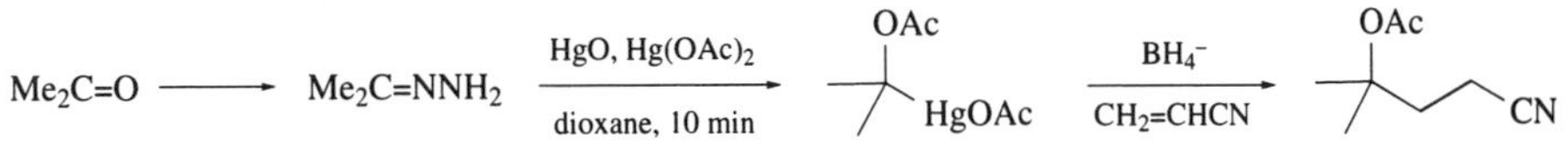

Scheme 25

If the intermediate mercury compound is isolated, the overall yield is 50%, but the various steps can be combined into a one-pot procedure, which gives a yield of 40%.[78]

3.2.7 Miscellaneous Methods

Mercury(II) fluorosulfonate in liquid SbF_5 at 100 °C reacts with carbon monoxide at 0.8–0.9 atm pressure to give the first carbonyl derivative of mercury (Equation (38)).

$$Hg(SO_3F)_2 + 2\,CO + 8\,SbF_5 \longrightarrow [Hg(CO)_2]^{2+}[Sb_2F_{11}^-]_2 + 2\,Sb_2F_9(SO_3F) \tag{38}$$

The $[Hg(CO)_2]^{2+}$ salt was isolated as a white solid, but it was stable only in $HSO_3F \cdot SbF_5$ solution; in HSO_3F or liquid SO_2 it decomposed with the evolution of CO.[79]

The mercury(I) salt $[Hg_2(CO)_2]^{2+}[Sb_2F_{11}^-]_2$ was obtained similarly as a pale-yellow solid in nearly quantitative yield, but it is less stable than the mercury(II) salt. Both compounds were characterized by infrared and ^{13}C and ^{199}Hg NMR spectroscopy. The $[Hg(CO)_2]^{2+}$ ion shows a carbonyl stretching vibration at 2281 cm^{-1}, which is the highest frequency yet reported for a metal carbonyl, and the $[Hg_2(CO)_2]^{2+}$ ion shows an intense Raman line at 169 cm^{-1}, which is ascribed to the Hg–Hg stretch. There seems little doubt that both ions have linear structures, OC–Hg–CO and OC–Hg–Hg–CO. These appear to be the first mercury carbonyls, and the first organomercury(I) compounds to be satisfactorily characterized.[79]

Benzyl and allyl halides will react with metallic mercury to give the corresponding organomercury halides in a process which is the formal equivalent of the formation of a Grignard reagent. The reaction of benzyl chlorides with mercury is best carried out in DMF solvent, with Br$^-$ catalysis, and the rate is enhanced by the presence of bromine substituents in the aromatic ring.[80] If propargylic bromides and iodides are shaken with mercury in sunlight, the corresponding propargylic or allenic mercury halides are formed in high yield (e.g., Equations (39) and (40)).[81]

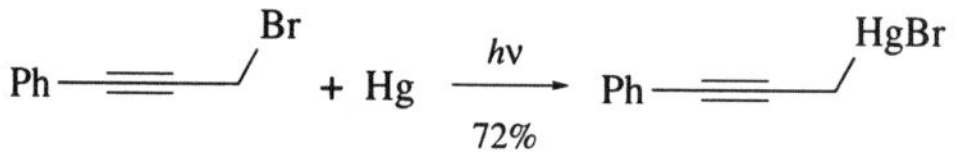

$$\tag{39}$$

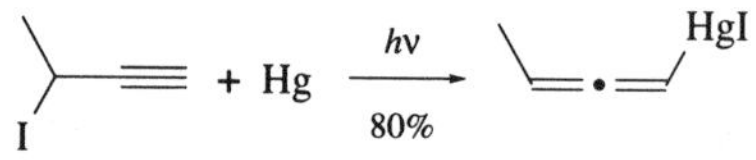

$$\tag{40}$$

The bistriethylsilyl- and bistriethylgermyl-mercury compounds $(Et_3M)_2Hg$ (M = Si or Ge) react with the alkyl chlorides $Me_{3-n}Cl_nSiCH_2Cl$ to give the corresponding dialkylmercurials in 85–90% yield (e.g., Equation (41)). The reaction shows the characteristics of a radical chain reaction, and appears to follow the mechanism shown in Equations (42) and (43), which includes (Equation (43)) a step involving bimolecular homolytic substitution (S_H2) at mercury.[82]

$$(Et_3Si)_2Hg + TMS\text{-}CH_2Cl \xrightarrow{h\nu} (TMS\text{-}CH_2)_2Hg + 2\,Et_3SiCl \tag{41}$$

$$Et_3M\bullet + RCl \longrightarrow Et_3MCl + R\bullet \tag{42}$$

$$R\bullet + (Et_3M)_2Hg \longrightarrow RHgMEt_3 + Et_3M\bullet,\ etc. \tag{43}$$

Bis(trimethylsilyl)mercury reacts with dialkynylmercurials to give quantitative yields of the alkynylsilylmercurials, which have been characterized by ^{13}C, ^{29}Si and ^{199}Hg NMR spectroscopy (e.g., Equation (44)).[83] These reactions can been regarded as transmetallation reactions in which both of the metals are mercury.

$$\left(Bu\!\!-\!\!\equiv\!\!-\right)_2 Hg + Hg(TMS)_2 \longrightarrow 2\,Bu\!\!-\!\!\equiv\!\!-Hg\text{-}TMS \tag{44}$$

3.3 STRUCTURES AND PROPERTIES OF ORGANOMERCURY COMPOUNDS

3.3.1 Methylmercury(II) Compounds

The isotope effects of ^{1}H, ^{2}H, ^{12}C and ^{13}C on the values of δ ^{1}H, ^{13}C and ^{199}Hg in the NMR spectra of the isotopomers CH_3HgCH_3, CHD_2HgCH_3, CD_3HgCH_3, CHD_2HgCD_3 and CD_3HgCD_3 have been determined.[84] The ^{1}H, ^{13}C, ^{14}N and ^{199}Hg NMR spectra of the mercury fulminates $Hg(CNO)_2$ and RHgCNO have been reported; in THF solution, the mercury in $Hg(CNO)_2$ is coupled to two nitrogen atoms, showing that dissociation does not occur.[85] The fulminates undergo cycloadditions with activated alkynes to give 3-(organomercurio)isoxazoles, which readily undergo cycloelimination to give 2-cyanoenolates (Scheme 26).[86]

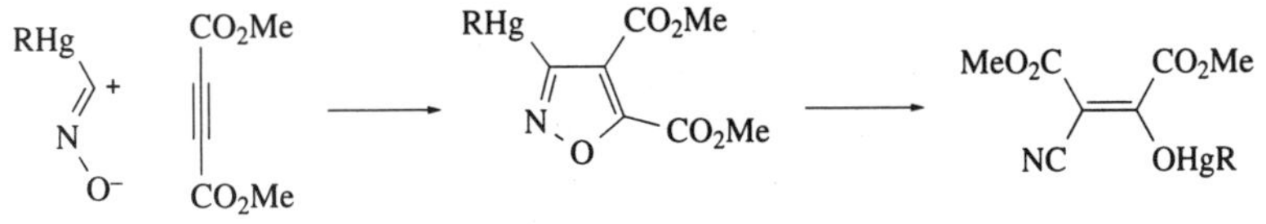

Scheme 26

Bis(trimethylmercury) oxide and sulfide react with methylmercury perchlorate to give the onium salts $(MeHg)_3O^+ClO_4^-$ and $(MeHg)_3S^+ClO_4^-$. The sulfonium cation is pyramidal at sulfur (**14**), with Hg–S–Hg 95.5(3)–106.5(4)°, and close to linear about mercury, with Me–Hg–S 172.3(14)–179.5(12)°.[87] Two further examples of essentially linear C–Hg–S structures are shown in (**15**) (177.7° and 178.8° in two independent molecules)[88] and (**16**).[89]

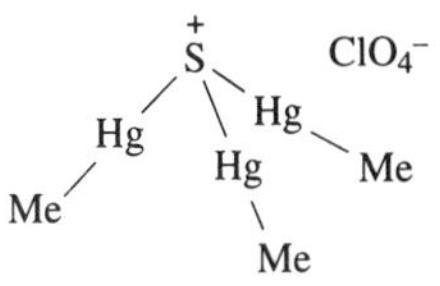

(**14**) 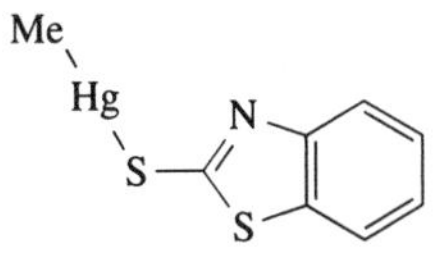　　　　　(**15**) 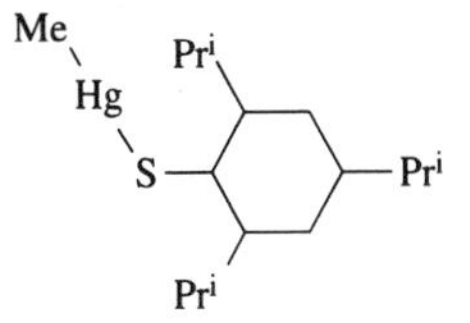　　　　　(**16**)

The carbon–mercury bond is sufficiently inert to allow modifications to be made to functional groups in the alkyl ligands. For example, di(bromomethyl)mercury reacts with thallium(I) 2,4,6-trichlorophenoxide to give the bis(trichlorophenoxymethyl)mercurial, which has a strictly linear, centrosymmetric structure (Equation (45)).[90]

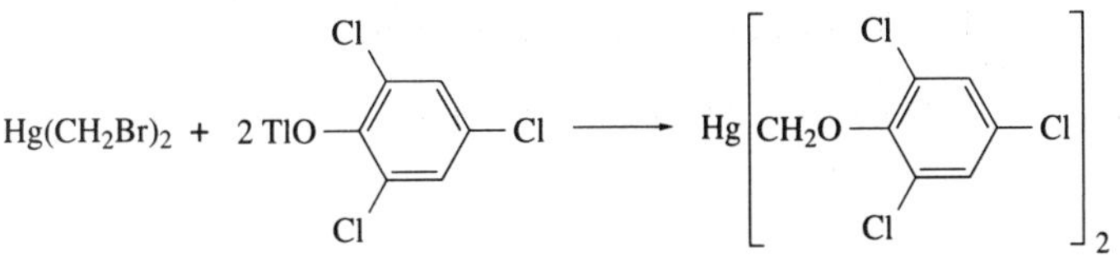

$$Hg(CH_2Br)_2 + 2\,TlO\text{—}\underset{Cl}{\overset{Cl}{\bigcirc}}\text{—}Cl \longrightarrow Hg\left[CH_2O\text{—}\underset{Cl}{\overset{Cl}{\bigcirc}}\text{—}Cl\right]_2 \tag{45}$$

The behaviour of methylmercury compounds as Lewis acids is important in connection with their toxicology. The solvation thermodynamics of the dissolution of Me_2Hg and MeHgX (X = Cl, Br and I) in water have been determined from the heats of solution,[91] and the interaction with pyridines in CH_2Cl_2, $MeNO_2$ and MeOH has been investigated by NMR.[92] The structure of the pyridine complex of methylmercury trifluoroacetate is shown in (**17**).[92] The methyl and pyridine groups are essentially collinear about mercury, and there are further weak interactions with one oxygen in each of two trifluoroacetate anions. The stability constants of the complexes which $MeHg^{II}X$ (X = Cl, Br or I) forms in pyridine solution increase in the sequence $Cl^- < Br^- < I^-$. Large-angle x-ray scattering of pyridine solutions shows that the molecules MeHgX are close to linear, with two asociated pyridine molecules.[93] The DMSO complex (**18**) is similarly nearly linear about mercury,[94] as is the *N*-methylimidazoline-2-thione complex (**19**) (C–Hg–S 176.1(4)°).[95]

The diastereoisomeric tripyridyl complexes (**20**) have the structure shown, and the *meso* and racemic isomers can be separated by crystallization. The central nitrogen forms the strongest bond (0.2283(9) nm), with a C–Hg–N angle of 166.3(5)°; the two other N–Hg distances are 0.2546 and 0.2595 nm, with C–Hg–N angles 113.3° and 113.7°.[96]

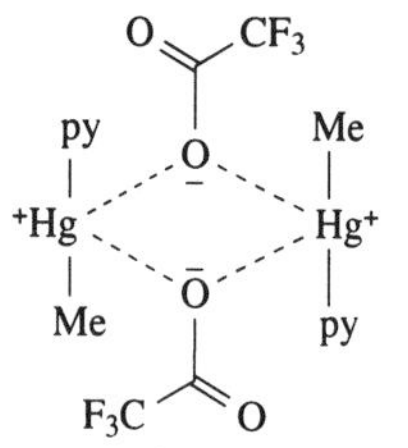

(17) 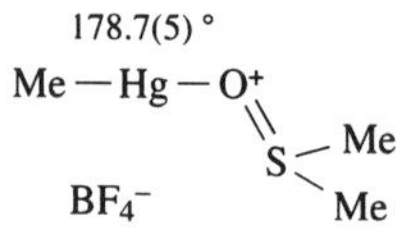   (18) 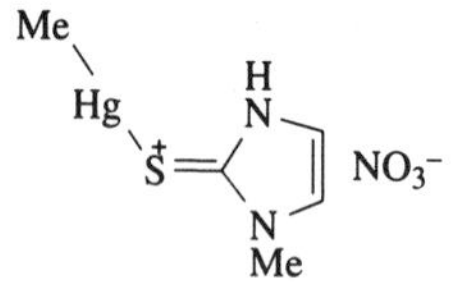   (19)

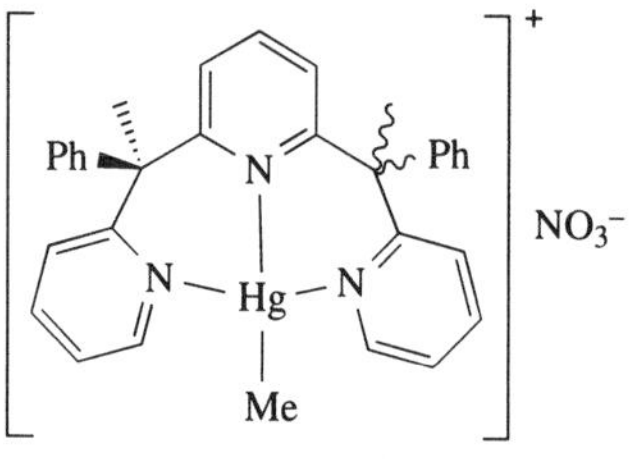

(20)

Similar studies have been made of the complexing of MeHg$^+$ by (pyridyl)/(*N*-methylimidazolyl)-methanols (**21**) (n = 0–3); the mercury is complexed preferentially by the *N*-methylimidazolyl group. In the tripyridylmethanol complex (**21**, n = 3), the largest C–Hg–N angle is 150(1)°, emphasizing that the bonding about mercury is not necessarily linear.[97]

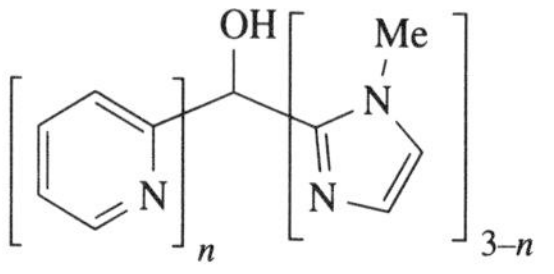

(21)

A ^{199}Hg NMR study of the complexation of MeHg$^+$, aimed at finding a therapeutic agent for MeHg^{II}, showed that simple sulfides and disulfides have a surprisingly low affinity and an absence of any macrocyclic ligand effect. K_f values in CH_2Cl_2 are RS$^-$ 10^{14}–10^{16} and R$_2$S 0.07, (**22**) 45, (**23**) 0.25 and (**24**) 0.36.[98]

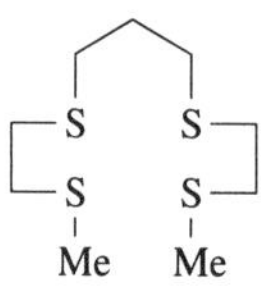

(22) 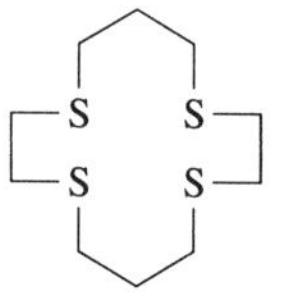   (23) 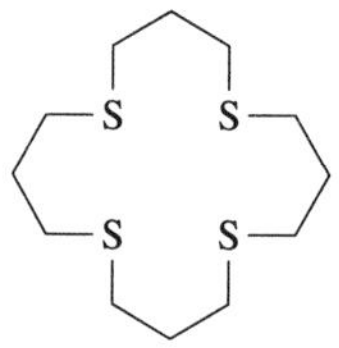   (24)

The fact that highly toxic methylmercury(II) compounds can be generated in the environment by the biological methylation of mercury(II) salts has stimulated extensive studies of the interaction of methylmercury(II) compounds with biologically important molecules, particularly the bases of nucleic acids. A complete list of the structures which have been determined by x-ray diffraction is given in Volume 9 of *COMC-I* and Volume 13 of the present edition, and some representative examples are shown in formulae (**25**)–(**32**).

Alanine reacts with the methylmercury cation to form the complex (**25**), in which the C–Hg–N angle is 175.6(9)°, and there is weak orthogonal interaction with oxygen of the carboxylate group.[99] Glycylglycine interacts with one MeHg$^+$ through the terminal amino group (**26**) (C–Hg–N 176(1)°), then a second can be added at the carboxylate group (**27**) (C–Hg–N 177(1)°).[100]

Xanthine gives the 3,7,9-trimercuriated cation (**28**), then the 1,3,7,9-tetramercuriated cation (**29**) (average C–Hg–N 176°).[101] Adenine bonds MeHg$^+$ initially by displacement of the proton at N-9, then a further group is introduced at N-3 without proton loss, and then two more by proton displacement at N-6 to give (**30**) (average C–Hg–N 177°).[102] 8-Azaadenine similarly reacts initially at N-9 to give (**31**).

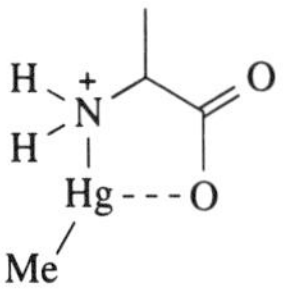

(25)

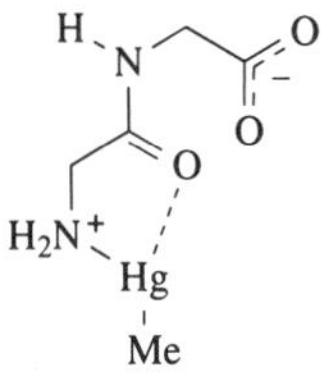

(26)

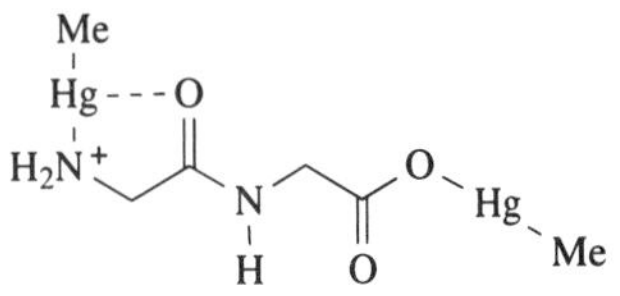

(27)

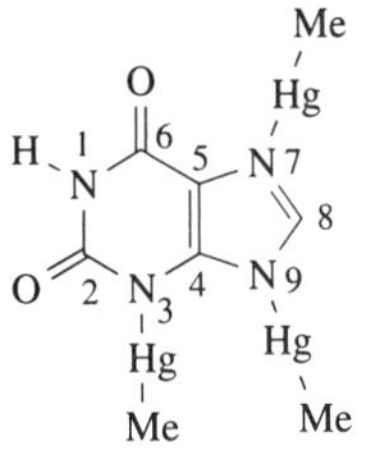

(28)

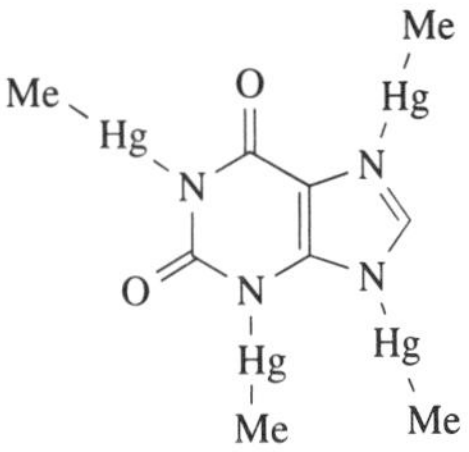

(29)

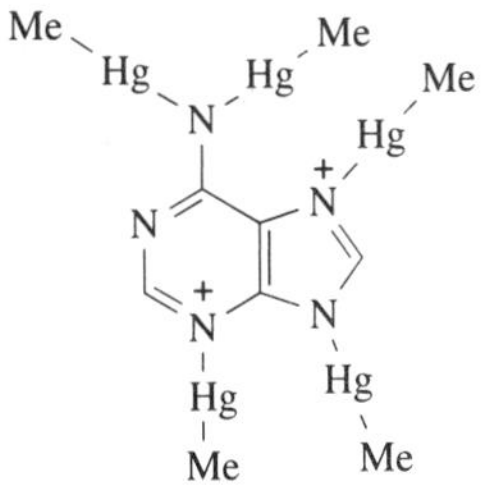

(30)

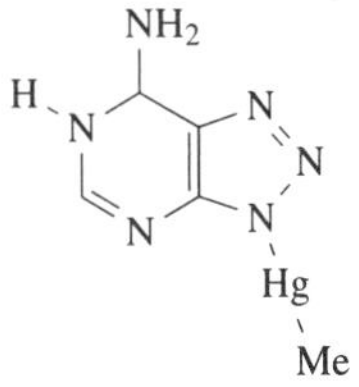

(31)

The structure of the complex between Hg^{2+} and two thymidine or guanine, or one thymidine and one guanine molecule, as a model for the putative mercury(II) interstrand cross-linking of DNA, has been studied by NMR spectroscopy. The thermodynamic stability decreases in the sequence [Thy–Hg–Thy] > [Guo–Hg–Guo] > [Thy–Hg–Guo], and the linking with the Thy–Guo pair is as shown in (32).[103]

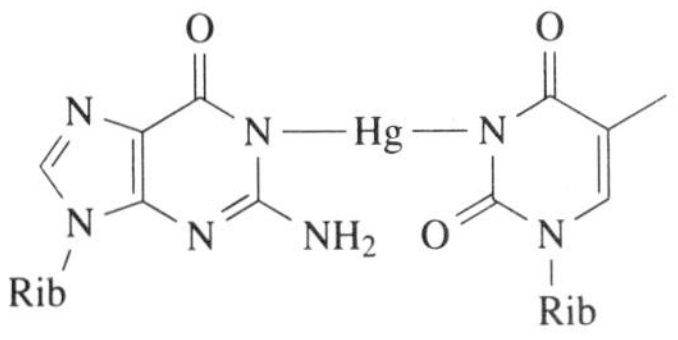

Rib = ribose

(32)

3.3.2 α-Mercuriated Carbonyl Compounds

The nature of the products from the reaction of acetaldehyde with mercury(II) nitrate depends on the mass fraction of nitric acid which is present, as shown in Table 5.[104]

The simplest structure is that of (35), which consists of infinite chains of dimercuriated hydroxonium ions HO^+. Compound (34) is built up of an infinite network of 12-membered $(OHgCHg)_3$ rings with trimercuriated oxonium ions, O^+. Compound (33) can be obtained from either acetaldehyde or ethyne with aqueous mercury(II) nitrate, and has the most complicated structure, in which adjacent rings from structure (34) are joined cyclically to give polymeric columns with a three-ring pitch. Crystals of adequate quality have not been obtained for the structure of $(NO_3Hg)_3CCHO$ (36) to be determined, but propionaldehyde and butyraldehyde react with aqueous mercury(II) chloride in the presence of sodium acetate to give the dimercuriated products $(ClHg)_2CMeCHO$ and $(ClHg)_2CEtCHO$, respectively, and the latter compound has been shown to consist of discrete molecules as shown in (37).[105]

Table 5 Products of the mercurideprotonation of acetaldehyde.

Mass fraction HNO$_3$ (%)	*Products*	*Formula*
<3	$[OHg_3CCHO]NO_3 \cdot H_2O$	**(33)**
3.5	$2\,[OHg_3CCHO]NO_3 \cdot HNO_3$	**(34)**
5–60	$[HOHg(NO_3Hg)CCHO]NO_3$	**(35)**
60	$(NO_3Hg)_3CCHO$	**(36)**

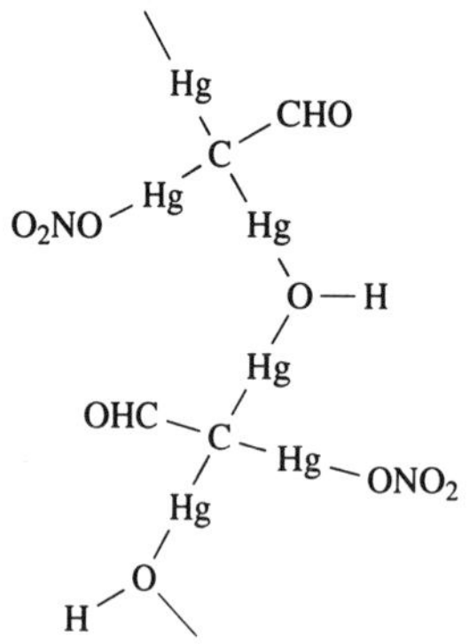

(33)

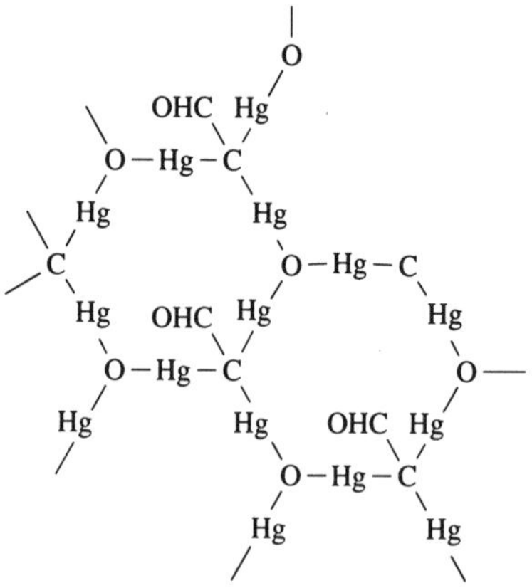

(34)

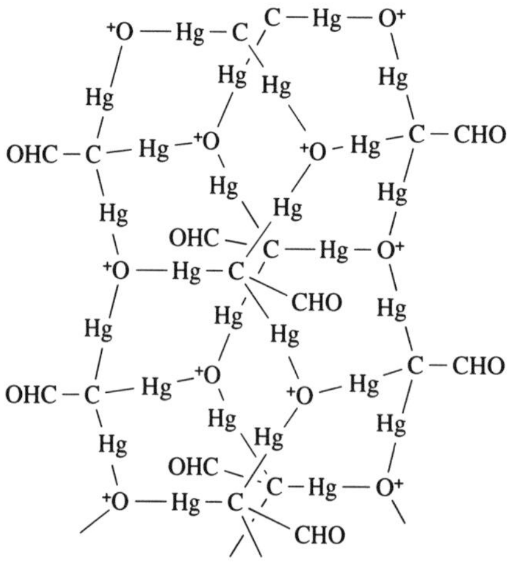

(35)

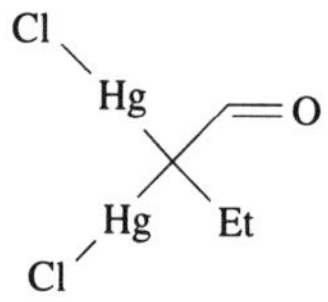

(37)

A variety of deuteriated acetones were prepared by reactions such as Equation (46) for study by NMR spectroscopy.[84]

$$CD_3MgI + MeHgCl \longrightarrow MeHgCD_3 \qquad (46)$$

Acetic acid reacts with mercury(II) acetate to give a condensation polymer of the monomercuriated acid (Scheme 27). This is depolymerized by nitric acid to give the monomer with the structure (**38**).[106]

$$MeCO_2H \xrightarrow{Hg(OAc)_2} \underset{n \approx 10}{AcO[HgCH_2CO_2]_nH} \xrightarrow[\text{ii, } CO_2]{\text{i, NaOH}} HO[HgCH_2CO_2]_nH \bullet H_2O \xrightarrow{HNO_3} \underset{\mathbf{(38)}}{NO_3HgCH_2CO_2H}$$

Scheme 27

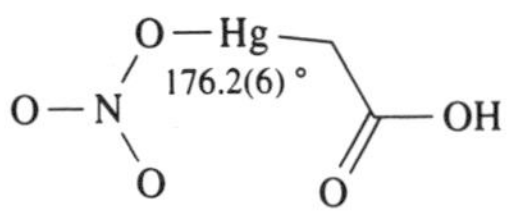

(38)

If acetic acid, acetaldehyde or ethanol is treated with mercury(II) nitrate in the presence of nitric acid, two trimercuriated acids can be isolated with the compositions $(O_3NHg)_3CCO_2H$ (39) and $[HgC(HgNO_3)(OHg)CO_2]_nNO_3$ (40). The former compound consists of discrete molecules, but the cation of the latter is an infinite chain.[107]

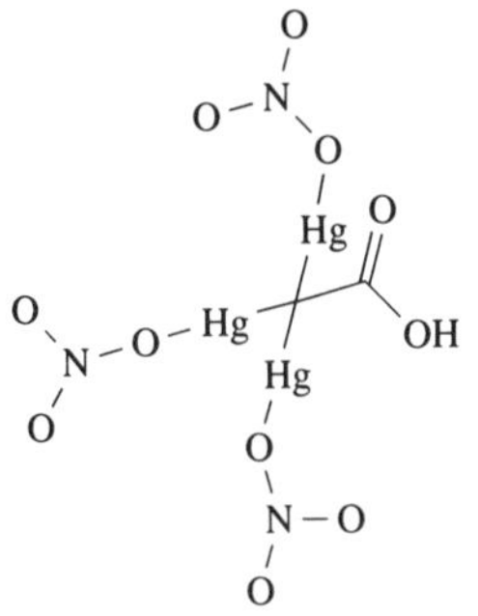

(39)

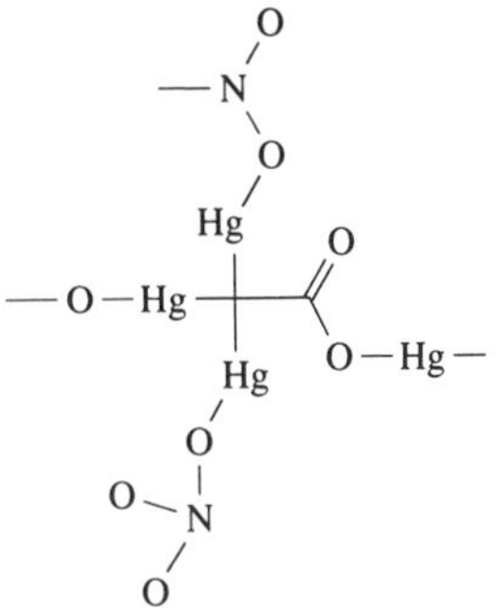

(40)

Higher acids undergo dimercuriation at the α-methylene group to give products of uncomplicated structure (Scheme 28),[108] and dimethyl or diethyl malonate gives the monomercuriated species (e.g., (41)).[109]

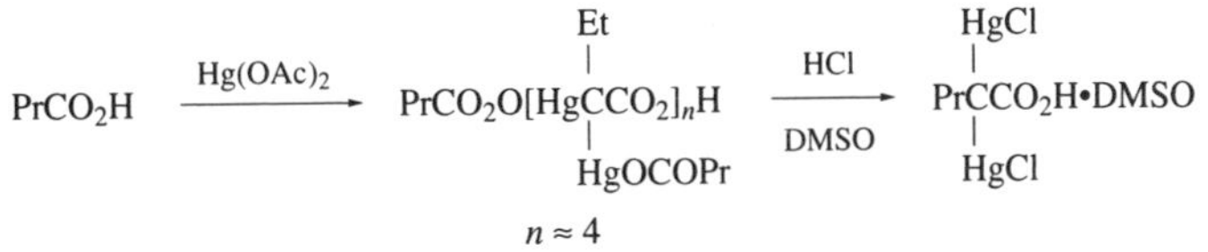

Scheme 28

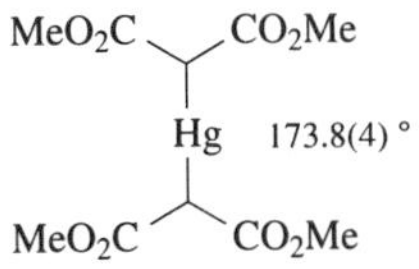

(41)

3.3.3 Arylmercury(II) Compounds

In the crystal, phenylmercury(II) acetate and trifluoroacetate have essentially the same structure; the C–Hg–O atoms are almost collinear (176.6°) and the mercury shows two further long-range interactions (0.2819–0.2952 nm) with two oxygen atoms.[110] In solution, phenylmercury acetate reacts with Lewis bases such as tertiary amines or phosphines to give three-coordinate complexes, but it is reluctant to increase its coordination number further.[111]

Bis{2-[(dimethylamino)methyl]phenyl}mercury has the structure shown in (42). It is near-planar, with C–Hg–C and N–Hg–N units linear (N–C 0.296 nm), and intersecting at 71°, which appears to require either *ds* orbital mixing or donation into a single acceptor orbital.[112]

The triphenylphosphine complex of phenylmercury(II) nitrate has the structure of a distorted T;[113] the C–Hg–P angle is 167.5°, and there is a weak perpendicular Hg···O interaction with oxygen of the nitrate anion. The behaviour of oligomeric 1,2-dimercuriarenes as polydentate Lewis acids is discussed in Section 3.3.5.

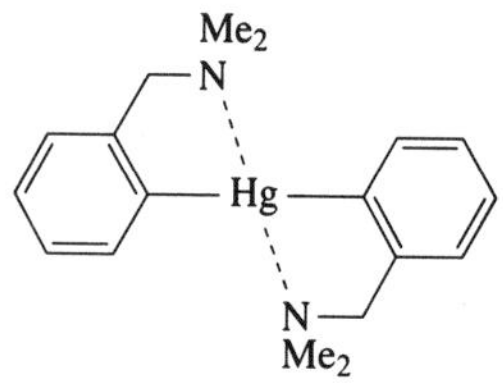

(42)

The naphthalene-derived Schiff's base 2-(1-ClHgC$_{10}$H$_6$)N=CHC$_6$H$_4$-4-NO$_2$, has a simple structure with two-coordinate mercury,[114] but the azo compound 2-ClHgC$_6$H$_4$N=NPh is dimerized through the HgCl group (43),[115] and 2-(2-pyridyl)phenylmercury(II) chloride (44) has an intramolecular N–Hg bond; then these units associate into a tetramer.[116]

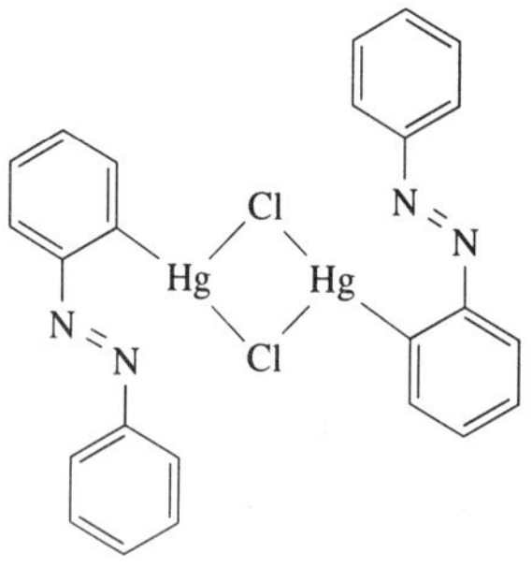

(43)   (44)

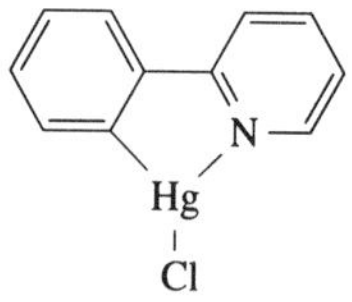

Phenylmercury(II) hydroxide or oxide reacts with strong acids HX to form salts (PhHg)$_2$OH$^+$ X$^-$ (whereas methylmercury hydroxide gives only the complex (MeHg)$_3$O$^+$ X$^-$). The x-ray crystal structure of [(PhHg)$_2$OH]$^+$ [BF$_4$$^-$]·H$_2$O shows that the Hg–O–Hg angle is 126°, but the position of the proton was not determined.[117]

Electrophilic displacement of mercury from arylmercury compounds is covered in Taylor's book.[23] Fluorodemercuriation of aromatic mercury compounds which do not carry deactivating substituents (e.g., 4-MeOC$_6$H$_4$HgOAc to 4-FC$_6$H$_4$HgOAc in 65% yield) can be achieved with acetyl hyperfluorite, but an electron-transfer mechanism has been suggested (Scheme 29).[118]

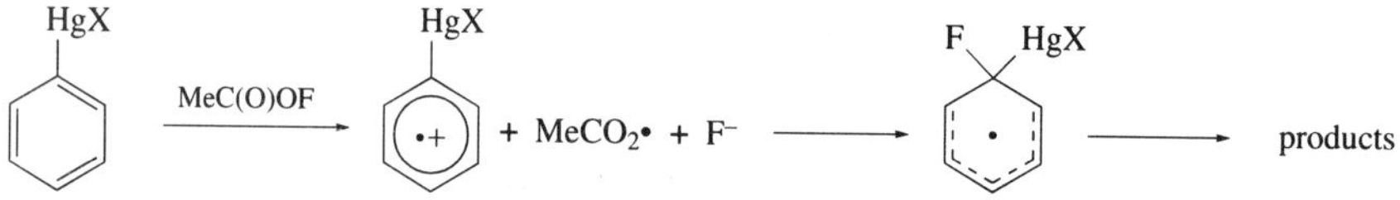

Scheme 29

3.3.4 Cyclopentadienylmercury Compounds

Since the mid-1980s, reports have appeared on the determination of the first structures of cyclopentadienylmercury compounds by single-crystal x-ray diffraction. Dicyclopentadienylmercury (Cp$_2$Hg) itself has the monohapto structure (45),[119] in which the two rings are nearly coparallel, subtending an angle of 8.0(8)°, and it is fluxional in solution.

On the other hand bis(pentamethylcyclopentadienyl)mercury, (Cp*$_2$Hg), prepared from Cp*Li and HgCl$_2$, had been reported to show separate NMR signals for the methyl groups in the intensity ratio 2:2:1 even at room temperature, with no sign of fluxionality.[120] It now seems likely that the material which was examined was in fact Cp*Cp* (46),[121] formed by decomposition of Cp*$_2$Hg, perhaps by way of extrusion of Hg or via the Cp* radical (Scheme 30).[122]

At −78 °C in THF, Cp*HgCl was obtained in 70% yield as yellow crystals from Cp*Li and HgCl$_2$. It is fluxional in solution.[121] The crystal structure shows that it has the novel 'battlement' structure (47). The C–Hg–Cl atoms are almost collinear (174° and 177°), and the molecules are joined by weak Hg···Cl interactions (0.310–0.324 nm) to give a double chain.[123]

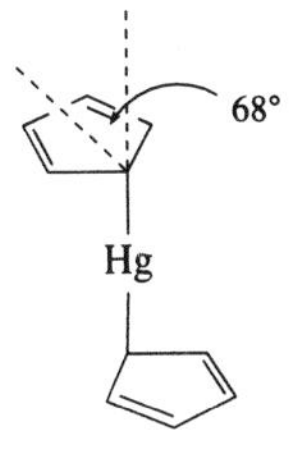

(45)

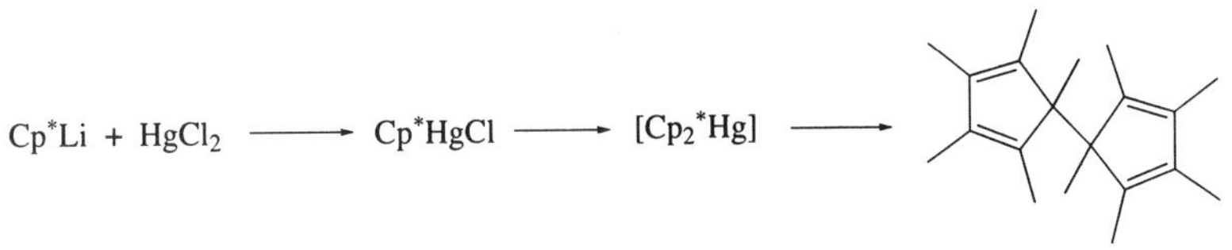

$$Cp^*Li + HgCl_2 \longrightarrow Cp^*HgCl \longrightarrow [Cp_2^*Hg] \longrightarrow$$

(46)

Scheme 30

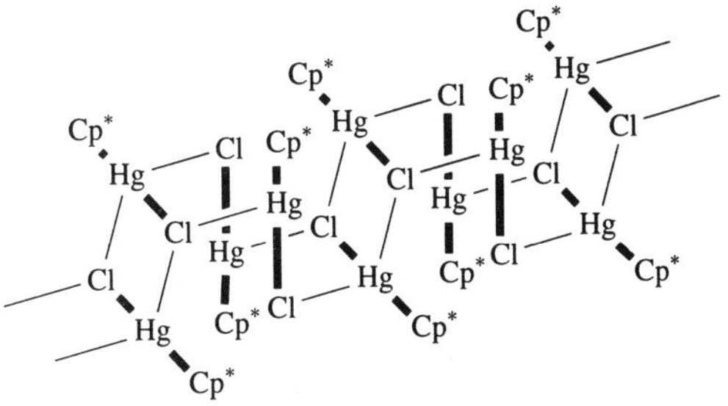

(47)

If cyclopentadienylmercury compounds are subjected to UV irradiation in solution, homolysis of the cyclopentadienyl–mercury bond occurs (Equation (47)), and a strong ESR spectrum of the corresponding cyclopentadienyl radical can be observed. This behaviour is common to the derivatives of a number of other metals, particularly tin and lead, but the ease of preparation of the mercury compounds and their special sensitivity to light have made them the reagent of choice for generating substituted cyclopentadienyl radicals for ESR studies.[124]

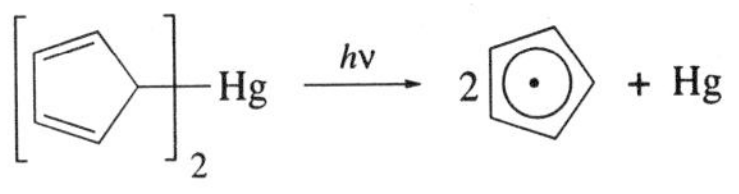

$$\left[\bigcirc \right]_2 Hg \xrightarrow{h\nu} 2 \bigcirc\!\!\cdot + Hg \qquad (47)$$

The appropriate cyclopentadiene is lithiated in solution with butyllithium, then mercury(II) chloride is added and the cyclopentadienyl compound which is formed is photolysed in solution without further purification. By this technique, the ESR spectra have been observed of the substituted cyclopentadienyl radicals $R_nC_5H_{5-n}$, where R_n = Me, 1,2-Me$_2$, 1,3-Me$_2$, 1,2,3-Me$_3$, 1,2,4-Me$_3$, Me$_4$ and Me$_5$.[124]

The structure of $Cl_5C_5HgC_6H_5$ is shown in (**48**); it is monohapto-bonded, and shows no sign of the Cl$\cdots$Hg interactions which were inferred from the ^{35}Cl NQR spectrum.[125] NMR spectra show that in solution the fluxionalities lie in the sequence $(Cl_5C_5)_2Hg < (Cl_5C_5)HgCl \approx (Cl_5C_5)HgPh$, but in the solid state (by ^{13}C NMR and ^{35}Cl NQR) the order is $(Cl_5C_5)Hg(C_6Me_5) > (Cl_5C_5)HgPh > (Cl_5C_5)_2Hg$.

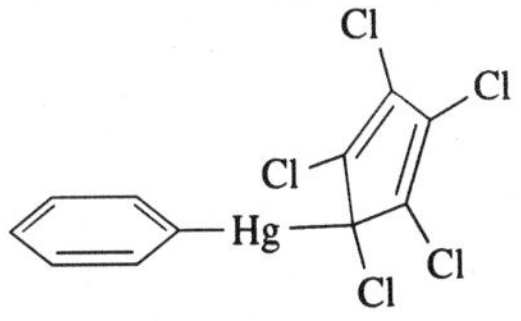

(48)

The cyclopentadienyl groups in ferrocene can be mercuriated with mercury(II) acetate. Ferrocenylmercury(II) chloride disproportionates (Scheme 31) when it is treated with L-cysteine to give di(ferrocenyl)mercury with the structure shown in (49).[126]

$$Cp_2Fe \xrightarrow[\text{ii, Cl}^-]{\text{i, Hg(OAc)}_2} Cp_2Fe(C_5H_4HgCl) \xrightarrow{CysSH^-} [Cp_2Fe(C_5H_4)]_2Hg{\cdot}H_2O + (CysS)_2Hg$$

Scheme 31

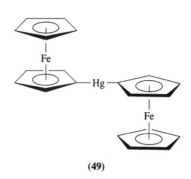

(49)

Ruthenocenes will undergo permercuriation. Mercury(II) acetate reacts with Cp*RuCp to give Cp*Ru[C_5(HgOAc)_5], which with KX_3^- (X = Cl or Br) gives Cp*Ru(C_5X_5) (Scheme 32). Under similar conditions Cp_2Ru gives $[C_5(HgOAc)_5]_2Ru$ and thence $(C_5X_5)_2Ru$ (Scheme 33), suggesting that this may provide a general method for preparing per-substituted derivatives.[127]

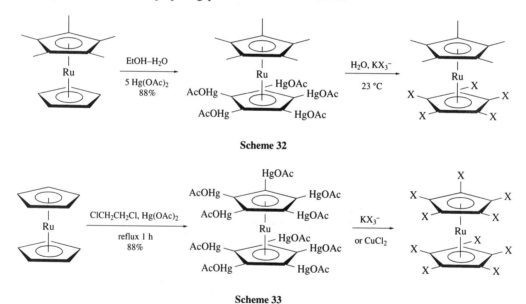

Scheme 32

Scheme 33

3.3.5 Polydentate Lewis Acids

The behaviour of simple organomercury compounds R_2Hg and RHgX as Lewis acids is well established.[1] Dialkylmercurials do not form isolable complexes $R_2Hg{\cdot}L_n$ unless the alkyl group contains electron-attracting substituents, but diphenylmercury gives a weak complex, $Ph_2Hg{\cdot}2$ phen, with 1,10-phenanthroline. The monoorganomercury compounds RHgX, where X is an electronegative ligand such as halide or carboxylate, are stronger acceptors, and a large number of complexes RHgX·L are known. These usually have a near-linear $R–Hg–L^+$ cation, and any interaction with X^- is relatively weak and is orthogonal to the RHgX line, giving a T-shape at mercury.

Examples of such complexes are given in the preceding sections. Our concern here is with compounds containing two or more mercury atoms per molecule, which have the potential of behaving as polydentate Lewis acids, and in a cyclic structure might be expected to act as 'anticrown' reagents towards anions and other Lewis bases.

Cooperative binding is apparent in 1,2-di(chloromercuri)benzene. Phenylmercury chloride does not form an isolable complex with chloride anion, but o-C$_6$H$_4$(HgCl)$_2$ reacts with Ph$_4$P$^+$ Cl$^-$ to give a 2:1 complex with the structure shown in (50).[128] There is little structural change in the organomercury unit, but there are four new Hg–Cl bonds with an average length of 0.302 nm.

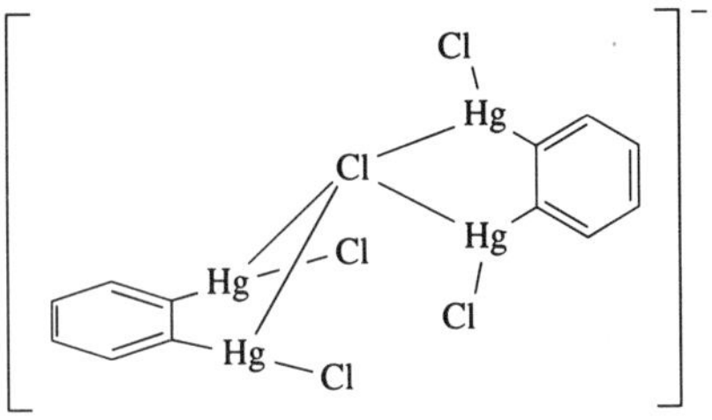

(50)

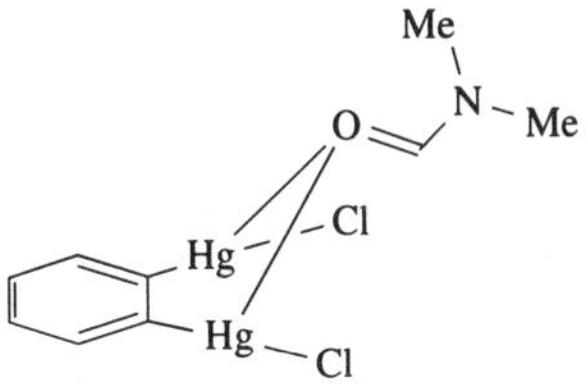

(51)

With DMF, o-C$_6$H$_4$(HgCl)$_2$ forms a 1:1 complex (51) in which the oxygen of the DMF bridges the two mercury atoms (Hg–O distances 0.2777(13) and 0.2681(13) nm), and the plane of the DMF molecule lies close to the plane bisecting the benzene ring.[129] Di-p-methoxybenzothiophenone similarly forms a complex with the trifluoroacetate, o-C$_6$H$_4$(HgOCOCF$_3$)$_2$·S=C(C$_6$H$_4$OMe)$_2$, which activates the thione towards reduction by nucleophilic attack.[130] If the aromatic ring of the mercurial carries four methyl substituents, it forms a 2:3 complex with dimethyl- or diethylformamide, [C$_6$Me$_4$(HgOCOCF$_3$)$_2$][O=CHNR$_2$], in which the oxygen of one amide molecule is bound to four mercury atoms, as represented diagramatically in (52).[131]

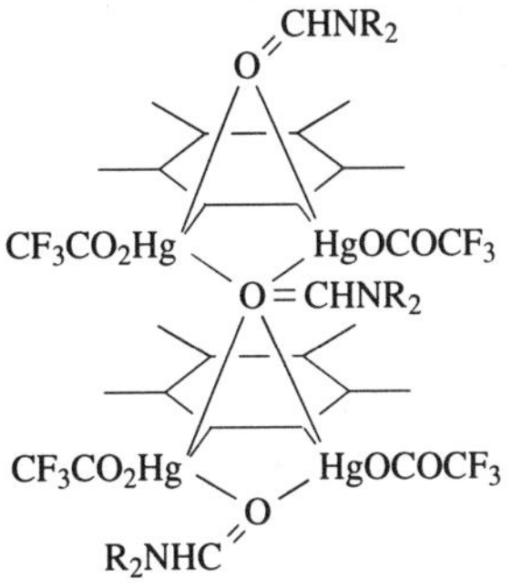

(52)

The cyclic tetraester which is formed with perfluoroglutaric acid gives a crystalline 2:1 complex (53) with THF, in which the two independent THF molecules are placed one above and one below the plane of the 22-membered ring.[132]

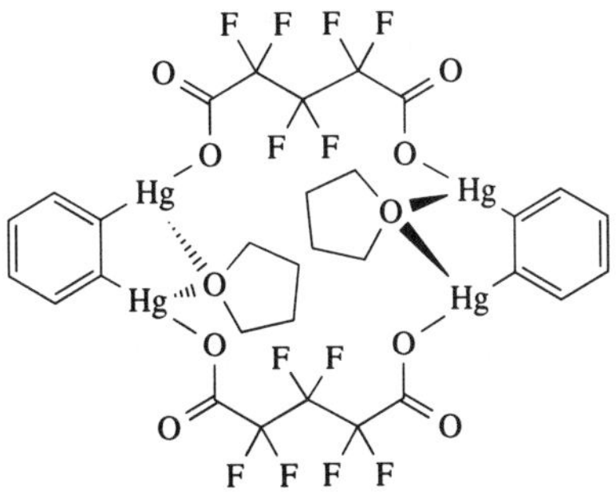

(53)

1,8-Di(chloromercuri)naphthalene binds one DMSO molecule on the plane bisecting the naphthalene ring as shown in (54).[11]

The cyclic trimer (o-C$_6$F$_4$Hg)$_3$ gives both 1:1 and 3:2 complexes with Ph$_3$PMe$^+$ Cl$^-$, Br$^-$ and I$^-$, and 1:1, 1:2 and 1:3 solvates with a variety of solvents. The 1:1 complex (o-C$_6$F$_4$Hg)$_3$·X$^-$ has the polynuclear

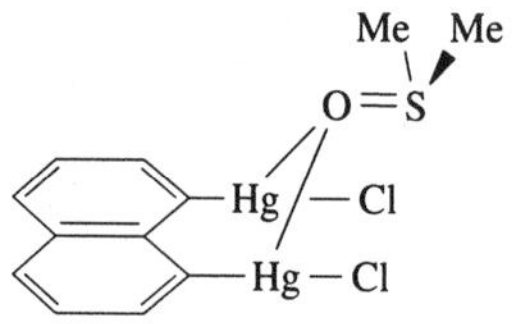

(54)

structure shown in (55), in which each Br^- is coordinated to six mercury atoms,[133] and the 3:2 complex may have the structure of a triple-decker sandwich. Complexes with a 1:2 structure have been identified from $[(CF_3)_2CHHg]_5$, and the structure of $[(CF_3)_2CHHg]_5Cl_2{}^{2-}[PPh_4]^+{}_2$ has been shown to be a pentagonal bipyramid (56), each chlorine complexed to five mercury atoms with average Hg—Cl distances of 0.3284 nm and 0.3221 nm.[134]

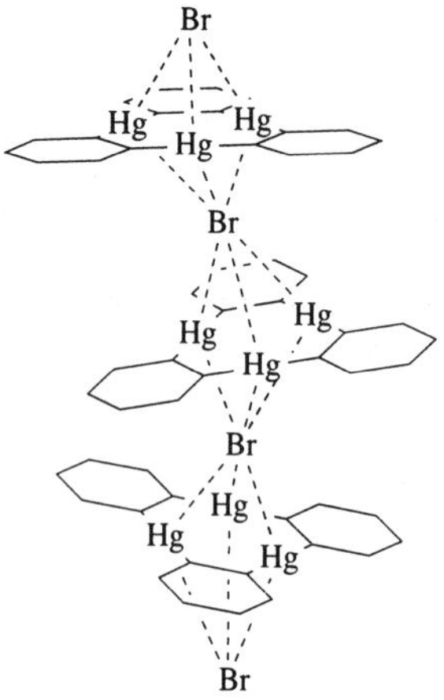

(55) 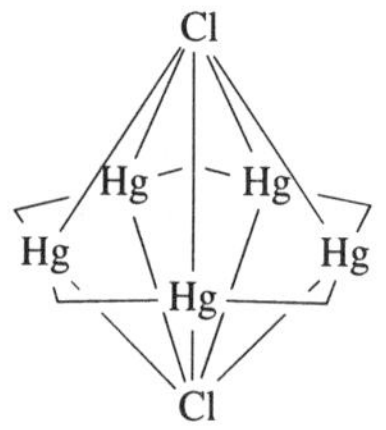   **(56)**

The cyclic trimeric and tetrameric mercuricarboranes $(B_{10}H_{10}C_2-Hg)_n$ ($n = 3$ or 4) (Scheme 1) similarly form complexes with halide ions and solvents. The $(B_{10}H_{10}C_2-Hg)_4\cdot Cl^-$ ion has the structure shown in (57), in which the chloride ion is 0.2944(2) nm from each of the four mercury atoms, and displaced out of their plane by 0.0383 nm; the C—Hg—C angle is 162.3°. The corresponding bromide has a similar structure, with an average Hg—Br distance of 0.3063(3) nm and displacement of 0.0960 nm.[14]

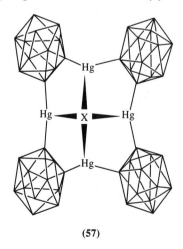

(57)

$(B_{10}H_{10}C_2-Hg)_4\cdot I_2$ has the structure shown in (58), in which the two I^- ions are located above and below the plane of the four mercury atoms by 0.1962(1) nm, equidistant (0.3304(1) nm and 0.3306(1) nm) from two 'para' mercury atoms, but nearer (0.3277(1) nm) to one of the remaining mercury atoms than to the other (0.3774(1) nm). The C—Hg—C angles are 152.6(4)° and 158.1(4)° and represent the largest deviation yet recorded for an R_2Hg molecule; this is presumably induced by strain in the tetrameric ring.[135]

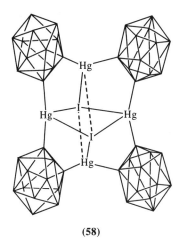

(58)

Similar complexes are fomed by 9,12-diethylmercuracarborane.[14] The corresponding tetra-3-phenylmercuracarborane forms a 1:1 complex with iodide anion, $(PhB_{10}H_9C_2-Hg)_4 \cdot I^- Li^+$.[136] The central ring is saddle-shaped as shown in (59), and only one of the four possible isomers is present, the four phenyl groups alternating above and below the ring. The four mercury atoms are located in a perfect square, and the iodide anion is displaced 0.125 nm above the centre of the square, 0.3125(5) nm from each mercury atom.

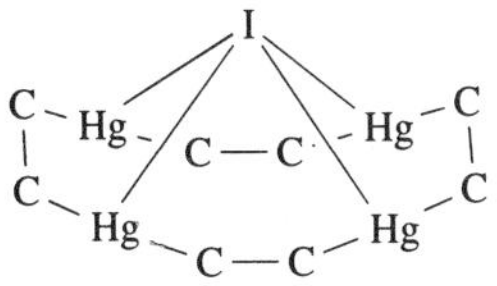

(59)

The trimer $(B_{10}H_{10}C_2-Hg)_3$ forms a 1:1 complex with chloride ion, $(B_{10}H_{10}C_2-Hg)_3 \cdot Cl^- Li^+$, with the structure shown in (60), where the chloride anion is symmetrically located over the triangle of mercury atoms.[6] The trimer also forms 1:3 and 1:5 acetonitrile complexes which cocrystallize in the ratio of 1:1. The binding of the nitrile molecules is unusual, and is represented diagrammatically in (61) and (62). In each complex, there is one nitrile molecule located above and below each ring, interacting with all three mercury atoms, and the remaining one or three nitrile molecules bind each to one mercury.[6]

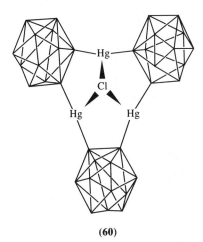

(60)

3.3.6 Organomercury Hydrides

When mercury compounds are used in organic synthesis, the final step often involves hydridodemercuration with a reagent such as sodium tetrahydroborate. The reactions occur through

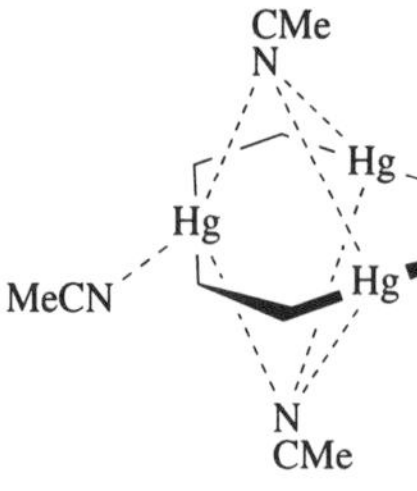

(61)

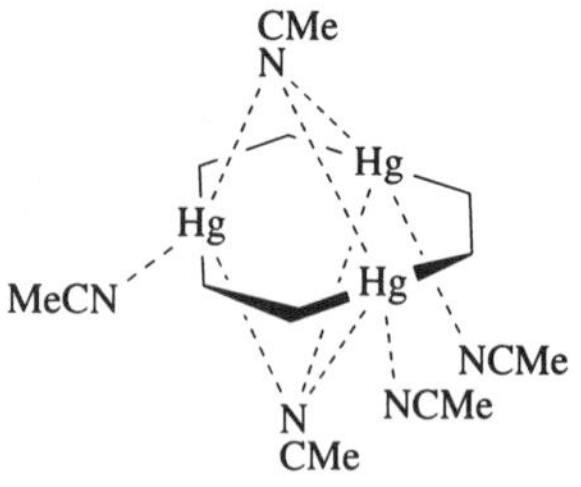

(62)

homolytic decomposition of intermediate mercury hydrides, and under some conditions the intermediate organic radicals can be trapped (Scheme 34, Equations (48) and (49)) (see Section 3.3.7).

$$RHgX \xrightarrow{\text{NaBH}_4} RHgH \longrightarrow RHg\bullet$$

Scheme 34

$$RHg\bullet \longrightarrow R\bullet + Hg^0 \tag{48}$$

$$R\bullet + RHgH \longrightarrow RH + RHg\bullet \tag{49}$$

It has generally been thought that the intermediate hydrides were too unstable to be observed, but three groups[137-9] have shown that they can be isolated and purified by gas–liquid chromatography and characterized by IR, NMR and mass spectrometry.

The reduction of methylmercury chloride with sodium tetrahydroborate in water gives methylmercury(II) hydride, which can be aspirated in an inert gas stream into benzene or toluene, and isolated in a cold trap after chromatography. In the ^{1}H NMR spectrum it shows a doublet for the methyl group, δ 0.10, $J(1H) = 4.2$ Hz, and a second signal for one hydrogen at δ 17.2, $J(3H) = 4.2$ Hz. In the IR spectrum it shows strong absorbances at 1969 cm^{-1} and 1943 cm^{-1} for the Hg–H stretch, which shifts to 1418 cm^{-1} and 1392 cm^{-1} in MeHgD. The mass spectrum shows a cluster of peaks at m/z 212–220 for MeHgH, which moves one mass unit higher in MeHgD. The half-life is about 2 h in water at ambient temperature.[137,138] The compounds EtHgH, PhHgH, C_6F_5HgH, EtHgD, PhHgD and C_6F_5HgD have been prepared in a similar way.[140]

A number of organomercury halides have also been reduced at $-70\,°C$ with lithium aluminum hydride or tributyltin hydride, and the NMR spectra of the products are given in Table 6.[139] The stability of these hydrides in solution appears to be strongly influenced by their concentration and by the nature of the solvent.

Support for the above conclusions comes from the preparation of methylmercury hydride and perdeuterio(methylmercury hydride) by a matrix-isolation technique. If a mixed of argon and methane (or CD$_4$) containing mercury is deposited at 22 K or 5.7 K, and is irradiated with light of 249 nm wavelength, the photoexcited ($^3P_1$) mercury reacts with the methane to give CH$_3$HgH (or CD$_3$HgD), which shows in the IR spectra bands at about 1956 cm^{-1} or 1404 cm^{-1} for the Hg–H or Hg–D stretch, respectively. These frequencies are close to those found in HgH$_2$ and HgD$_2$, which can be prepared in a similar way by irradiation of a solid solution of mercury in an argon–hydrogen matrix.[141]

3.3.7 Radicals and Radical Reactions

The weakness of the C–Hg and Hg–H bonds and the low ionization energy of organomercury compounds render organomercury chemistry an important field for radical and radical ion processes, which provide the basis for a number of important synthetic procedures.

A standard technique for generating radical cations for studies by EPR spectroscopy is to irradiate a dilute solution of the substrate in a frozen Freon matrix with γ-rays, and a number of organomercury compounds have been examined in this way. Dialkyl-, dibenzyl- and diphenyl-mercury compounds show spectra of the radical cations R$_2$Hg$^{\bullet+}$ which indicate that an electron has been lost from the C–Hg–C σ-bond system. If the matrix is allowed to warm, the dialkyl- and dibenzylmercury radical cations dissociate into R$\bullet$ and RHg$^+$.[142,143]

Table 6 NMR spectra of organomercury hydrides.

RHgH	$\delta\ ^{199}Hg$	$\delta\ ^1H$	$^1J(Hg–H)$ (Hz)
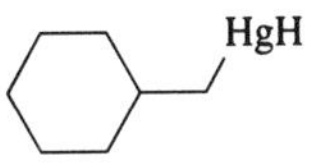 [cyclohexyl–CH₂HgH structure]	–508	+17.3	2441
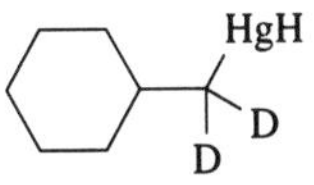 [cyclohexyl–CHD with HgH and D structure]	–500	+17.6	2409
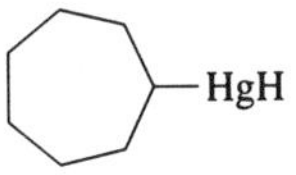 [cycloheptyl–HgH structure]	–640	+17.1	2308
MeHgH	–456	+16.8	2660
PhHgH	–830	+13.3	2936
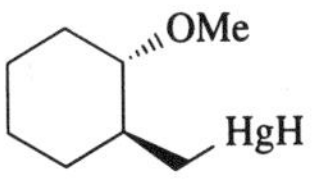 [cyclohexyl with OMe and HgH structure]	–725	+17.5	2314
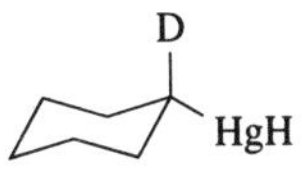 [cyclohexane with D and HgH structure]	–756	+16.1	2302
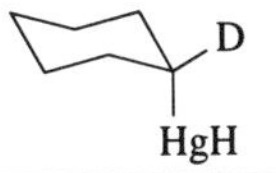 [cyclohexane with D and HgH structure]	–632	+17.3	2307

In arylmercurials, Ar$_2$Hg or ArHgX, if the ionization energy of the aryl group is lowered by substituents or by extended conjugation (e.g., in di(*o*-allyloxyphenyl)mercury or 2-biphenylenylmercury trifluoroacetate),[27] the electron is lost instead from the aromatic π-system, and the EPR spectra can be observed in fluid solution, often at room temperature. The preparation of arylmercury radical cations in this way by UV irradiation of a solution of the arene in trifluoroacetic acid containing mercury(II) trifluoroacetate is described in Section 3.2.3. The replacement of a proton in the aromatic system by mercury has a negligible effect on the hyperfine coupling constants of the remaining protons, implying that the mercury substituent does not alter the π-electron distribution, and it is found that, in the radical cations of the arene and corresponding arylmercury trifluoroacetate, ArH$^{•+}$ and ArHgOCOCF$_3^{•+}$, the ratio of the hyperfine coupling constants $a(^{199}Hg)/a(^1H)$ is approximately constant at ca. 20.6.[27]

Organomercury salts can also form radical anions RHgX$^{•-}$, and the compounds RHgX can take part in organic reactions by mechanisms involving neutral radicals, radical cations or radical anions, depending on the electron demands of the substrates; in the last two mechanisms, the organomercury compound reacts as an electron donor or as an electron acceptor, respectively. This is illustrated by the photoinitiated reaction of *t*-butylmercury(II) chloride with the 1,1-diarylethenes Ar$_2$C=CH$_2$, where Ar = Ph, 4-MeOC$_6$H$_4$ or 4-NO$_2$C$_6$H$_4$.[144] In each case, the *t*-butyl radical which is generated adds to the alkene to give the corresponding diphenylmethyl radical (Equations (50)–(52)).

$$\text{Bu}^t\text{HgCl} \xrightarrow{h\nu} \text{Bu}^t\text{•} + \text{•HgCl} \tag{50}$$

$$\text{Bu}^t\text{HgCl} + \text{•HgCl} \longrightarrow \text{Bu}^t\text{•} + \text{HgCl}_2 + \text{Hg}^0 \tag{51}$$

$$\text{Bu}^t\text{•} + \text{CH}_2=\text{CAr}_2 \longrightarrow \text{Bu}^t\text{CH}_2\overset{•}{\text{C}}\text{Ar}_2 \tag{52}$$

When Ar = Ph, this radical disproportionates to give the corresponding alkene and alkane (Equation (53)).

$$2\ \text{Bu}^t\text{CH}_2\overset{•}{\text{C}}\text{Ph}_2 \longrightarrow \text{Bu}^t\text{CH}_2\text{CHPh}_2 + \text{Bu}^t\text{CH}=\text{CPh}_2 \tag{53}$$

When Ar = 4-MeOC$_6$H$_4$, the radical donates an electron to the mercury(II) halide to give the carbon cation and alkylmercury radical anion, and thence the alkene, mercury, chloride ion and *t*-butyl radical, establishing a chain reaction (Equations (54)–(56)).

$$Bu^tCH_2\dot{C}(C_6H_4OMe\text{-}4)_2 + Bu^tHgCl \longrightarrow Bu^tCH_2\overset{+}{C}(C_6H_4OMe\text{-}4)_2 + Bu^tHgCl^{\bullet -} \tag{54}$$

$$Bu^tHgCl^{\bullet -} \longrightarrow Bu^t\bullet + Hg^0 + Cl^- \tag{55}$$

$$Bu^tCH_2\overset{+}{C}(C_6H_4OMe\text{-}4)_2 \xrightarrow{-H^+} Bu^tCH=C(C_6H_4OMe\text{-}4)_2 \tag{56}$$

On the other hand, when Ar = 4-NO$_2$C$_6$H$_4$, an electron is donated to the radical by the alkylmercury halide, to give the carbanion and alkylmercury radical cation, and thence the alkane, and *t*-butyl radical, again setting up a chain process (Equations (57)–(59)).

$$Bu^tCH_2\dot{C}(C_6H_4NO_2\text{-}4)_2 + Bu^tHgCl \longrightarrow Bu^tCH_2\overset{-}{C}(C_6H_4NO_2\text{-}4)_2 + Bu^tHgCl^{\bullet +} \tag{57}$$

$$Bu^tHgCl^{\bullet +} \longrightarrow Bu^t\bullet + HgCl^+ \tag{58}$$

$$Bu^tCH_2\overset{-}{C}(C_6H_4NO_2\text{-}4)_2 \xrightarrow{+H^+} Bu^tCH_2CH(C_6H_4NO_2\text{-}4)_2 \tag{59}$$

Most of the reactions that have achieved preparative importance involve neutral radicals,[145-8] and can be generalized as in Scheme 35; the formation of the organomercury hydrides under these conditions has been confirmed (see Section 3.3.6).

$$RHgX \xrightarrow[\text{or Bu}_3\text{SnH}]{NaBH_4} RHgH \longrightarrow R\bullet$$

$$R\bullet + Z \longrightarrow RZ\bullet$$

$$RZ\bullet + RHgH \longrightarrow RZH + R\bullet + Hg^0$$

Scheme 35

Commonly the reagent Z is an alkene when the reaction cycle is as illustrated in Scheme 36. These are 'one-pot' reactions which can be carried out without irradiation at room temperature, and are usually complete in a few minutes. They are tolerant towards a variety of functional groups, which can be introduced into the α-position of the mercurial by a carbene insertion, or into the β- or γ-positions by solvomercuriation of an alkene or a cyclopropane (see Sections 3.2.6, 3.2.5.1 and 3.2.5.2). One limitation is that, as the alkylmercury hydride is a very good hydrogen donor, the alkene must be activated towards reaction with the radical R•, or R• will react with the hydride to give RH. These reactions provide a novel method of forming the C–C bond and they have been developed, particularly by Giese's group, into useful synthetic procedures. Some examples are given in Scheme 37,[149] Equation (60)[150] and Schemes 38–40.[151-3]

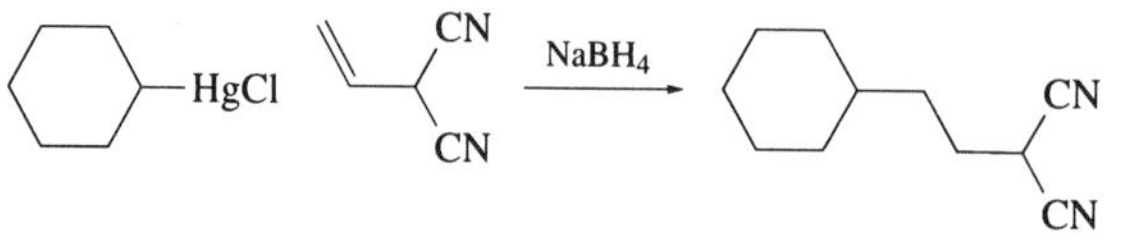

$$\text{(cyclohexyl)}-HgCl + \overset{CN}{\underset{CN}{\diagup\!\!\diagdown}} \xrightarrow{NaBH_4} \text{(cyclohexyl)}-CH_2CH_2CH\overset{CN}{\underset{CN}{}} \tag{60}$$

If the alkene carries a suitable substituent Y (e.g., Bu$_3$Sn, HgCl, PhC≡CHg, I, SO$_2$Ph), the addition of the alkyl radical can be followed by a geminal elimination of that substituent, resulting in overall substitution (Scheme 41). These reactions are usually initiated photolytically (e.g., a 275 W sunlamp at 35–40 °C).[154] An example is given in Equation (61).

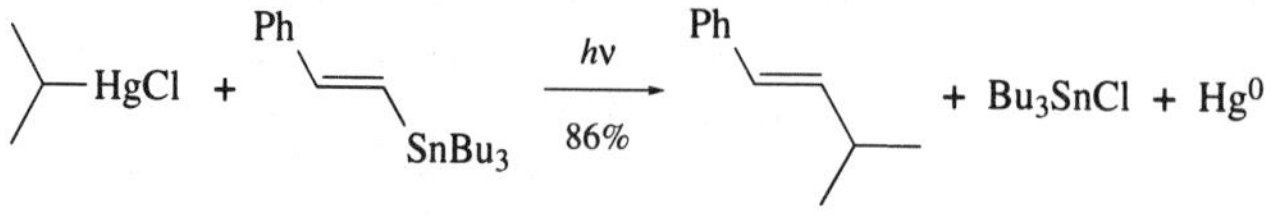

$$\text{(isopropyl)}-HgCl + \overset{Ph}{\underset{SnBu_3}{\diagup\!\!\diagdown}} \xrightarrow[86\%]{h\nu} \overset{Ph}{\diagup\!\!\diagdown}\text{(isopropyl)} + Bu_3SnCl + Hg^0 \tag{61}$$

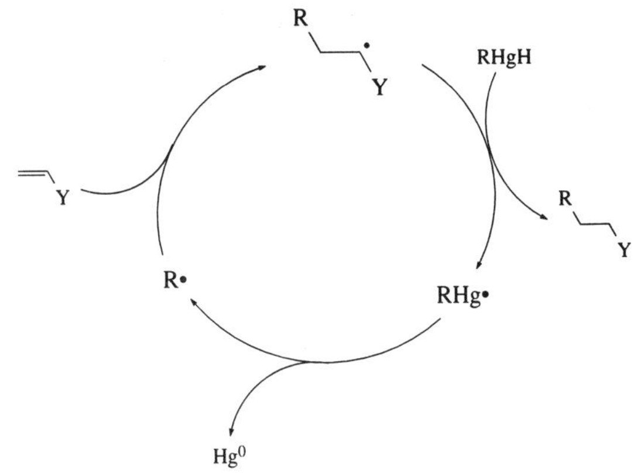

Scheme 36

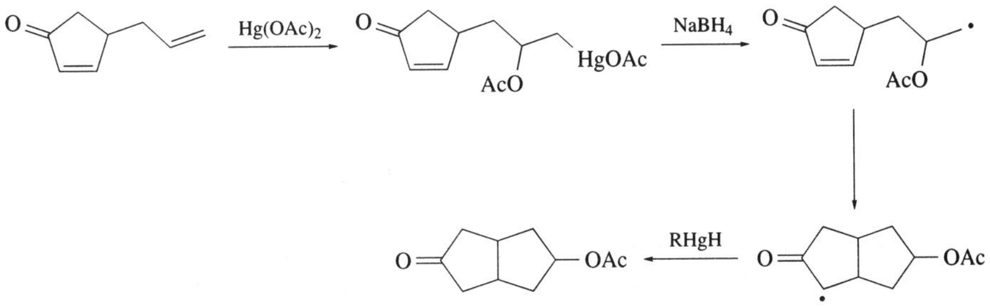

Scheme 37

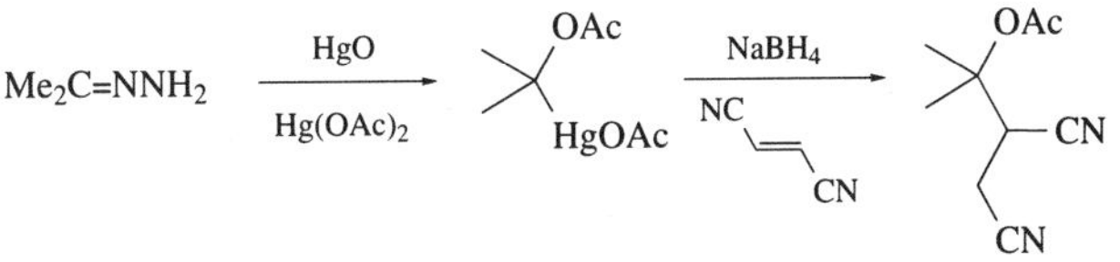

Scheme 38

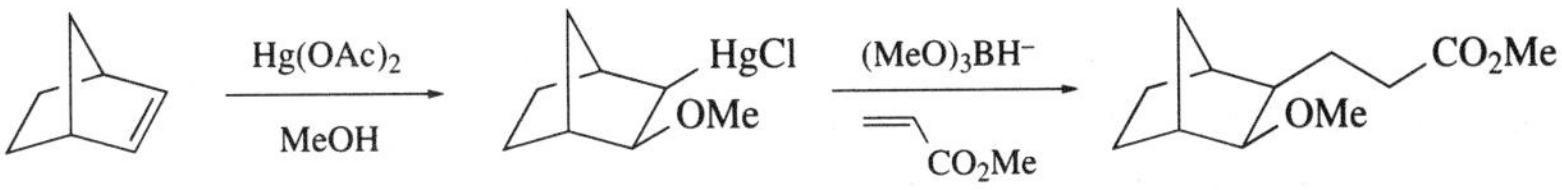

Scheme 39

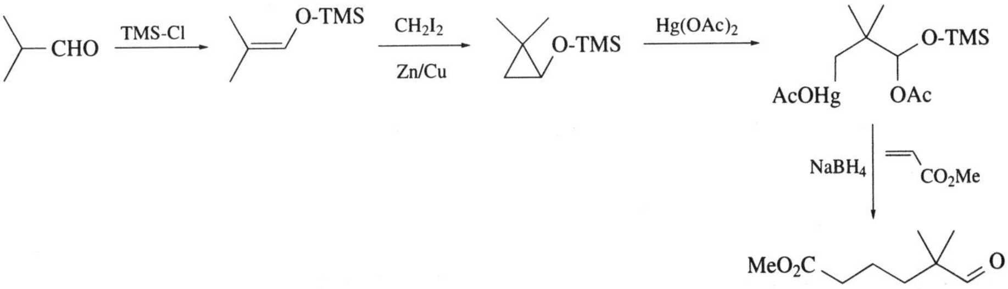

Scheme 40

Under similar photolytic conditions, or with initiation with AIBN (2,2'-azobisisobutyronitrile) at 80 °C, direct homolytic substitution (S_H2) can occur at a heteroatom Y in compounds QY, for example, RSSR, PhSeSePh, PhTeTePh, ArSO$_2$Cl, 4-MeC$_6$H$_4$SO$_2$Cl (Scheme 42 and Equation (62)).[155]

Russell has also established $S_{RN}1$ chain reactions which involve intermediate organomercury radical anions as shown in Scheme 43. Two examples are given in Equations (63) and (64).[156]

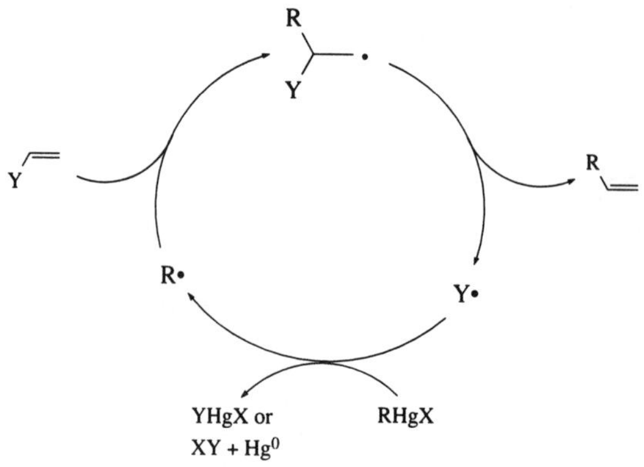

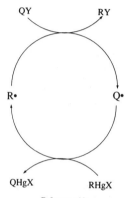

Scheme 41

Scheme 42

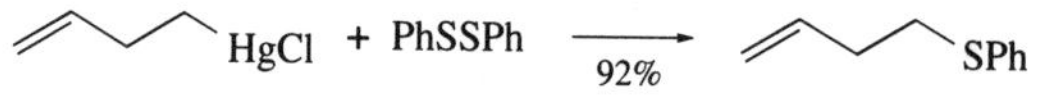

$$\text{(alkenyl)}HgCl + PhSSPh \xrightarrow[92\%]{} \text{(alkenyl)}SPh \qquad (62)$$

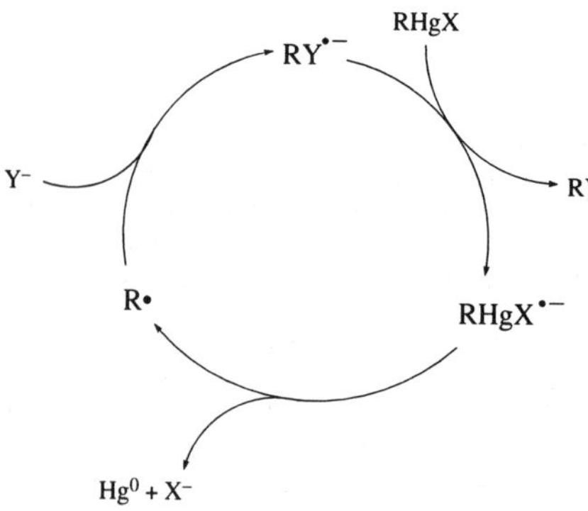

Scheme 43

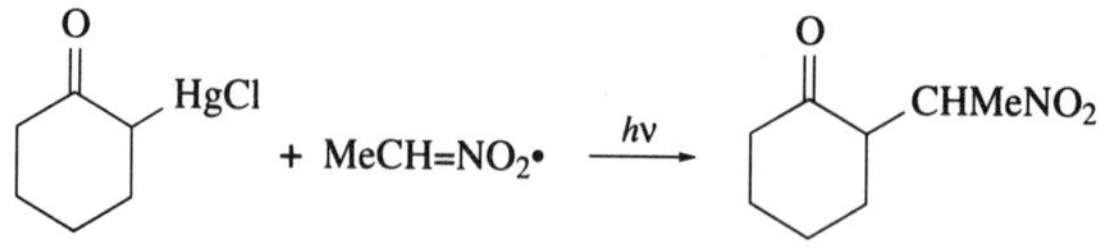

$$\text{(cyclohexanone-HgCl)} + MeCH=NO_2\bullet \xrightarrow{h\nu} \text{(cyclohexanone-CHMeNO_2)} \qquad (63)$$

$$Bu^tHgCl + \text{(fluorenyl)}^- K^+ \xrightarrow[44\%]{h\nu} \text{(9-Bu^t-fluorene)} \qquad (64)$$

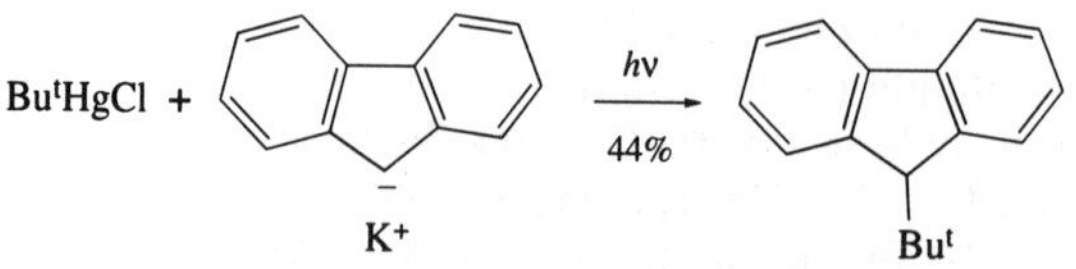

3.4 REFERENCES

1. J. L. Wardell, in 'COMC-I', vol. 2, p. 863.
2. A. Sebald, 'NMR: Basic Principles and Progress', Springer, Berlin, 1994, vol. 31, p. 91 (*Chem. Abstr.*, 1994, **121**, 219 702f).
3. R. C. Larock, 'Solvomercuration/Demercuration Reactions in Organic Synthesis', Springer, Berlin, 1986.
4. J. L. Wardell, in 'Organometallic Chemistry', Royal Society of Chemistry, London, 1983–1993, vols. 13–23.
5. S. M. S. V. Doidge-Harrison and J. L. Wardell, in 'Dictionary of Organometallic Compounds', 2nd edn., Chapman and Hall, London, 1994, vol. 2, p. 1789.
6. X. Yang, Z. Zheng, C. B. Knobler and M. F. Hawthorne, *J. Am. Chem. Soc.*, 1993, **115**, 193.
7. F. I. Aigbirhio, S. S. Al-Juaid, C. Eaborn, A. Habtemariam, P. B. Hitchcock and J. D. Smith, *J. Organomet. Chem.*, 1991, **405**, 149.
8. C. Eaborn, K. L. Jones, J. D. Smith and K. Tavakkoli, *J. Chem. Soc., Chem. Commun.*, 1989, 1201.
9. S. Rapsomanikis and P. J. Craig, *Anal. Chim. Acta*, 1991, **248**, 563.
10. H. C. Brown, R. C. Larock, S. K. Gupta, S. Rajagopalan and N. G. Bhat, *J. Org. Chem.*, 1989, **54**, 6079.
11. H. Schmidbaur, H.-J. Öller, D. L. Wilkinson, B. Huber and G. Müller, *Chem. Ber.*, 1989, **122**, 31.
12. E. Negishi, K. P. Jadhav and N. Daotien, *Tetrahedron Lett.*, 1982, **23**, 2085.
13. M. J. Rozema, D. Rajagopal, C. E. Tucker and P. Knochel, *J. Organomet. Chem.*, 1992, **438**, 11.
14. X. Yang, C. B. Knobler, Z. Zheng and M. F. Hawthorne, *J. Am. Chem. Soc.*, 1994, **116**, 7142.
15. A. Naganuma, T. Urano and N. Imura, *J. Pharmacobio-Dyn.*, 1985, **8**, 69.
16. G. N. Howell, M. J. O'Connor, A. M. Bond, H. A. Hudson, P. J. Hanna and S. Strother, *Aust. J. Chem.*, 1986, **39**, 1167.
17. P. E. Eaton and R. M. Martin, *J. Org. Chem.*, 1988, **53**, 2728.
18. P. E. Eaton, G. T. Cunkle, G. Marchioro and R. M. Martin, *J. Am. Chem. Soc.*, 1987, **109**, 948.
19. N. A. A. Al-Jabar and A. G. Massey, *J. Organomet. Chem.*, 1984, **275**, 9.
20. F. A. Johnson and W. D. Perry, *Organometallics*, 1989, **8**, 2646.
21. L. Gorrichon-Guigon, P. Maroni, R. Meyer and J. Corset, *J. Organomet. Chem.*, 1982, **228**, 15.
22. K. König, W. Weiss and H. Musso, *Chem. Ber.*, 1988, **121**, 1271.
23. R. Taylor, 'Electrophilic Aromatic Substitution', Wiley, Chichester, 1990.
24. G. B. Deacon, M. J. O'Connor and G. N. Stretton, *Aust. J. Chem.*, 1986, **39**, 953.
25. W. Lau and J. K. Kochi, *J. Org. Chem.*, 1986, **51**, 1801.
26. A. G. Davies, *Chem. Soc. Rev.*, 1993, **22**, 299.
27. A. G. Davies and D. C. McGuchan, *Organometallics*, 1991, **10**, 329.
28. J. L. Courtneidge, A. G. Davies, D. C. McGuchan and S. N. Yazdi, *J. Organomet. Chem.*, 1988, **341**, 63.
29. R. Eujen, *Inorg. Synth.*, 1986, **24**, 52.
30. G. B. Deacon, M. F. O'Donoghue, G. N. Stretton and J. M. Miller, *J. Organomet. Chem.*, 1982, **233**, C1.
31. V. F. Chertskov, M. V. Galakhov, S. R. Sterlin, L. S. German and I. L. Knunyants, *Bull. Acad. Sci. USSR*, 1983, **32**, 1095.
32. G. B. Deacon and M. F. O'Donoghue, *Inorg. Chim. Acta*, 1986, **118**, L41.
33. G. B. Deacon and G. N. Stretton, *Aust. J. Chem.*, 1985, **38**, 419.
34. A. G. Massey, N. A. A. Al-Jabar, R. E. Humphries and G. B. Deacon, *J. Organomet. Chem.*, 1986, **316**, 25.
35. M. Y. Antipin, Y. T. Struchkov, A. Yu. Volkonskii and E. M. Rokhlin, *Bull. Acad. Sci. USSR*, 1983, **32**, 410.
36. G. B. Deacon, B. M. Gatehouse, L. W. Guddat and S. C. Ney, *J. Organomet. Chem.*, 1989, **375**, C1.
37. J. Barluenga, C. Jimenez, C. Nájera and M. Yus, *J. Chem. Soc., Perkin Trans. 1*, 1983, 591.
38. W. Carruthers, M. J. Williams and M. T. Cox, *J. Chem. Soc., Chem. Commun.*, 1984, 1235.
39. J. Barluenga, C. Jimenez, C. Nájera and M. Yus, *J. Chem. Soc., Perkin Trans. 1*, 1984, 721.
40. J. Perthius and P. Poisson, *Bull. Soc. Chim. Fr.*, 1985, 75.
41. K. Inomata, T. Kobayashi, S. Sasaoka, H. Kinoshita and H. Kotake, *Chem. Lett.*, 1986, 289.
42. A. J. Bloodworth and P. N. Cooper, *J. Chem. Soc., Chem. Commun.*, 1986, 709.
43. V. R. Kartashov, T. N. Sokolova, E. V. Skorobogatova, Y. K. Grishin, D. V. Bazhenov and N. S. Zefirov, *J. Org. Chem. USSR*, 1991, **27**, 223.
44. V. R. Kartashov, E. V. Gal'yanova, E. V. Skorobogatova, A. N. Chernov and N. S. Zefirov, *J. Org. Chem. USSR*, 1984, **20**, 2389.
45. V. R. Kartashov *et al.*, *J. Org. Chem. USSR*, 1991, **27**, 1080.
46. A. J. Bloodworth and M. D. Spencer, *J. Organomet. Chem.*, 1990, **386**, 299.
47. J. Barluenga, J. M. Martínez-Gallo, C. Nájera and M. Yus, *J. Chem. Soc., Chem. Commun.*, 1985, 1422.
48. J. Einhorn, C. Einhorn and J. L. Luche, *J. Org. Chem.*, 1989, **54**, 4479.
49. H. C. Brown, J. T. Kurek, M.-H. Rei and K. L. Thompson, *J. Org. Chem.*, 1984, **49**, 2551.
50. A. J. Bloodworth, C. J. Cooksey and D. Korkodilos, *J. Chem. Soc., Chem. Commun.*, 1992, 926.
51. A. J. Bloodworth, R. J. Curtis, M. D. Spencer and N. A. Tallant, *Tetrahedron*, 1993, **49**, 2729.
52. W. Kitching, J. A. Lewis, M. T. Fletcher, J. J. de Voss, R. A. I. Drew and C. J. Moore, *J. Chem. Soc., Chem. Commun.*, 1986, 855.
53. T. S. Kuznetsova *et al.*, *J. Org. Chem. USSR*, 1991, **27**, 67.
54. A. J. Bloodworth and N. A. Tallant, *J. Chem. Soc., Chem. Commun.*, 1992, 428.
55. J. Cai and A. G. Davies, *J. Chem. Soc., Perkin Trans. 1*, 1992, 3383.
56. M. Bassetti, M. P. Trovato and G. Bocelli, *Organometallics*, 1990, **9**, 2292.
57. Y. K. Grishin *et al.* *Tetrahedron Lett.*, 1988, **29**, 4631.
58. J. Barluenga, F. Aznar, R. Liz and R. Rodes, *J. Chem. Soc., Perkin Trans. 1*, 1983, 1087.
59. J. Barluenga, F. Aznar and R. Liz, *J. Chem. Soc., Perkin Trans. 1*, 1983, 1093.
60. R. C. Larock and C.-L. Liu, *J. Org. Chem.*, 1983, **48**, 2151.
61. R. C. Larock *et al.*, *Organometallics*, 1987, **6**, 1780.
62. V. R. Kartashov, T. N. Sokolova, Y. K. Grishin, D. V. Bazhenov and N. S. Zefirov, *Organomet. Chem. USSR*, 1992, **5**, 472.
63. M.-A. Boaventura, J. Drouin, F. Theobald and N. Rodier, *Bull. Soc. Chim. Fr.*, 1987, 1006.
64. R. C. Larock and L. W. Harrison, *J. Am. Chem. Soc.*, 1984, **106**, 4218.

65. A. J. Bloodworth and C. J. Cooksey, *J. Organomet. Chem.*, 1985, **295**, 131.
66. J. B. Lambert, E. C. Chelius, R. H. Bible and E. Hajdu, *J. Am. Chem. Soc.*, 1991, **113**, 1331.
67. S. G. Bandaev, Y. K. Eshnazarov, I. M. Nasyrov, S. S. Mochalov and Y. S. Shabarov, *J. Org. Chem. USSR*, 1988, **24**, 659.
68. A. J. Bloodworth and D. Korkodilos, *Tetrahedron Lett.*, 1991, **32**, 6953.
69. A. J. Bloodworth, K. H. Chan and C. J. Cooksey, *J. Org. Chem.*, 1986, **51**, 2110.
70. A. J. Bloodworth and G. M. Lampman, *J. Org. Chem.*, 1988, **53**, 2668.
71. N. M. Abramova, S. V. Zotova and O. A. Nesmeyanov, *Bull. Acad. Sci. USSR*, 1982, 961.
72. A. J. Bloodworth, K. H. Chan, C. J. Cooksey and N. Hargreaves, *J. Chem. Soc., Perkin Trans. 1*, 1991, 1923.
73. D. B. Collum, W. C. Still and F. Mohamadi, *J. Am. Chem. Soc.*, 1986, **108**, 2094.
74. J. M. Coxon, P. J. Steel, B. I. Whittington and M. A. Battiste, *J. Am. Chem. Soc.*, 1988, **110**, 2988.
75. B. E. Kogai, U. A. Sokolenko, P. V. Petrovskii and V. I. Sokolov, *Bull. Acad. Sci. USSR*, 1982, 1464.
76. H. G. M. Edwards, *Spectrochim. Acta, Part A*, 1986, **42**, 427.
77. H. G. M. Edwards, *J. Organomet. Chem.*, 1986, **314**, 13.
78. B. Giese and U. Erfort, *Angew. Chem., Int. Ed. Engl.*, 1982, **21**, 130.
79. H. Willner, M. Bodenbinder, C. Wang and F. Aubke, *J. Chem. Soc., Chem. Commun.*, 1994, 1189.
80. K. B. Butin, A. A. Ivkina and O. A. Reutov, *Bull. Acad. Sci. USSR*, 1985, **34**, 613.
81. R. C. Larock and M.-S. Chow, *Tetrahedron Lett.*, 1984, **25**, 2727.
82. L. I. Rybin *et al.*, *J. Gen. Chem. USSR*, 1986, **54**, 1807.
83. A. Sebald and B. Wrackmeyer, *Spectrochim. Acta, Part A*, 1986, **42**, 1107.
84. Y. K. Grishin, V. A. Rozyatovskii, V. N. Torocheshnikov and Y. A. Ustynyuk, *Organomet. Chem. USSR*, 1991, **4**, 522.
85. F. de Sarlo, A. Brandi, L. Fabrizi, A. Guarna and N. Niccolai, *Org. Magn. Reson.*, 1984, **22**, 372.
86. F. de Sarlo, A. Guarna, A. Goti and A. Brandi, *J. Organomet. Chem.*, 1984, **269**, 115.
87. B. Kamenar, B. Kaitner and S. Pocev, *J. Chem. Soc., Dalton Trans.*, 1985, 2457.
88. J. Bravo *et al.*, *Inorg. Chem.*, 1985, **24**, 3435.
89. E. Block, M. Brito, M. Gernon, D. McGowty, H. Kang and J. Zubieta, *Inorg. Chem.*, 1990, **29**, 3172.
90. G. B. Deacon, B. M. Gatehouse and S. C. Ney, *J. Organomet. Chem.*, 1988, **348**, 141.
91. Å. Iverfeldt and I. Persson, *Inorg. Chim. Acta*, 1985, **103**, 113.
92. R. D. Bach, H. B. Vardhan, A. F. M. M. Rahman and J. P. Oliver, *Organometallics*, 1985, **4**, 846.
93. Å. Iverfeldt and I. Persson, *Inorg. Chim. Acta*, 1986, **111**, 171.
94. H. Schmidbaur, H.-J. Öller, S. Gamper and G. Müller, *J. Organomet. Chem.*, 1990, **394**, 757.
95. A. R. Norris, S. E. Taylor, E. Buncel, F. Belanger-Gariépy and A. L. Beauchamp, *Can. J. Chem.*, 1983, **61**, 1536.
96. A. J. Canty, N. J. Minchin, B. W. Skelton and A. H. White, *J. Chem. Soc., Dalton Trans.*, 1986, 2201.
97. A. J. Canty, J. M. Patrick and A. H. White, *Inorg. Chem.*, 1984, **23**, 3827.
98. R. D. Bach and H. B. Vardhan, *J. Org. Chem.*, 1986, **51**, 1609.
99. M.-C. Corbeil, A. L. Beauchamp, S. Alex and R. Savoie, *Can. J. Chem.*, 1986, **64**, 1876.
100. S. Alex, R. Savoie, M.-C. Corbeil and A. L. Beauchamp, *Can. J. Chem.*, 1986, **64**, 148.
101. F. Allaire and A. L. Beauchamp, *Can. J. Chem.*, 1984, **62**, 2249.
102. J.-P. Charland and A. L. Beauchamp, *Inorg. Chem.*, 1986, **25**, 4870.
103. E. Buncel, C. Boone and H. Joly, *Inorg. Chim. Acta*, 1986, **125**, 167.
104. D. Grdenić, D. Matković-Čalogović and M. Sikirica, *J. Organomet. Chem.*, 1987, **319**, 1.
105. B. Korpar-Čolig, Z. Popović and M. Sikirica, *Croat. Chem. Acta*, 1984, **57**, 689.
106. D. Grdenić, B. Korpar-Čolig, D. Matković-Čalogović, M. Sikirica and Z. Popović, *J. Organomet. Chem.*, 1991, **411**, 19.
107. D. Grdenić, M. Sikirica and D. Matković-Čalogović, *J. Organomet. Chem.*, 1986, **306**, 1.
108. B. Korpar-Čolig, Z. Popović, M. Sikirica and D. Grdenić, *J. Organomet. Chem.*, 1991, **405**, 59.
109. D. Matković-Čalogović, *Acta Crystallogr., Sect. C*, 1987, **43**, 1473.
110. B. Kamenar, M. Penavic and A. Hergold-Brundic, *Croat. Chem. Acta*, 1984, **57**, 145.
111. Y. Farhangi, *Inorg. Chim. Acta*, 1987, **129**, 11.
112. J. L. Atwood, D. E. Berry, S. R. Stobart and M. J. Zaworotko, *Inorg. Chem.*, 1993, **22**, 3480.
113. T. S. Lobana, M. K. Sandhu, D. C. Povey and G. W. Smith, *J. Chem. Soc., Dalton Trans.*, 1988, 2913.
114. K. Ding, Y. Wu, H. Hu, L. Shen and X. Wang, *Organometallics*, 1992, **11**, 3849.
115. M. Ali, W. R. McWhinnie and T. A. Hamor, *J. Organomet. Chem.*, 1989, **371**, C37.
116. E. C. Constable, T. A. Leese and D. A. Tocher, *J. Chem. Soc., Chem. Commun.*, 1989, 570.
117. B. K. Nicholson and A. J. Whitton, *J. Organomet. Chem.*, 1986, **306**, 139.
118. G. W. M. Visser, C. N. M. Bakker, B. W. v. Halteren, J. D. M. Herscheid, G. A. Brinkman and A. Hoekstra, *J. Org. Chem.*, 1986, **51**, 1886.
119. B. Fischer, G. P. M. van Mier, J. Boersma and G. van Koten, *Recl. Trav. Chim. Pays-Bas*, 1988, **107**, 259.
120. B. Floris, G. Illuminati and G. Ortaggi, *J. Chem. Soc., Chem. Commun.*, 1969, 492.
121. A. Razavi, M. D. Rausch and H. G. Alt, *J. Organomet. Chem.*, 1987, **329**, 281.
122. A. G. Davies and J. Lusztyk, *J. Chem. Soc., Perkin Trans. 2*, 1981, 692.
123. J. Lorberth, T. F. Berlitz and W. Massa, *Angew. Chem., Int. Ed. Engl.*, 1989, **28**, 611.
124. A. G. Davies, E. Lusztyk and J. Lusztyk, *J. Chem. Soc., Perkin Trans. 2*, 1982, 729.
125. A. G. Davies, J. P. Goddard, M. B. Hursthouse and N. P. C. Walker, *J. Chem. Soc., Dalton Trans.*, 1985, 471.
126. L. Zhu, L. M. Daniels, L. M. Peerey and N. M. Kostić, *Acta Crystallogr., Sect. C*, 1988, **44**, 1727.
127. C. H. Winter, Y.-H. Han, R. L. Ostrander and A. L. Rheingold, *Angew. Chem., Int. Ed. Engl.*, 1993, **32**, 1161.
128. J. D. Wuest and B. Zacharie, *Organometallics*, 1985, **4**, 410.
129. A. L. Beauchamp, M. J. Olivier, J. D. Wuest and B. Zacharie, *Organometallics*, 1987, **6**, 153.
130. J. D. Wuest and B. Zacharie, *J. Am. Chem. Soc.*, 1985, **107**, 6121.
131. M. Simard, J. Vaugeois and J. D. Wuest, *J. Am. Chem. Soc.*, 1993, **115**, 370.
132. J. D. Wuest and B. Zacharie, *J. Am. Chem. Soc.*, 1987, **109**, 4714.
133. V. B. Shur *et al.*, *J. Organomet. Chem.*, 1991, **418**, C29.
134. V. B. Shur *et al.*, *J. Organomet. Chem.*, 1993, **443**, C19.
135. X. Yang, C. B. Knobler and M. F. Hawthorne, *J. Am. Chem. Soc.*, 1992, **114**, 380.
136. Z. Zheng, X. Yang, C. B. Knobler and M. F. Hawthorne, *J. Am. Chem. Soc.*, 1993, **115**, 5320.

137. M. Filippelli, F. Baldi, F. E. Brinckmann and G. J. Olson, *Environ. Sci. Technol.*, 1992, **26**, 1457.
138. P. J. Craig, D. Mennie, M. Needham, N. Oshah, O. F. X. Donard and F. Martin, *J. Organomet. Chem.*, 1993, **447**, 5.
139. K. Kwetkat and W. Kitching, *J. Chem. Soc., Chem. Commun.*, 1994, 345.
140. P. J. Craig, H. Garraud, S. H. Laurie, D. Mennie and G. H. Stojak, *J. Organomet. Chem.*, 1994, **468**, 7.
141. N. Legay-Sommaire and F. Legay, *Chem. Phys. Lett.*, 1994, **217**, 97.
142. J. Rideout and M. C. R. Symons, *J. Chem. Soc., Chem. Commun.*, 1985, 129.
143. C. J. Rhodes, C. Glidewell and H. Agirbas, *J. Chem. Soc., Faraday Trans.*, 1991, **87**, 3171.
144. G. A. Russell, R. K. Khanna and D. Guo, *J. Chem. Soc., Chem. Commun.*, 1986, 632.
145. B. Giese, *Angew. Chem., Int. Ed. Engl.*, 1985, **24**, 553.
146. M. Regitz and B. Giese, in 'Methoden der Organische Chemie (Houben–Weyl)', Thieme, Stuttgart, 1989, Band E19a.
147. B. Giese and J. A. Gonzales-Gomez, *Ber. Dtsch. Chem. Ges.*, 1986, **119**, 1291.
148. G. A. Russell, *Acc. Chem. Res.*, 1989, **22**, 1.
149. S. Danishefsky, S. Chackalamannil and B.-J. Uang, *J. Org. Chem.*, 1982, **47**, 2231.
150. B. Giese, H. Harnisch and S. Lachhein, *Synthesis*, 1983, 733.
151. B. Giese and U. Erfort, *Chem. Ber.*, 1983, **116**, 1240.
152. B. Giese, K. Hueck, H. Lenhardt and U. Luning, *Ber. Dtsch. Chem. Ges.*, 1984, **117**, 2132.
153. B. Giese and H. Horler, *Tetrahedron*, 1985, **41**, 4025.
154. G. A. Russell and P. Ngoviwatchai, *Tetrahedron Lett.*, 1986, **27**, 3479.
155. G. A. Russell and H. Tashtoush, *J. Am. Chem. Soc.*, 1983, **105**, 1398.
156. G. A. Russell and R. K. Khanna, *J. Am. Chem. Soc.*, 1985, **107**, 1450.

4
Cadmium and Zinc

PAUL O'BRIEN
Imperial College of Science, Technology and Medicine, London, UK

4.1 GENERAL INTRODUCTION	175
4.2 ORGANOZINC COMPOUNDS	176
4.2.1 Simple Diorganozinc Compounds	176
4.2.2 Adducts of Simple Alkyls	177
4.2.3 Simple Zincates	179
4.2.4 Bulky Silyl Ligands	179
4.2.5 Organometallic Species Involving Unsaturated Ligands	180
4.2.6 Zinc Enolates	180
4.2.7 Organoamides and Related Species	181
4.2.8 Organozinc Compounds Containing a Zinc–Halide Bond	182
4.2.9 Organozinc Compounds with R–Zn–O Bonds	184
4.2.10 Compounds Containing R–Zn–S and R–Zn–Se Bonds	186
4.2.11 Heterometallic Zinc Organometallic Complexes	187
4.2.12 Theoretical and Physical Studies of Organozinc Species	188
4.2.13 The Use of Organozinc Reagents in Organic Synthesis	188
4.2.13.1 Catalytic asymmetric synthesis	189
4.2.13.2 The reaction pathway with aldehydes	190
4.2.13.3 Chiral amplification and enantiomer recognition	192
4.2.13.4 The Simmons–Smith reaction	194
4.2.13.5 The Reformatsky reaction	195
4.3 ORGANOCADMIUM CHEMISTRY	196
4.3.1 Diorganocadmium Compounds	196
4.3.2 Adducts of Organocadmium Compounds	197
4.3.3 Organocadmium Compounds Containing a Cadmium–Heteroatom Bond	198
4.3.4 Theoretical and Physical Studies of Organocadmium Species	200
4.4 THE USE OF METAL ALKYLS IN THE ELECTRONICS INDUSTRY	200
4.4.1 Simple Alkyls	200
4.4.2 Adducts as Precursors	200
4.4.3 Nature of the Effect of the Adducts on Growth	201
4.4.4 The Synthesis of Highly Dispersed Materials	203
4.5 REFERENCES	203

4.1 GENERAL INTRODUCTION

The aim of this chapter is to provide a comprehensive introduction to developments in the organometallic chemistry of zinc and cadmium that have occurred since the mid-1970s. It should be seen as a complement or supplement to the chapter by Boersma in *COMC-I*.[1] No attempt has been made to cover the historical development of this chemistry, an important area because of the early development of organozinc compounds by Frankland.[2] Similarly, anyone trying to gain an entry to this area will find that general methods of preparation and physical properties are well covered by Boersma.[1]

The structure of the chapter is similar to that used in *COMC-I*, which will make it easy for readers to use the two chapters in combination. However, in surveying the chemistry of zinc and cadmium it became clear that several areas of research have gained greater impetus in the last few years. Specifically, these areas are:

(i) the use of bulky ligands, such as trimethylsilyl, to control the structure and reactivity of simple alkyls;
(ii) the use of zinc reagents in organic synthesis, especially in catalytic asymmetric synthesis; and
(iii) the use of metal alkyls in materials chemistry, especially the deposition of thin films by metal organic chemical vapour deposition (MOCVD).

Preliminary literature searches indicated that in the time since the review by Boersma over 4000 publications have appeared in which compounds with direct metal to zinc or cadmium bonds are discussed. Clearly the limitations of space and time, and consideration for the reader, mean that this chapter cannot be truly comprehensive. I have tried to give references to recent key papers in each main area to provide an entry to the literature, to cite major papers and where possible to provide references to reviews. I hope the chapter will help new people to develop an interest in this area and that those working in the field will find it a useful summary. Main group organometallic chemistry has become more popular in recent years; much of the work having been driven by the quest for precursors for advanced materials. I hope that someone will be inspired to take up work in this area as a result of reading this chapter.

4.2 ORGANOZINC COMPOUNDS

4.2.1 Simple Diorganozinc Compounds

There have been relatively few innovations in the methods for the synthesis of metal alkyls since the first volume of this publication. A reliable procedure for the preparation of dimethylzinc using commercially available trimethylaluminum as an alkylating agent has been reported.[3] Metal vapours of cadmium or zinc have been reacted with trifluorosilyl and trifluoromethyl radicals to produce several new metal alkyls.[4] A low-temperature radio-frequency glow discharge or plasma of hexafluoroethane or hexafluorodisilane was used to generate radicals. Bis(trifluorosilyl)cadmium, bis(trifluorosilyl)zinc, bis(trifluoromethyl)cadmium and bis(trifluoromethyl)zinc were isolated. The compounds were unstable at room temperature with the exception of bis(trifluorosilyl)cadmium, which was marginally stable at room temperature. The reaction of CF_3I or C_6F_5I with dialkylzinc in the presence of a Lewis base results in the quantitative synthesis of $[Zn(CF_3)_2]$ and $[Zn(C_6F_5)_2]$ complexes; however, the analogous reactions with C_2F_5I or $(CF_3)_2CF_2$ do not yield the pure compounds.[5]

Aliphatic organoboron derivatives have been transmetallated to the corresponding dialkylzincs using diethyl- or dimethylzinc, which allows access to zinc reagents otherwise not readily available, for example didecylzinc.[6] The NMR spectra of ternary systems have been studied.[7] The ^{13}C NMR spectrum of $[Zn(Me_2CH)_2]$ showed a single environment for the methyl groups, whereas that of $[Zn(MeEtCH)_2]$ was diastereomeric. Alkyl exchange for the chiral system had an activation energy of >72 kJ mol^{-1}. The ^{13}C spectrum of a mixture of the alkyls showed a set of signals due to $[Zn(Me_2CH)(CHMeEt)]$, accounting for 33% of the zinc. Mixed organozinc compounds have been prepared by exchange reactions of diorganozinc compounds.[8] The exchange of organo groups between various diorganozinc compounds R^1_2Zn (R^1 = CHMe$_2$, CH$_2$(CH$_2$)$_2$Me, CH$_2$CMe:CH$_2$, CH$_2$CH:CHMe) and R^2_2Zn (R^2 = Et, CHMeEt, CH$_2$(CH$_2$)$_3$Me, CMe$_3$, menthyl, Ph, cyclohexyl), has been followed by mass spectroscopy; molecular ions for the mixed compounds R^1R^2Zn were observed. Ethyl(4-dimethylamino-1-butynyl)zinc was prepared[9] by metallation of 1-dimethylamino-3-butyne with Et$_2$Zn in benzene. Bis(4-dimethylamino-1-butynyl)zinc was synthesized similarly from Ph$_2$Zn and two equivalents of the alkynic amine.

Simple alkyls formed from the unsaturated systems bis(alk-2-enyl)zinc (**1**) have been synthesized from organomagnesium halides, for example $[Zn(R^1_2R^2C:CR^1CH_2)_2]$ ($R^1 = R^2 = H$, Me; $R^2 = H$, Me, Ph) were prepared[10] (68–94% yield) by reaction of the Grignard reagent with ZnCl$_2$. An electron diffraction study of the linear di-3-butenylzinc has been reported (**2**).[11] The crystal structure of $[Zn\{2,4-(t-C_4H_9)_2C_5H_3\}_2]$ (**3**) has been determined.[12]

A number of studies have been reported concerning the ability of pendant, potentially coordinating, groups in an alkyl fragment to coordinate to the metal centre. One of the most thorough involved bis[3-(dimethylamino)propyl-*C,N*]zinc (**4**), for which both electron diffraction (vapour) and x-ray diffraction structures have been reported. There is a weak contact, around 0.23 nm, between the nitrogen and the

[MeCH=CHCH$_2$ZnCH$_2$CH=CHMe]

(1)

[Zn(H$_2$C=CHCH$_2$CH$_2$)$_2$]

(2)

Bu^t ⎯ Zn ⎯ Bu^t
Bu^t Bu^t

(3)

metal in both phases.[13] Related although less detailed studies have been reported on phosphorus- and sulfur-containing ligands.[14] Similar interactions have been observed in dimethylaminophenyls[15] (**5**) and the highly sterically hindered complex [M{NC$_5$H$_4$C(TMS)$_2$}$_2$] (M = Zn or Cd) has been reported and characterized crystallographically.[16] In the zinc compound there are weak interactions between the nitrogen and the metal centre.

[Zn(Me$_2$NCH$_2$CH$_2$CH$_2$)$_2$]

(4)

Me Me
\ /
N
Zn
N
/ \
Me Me

(5)

4.2.2 Adducts of Simple Alkyls

Until recently the nature of the adducts formed by group 12 alkyls with bidentate amines had not been confirmed crystallographically. A number of relevant structures have appeared in the last few years, including those of the methyls and neopentyls of zinc or cadmium with tetramethylethylenediamine[17] or (−)-sparteine.[18] Some important bond lengths are summarized in Table 1. As is expected, the structures of these compounds show that the linear C–M–C backbone has been distorted by coordination to the amine (**6**) (Scheme 1). The dimethylzinc adduct of the natural product sparteine has been studied; the compound is interesting in that it is almost air stable and still functions as an alkylating agent in organic syntheses.[18] The coordination chemistry of the amine 1,3,5-trimethylhexahydro-1,3,5-triazine is interesting as stable 2:1 adducts are formed with dimethylzinc and dimethylcadmium[19] [M({CH$_2$NMe}$_3$)$_2$Me$_2$]. The dimethylzinc adduct was characterized by single-crystal x-ray methods. The coordination mode of the triazine is interesting as only one of the three nitrogens is coordinated to the zinc atom (**7**).

Table 1 Important bond lengths and angles in some adducts of group 12 alkyls and related compounds.

	Bond length (nm)		*Bond angle (°)*		
	M–C	M–N	C–M–C	N–M–N	*Ref.*
[Zn{(CH$_2$NMe)$_3$}$_2$Me$_2$]	0.1987	0.2410	145.1	105.6	19
[Zn{Me$_2$N(CH$_2$)$_2$NMe$_2$}Me$_2$]	0.1975 (av.)	0.2269 (av.)	136.7	80.0	17
[Zn{Me$_2$N(CH$_2$)$_3$}$_2$]	0.1984	0.2307	156.4	109.7	13
[Zn{Me$_2$N(CH$_2$)$_2$NMe$_2$}(Me$_3$CCH$_2$)$_2$]	0.2000	0.2411	148.3	77.3	19
[Cd{Me$_2$N(CH$_2$)$_2$NMe$_2$}Me$_2$]	0.2163 (av.)	0.2564 (av.)	153.6	70.8	19
[Cd{Me$_2$N(CH$_2$)$_2$NMe$_2$}(Me$_3$CCH$_2$)$_2$]	0.2160	0.2642	157.0	71.2	19

The Grignard reaction of RC$_6$H$_4$CH$_2$MgX (X = Cl, Br) with ZnCl$_2$ gave [Zn(RC$_6$H$_4$CH$_2$)$_2$] (R = H, *o*-F, *o*-Cl, *o*-Br, *p*-Cl, *o*-MeO), which were initially isolated as dioxane complexes, and then reacted with bipyridyl to give the bipyridyl complexes.[20] The first fully characterized structure for a fluorinated alkyl zinc has been reported, namely the pyridine adduct [Zn(CF$_3$)$_2$(C$_5$H$_5$N)$_2$]; the zinc atom is in a distorted tetrahedral configuration.[21,22] An extremely unusual structure has been reported for the adduct formed between dimethylzinc and 18-crown-6; the alkyl is threaded through the macrocycle and coordinated equatorially by the oxygen atoms.[23] The reaction of bis[bis(trimethylsilyl)methyl]zinc with 2,2'-bipyridine gives a 1:1 adduct, which crystallizes from cyclopentane. The zinc atom is surrounded by the two nitrogen donors in a distorted tetrahedral fashion; the C–Zn–C angle (126.4°) is remarkably large.[24]

The bis(dienyl)zinc [(CH$_2$:CHCH:CHCH$_2$)$_2$Zn] compound was isolated as crystals of the tetramethylethylenediamine adduct. The compound exists as a (*Z*),(*E*)-mixture of terminally σ-bonded structures.[25]

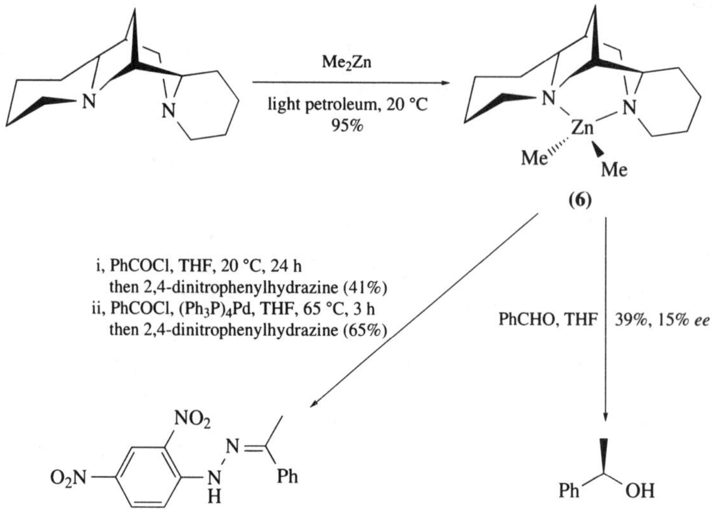

Scheme 1

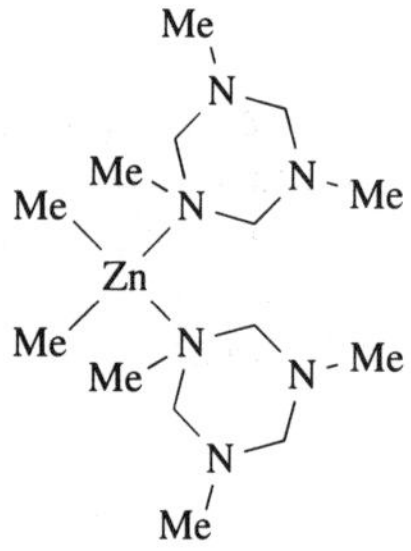

(7)

Adducts of the group 12 metals have also found use in the purification of these compounds. Those with two nitrogen atoms, such as 4,4'-bipyridyl, which cannot for geometrical reasons bind to the same metal atom, give involatile polymeric adducts which on dissociation with heating can give rise to purified alkyl.[26] These purified alkyls may be useful to the electronic industry and several detailed procedures have been reported.[27] The vapour pressures of a range of σ-bonded adducts of dimethylzinc with nitrogen-containing ligands have been studied in detail because these compounds are potentially useful as precursors for the deposition of semiconducting materials.[28]

The interaction of diphenylzinc with crown ethers has been studied[29] and produces compounds in which zinc is four-coordinate in the solid state (8); with diglyme a five-coordinate adduct that persists in solution (9) is formed.[29]

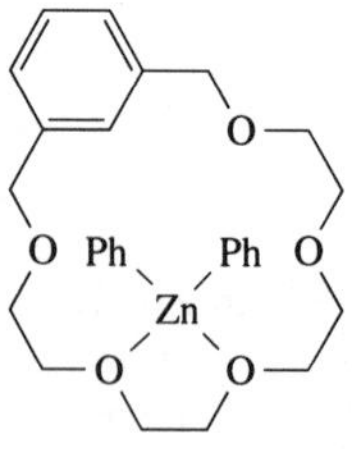

(8) 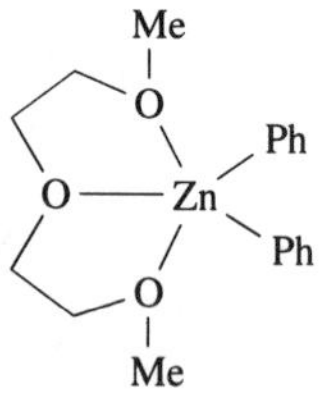   (9)

Finally, the reaction of dimethylzinc with Lewis bases containing acidic hydrogens such as NH_3, PH_3, AsH_3, H_2O, H_2S and H_2Se has been studied in argon matrices and cryogenic thin films.[30] Compounds of both 1:1 and 2:1 stoichiometry were formed. The reaction of these was rapid with H_2S and H_2Se, but slow with a low yield of product for PH_3 and AsH_3. The most sensitive band in the infrared spectrum was the ZnC_2 antisymmetric stretch near 600 cm^{-1}. This mode shifted between 8 and 34 cm^{-1} to the red in the different complexes.

4.2.3 Simple Zincates

The sodium and potassium trialkylzincates, [MZn(CH$_2$Bu^t)$_3$] (M = Na or K) and [KZn(CH$_2$(TMS))$_3$] were isolated from the reactions between the alkali metal and alkyl in benzene. Metallation of benzene led to the isolation of [KZn(CH$_2$(TMS))$_2$Ph] and crystal structures were determined for all four zincates. The simple zincates crystallize as discrete dimers.[31]

Heteroleptic diorganozincate compounds of the type [ZnRC(TMS)$_3$] have been isolated from the reaction of lithium trichlorobis{[tris(trimethylsilyl)methyl]zincate}.[32] The compounds are monomeric in the solid state and solution and no evidence for ligand exchange reactions was observed. The crystal structures of bis(trimethylsilyl)methyl[tris(trimethylsilyl)methyl]zinc and phenyl[tris(trimethylsilyl)-methyl]zinc were determined; rather short Zn–C–bond lengths of ca. 0.1935 nm are seen. The reaction of lithium bis(trimethylsilyl)methanide and TMEDA in ether gave the zincate Li[ZnR$_3$]·TMEDA·2Et$_2$O (R = bis(trimethylsilyl)methyl).[33] Adducts have also been well characterized for such zincates. Bis[bis(trimethylsilyl)methyl]zinc and the amine 1,3,5-trimethyl-1,3,5-triazinane form a 1:1 adduct; as in the adducts with dimethylzinc the amine binds as a unidentate ligand. The crystal structure contains a nearly T-shaped zinc environment, (C–Zn–C, 157°, Zn–C 0.199 nm, Zn–N 0.239 nm).[34] The formation of organozinc anions and cations has been observed on the addition of diethylzinc to crown ethers or cryptands in benzene solution.[35]

4.2.4 Bulky Silyl Ligands

An area of chemistry that has opened up in the last few years is the use of extremely bulky groups such as trimethylsilyl on alkyl ligands to produce novel organometallic complexes that often have interesting properties. This section is concerned with a survey of such ligands but reference to compounds with simple structures will also be found in other sections of the chapter.

The remarkable properties of these compounds are well illustrated in the work of Eaborn.[36] Di[tris(trimethylsilyl)methyl]zinc or -cadmium [M{(TMS)$_3$C}$_2$] were prepared by treating [LiC(TMS)$_3$] with the anhydrous metal chloride. The compounds decompose above 300 °C to give (TMS)$_3$CH and are stable to H$_2$O in refluxing THF. The zinc compound can be steam-distilled without decomposition. The properties of the trimethylsilyl-substituted derivatives of dimethylzinc [Zn{(TMS)$_n$CH$_{3-n}$}$_2$] (n = 1–3) are strongly dependent upon the level of substitution;[37] for n = 0–2 the compounds are pyrophoric liquids, whereas bis[tris(trimethylsilyl)methyl]zinc (n = 3) has a m.p. of about 300 °C and is air and water resistant. The crystal structure shows Zn–C of 0.198 nm, and the C–Zn–C unit is linear. The –CSi$_3$ fragment is flattened but not as significantly as NMR results would suggest. Heteroleptic diorganozinc compounds of the type[32] [ZnRC(TMS)$_3$] have been prepared by the equimolar reaction of lithium tris(trimethylsilyl)methanide with zinc(II) chloride to give lithium trichlorobis{[tris(trimethyl-silyl)methyl]zincate}. The addition of organolithium leads to the formation of the heteroleptic organo[tris(trimethylsilyl)methyl]zinc. No ligand exchange reactions can be seen for these compounds. The crystal structures of bis(trimethylsilyl)methyl[tris(trimethylsilyl)methyl]zinc and phenyl-[tris(trimethylsilyl)methyl]zinc have been determined; rather short Zn–C bonds (0.1935 nm) are found in the phenyl compound.

Zincates of heteroleptic triorganylzincate anions have been reported;[38] bis[(trimethylsilyl)methyl]-, bis[bis(trimethylsilyl)methyl]- and bis(2,2,4,4,6,6-hexamethyl-2,4,6-trisilacyclohexyl)zinc react with methyl- or phenyllithium in the presence of the tridentate 1,3,5-trimethyl-1,3,5-triazinane (TMTA) to yield zincates of the type LiZnR^{1}2R^2·2TMTA. The reaction of lithium bis(trimethylsilyl)amide with bis[(trimethylsilyl)methyl]zinc yields lithium [bis(trimethylsilyl)amino]bis[(trimethylsilyl)methyl]-zincate·TMTA. The structure was confirmed by x-ray diffraction. The bridging bis(trimethylsilyl)amino ligands have long Zn–N and Li–N distances of 0.213 and 0.208 nm, respectively.

Ternary species involving a second functional group have been prepared. Halides [MClC(PhMe$_2$Si)$_3$] (M = Zn or Cd) have been prepared by the reaction of [LiC(PhMe$_2$Si)$_3$] with the metal chloride.[39] The compounds were isolated via lithium-containing intermediates thought to be of the form [M(μ-Cl)$_2$Li(THF)$_2$C(PhMe$_2$Si)$_3$]. The zinc and cadmium chlorides are present in the crystal as the dimers [M(μ-Cl)$_2$C(PhMe$_2$Si)$_3$MC(SiMe$_2$Ph)$_3$]. The metals are in three-coordinate planar environments.

Another theme is the functionalization of the silyl ligand. The x-ray structures of [Zn{C(TMS)$_2$(SiMe$_2$O$_2$CCF$_3$)}$_2$] and [Cd{C(TMS)$_2$(SiMe$_2$OMe)}$_2$] have been reported[40] (Scheme 2). Several compounds of the form [Zn{C(TMS)$_2$(SiMe$_2$X)}$_2$] (X = Cl, Br, I, F, OH, OMe, O$_2$CCF$_3$, O$_2$CH or NCS), were obtained by substitutions at silicon without cleavage of Zn–C bonds. The compounds [Cd{C(TMS)$_2$(SiMe$_2$X)}$_2$] (X = H, OMe or Ph) were obtained similarly, but subsequent substitutions at silicon were accompanied by reactions at cadmium. The crystal structure of

[Zn{C(TMS)$_2$(SiMe$_2$O$_2$CCF$_3$)}$_2$] shows that the molecules are centrosymmetric with the O$_2$CCF$_3$ groups forced outwards by steric hindrance away from the metal centre. The compound [Zn{(HOMe$_2$Si)(TMS)$_2$C}$_2$] has been prepared.[41] The x-ray structure shows that the compound is centrosymmetric, with the OMe group directed towards the metal (Zn–O, 0.293 nm). There is hydrogen bonding between the two oxygen atoms within each molecule, and hydrogen bonding between two molecules to give discrete dimers; the C–Zn–C linkage is slightly bent (175.9°).

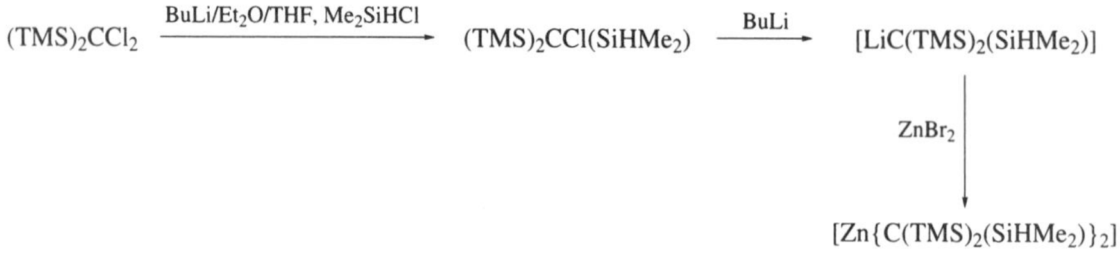

$$(TMS)_2CCl_2 \xrightarrow{\text{BuLi/Et}_2\text{O/THF, Me}_2\text{SiHCl}} (TMS)_2CCl(SiHMe_2) \xrightarrow{\text{BuLi}} [LiC(TMS)_2(SiHMe_2)]$$

$$\Big\downarrow ZnBr_2$$

$$[Zn\{C(TMS)_2(SiHMe_2)\}_2]$$

Scheme 2

4.2.5 Organometallic Species Involving Unsaturated Ligands

Carbenes and carbynes have been studied by Fourier transform infrared spectroscopy at 12 K. Monatomic zinc, CH$_2$N$_2$ and argon on a cold finger were converted to ZnCH$_2$ by photolysis; a carbene to carbyne rearrangement (ZnCH$_2$ to HZnCH) is observed during UV photolysis.[42]

The dialkynylzinc compounds[43,44] [Zn{MeC≡C(CH$_2$)$_3$}$_2$] and [Zn{MeC≡C(CH$_2$)$_4$}$_2$] have been prepared. The interation between zinc and the C–C triple bond in dihexyn-4-ylzinc was confirmed by ^{13}C and ^{1}H NMR and Raman spectroscopies. Metallation of HC≡C(CH$_2$)$_n$NMe$_2$ with Et$_2$Zn or Ph$_2$Zn in benzene (at 60 °C, 50 h) gives [EtZnC≡C(CH$_2$)$_n$NMe$_2$] or Zn[C≡C(CH$_2$)$_n$NMe$_2$]$_2$ (n = 1–4).[45] All the compounds show oligomeric or polymeric association in benzene. The presence of coordinating dimethylamino groups and/or bridging alkynyl groups was established by IR and NMR spectroscopy. The reaction of methylenetriphenylarsorane,[46] Ph$_3$As:CH$_2$, with ZnCl$_2$ in a 2:1 ratio gives the monomeric complex [AsCH$_2$MCH$_2$Ph$_3$As]Cl$_2$. Adducts of dialkylzinc with nucleophilic carbenes have been prepared and characterized by the reaction of the stable nucleophilic carbenes 1,3-dimesitylimidazol-2-ylidene or 1,3-di(1-adamantyl)imidazol-2-ylidene and the metal alkyl.[47] The x-ray crystal structure of the monomeric 1,3-di(1-adamantyl)imidazol-2-ylidenediethylzinc adduct has been reported.

Structural studies of zincocene have appeared. In the solid state there are infinite chains of zinc atoms with bridging cyclopentadienyl groups. Both σ- and π-type interactions contribute to the cyclopentadiene–zinc bonding.[48] Decamethylzincocene (**10**) has a different structure, with both η^1 and η^5 donation to the metal from the ligand, and has been studied in both the solid state and as a vapour by electron diffraction.[49] A novel cyclopentadienyl-bridged structure has been observed[50] for [(ZnCp)$_2$(μ-Cp){μ-N(TMS)$_2$}]. The compound is the first example of a cyclopentadienyl group that bridges between nonbonded metal atoms located on the same side of the ring plane. The coordination at zinc suggests that both σ- and π-type interactions contribute to the cyclopentadiene–zinc bonding. In solution an intramolecular exchange of cyclopentadienyl groups occurs at elevated temperature, presumably via opening of the cyclopentadienyl bridge.

4.2.6 Zinc Enolates

Alkylzinc enolates have attracted particular attention because of their potential for use in organic synthesis (see Section 4.2.13). Several structural studies have appeared and these are collected together in this short section. The structure of [{ZnOMeEt·Zn(Pac)$_2$}$_2$] (HPac = pivaloylacetone, Bu^{t}COCH$_2$COMe), isolated in 20% yield from the reaction of [ZnOMeEt] with MeCO$_2$C(Bu^t):CH$_2$, was determined[51] (**11**). The reactivity of the zinc enolates suggests that they contain both zinc–carbon and zinc–oxygen bonds. They are assumed to have a cyclic structure which resembles that of the Reformatsky reagent. The product of the reaction between cyclopentadienylethylzinc and ethyl N,N-diethylglycinate is an unexpected enolate of 1-(2,4-cyclopentadienyl)-2-(diethylamino)-1-ethanone, the structure of which has been determined.[52] The reaction of N-(ethylzincio)diisopropylamine with N,N-dialkylglycine esters gives pure ethylzinc ester enolates in almost quantitative yields.[53] The ester enolate anions are chelate-bonded to the metal centre through covalent Zn–O and dative Zn–N bonds. The strength of the Zn–N bond, and consequently the thermostability of the zinc enolates, is influenced by

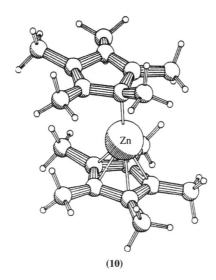

(10)

the electronic properties of the amino-nitrogen substituents. The structure of [EtZnOC(OMe):C(H)N(Bu^t)Me]$_4$ shows a tetrameric association of four independent ZnOCCN units (av. Zn–O = 0.2065 nm), which are linked via Zn–O–Zn bridges (av. Zn–O = 0.2032 nm). Pure ethylzinc α-amino ester enolates (R = Me, Bu^t) have been prepared from RMeNCH$_2$CO$_2$Me and [ZnN(Me$_2$CH)$_2$Et]. An x-ray diffraction study of the neopentyl compound again shows a tetrameric species,[54] in which the four independent zinc enolate units are interconnected via covalent Zn–O–Zn bridges. As a result of the Zn–N coordination, all the enolate moieties have the Z-configuration (**12**). The preparation and reactivity of ethylzinc enolate [(Z)-Bu^{t}C(OZnEt):CHMe] has been reported;[55] the compound is less reactive than the corresponding lithium enolate but undergoes many of the same reactions, for example, aldol condensation with PhCHO to give hydroxy ketones and Michael addition with PhCH:CHCOPh. The metal exchange reaction used to prepare the enolate is not a general method for other alkylzinc enolates.

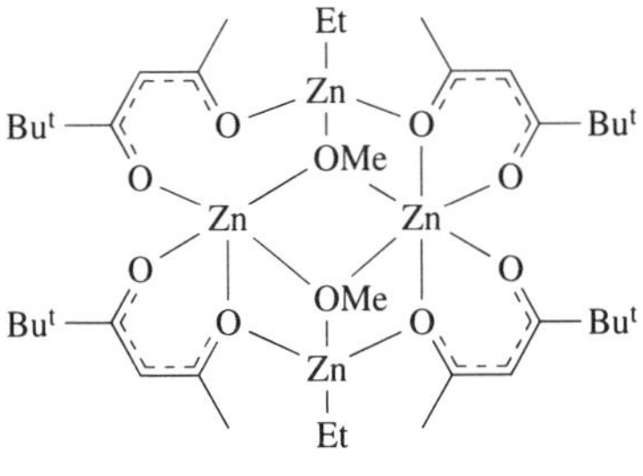

(11)

4.2.7 Organoamides and Related Species

Simple amides that have been reported include ethylzinc diisopropylamide[55] EtZnN(CHMe$_2$)$_2$ and the silylamide[56] [ZnN(TMS)$_2$Et]. (Diphenylaminato)methylzinc, which has been known for some time, has been shown by crystallography[57] to exist as a dimer (**13**).

The reaction of ZnEt$_2$ with α-imino amides gave the products of *N*-addition as well as those of N–C reduction.[58] Reduction in this system was confirmed by the isolation and crystallographic characterization of the dimeric amide [{EtZnN(CMe$_3$)CH$_2$C(O)NEt$_2$}$_2$] (**14**), which contains a central four-membered Zn$_2$N$_2$ ring. Another dimeric silylamide is the cyclopentadienyl compound (**15**), which has both η^1 and η^{2,3} cyclopentadienyl groups and a bridging disilylamide.[59] The zinc amides [(ZnEt)$_2$(mpsa)$_2$] (mpsa = 2-N(TMS)C$_5$H$_3$N$_6$Me), [ZnCl(mpsa)] and [Zn(mpsa)$_2$] were prepared from lithium reagents and ZnCl$_2$. In [(ZnEt)$_2$(mpsa)$_2$] the *N*-amido groups bridge the metal centres.[60]

The reactivity of Zn–C and Mg–C in some pyrazolyl borate[61] species [M{η^3-HB(3-Bu^{t}pz)$_3$}R] (3-Bu^{t}pz = 3-C$_3$N$_2$Bu^{t}H$_2$, R = Me, Et) have been compared. The Mg–C bonds in this well-defined environment are found to be more reactive than those between zinc and carbon (**16**). A study of the

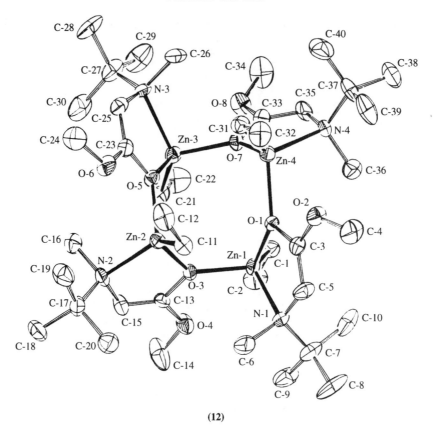

(12)

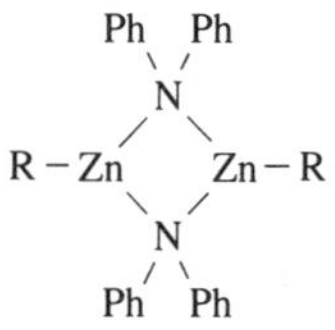

$$
\begin{array}{c}
\text{Ph} \quad \text{Ph} \\
\diagdown \diagup \\
\text{N} \\
\diagup \quad \diagdown \\
\text{R} - \text{Zn} \qquad \text{Zn} - \text{R} \\
\diagdown \quad \diagup \\
\text{N} \\
\diagup \quad \diagdown \\
\text{Ph} \quad \text{Ph}
\end{array}
$$

R = Me
(13)

related compounds such as [Zn{η^2-HB(3-Bu^{t}pz)$_3$}CMe$_3$] (**17**) exhibited the two different reaction pathways shown in Scheme 3 with protic acids, but the B–H bond reacts with ketones and aldehydes (Scheme 3).[62] A triangular planar zinc environment is found[63] in the phosphonimide (**18**).

4.2.8 Organozinc Compounds Containing a Zinc–Halide Bond

The 2,2'-bipyridyl adducts of a number of organozinc(II) halides [ZnXR] were prepared[64] by the electrochemical oxidation of zinc metal in the presence of organic solutions of various alkyl halides (RX, R = Me, Et, CF$_3$, C$_3$H$_3$, Ph, C$_6$F$_5$, Bz; X = Cl, Br); similar methods were used to prepare anionic species [RZnX$_2$]$^-$. The reaction of zinc dust with primary or secondary benzylic halides, for example, BzBr, gives species such as [ZnBrBz] in high yields.[65] The chloro complex of the chelating diethylaminopropyl ligand (**19**)[66] and the zincate derived from diethylamino-1-ethoxyethanolate and ethylzinc chloride have been reported[67] (**20**).

The fluoroalkyl zinc bromide species [ZnBr(CF$_3$)·2L] (L = DMF, MeCN) can easily be prepared by the reaction of CBrF$_3$ with elemental zinc.[68] The reaction of this compound with iodine monochloride in DMF provides a convenient route for the synthesis of CF$_3$I. The first x-ray structure for an (iodomethyl)zinc compound has been reported,[69] the zinc having complexed with a glycol-ether (**21**).

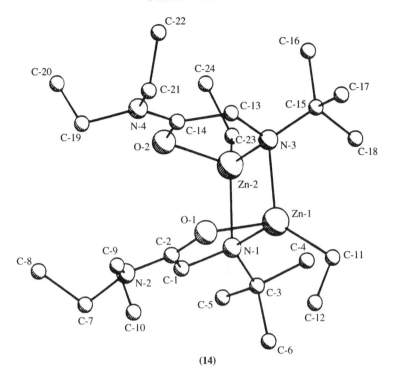

(14)

TMS   TMS

(15)

(16)

$\{\eta^2\text{-}H_2B(Bu^tpz)_2\}Zn(\eta^2\text{-}O_2CMe)$

MeCO$_2$H $\mid$ –CH$_4$

$\xrightarrow{\begin{array}{c} R^2_2C=O \\ R^2_2 \left\{\begin{array}{l} H_2 \\ (Me)H \\ (Me)_2 \end{array}\right. \end{array}}$ $\{HB(OCHR^2_2)(Bu^tpz)_2\}ZnR^1$

(17)

H$_2$O $\mid$ –CH$_4$

$[\{\eta^2\text{-}H_2B(Bu^tpz)_2\}Zn(\mu\text{-}OH)]_3$

Scheme 3

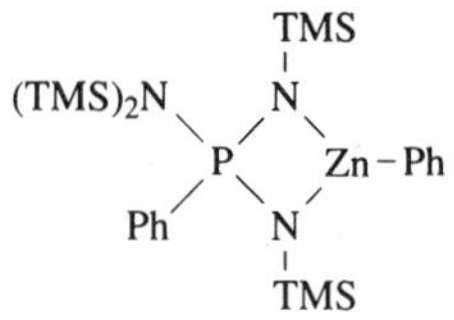

(18)

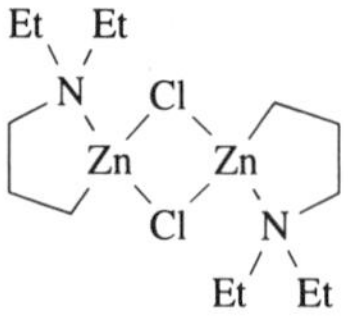

(19) 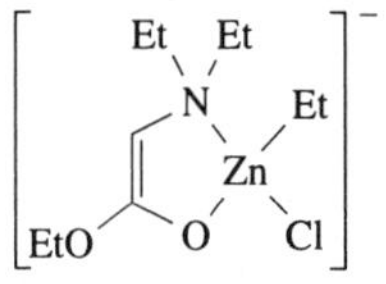   (20)

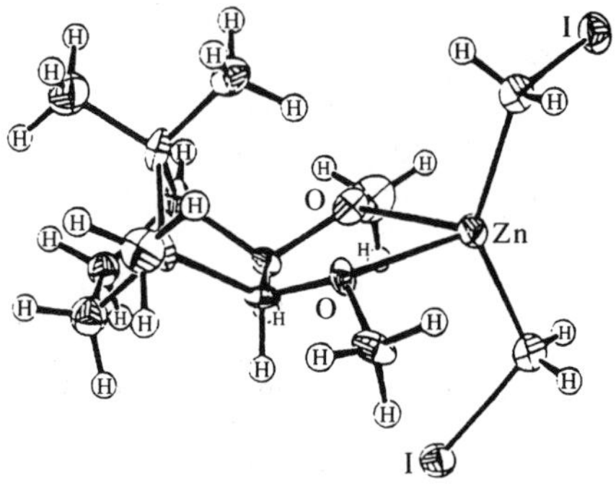

(21)

4.2.9 Organozinc Compounds with R–Zn–O Bonds

Ethylhydroxyzinc (**22**) is reported to be dimeric in dioxane and is a useful catalyst for polymerizations.[70] The use of bulky ligands, as is so often the case, leads to interesting compounds. A stable alkyl metal hydroxide of zinc, [{Zn(OH)C(Me$_2$PhSi)$_3$}$_2$], prepared from (**23**), is based on four-membered H$_2$O$_2$ rings.[71]

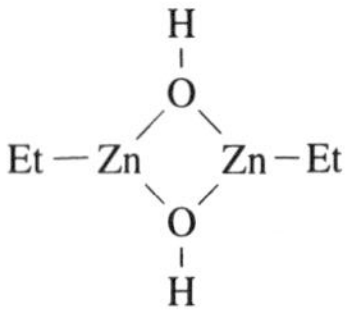

(22)

The reactions of Et$_2$Zn with RC$_6$H$_4$OH (R = H, Et, Cl) were carried out in THF and 1,4-dioxane. Monomeric [Zn(OPh)Et] was formed from the stoichiometric reaction of diethylzinc and phenol in 1,4-dioxane.[72] The reaction of Et$_2$Zn with 1,2,3-(HO)$_3$C$_6$H$_2$R (R = H(I), 4-Et, 4-Ac) in a 2:1 molar ratio in THF or 1,4-dioxane solutions led to oligomeric and polymeric products. The products contain zinc atoms bonded to oxygen atoms of the trihydroxybenzene; the other coordination sites are filled with solvent molecules or alkyl groups.[73] The diethylzinc derivatives with pyrocatecholate and pyrogallolate, obtained in the reaction of [ZnEt$_2$] with the appropriate phenol, were subjected to reaction with phenol, which resulted in the formation of [Zn(OPh)$_2$], ethylzinc phenoxyzinc pyrocatecholate, and bis(ethylzinc)phenoxyzinc pyrogallolate.[74] The x-ray single-crystal structure of a cubane-type complex [MeZnOBu^t]$_4$ has been reported.[75]

Complexes of the general formulae Et$_3$SiOMEt (M = Zn or Cd) were prepared by the stoichiometric reaction of [MEt$_2$] with Et$_3$SiOH in a hydrocarbon solvent at room temperature.[76] The structure of bis[(dimethylisopropoxysilyl)methyl]zinc has been determined by x-ray diffraction. The structure shows the internal coordination of an oxygen atom to the zinc atom and is a helical polymer. Coordination at zinc is trigonal planar.[77]

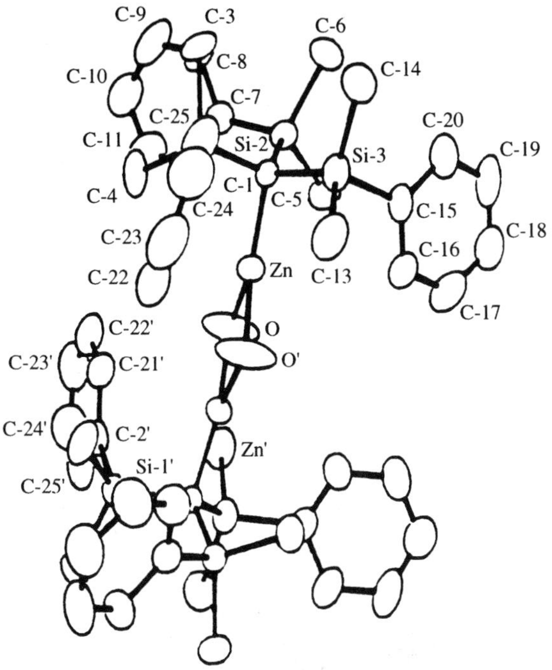

(23)

Dimeric chelated complexes with diatomic oxygen bridges (**24**) [EtZ̄n(Et)(Bu^t)NCH=CMeŌ]$_2$ have been synthesized.[78] An important intermediate in catalytic enantioselective alkylation (*vide infra*) is thought to be another dimeric compound[79] (**25**). Several alkylzinc alkoxides and aryloxides have been prepared[80] by the reaction of [Zn(CH$_2$TMS)$_2$] with ROH (such as R = 2,6-(Me$_2$CH)$_2$C$_6$H$_3$; 2,4,6-Bu^t$_3$C$_6$H$_2$). The products were typically dimers (**26**).

(24)

(R,R)-form

(25)

R^1 = H, R^2 = Pr^i

(26)

Attempts to synthesize carbamates by CO$_2$ insertion reactions with mixed alkyls/alkylamides of zinc have led to a range of novel compounds[81] (Equation (1)), the insoluble crystalline parent of which is the tetramer [Zn$_4$(OCNEt$_2$)$_6$Me$_2$]. The tetramer has an unusual structure (**27**) and analogous chemistry has been observed for the cadmium system. The tetramer can be reacted with stoichiometric quantities of dimethylzinc to give a second tetramer (**28**) [Zn$_4$(OCNEt$_2$)$_4$Me$_4$], which has a distorted cubane structure.[82]

$$[Zn(NEt_2)Me] + CO_2 \longrightarrow [Zn_4(OCNEt_2)_6Me_2] \tag{1}$$

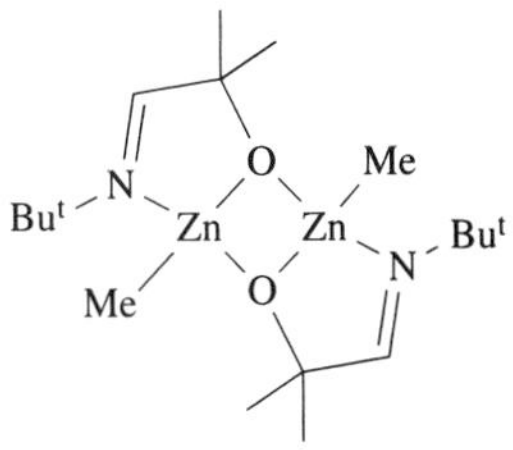

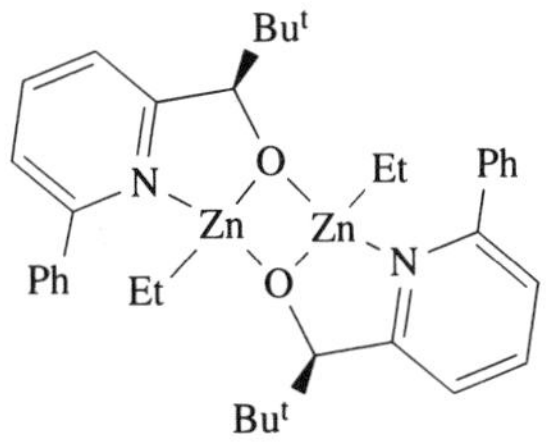

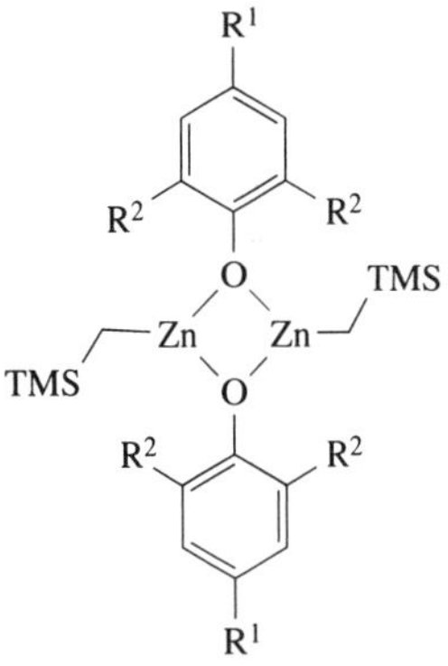

The reaction of Et$_2$Zn and Bu^{t}OK gave a product the x-ray structure of which showed it to consist of units of dimers of [Et$_2$ZnOBu^t]$^-$K$^+$. The zinc atoms share two bridging Bu^{t}O$^-$ groups to form a four-membered ring of alternating zinc and oxygen atoms; each zinc is bonded to two terminal ethyl groups.[83]

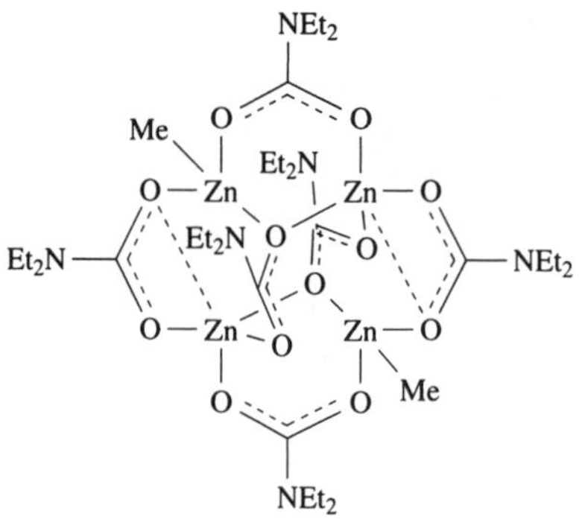

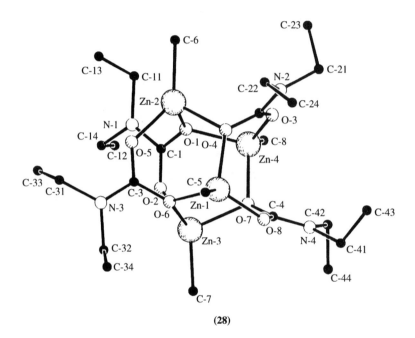

(27)

(28)

4.2.10 Compounds Containing R–Zn–S and R–Zn–Se Bonds

There appear to have been relatively few studies of these compounds to build upon the pioneering work of Coates, although the crystal structure of the pentamer[84] has been reported ([Zn(SBu^t)Me]$_5$, (29)). The synthesis and structural characterization of several thiolates has been undertaken,[80] for example [R^{1}ZnSR^2]$_n$ (R^1 = CH$_2$TMS, $n = 2$, R^2 = CPh$_3$; $n = 3$, R^2 = 2,4,6-(Me$_2$CH)$_3$C$_6$H$_2$, 2,4,6-Bu$^t_3$C$_6$H$_2$). The aggregate [Zn$_3${O(2,6-(Me$_2$CH)$_2$C$_6$H$_3$)}$_4$R$_2$] was also obtained. The trimers have a flattened, almost planar, Zn$_3$S$_3$C$_6$ array (30). The selenolate [Zn(CH$_2$TMS)SeR]$_3$·0.5C$_6$H$_{14}$ (R = 2,4,6-Bu$^t_3$C$_6$H$_2$) was synthesized from the reaction of Zn(CH$_2$TMS)$_2$ with HSeR, and has been shown to have a Zn$_3$Se$_3$ core.[85] Some species in this class have proved of synthetic use, for example, PhC(O)SCH$_2$I and PhSCH(R)Cl (R = H, Me, Pr, CH$_2$CN, CH$_2$CH$_2$CO$_2$Et), which gave zinc α-thioorganometallics under mild conditions in THF.[86] The x-ray structure of [Zn(Me){SC$_6$H$_4$((R)-CH(Me)NMe$_2$)$^{2-}$}]$_2$ (31) was determined.[87] The compound was prepared by the reaction of the arenethiolate [ArS-TMS] with [ZnClMe].

Mixed zinc–alkyl dithio- or diselenocarbamates were first prepared by Noltes[88] using an insertion reaction[89] (Equation (2)). However, a comproportionation often provides a more convenient synthesis for these compounds[90–2] (Equation (3)). The compounds are dimers (32) in the solid state and the parent dimeric structure has been confirmed for a wide range of compounds.[90–2] The metal dialkyls MR$_2$ (M = Zn, R = Me, Et, Pr, CHMe$_2$; M = Cd, R = Me) were reacted[93] with the sterically hindered selenol 2,4,6-(Me$_2$CH)$_2$C$_6$H$_2$SeH (Ar'SeH) to give complexes (33) [{M(SeAr')R}$_4$]. The less bulky mesityl selenol gave poorly soluble metal bis(selenolato) complexes.

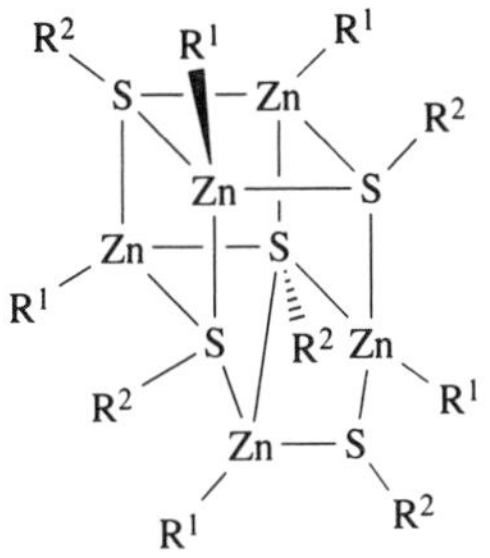

pentamer

$R^1 = Me$

$R^2 = Bu^t$

(29)

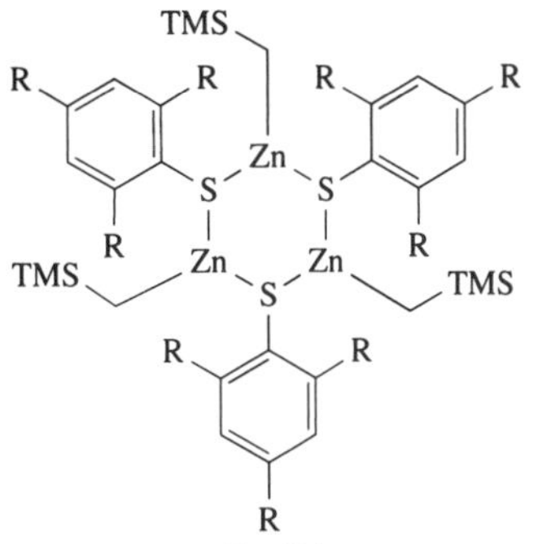

R = Pr^i

(30)

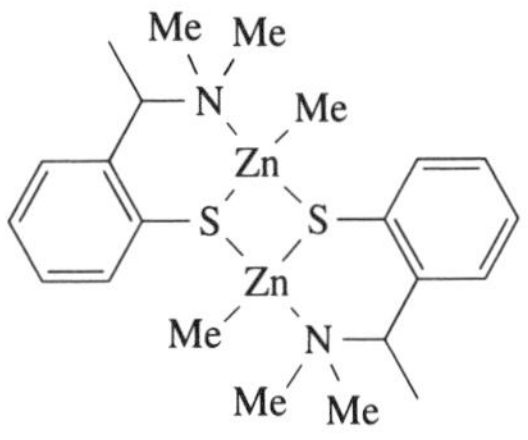

dimer

(31)

$$2\,[Zn(NEt_2)Me] + 2\,CS_2 \longrightarrow [\{ZnS_2C(NEt_2)Me\}_2] \tag{2}$$

$$[ZnMe_2] + [Zn(Se_2CNEt_2)_2] \longrightarrow [\{MeZnSe_2CNEt_2\}_2] \tag{3}$$

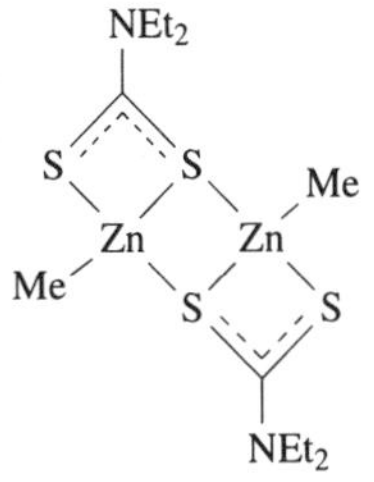

(32)

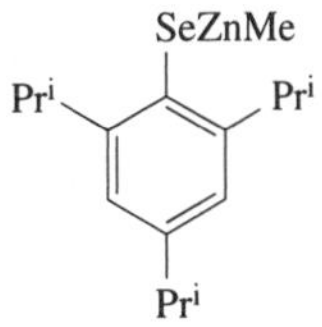

(33)

4.2.11 Heterometallic Zinc Organometallic Complexes

The reactions of zinc alkyls with several clusters have been studied. The reaction of $[Co_3(CO)_9\{\mu^3\text{-}C[1,4\text{-}C_6H_4(CH_2)_2CO_2H]\}]$ with half an equivalent of diethylzinc gives $[Zn\{(CO)_9Co_3CC_6H_4(CH_2)_2CO_2\}_2]$. In contrast, the reaction of $[Co_3(CO)_9(\mu^3\text{-}CCO_2H)]$ with one equivalent of diethylzinc gives $[Zn_4O\{(CO)_9Co_3CCO_2\}_6]$ in a high yield at room temperature.[94] The cluster $[Zn_4O\{(CO)_9Co_3(\mu^3\text{-}CCO_2)\}_6]$ is obtained in high yield from the reaction of diethylzinc with $[Co(CO)_9(\mu^3\text{-}CCO_2H)]$. A single-crystal structure determination shows an oxygen-atom-centred tetrahedral zinc(II) core with chelating clusters acting as ligands.[95]

The stabilities of some organozinc–transition metal complexes such as $[ZnCo(CO)_4Et]$ towards disproportionation have been investigated.[96] Simple alkyls and aryls disproportionate to such an extent that isolation is impossible. Ligands that can form chelate rings such as the dimethylaminopropyl group improve stability. The crystal structure of $[ZnWCp(CO)_3Me_2N(CH_2)_3]$ was determined. Zinc can be inserted in the Mo–C bond of $[\eta^7\text{-}C_7H_7Mo(CO)_2Me]$.[97]

A compound stabilized by a Sn–Zn interaction has been reported:[98] reaction of bis[3-(dimethylamino)propyl]zinc with bis(dibenzoylmethanato)tin(II) quantitatively gives zinc[bis[3-(dimethylamino)propyl-C,N]tin]bis(dibenzoylmethanato). The crystal structure shows an intramolecularly coordinated dialkyltin(II), with tin acting as an electron donor to the zinc β-diketonate

((**34**), Equation (4)). The reaction of Ni[salen] (salen = bis(salicylaldehyde)ethylenediiminato) with [ZnEt$_2$] in hydrocarbon solvents yields an adduct[99] the structure of which has been determined. The zinc atom is only weakly coordinated to the *cis* oxygen atoms of the ligand with an average Zn–O distance of 0.2321 nm.

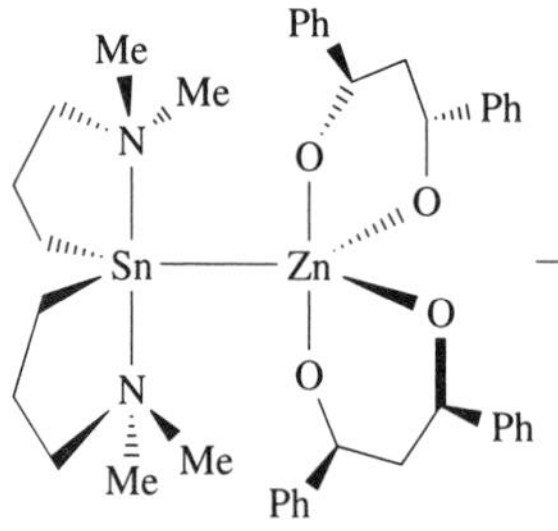

$$\xrightarrow{\text{TMEDA}} \quad [SnR_2]_n \ + \ [Zn(DBM)_2(TMEDA)]\downarrow \qquad (4)$$

R = (CH$_2$)$_3$NMe$_2$
DBM = dibenzoylmethanato

(**34**)

4.2.12 Theoretical and Physical Studies of Organozinc Species

The AM1 parameterization was used for a number of organometallic zinc species ZnR$^1_2$ (R^1 = Me, Et, Pr, Bu, F, Cl, Br, iodo) or ZnCpR^2 (R^2 = Me, Et, Ph, Cp).[100] The results are generally better than those given by MNDO calculations; for ZnCpR^2 (R^2 = Me, Cp) a half-sandwich structure is predicted to be the most stable geometry. *Ab initio* pseudopotential molecular orbital calculations have been used[101] to study species formed in thermal and photoinduced alkyl transfer reactions between metal alkyls R$^2_2$M (M = Zn, Mg; R^2 = alkyl, aryl) and 1,4-diaza-1,3-butadienes R^{1}N:CHCH:NR^1 (R^1-DAB). The x-ray structure of ZnMe$_2$·Bu^t-DAB was determined and its main features are well reproduced by modelling on systems with R^1 = H, Me. The absorption spectrum of the [ZnEt$_2$] has been reported, and is discussed in terms of the electronic transitions.[102]

There have been a number of studies of the photodissociation of simple alkyl zinc species. Dimethylzinc or -cadmium was photodissociated at 193 nm into methyl radicals and the monoalkyl, and further into methyl radicals and metal atoms.[103] The emission spectrum of the monoalkyl fragment in the visible region was reported. The photoabsorption cross-sections of zinc and cadmium were measured at wavelengths between 190 and 400 nm, based on measurements of the UV absorption spectra.[104] The spectra had intense absorption bands between 190 and 250 nm. *In situ* measurements of photolysis of these alkyls were made by monitoring UV absorption spectra in a vapour deposition reactor. [ZnMe$_2$], [[ZnEt$_2$] and [ZnPr$_2$] were studied at 248 nm by using laser-induced fluorescence to detect the monoalkyl and zinc atom products.[105] The monoalkylzinc is the primary photoproduct for each alkyl and is formed sufficiently hot that it spontaneously dissociates to an alkyl radical and a zinc atom. Time-of-flight mass spectrometry has been used to study[106] the photodecomposition of dimethylzinc on a quartz substrate; methyl fragments were the main species detected.

An excimer laser was used[107] in the *in situ* monitoring of the room-temperature photodecomposition of dimethylzinc in a low-pressure MOCVD reactor. The laser-induced luminescence intensity from an upper level of a zinc species is linearly dependent on the zinc and scales quadratically with laser intensity. The decomposition of [ZnMe$_2$] has been studied in an atmospheric-pressure MOCVD system,[108] sampling both in the reactor and the exhaust. The concentrations of [ZnMe$_2$] and methane produced were followed. If it was assumed that decomposition was a first-order reaction, the activation energy became dependent on reactor conditions and gave a value of close to 100 kJ mol^{-1}, about half that expected.

4.2.13 The Use of Organozinc Reagents in Organic Synthesis

Organozinc reagents are less reactive than the widely used Grignard reagents, a property that is a distinct advantage when more gentle reactions are required. Recent developments in the use of such reagents have been surveyed.[109–11]

4.2.13.1 Catalytic asymmetric synthesis

Organozinc compounds, unlike Grignard reagents, do not react with aldehydes at or below room temperature. They are hence useful as alkyl donors in catalytic asymmetric alkylations. Addition of a ligand or auxiliary to the organozinc reagent distorts the linear geometry of the zinc–alkyl bond, and accelerates the alkyl transfer reaction.[1] An early example of this type of reaction was the reaction of benzaldehyde and diethylzinc catalysed by 2 mol.% (S)-leucinol to give (R)-1-phenylpropan-1-ol (96% yield, 49% ee).[112]

The chiral auxiliary must have a three-dimensional structure that discriminates between the diastereoisomeric transition states involved in the alkylation step. Several kinetic conditions must be met before catalytic asymmetric induction is achieved. The rate of alkylation by the chiral zinc complex (**35**) must exceed that of the original alkyl. The chiral ligand must be labile with the metal alkoxide (**36**) to establish catalysis (Equation (5)).

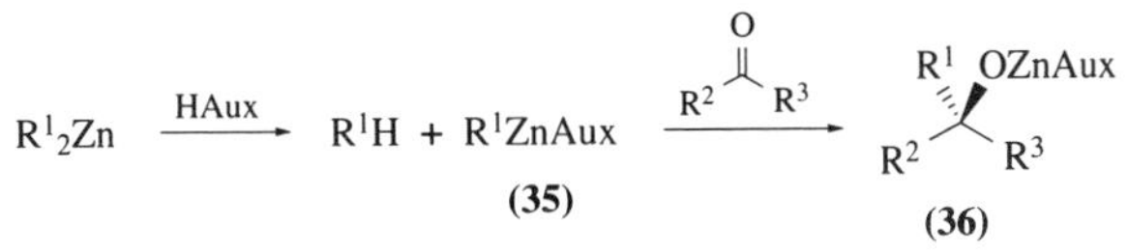

$$R^1_2Zn \xrightarrow{\text{HAux}} R^1H + R^1ZnAux \xrightarrow{\underset{R^2}{\overset{O}{\parallel}}\underset{R^3}{}} \underset{R^2\quad R^3}{\overset{R^1\ OZnAux}{}} \tag{5}$$

(**35**)   (**36**)

β-Amino alcohol chiral auxiliaries and aprotic chiral auxiliaries, such as N,N,N'N'-tetramethylethylenediamine derivatives, have been successfully used. N,N-Dialkyl-β-amino alcohols are better promoters than the corresponding N-monoalkyl or nonalkylated compounds. Some cyclic compounds and compounds with bulky α-substituents are found to be 10–100 times more reactive than simple α-amino acid derived alcohols.[113] Alkylation using (−)-3-exo-(dimethylamino)isoborneol (daib) (**37**) is highly enolic selective;[114] related auxiliaries have now been used. Soai et al. have shown (**38**) to be a highly efficient catalyst with a wide variety of aldehydes and dialkylzinc reagents.[115]

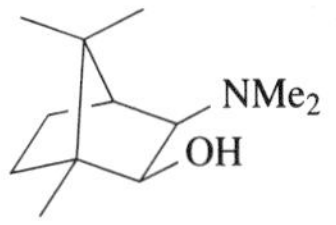

(**37**) 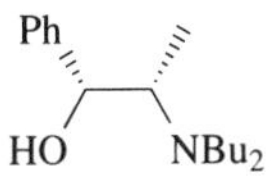   (**38**)

A range of immobilized systems have been developed including N-alkylnorephedrines bound to polystyrene,[116,117] alumina or silica gel.[118] Fréchet and co-workers have attached daib to polystyrene.[119] Polymer-bound catalysts, with a methylene spacer arm between the polymer and the catalyst, are more efficient asymmetric catalysts than those that are directly bound to the polymer. The spacer allows the chiral auxiliary the freedom to form and imitate monomeric auxiliaries. The catalyst (**39**) is particularly effective and in the reaction of diethylzinc with benzaldehyde the corresponding alcohol is obtained in 91% yield and 82% ee.[120] Oppolzer[121] has used divinylzinc, in the presence of the daib-derived auxiliary (**40**), to ethenylate benzaldehyde (96% yield and 87% ee) (Equation (6)).

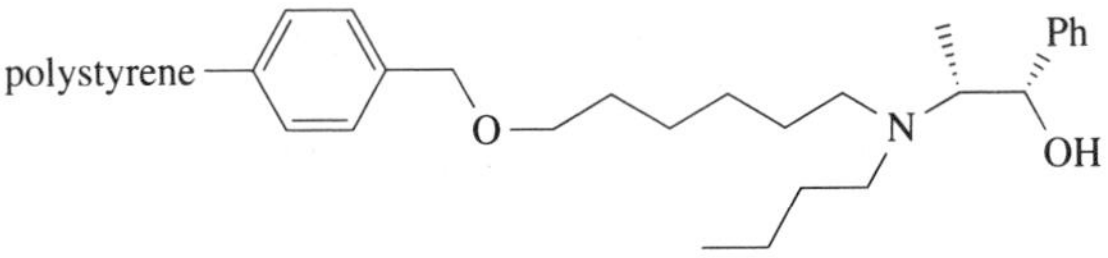

(**39**)

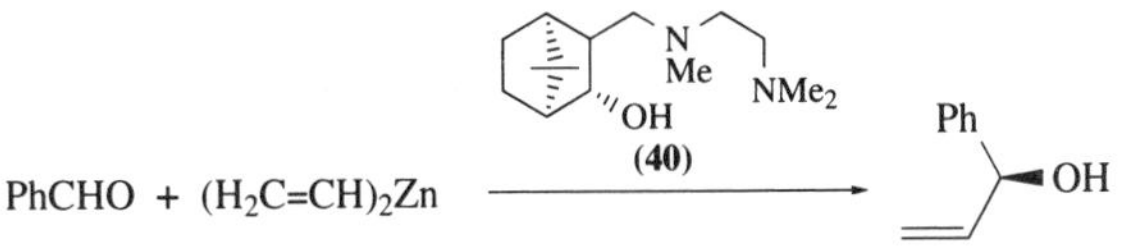

$$\text{PhCHO} + (\text{H}_2\text{C=CH})_2\text{Zn} \xrightarrow{\quad (\textbf{40}) \quad} \tag{6}$$

A series of dialkynylzinc reagents react with aldehydes in the presence of catalytic quantities of dbne (**38**). Higher enantioselectivities are obtained when dialkylzinc reagents are reacted with alkynyl aldehydes in the presence of the auxiliary (**41**)[122,123] (Equation (7)).

Lithium, titanium, boron, chromium and iron complexes of diamines, diols and β-amino alcohols (although the iron and the chromium atoms are not actually complexed to the β-amino alcohol) have also been used to good effect as chiral auxiliaries. Their use is summarized in Table 2.

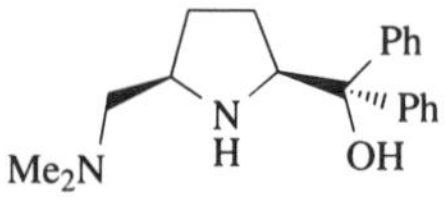

(41)

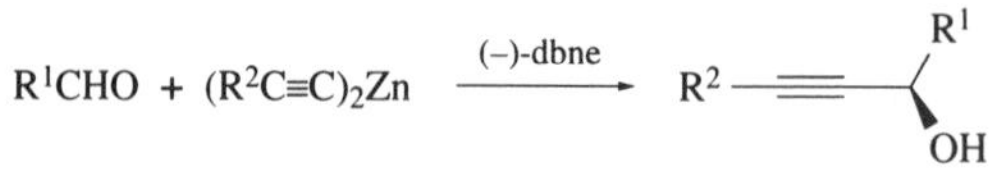

$$R^1CHO + (R^2C{\equiv}C)_2Zn \xrightarrow{\text{(−)-dbne}} R^2 {=\!\!=} \overset{R^1}{\underset{OH}{\diagup}}$$

(7)

The reaction between diethylzinc and terephthalic aldehyde (42) or isophthalic aldehyde (43) in the presence of the chiral spirotitanate (44) (Table 2) gives excellent diastereoisomeric and enantiomeric excesses (Equations (8) and (9)).[124–9]

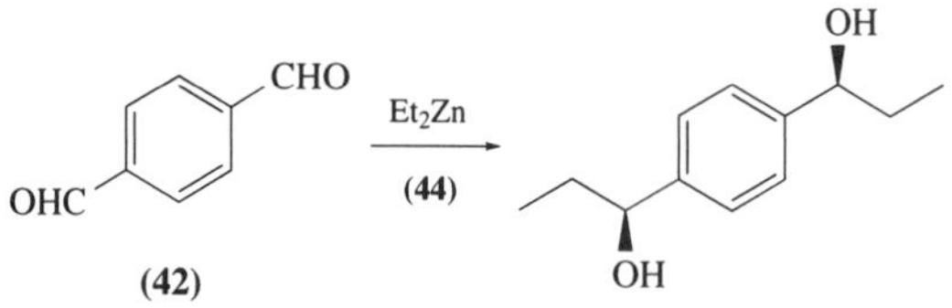

>99% *de*, 98% *ee*

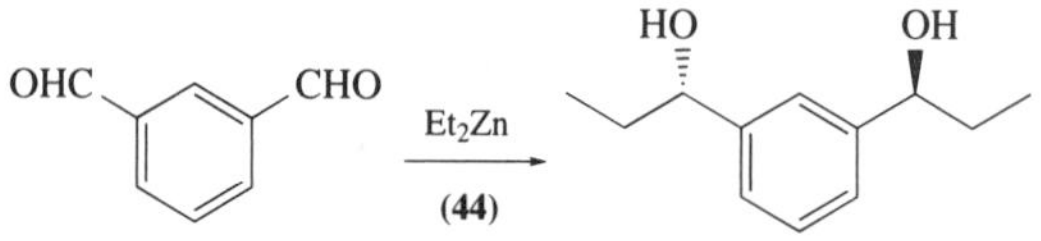

(8)

93% *de*, >99% *ee*

(9)

4.2.13.2 The reaction pathway with aldehydes

The action of dialkylzinc with an auxiliary (e.g., daib) leads to the formation of the dimeric species (45). The structure of this compound has been determined by x-ray crystallography.[130] The dimer (45) does not alkylate aldehydes (despite having two available alkyl groups) unless there is an excess of the dialkylzinc reagent present. This observation is consistent with the alkylating species being a product of the reaction between complex (45) and dialkylzinc (Equation (10)).

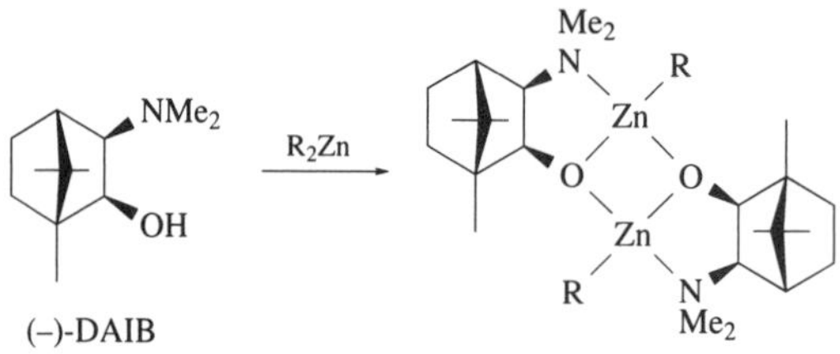

(−)-DAIB

(37)   (45)

(10)

Noyori has proposed a catalytic cycle for the addition of dialkylzincs to aldehydes in the presence of (−)-daib (37).[131] Evidence to support the theory of there being a bridging alkyl group in complex (46) comes from the fact that in dialkylzincs self-exchange occurs via alkyl-bridged species.[132] The bridging alkyl group is the one transferred. However, the postulation of (46) as the key intermediate does not preclude the existence of an intermediate species such as complex (47). In this case a six-centre transition state would result, which may be more plausible than the four-centre transition state depicted in complex (46).

Both Fréchet[119] and Oppolzer[133] have proposed mechanistic details and in neither paper is there any reference to a bridging alkyl species such as in complex (46). Using the auxiliary (48), Oppolzer has postulated the existence of complex (49) as a possible intermediate containing a tetrahedrally coordinated zinc(II) species (Equation (11)).

Table 2 Enantioselective addition of diethylzinc to aldehydes using complexes of lithium, titanium, boron, iron and chromium as chiral auxiliaries.

Chiral auxiliary	Aldehyde	Yield (%)	ee (%)	Configuration	Ref.
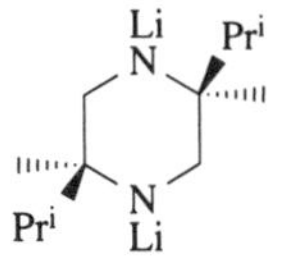	PhCHO	68	92	(R)	124
	p-MeO-C$_6$H$_4$CHO	65	79	(R)	
	p-Cl-C$_6$H$_4$CHO	77	98	(R)	
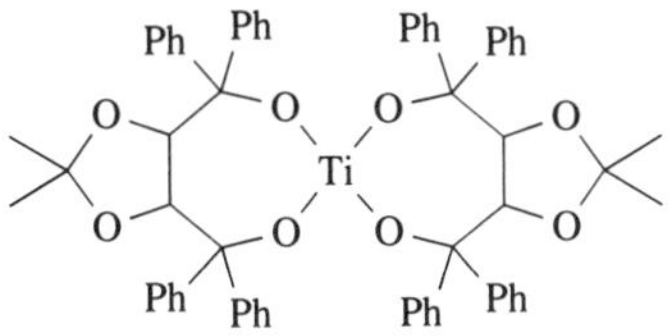	PhCHO	75	99	(S)	125
	p-MeO-C$_6$H$_4$CHO	86	94	(S)	
	PhCH$_2$CH$_2$CHO	85	82	(S)	
	CyCHO	67	82	(S)	
	(E)-PhCH=CHCHO	89	96	(S)	
	p-BrC$_6$H$_4$CHO	88	90	(S)	
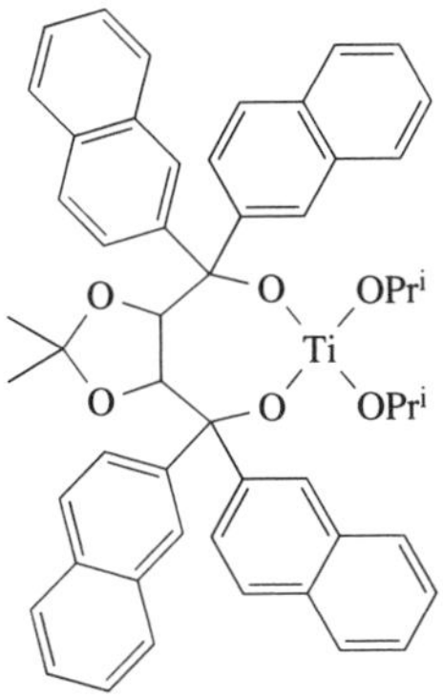 (44)	PhCHO	76	99	(S)	126
	n-C$_6$H$_{13}$CHO	70	97	(S)	
	PhCH$_2$CH$_2$CHO	87	99	(S)	
	CyCHO	77	99	(S)	
	(E)-PhCH=CHCHO	87	91	(S)	
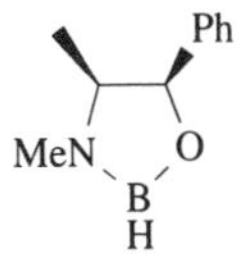	PhCHO	>95	95	(R)	127
	p-MeO-C$_6$H$_4$CHO		93	(R)	
	p-Cl-C$_6$H$_4$CHO		96	(R)	
	n-C$_6$H$_{13}$CHO		52	(R)	
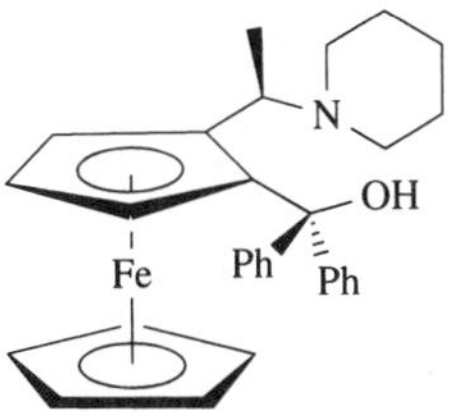	PhCHO	99	99	(S)	128
	p-Cl-C$_6$H$_4$CHO	100	100	(S)	
	p-MeO-C$_6$H$_4$CHO	97	90	(S)	
	(E)-PhCH=CHCHO	90	100	(S)	
	n-C$_6$H$_{13}$CHO	93	59	(S)	
	n-C$_6$H$_{13}$CHO	92	>98	(S)	
	Me$_2$CHCH$_2$CHO	70	62	(S)	
	PhCH$_2$CH$_2$CHO	88	63	(R)	
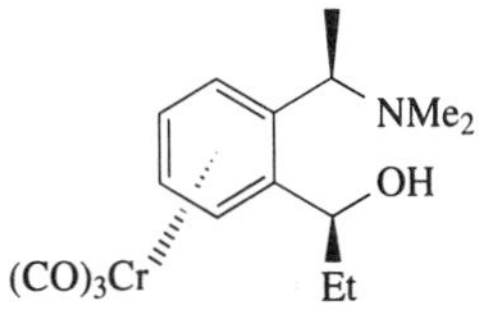	PhCHO	75	64	(S)	129
	p-MeO-C$_6$H$_4$CHO	18	10	(S)	
	p-PhC$_6$H$_4$CHO	23	17	(S)	
	p-Cl-C$_6$H$_4$CHO	28	54	(S)	

Evidence to support an unbridged intermediate has been given by Soai[115,134] and Corey.[135,136] Using lithium salts of chiral auxiliaries they have alkylated aldehydes with dialkylzinc reagents, giving the secondary alcoholic products in good yields and high enantiomeric excess. Soai has mirrored the work

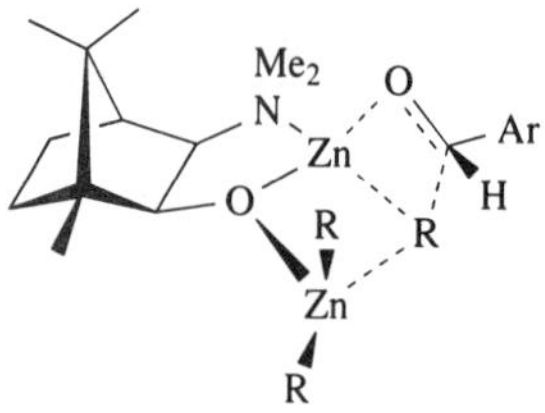

(46) 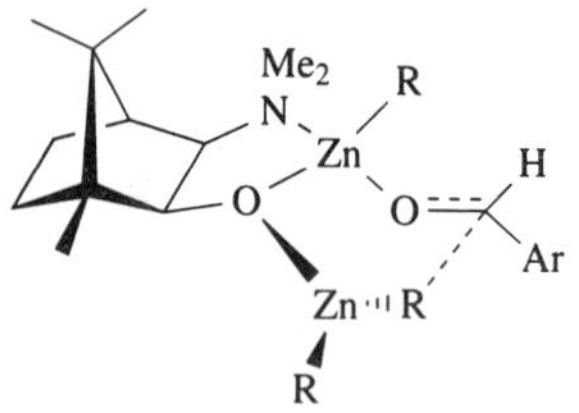   (47)

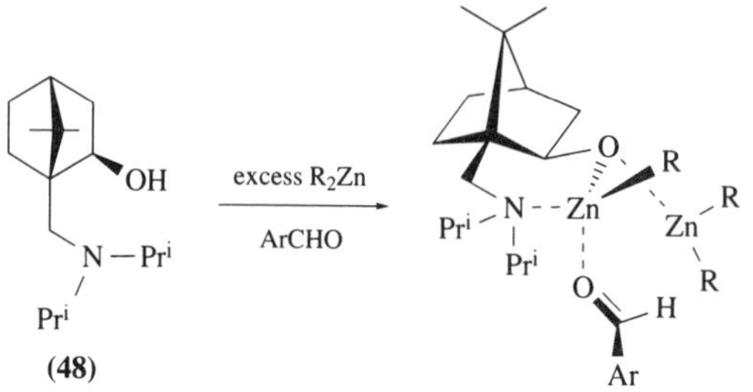

(11)

(48)   (49)

of Noyori using as catalyst (+)-dbne (**38**), which has similar directing properties to (−)-daib (**37**). The lithium salt of dbne (**50**) may react with dialkylzinc in the presence of an aldehyde to form (**51**), in which a six-centre transition state may be observed (Equation (12)). This complex may then alkylate the aldehyde.

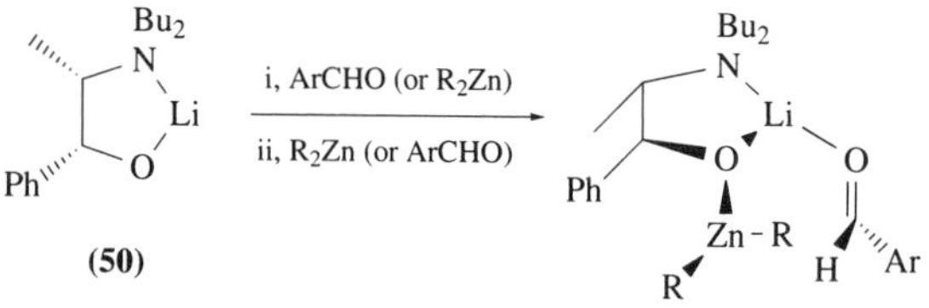

(12)

(50)   (51)

Complex (**51**) is similar to complex (**47**). As there can be no alkyl group attached to the lithium in (**51**) the alkylation must occur from the complexed dialkylzinc reagent. Therefore it is logical to assume that a similar process occurs in (**47**). Corey has carried out a similar reaction using chiral tertiary amino phenolic alcohols as chiral auxiliaries.[136] The ligand (**52**) reacts with dialkylzinc to give the complex (**53**). Again, this intermediate is similar to complex (**47**) with no alkyl group on the centrally complexed zinc ion. Treatment of (**53**) with an excess of dialkylzinc and aldehyde again leads to alkylation of the aldehyde via the intermediate (**54**) (Scheme 4).

There is some evidence to support the postulation of both the bridged complex (**46**) and the nonbridged complex (**47**) as the possible intermediate that leads to the subsequent alkylation of the aldehyde. From the work of Soai and Corey it would seem that complex (**47**) would be the preferred one, but the results of the alkyl scrambling experiments and variable-temperature ^{1}H NMR spectroscopic experiments indicate that there is facile alkyl exchange between different zinc atoms. The results of the alkyl scrambling experiment indicate the fluxional nature of the intermediates and this would tend to support the postulation of complex (**46**) as the key intermediate.

4.2.13.3 *Chiral amplification and enantiomer recognition*

Chiral amplification is an interesting phenomenon; for example the reaction of diethylzinc and benzaldehyde in the presence of (−)-daib of only 15% *ee* affords (*S*)-1-phenylpropan-1-ol in 92% yield and 95% *ee*,[130] only slightly less than when optically pure (−)-daib is used for the same reaction (cf. 97% yield, 98% *ee*).[137]

The explanation for this phenomenon is believed to lie in the formation of ternary *meso* complexes. If enantiomerically pure (−)-daib is reacted with an equivalent amount of dialkylzinc, the dimer (**45**) is formed. X-ray crystallography shows complex (**55**) to have chiral C_2 symmetry; ^{1}H NMR spectroscopic

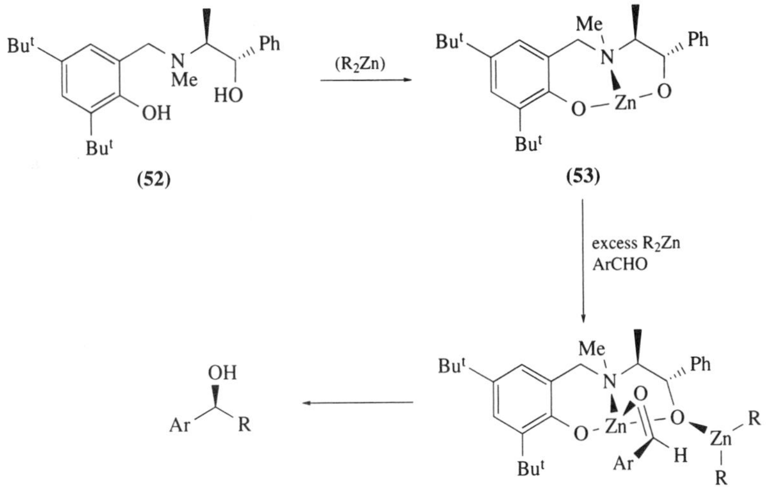

Scheme 4

experiments indicate that the two daib species are magnetically equivalent. However, when (±)-daib is treated with either a whole or a half equivalent of dialkylzinc, a different dimer (**57**) is formed (Equation (13)). The two five-membered rings are *endo*-fused to the Zn_2O_2 moiety, but the central tricyclic system has *anti* geometry. This is opposite to that observed for the enantiomerically pure daib (**55**).

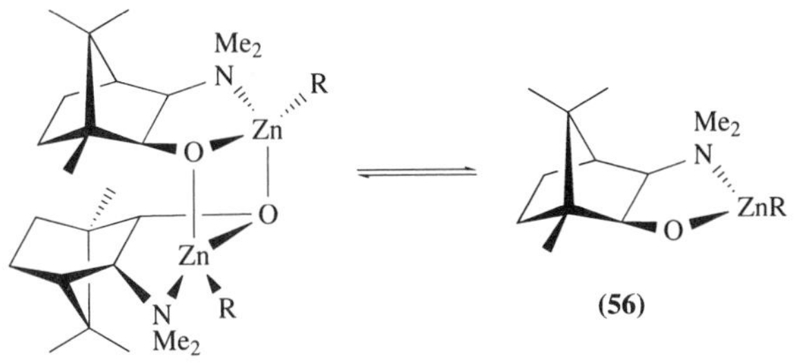

(13)

Cryoscopic molecular weight measurements indicate that complex (**55**) dissociates to the monomeric form (**56**). Complex (**57**) is unable to alkylate aldehydes in the presence of equimolar amounts of aldehyde and dialkylzinc because it is more stable. Oguni[138] has also demonstrated this chiral amplification effect in the alkylation of benzaldehyde with excess diethylzinc and 1-piperidino-3,3-dimethylbutan-2-ol.

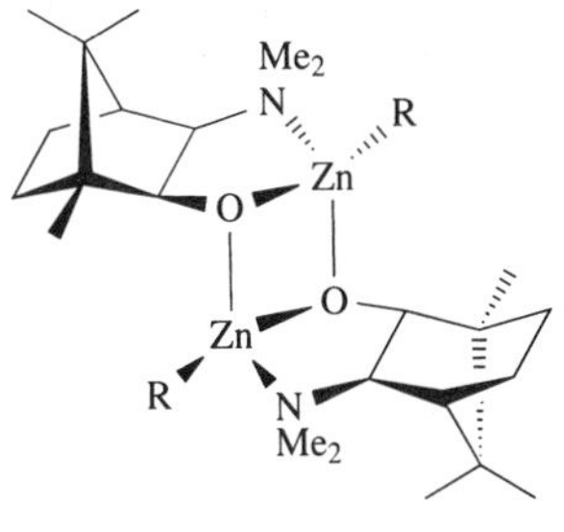

(**57**)

Kagan and co-workers have provided a detailed discussion of the origins of nonlinear effects in asymmetric catalysis.[139] A large number of results in the literature[140-2] can be explained using the models evolved by these authors.

4.2.13.4 The Simmons–Smith reaction

The Simmons–Smith reaction involves the reaction of diiodomethane, a zinc–copper couple and alkenes to give cyclopropanes (Equation (14)). A variation of this reaction is the Furukawa cyclopropanation reaction,[143] which involves the treatment of alkenes with diethylzinc and diiodomethane. Another variation is the Wittig–Schwarzenbach cyclopropanation involving the use of diazomethane and zinc iodide to cyclopropanate alkenes.[144]

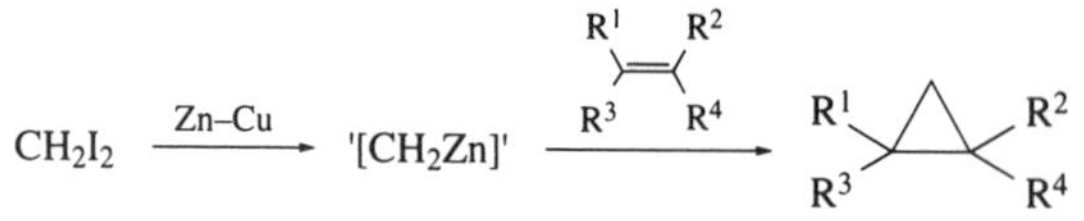

$$CH_2I_2 \xrightarrow{Zn-Cu} {}'[CH_2Zn]' \longrightarrow \quad (14)$$

As two new chiral centres are manufactured stereospecifically from the *syn* addition of a carbene unit to the alkene, an asymmetric synthesis should be realizable. The addition of the carbene unit goes via the three-centre transition state[145] (**58**) rather than the simple addition of the reagent as an electrophile to give transition state (**59**) (Scheme 5).

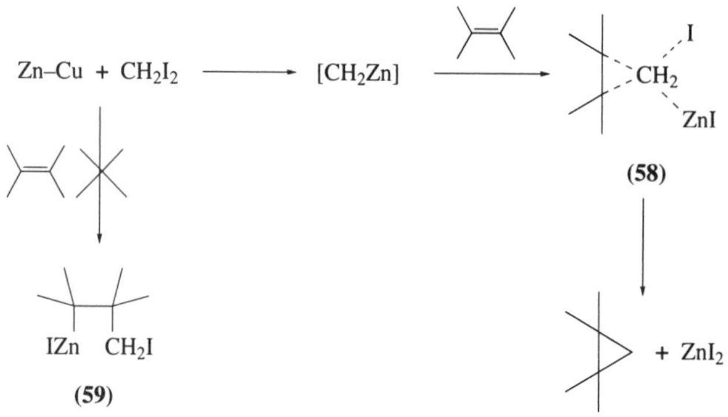

Scheme 5

It has been shown that in some alkenic alcohols high diastereoselectivities are obtained, and this has been ascribed to the complexing of the zinc to the alcohol and the subsequent directing effect this has. In the reaction of the chiral alkenic alcohol (**60**) with the Simmons–Smith reagent or the Furukawa variant, high diastereoselectivities are observed.[146]

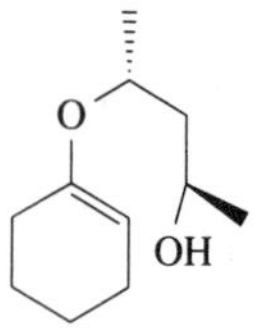

(**60**)

Alkenes to which other chiral auxiliaries have been attached, have also been seen to achieve a diastereoselection in the cyclopropanation of large-ring α,β-unsaturated ketals derived from 1,4-di-*O*-benzyl-L-threitol (**61**) (Equation (15)).

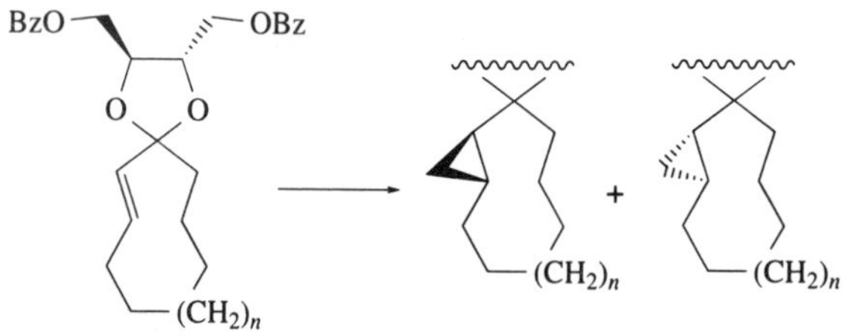

(**61**)

$$(15)$$

The diastereoselectivity of the reaction has been ascribed to a conformational effect. An enantioselective Simmons–Smith reaction uses stoichiometric amounts of (1*R*,2*R*)-tartaric acid esters.[147] (*E*)-3-Phenylprop-2-en-1-ol (**62**) reacts with diethylzinc to produce an alkoxide (**63**). The alkoxide is then added to a stoichiometric amount of (1*R*,2*R*)-tartaric acid ester (**64**) which, in the presence of diiodomethane and excess diethylzinc, cyclopropanates the double bond (Scheme 6).

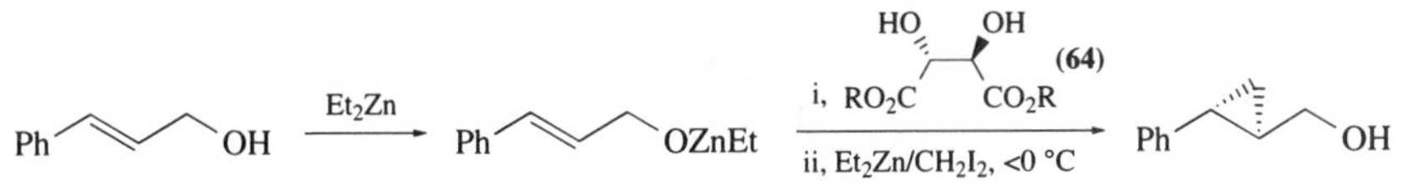

Scheme 6

An effective asymmetric Simmons–Smith reaction using a catalytic amount of chiral auxiliary,[148] reaction of compound (**65**) with diethylzinc, leads to the formation of a chiral Lewis acid complex (**66**); reaction of this complex with diiodomethane, diethylzinc and an allylic alcohol (**67**) gives an enantioselective synthesis of the cyclopropane derivative (**68**) (Scheme 7).

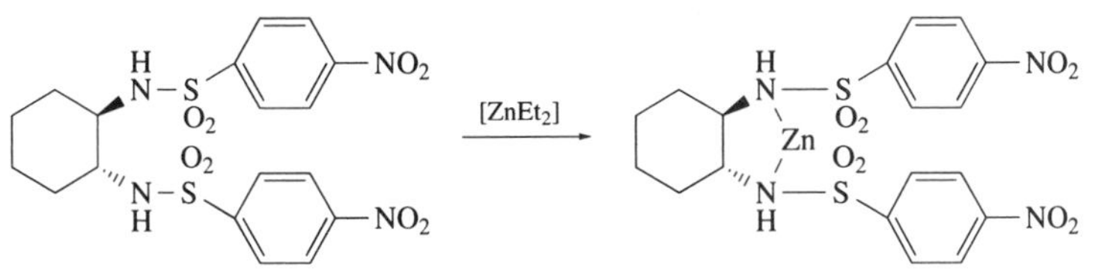

Scheme 7

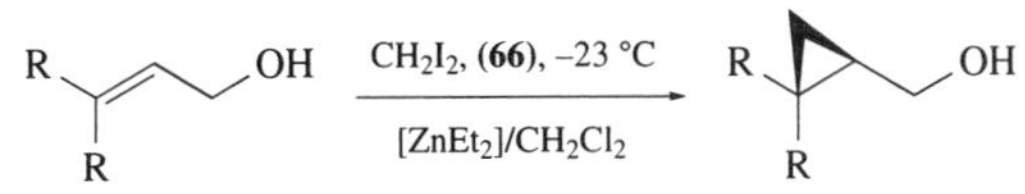

4.2.13.5 The Reformatsky reaction

The reaction between activated zinc metal and an α-bromoester to give an organozinc reagent is termed the Reformatsky reaction. The reagent reacts with carbonyl compounds to give β-hydroxyesters. The actual structure of the intermediate has been much debated, with suggestions for an enolate-type structure (**69**)[149] or alternatively a carbon-metallated species (**70**).[150]

The situation has been clarified by Boersma, who has shown the actual structure of the intermediate to be dimeric, both in solution (molecular weight measurements) and in the solid state (x-ray crystal structure).[151] An ORTEP drawing of the crystalline Reformatsky intermediate derived from *t*-butyl bromoacetate is shown in Structure (**71**). The crystal structure shows a shortening of the C–C bond (C-22–C-23a 0.141 nm). A corresponding lengthening of the carbonyl bond (C-22–O-21 0.131 nm) is also observed. The intermediate exists neither as a *C*-metallated species nor as an enolate, but rather as a combination of the two.

Very little work has been done to develop the Reformatsky reaction enantioselectively. A diastereoselective Reformatsky reaction, the cleavage of oxazolidine (**72**) reagents leading to the formation of protected β-amino acids (**73**) (Equation (16)), has been reported.[152] A recent report of an enantioselective Reformatsky reaction uses (*S*)-(+)-dpmpm as chiral catalyst and *t*-butyl bromoacetate.[153]

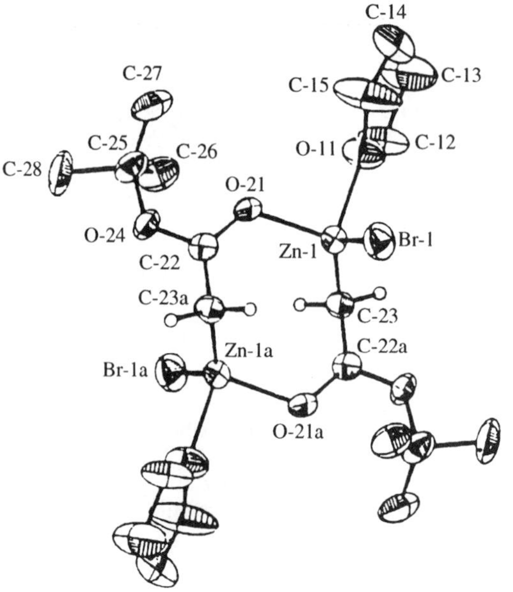

(71)

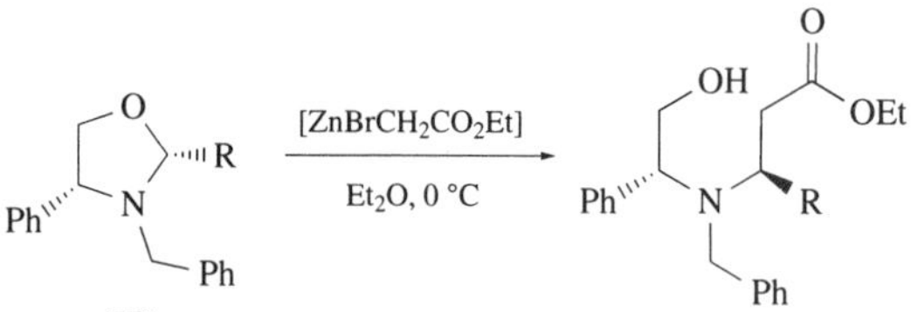

(16)

(72)   (73)

4.3 ORGANOCADMIUM CHEMISTRY

4.3.1 Diorganocadmium Compounds

Dimethylcadmium has been prepared[154] by the condensation of Me and the metal on a cold finger at −196 °C. The equilibrium between ethyl- and methylcadmium was studied in an attempt to prepare a pure sample of EtMeCd; only equilibrium mixtures were obtained.[155] The synthesis and spectroscopic characterization of bis(γ-methoxypropyl)cadmium has been reported; the compound is a colourless oil and there is no evidence of internal coordination by the oxygen atom.[156]

The compound [Cd((TMS)CH$_2$)$_2$] was prepared by treating CdI$_2$ with (TMS)CH$_2$MgCl in Et$_2$O.[157] The compound has high thermal stability but reacts immediately with molecular oxygen to give peroxo-derivatives and with water to give [Cd(OH)$_2$] and SiMe$_4$. A simple adduct is formed with 1,10-phenanthroline, whereas with 2,2'-bipyridyl (2,2'bipy)Cd(CH$_2$TMS)$_2$ 5(2,2'bipy) is formed. In another report the trimethylsilylmethyl and bis(trimethylsilyl)methyl derivatives of cadmium are detailed.[158] In continuing studies of the interaction of metals with bulky silicon-containing ligands (see also Section 4.2.4) the linear cadmium species [Cd((TMS)$_2$(CH$_2$:CHMe$_2$Si)C)$_2$] has been prepared.[159]

A considerable number of papers are concerned with the fluoroalkyls of cadmium. Bis(perfluoroorgano)cadmium compounds were prepared[160] by a simple method in which perfluoro-organoiodides RI and dialkylcadmium R$_2$Cd (R = Me, Et) were reacted in the presence of a Lewis base. The exchange reaction proceeded in steps via the ternary species. An extremely simple synthesis[161] using (trifluoromethyl)cadmium reagents has been reported, in which CF$_2$X$_2$ (X = a halide) reacts with cadmium or zinc powder in DMF to give CF$_3$MX (M = Cd, Zn) in excellent yield. A mechanism for the conversion of the CF$_2$ group into a CF$_3$ group was proposed involving the formation of CF$_3^-$ ion from the capture of F$^-$. The perfluoroalkyl cadmium compounds [Cd(C$_6$F$_{13}$)$_2$] and [Cd(C$_8$F$_{17}$)$_2$] have been isolated[162] along with their adducts with DMF, MeCN, GLYME, and DIGLYME. Perfluoroisopropyl-cadmium has been prepared[163] in excellent yield (98%) by the reaction of (CF$_3$)$_2$CFI with activated cadmium powder in DMF. The compound decomposes into a mixture of dimers and trimers of hexafluoroprene. Sulfur dioxide can be inserted into the cadmium–carbon bond. There are a number of reports of chelate-supported organometallics of cadmium, of which (74)–(76) are good examples.[164-6]

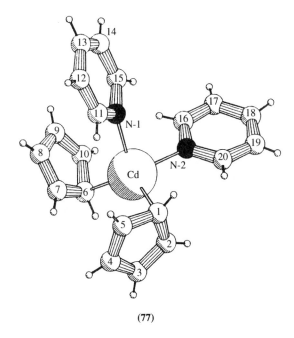

(74)   (75)   (76)

The crystal structure of bis(cyclopentadienyl) bis(pyridine) cadmium(II) (77) has been reported,[167,168] as has the structure of the pyridine complex of bis(pentamethylcyclopentadien-1-yl)cadmium(II).[169]

(77)

4.3.2 Adducts of Organocadmium Compounds

The isolation and characterization of bis(2,2-dimethylpropyl)cadmium(II), [bis(neopentyl)-cadmium(II)] and its 1:1 adduct with 2,2'-bipyridyl have been reported.[170] Adducts between tetramethylethylenediamine and methylcadmium or neopentylcadmium have been reported[17] (78) (see Table 1). An electron diffraction study of the tetramethylenediamine adduct of dimethylzinc has also been reported;[171] the structure is similar to that in the solid state. However, an infrared study of the same compound at 60 °C gave evidence for extensive dissociation. It is clear that the extent of dissociation of such alkyls is highly temperature dependent; there was also no evidence for the adduct in the mass spectrum.

The 1:1 adduct of dimethylcadmium and 2,2'-bipyridyl has been studied.[172] Cadmium is four-coordinate and the C-13–Cd–C-14 angle is 148.4°. The crystal structure of the 1:1 adduct formed between dimethylcadmium and 1,4-dioxane has been reported (79).[173] The cadmium atom is four-coordinate and bound to two methyl groups (Cd–C 0.209 nm) and two oxygen atoms (Cd–O 0.288 nm and 0.275 nm) from different dioxanes, giving rise to an unusual one-dimensional polymeric structure.

The properties of a number of adducts of dimethylcadmium with group 5 and 6 donors have been studied.[174] Volatile monomeric adducts are formed with chelating bases such as 1,2-bis-(dimethylamino)ethane, 1,4-dioxane and 2,2'-bipyridyl. Crystallizable adducts, which dissociate on heating but are themselves nonvolatile, are formed with 4,4'-bipyridyl and 1,4-bis(dimethylamino)-benzene. These adducts may be suitable for the purification of dimethylcadmium. The heat of dissociation of the 4,4'-bipyridyl adduct is 54 kJ mol^{-1}.

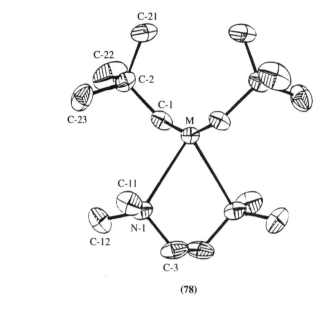

(78)

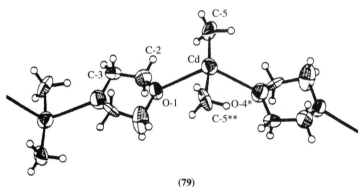

(79)

4.3.3 Organocadmium Compounds Containing a Cadmium–Heteroatom Bond

The reaction between $CdBr_2$ and $[Li\{C(TMS)_3\}]$ gives the tricadmate $[Li(THF)_4][\{Cd[C(TMS)_3]\}_3(\mu$-$Br)_3(\mu^3$-$Br)]\cdot0.5C_6H_{12}$, with a structure based on a Cd_3Br_4 cube with one corner missing (**80**).[175] In the presence of traces of moisture, the neutral compound $[\{Li(THF)\}_3(\mu^3$-$Br)_3(\mu^3$-O-$TMS)Cd\{C(TMS)_3\}]$, which also has a cage structure, is obtained (**81**). These compounds were converted into $[CdBr\{C(TMS)_3\}]$, and $[Cd(OH)\{C(TMS)_3\}]$. The chloride $[CdCl\{C(TMS)_3\}]$ is dimeric in the gas phase. The reaction between tris(dimethylphenylsilyl)methyllithium and $CdBr_2$ yields the cadmate $[Li(THF)_2CdBr_2\{C(SiMe_2Ph)_3\}]$, which was converted into $[Cd(OH)\{C(TMS)_3\}]$, $[CdBr\{C(SiMe_2Ph)_3\}]$, or $[Cd(OMe)\{C(SiMe_2Ph)_3\}]$.

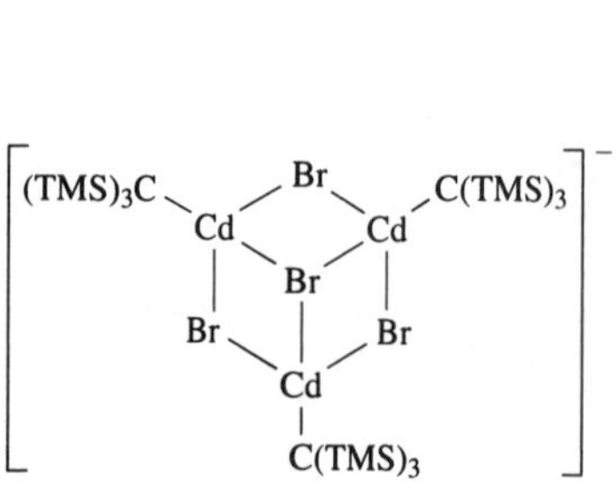

(80) 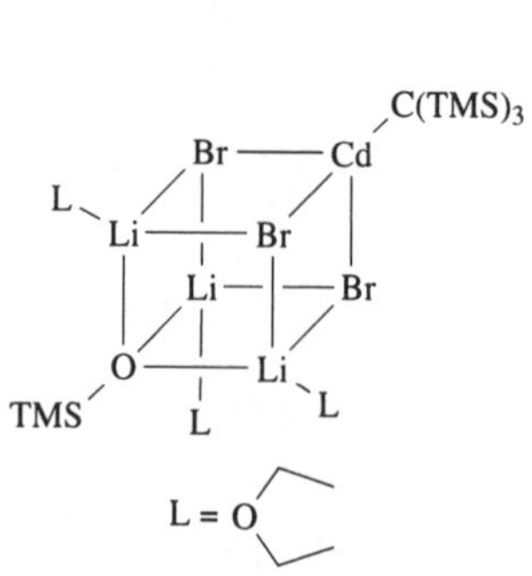   **(81)**

The electrochemical oxidation of cadmium in MeCN solution has been used to synthesize bromo(2-cyanophenyl)cadmium, which has been isolated as a complex with 2,2'-bipyridyl (**82**).[176] The hydroxy(pentafluorophenyl)cadmium (**83**) is a tetramer[177] in the solid state. The dimer formed, by Cp*CdN(TMS)$_2$ has been prepared.[169] Several diseleno and dithiocarbamate complexes of methylcadmium have been reported; the crystal structure of the neopentyl compound shows the expected dimers[178] (see also Section 4.2.11). The three-coordinate diorganophosphide of cadmium [{MeCd(μBu$^t_2$P)}$_3$], the first cadmium diorganophosphide, has been reported.[179] The reaction between [Ga{(Bu$^t_2$P)$_3$}] and Me$_2$M (M = Zn, Cd, Hg) gives trimeric phosphido complexes. For cadmium, the structure has been determined by crystallography as shown in (**84**); it contains planar three-coordinate cadmium(II) atoms in a six-membered Cd$_3$P$_3$ ring.

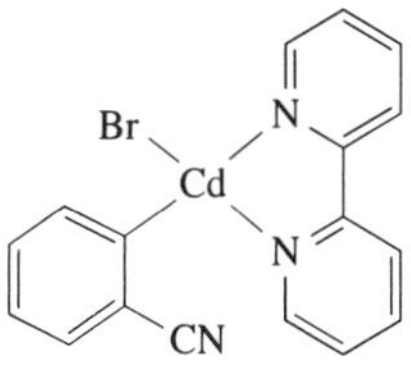

(82)

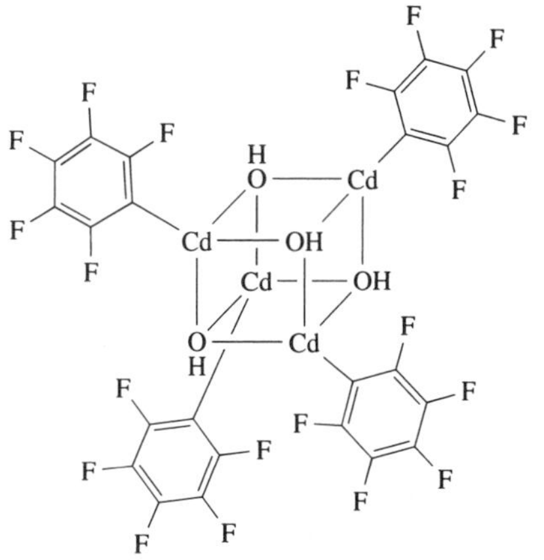

(83)

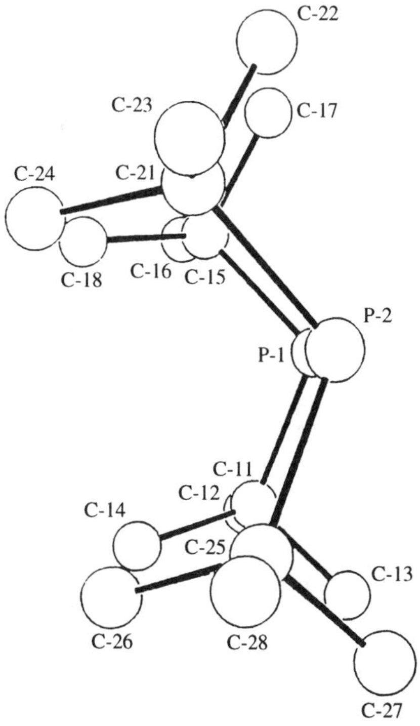

(84)

4.3.4 Theoretical and Physical Studies of Organocadmium Species

The photochemistry of dimethylcadmium on quartz and silicon Si(111) surfaces has been studied[180] at 193 nm and 248 nm using a rare-gas fluoride excimer laser. The desorbed gas products detected included: CH_2, Me, CH_4, C_2H_4, Et, C_2H_6, Cd, MeCd, and Me_2Cd.

The dynamics of adsorption of laser-induced resonance photoprocesses in physisorbed dimethylcadmium at 150 K have been studied. The desorption of dimethylcadmium, MeCd, cadmium and Me fragments in both neutral and ionic forms was followed.[181] Similarly, the UV laser chemistry of dimethylcadmium either chemisorbed at 297 K on *n*-type Si(100) with native oxide or physisorbed at 150 K on a photodeposited cadmium film was studied by using mass spectrometry of desorbed species.[182] A XeCl laser induced heterogeneous fragmentation of the sorbed molecules was observed.

$[Cd(SiF_3)_2]$ and $[Cd(CF_3)_2]$ were prepared by cocondensation of cadmium metal vapour and the organic radical.[183] Fluorine-19 NMR spectroscopy showed that $[Cd(SiF_3)_2]$ slowly decomposes at room temperature but $[Cd(CF_3)_2]$ decomposes at 5 °C.

4.4 THE USE OF METAL ALKYLS IN THE ELECTRONICS INDUSTRY

There is considerable current interest in the use of cadmium and zinc alkyls for the deposition of chalcogenide-containing materials such as zinc selenide or cadmium sulfide by techniques such as MOCVD.[184-6] In this method volatile precursors for the deposition of the required material are passed over a heated substrate, where pyrolysis leads to the deposition of material.

4.4.1 Simple Alkyls

Deposition studies are dominated by the use of the methyls of cadmium or zinc, or an adduct of these; reviews of the use of these compounds are available.[186-8] Only relatively rarely are compounds other than dimethylzinc or dimethylcadmium used, but there can be distinct advantages in the use of a different alkyl. Jones, in collaboration with Thompson and Smith, has reported[189] the use of diethylcadmium for the deposition of cadmum mercury telluride. An advantage of this alkyl is a lower decomposition temperature, better matched to some of the more recently developed tellurium precursors. However, the higher alkyls of cadmium are particularly prone to photochemical decomposition and are hazardous materials, even by the standards of metal alkyls. This is probably due to particularly stable radical species, generated by β-elimination reactions, during the decomposition of such alkyls. Bis(neopentyl)cadmium(II) has successfully been used in combination with H_2S to grow films of CdS. The low growth rate obtained with the source at room temperature suggests that the alkyl is unsuitable for atmospheric-pressure MOCVD. Bis(neopentyl)cadmium may also have an application as a *p*-type dopant for InP and related III–VI alloys,[170] an alternative to controlling the vapour pressure of dimethyl-zinc by the use of an adduct.

4.4.2 Adducts as Precursors

Adducts of the alkyls of zinc and cadmium find application as precursors for the growth of wide-bandgap II–VI semiconductors by MOCVD. The dioxane,[190-2] thioxane,[192] triethylamine[193-9] and triazine[193] adducts of dimethylzinc successfully inhibit homogeneous reaction of chalcogenide and metal alkyl (prereaction) in the growth of ZnSe and related alloys by MOCVD. Adducts of dimethylcadmium have been used to grow CdSe and CdS.[200,201] The use of such adducts has several potential advantages:
  (i)   the vapour pressure of the metal alkyl is effectively reduced, eliminating the necessity to cool bubblers containing dimethylzinc;
  (ii) reaction before the precursors reach the hot zone of the reactor, so-called 'homogeneous pre-reaction', may be considerably limited; and
  (iii) layers may have improved electrical properties due to the purification of the alkyl during the preparation of the adduct.

Despite the increasing use of such compounds in MOCVD, little is known about the nature of the chemical species present in the gas phase in the MOCVD reactor.

Recent infrared studies indicate that many of the adducts of dimethylcadmium and dimethylzinc are close to fully dissociated in the vapour phase.[202] Totally dissociative vapourization is common for organometallic species; one of the simplest examples is trimethylindium, which is a tetramer in the solid state[203] and a monomer in the vapour phase.[204] The adduct of dimethylcadmium with the chelating ligand tetramethylethylenediamine shows some evidence of association at room temperature; under MOCVD conditions any association is likely to be extremely limited.[171]

A large number of adducts have been studied as potential precursors. By far the most useful seems to be the triethylamine adduct of dimethylzinc, a compound first prepared by Thiele in the 1960s. Several groups have now shown that this compound is useful both in the deposition of chalcogenides[193-8] and in the *p*-type doping of III–V materials with zinc.[205,206] One useful feature of this compound is that the 1:1 species is a eutectic mixture (Figure 1) and the stoichiometry of the precursor is consequently self-limiting.

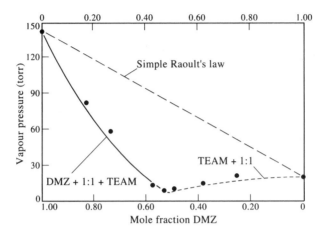

Figure 1 The variation of vapour pressure with composition for the system triethylamine:dimethylzinc at 0 °C. DMZ, dimethylzinc; TEAM, triethylamine.

4.4.3 Nature of the Effect of the Adducts on Growth

Crystal growth results indicate that there is a marked difference in the extent to which each adduct inhibits homogeneous 'prereaction' with the group 6 hydride during growth, a surprising observation in view of the similar, and largely dissociated, nature of the adducts in the vapour phase. The effectiveness of the adducts in inhibiting homogeneous reactions in MOCVD reactors had been tacitly interpreted (see, for example, Ref. 189) in terms of the 'blocking' of reaction by adduct formation in the vapour phase, the implication being that homogeneous reaction proceeds via an initial step involving the formation of an addition compound (adduct) between the group 2 source and the metal alkyl (a 'preequilibrium' process (Equation (17)), in chemical terminology an associative reaction), which is inhibited by excess of a 'stronger' Lewis base (Equation (18)). The kind of scheme envisaged is shown in Equations (17)–(19). Adducts with strong Lewis bases and group 2 alkyls are effectively totally dissociated in the vapour phase. Results suggest that the inhibition of homogeneous reaction ('prereaction') must occur by some other mechanism.

$$[ZnMe_2] + H_2X \longrightarrow [ZnMe_2XH_2] \qquad (17)$$

$$[ZnMe_2XH_2] \longrightarrow ZnX + \text{volatile products (CH}_4 \text{ etc.)} \qquad (18)$$

X is a chalcogen

$$[ZnMe_2] + L \longrightarrow [ZnMe_2L] \qquad (19)$$

L is a Lewis base

The extreme alternative mechanism to an associative process, as described above, is a dissociative mechanism. Free radicals (Me·, etc.) are now well-established components of the reactive vapour phase in MOCVD reactors containing group 3 metal-containing precursors.[207] An alternative mechanism for homogeneous reaction could invoke metal–carbon bond homolysis (Equation (20)). The highly reactive intermediates thus formed could react with other molecules in the vapour phase. In the absence of a better electron donor the metal intermediates are likely to react with the chalcogen-containing precursor, leading to homogeneous reaction and 'snow'. However, in the case of adduct precursors the vapour will contain a considerable (stoichiometric or greater) quantity of the Lewis base. The bases are much better ligands than the chalcogen precursors and are likely to act as 'traps' for the reactive-metal-containing intermediates. Thus, the reactive-metal-containing intermediates could pass through the hot zone of the reactor without reaction proceeding far enough for the nucleation and growth of particulate material.

$$\text{Me–M–Me} \longrightarrow \text{Me}\bullet + \text{M–Me} \qquad (20)$$

It is interesting to consider the recent conclusions of Jackson concerning the homogeneous decomposition of the group 12 metal alkyls.[208] At the temperatures typically used for the preparation of II–VI materials, 400–500 °C, simple bond homolysis does not seem likely. However, radical and/or Lewis acid species could be generated during decomposition of the initial complex formed on the association of the alkyl and the chalcogen.

One area of chemistry of direct relevance to such work concerns the formation of discrete complexes by the reaction of thiols with metal alkyls such as dimethylzinc. Coates[209] reported the formation of a pentamer[56] [{Zn(SBu^t)Me}$_5$] from the reaction of dimethylzinc with *t*-butyl thiol, and that the compound reacted with amines, such as pyridine, to form stable adducts. A low-resolution x-ray single-crystal structure of the pentamer has been reported,[56] but no structural information was available for the adducts. In view of the fact that layers of zinc sulfide have been grown from dimethylzinc, or its adducts, in combination with *t*-butyl thiol, the adducts of the pentamer [{ZnSBu^{t}Me}$_5$] with a number of nitrogenous bases have been studied. Stable dimeric complexes are formed[210] by the reaction of the pentamer with pyridine or trimethyltriazine. The dimers formed have similar structures, in which the zinc is four-coordinate with the methyl groups, amines are monodentate and the thiolates form monatomic bridges (**85**). The coordination of the triazine is similar to that in the 2:1 adduct formed with dimethylzinc in that the amine behaves as a simple two-electron donor.

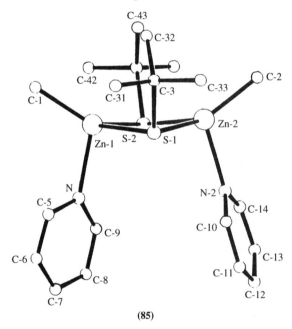

(**85**)

In summary, there are a number of ways in which adducts could be operating in controlling unwanted reactions.

(i)   Surface, possibly acid catalysed, reactions in the cool zone of the reactor are inhibited.

(ii) Unwanted reactions are often less apparent with high purity reagents, so the improved purity of adducted alkyls may be important.

(iii) With highly volatile precursors such as dimethylzinc, the adducts provide for better control of mass flow. Under equivalent flow conditions a lower partial pressure of the alkyl in the MOCVD reactor will be supported by an adduct compared to the pure alkyl. Homogeneous reactions are likely to proceed at rates proportional to the partial pressure of alkyls; a lower partial pressure will lead to a lower rate of homogeneous reaction.

(iv) The Lewis bases from which adducts are formed may play a role in trapping highly reactive intermediates generated during the decomposition of alkyls.

In considering factors (iii) and (iv) it is important to note that adducts do not need to totally stop the formation of precipitates from the vapour phase but only to sufficiently inhibit their formation so that they pass through the hot zone of the reactor before forming particulate material. As such reactions probably proceed by a process of nucleation and growth, inhibition of either of these steps could stop significant homogeneous reaction occurring; the growing nuclei of metal chalcogen material are likely to contain highly reactive (Lewis acidic) sites, which could be blocked by the Lewis base.

Much needs to be learned about the properties of the adducts of group 2 metals and the way they affect the quality of grown layers. More information concerning vapour pressures, properties in the vapour phase and, above all, the mechanisms of homogeneous decomposition is needed.

4.4.4 The Synthesis of Highly Dispersed Materials

During the past few years there has been considerable interest in the synthesis and characterization of semiconductor nanocrystallites.[211,212] Nanocrystallites, also known as Q-particles or quantum dots, are nanometric particles with a high surface/volume ratio and diameters of up to 10–20 nm; their optoelectronic properties are different from the bulk counterparts and new technological applications have been proposed for this type of material. Chalcogenides of the group 12 elements are a particularly popular class of compounds prepared in nanodispersed form. Many of the routes used to prepare such materials involve the controlled reaction of a metal alkyl with a source for the chalcogen, either in the vapour phase or in a high boiling point solvent. One particularly well-developed synthesis of CdSe uses the reaction of the trioctylphosphineselenide with dimethylcadmium in tri-octyl(n)phosphine oxide.[213,214] Highly monodispersed materials in the 1.2–11.5 nm range have been obtained. In related work well-defined whiskers of ZnS have been synthesized[215] by the reaction of [{Zn(SBu^t)Et}$_5$] with H$_2$S in CH$_3$CHCl$_2$.

ACKNOWLEDGEMENTS

It is a great pleasure to acknowledge the help of my group in writing this chapter, and in my work in general; especial thanks are due to Drs Azad Malik and John Walsh and Mr Tito Trindade. I would like to thank Drs Peter Wyatt and Andrew Robinson (Queen Mary and Westfield College) for a useful report on the organic chemistry of zinc alkyls and for several diagrams, Ms L. Humphries who helped with literature searches, Mrs Bernice Elliot Smith and Sylvia Reading (Queen Mary and Westfield College) who helped with the rapid production of this chapter, and finally, Kym for her understanding when I'm under pressure.

4.5 REFERENCES

1. J. Boersma, in 'COMC-I', vol. 2, p. 823.
2. E. Frankland, *Leibigs Ann. Chem.*, 1848, **71**, 171.
3. A. L. Galyer and G. Wilkinson, *Inorg. Synth.*, 1979, **19**, 253.
4. M. A. Guerra, T. R. Bierschenk and R. J. Lagow, *J. Am. Chem. Soc.*, 1986, **108**, 4103.
5. H. Lange and D. Naumann, *J. Fluorine Chem.*, 1984, **26**, 435.
6. F. Langer, J. Waas and P. Knochel, *Tetrahedron Lett.*, 1993, **34**, 5261.
7. R. Mynott, B. Gabor, H. Lehmkuhl and I. Döring, *Angew. Chem., Int. Ed. Engl.*, 1985, **24**, 335.
8. H. Nehl and W. R. Scheidt, *J. Organomet. Chem.*, 1985, **289**, 1.
9. A. J. de Koning, P. E. van Rijn, J. Boersma and G. J. M. van der Kerk, *J. Organomet. Chem.*, 1978, **153**, C37.
10. H. Lehmkuhl, I. Döring, R. McLane and H. Nehl, *J. Organomet. Chem.*, 1981, **221**, 1.
11. A. Haaland, H. Lehmkuhl and H. Nehl, *Acta Chem Scand., Ser. A.*, 1984, **38**, 547.
12. R. D. Ernst, J. W. Freeman, P. N. Swepston and D. R. Wilson, *J. Organomet. Chem.*, 1991, **402**, 17.
13. J. Dekker, J. Boersma, L. Fernholt, A. Haaland and A. L. Spek, *Organometallics*, 1987, **6**, 1202.
14. J. Dekker, J. W. Münninghoff, J. Boersma and A. L. Spek, *Organometallics*, 1987, **6**, 1236.
15. J. L. Atwood, D. E. Berry, S. R. Stobart and M. J. Zaworotzo, *Inorg. Chem.*, 1983, **22**, 3480.
16. M. Henderson, R. I. Papasergio, C. L. Raston, A. H. White and M. F. Lappert, *J. Chem. Soc., Chem. Commun.*, 1986, 672.
17. P. O'Brien, M. B. Hursthouse, M. Motevalli, J. R. Walsh, and A. C. Jones, *J. Organomet. Chem.*, 1993, **449**, 1.
18. M. Motevalli, P. O'Brien, A. J. Robinson, J. R. Walsh, P. B. Wyatt and A. C. Jones, *J. Organomet. Chem.*, 1993, **461**, 5.
19. (a) M. B. Hursthouse, M. Motevalli, P. O'Brien, J. R. Walsh and A. C. Jones, *J. Mater. Chem.*, 1991, **1**, 139; (b) M. B. Hursthouse, M. Motevalli, P. O'Brien, J. R. Walsh and A. C. Jones, *Organometallics*, 1991, **10**, 3196.
20. V. Weissig, R. Beckhaus, U. Banasiak and K.-H. Thiele, *Z. Anorg. Allg. Chem.*, 1980, **467**, 61.
21. J. Behm, S. D. Lotz and W. A. Herrmann, *Z. Anorg. Allg. Chem.*, 1993, **619**, 849.
22. E. K. S. Liu, *Inorg. Chem.*, 1980, **19**, 266.
23. A. D. Pajerski, G. L. BergStresser, M. Parvez and H. G. Richey, Jr., *J. Am. Chem. Soc.*, 1988, **110**, 4844.

24. M. Westerhausen, B. Rademacher and W. Schwarz, *J. Organomet. Chem.*, 1992, **427**, 275.
25. H. Yasuda, Y. Ohnuma, A. Nakamura, Y. Kai, N. Yasuoka and N. Kasai, *Bull. Chem. Soc. Jpn.*, 1980, **53**, 1101.
26. D. V. Shenai-Khatkhate, E. D. Orrell, J. B. Mullin, D. C. Cupertino and D. J. Cole-Hamilton, *J. Cryst. Growth*, 1986, **77**, 27.
27. Several patents have been filed, e.g. J. B. Mullin, D. J. Cole-Hamilton, E. D. Orrell, D. V. Shenai-Khatkhate and P. R. Jacobs, PCT, *Int. Appl.* WO 8606071 A1 861023; PRIORITY: GB 85-9055 850409 (*Chem. Abstr.*, **106**, 84850; see also ibid., **106**, 84850, and **99**, 212714).
28. P. O'Brien, J. R. Walsh, A. C. Jones, S. A. Rushworth and C. Meaton, *J. Mater. Chem.*, 1993, **3**, 739.
29. (a) P. R. Markies, G. Schat, O. S. Akkerman, F. Bickelhaupt, W. J. J. Smeets and A. L. Spek, *Organometallics*, 1991, **10**, 3538; (b) P. R. Markies, G. Schat, O. S. Akkerman, F. Bickelhaupt and A. L. Spek, *J. Organomet. Chem.*, 1992, **430**, 1; *Organometallics*, 1992, **11**, 1428.
30. H. Bai and B. S. Ault, *J. Phys. Chem.*, 1994, **98**, 6082.
31. A. P. Purdy and C. F. George, *Organometallics*, 1992, **11**, 1955.
32. M. Westerhausen, B. Rademacher, W. Schwarz, J. Weidlein and S. Henkel, *J. Organomet. Chem.*, 1994, **469**, 135.
33. M. Westerhausen and B. Rademacher, *Spec. Publ.-R. Soc. Chem.*, 131 (Chemistry of the Copper and Zinc Triads), 1993, **148**.
34. M. Westerhausen, B. Rademacher and W. Schwarz, *Z. Anorg. Allg. Chem.*, 1993, **619**, 675.
35. R. M. Fabicon, A. D. Pajerski and H. G. Richey, Jr., *J. Am. Chem. Soc.*, 1991, **113**, 6680.
36. C. Eaborn, N. Retta and J. D. Smith *J. Organomet. Chem.*, 1980, **190**, 101.
37. M. Westerhausen, B. Rademacher and W. Poll, *J. Organomet. Chem.*, 1991, **421**, 175.
38. M. Westerhausen, B. Rademacher, W. Schwarz, and S. Henkelonja, *Z. Naturforsch., Teil B*, 1994, **49**, 199.
39. S. S. Al-Juaid *et al.*, *J. Organomet. Chem.*, 1993, **462**, 45.
40. S. S. Al-Juaid, C. Eaborn, A. Habtemariam, P. B. Hitchcock and J. D. Smith, *J. Organomet. Chem.*, 1992, **437**, 41.
41. F. I. Aigbirhio, S. S. Al-Juaid, C. Eaborn, A. Habtemariam, P. B. Hitchcock and J. D. Smith, *J. Organomet. Chem.*, 1991, **405**, 149.
42. S.-C. Chang, R. H. Hauge, Z. H. Kafafi, J. L. Margrave and W. E. Billups, *J. Chem. Soc., Chem. Commun.*, 1987, 1682.
43. E. Okninska and K. B. Starowieyski, *J. Organomet. Chem.*, 1988, **347**, 1.
44. E. Okninska and K. B. Starowieyski, *J. Organomet. Chem.*, 1989, **376**, 7.
45. A. J. de Koning, P. E. van Rijn, J. Boersma and G. J. M. van der Kerk, *J. Organomet. Chem.*, 1979, **174**, 129.
46. Y. Yamamoto, *Bull. Chem. Soc. Jpn.*, 1984, **57**, 2835.
47. A. J. Arduengo III, H. V. R. Dias, F. Davidson and R. L. Harlow, *J. Organomet. Chem.*, 1993, **462**, 13.
48. P. H. M. Budzelaar, J. Boersma, G. J. M. van der Kerk, A. L. Spek and A. J. M. Duisenberg, *J. Organomet. Chem.*, 1985, **281**, 123.
49. B. Fischer *et al.*, *J. Organomet. Chem.*, 1989, **376**, 223; R. Blom *et al.*, *Acta Chem. Scand., Ser. A*, 1986, **40**, 113.
50. P. H. M. Budzelaar, J. Boersma, G. J. M. van der Kerk and A. L. Spek, *Organometallics*, 1984, **3**, 1187.
51. J. Dekker *et al.*, *J. Organomet. Chem.*, 1987, **320**, 1.
52. J. T. B. H. Jastrzebski, J. Boersma, G. van Koten, W. J. J. Smeets and A. L. Spek, *Recl. Trav. Chim. Pays-Bas*, 1988, **107**, 263.
53. F. H. van der Steen, J. Boersma, A. L. Spek and G. van Koten, *Organometallics*, 1991, **10**, 2467.
54. F. H. van der Steen, J. Boersma, A. L. Spek and G. van Koten *J. Organomet. Chem.*, 1990, **390**, C21.
55. M. M. Hansen, P. A. Bartlett and C. H. Heathcock, *Organometallics*, 1987, **6**, 2069; 2074.
56. P. H. M. Budzelaar, J. Boersma, G. J. M. van der Kerk and A. L. Spek, *Organometallics*, 1984, **3**, 1187.
57. N. A. Bell, H. M. M. Shearer and C. B. Spencer, *Acta Crystallogr. Sect. C*, 1983, **39**, 1182.
58. M. R. P. van Vliet, *et al.*, *J. Organomet. Chem.*, 1987, **319**, 285.
59. P. H. M. Budzelaar, J. Boersma, G. J. M. van der Kerk and A. L. Spek, *Organometallics*, 1984, 1187.
60. L. M. Engelhardt, G. E. Jacobsen, W. C. Patalinghug, B. W. Skelton, C. L. Raston and A. H. White, *J. Chem. Soc., Dalton Trans.*, 1991, 2859.
61. I. B. Gorrell, A. Looney and G. Parkin, *J. Chem. Soc., Chem. Commun.*, 1990, 220.
62. I. B. Gorrell, A. Looney and G. Parkin, *J. Am. Chem. Soc.*, 1990, **112**, 4068.
63. L. N. Markovskii, V. D. Romanenko, V. F. Shul'gin, A. N. Chemega, M. Yu Antipin and Yu. T. Struchkov, *J. Gen. Chem. USSR (Engl. Transl.)*, 1987, **57**, 1993.
64. J. J. Habeeb, A. Osman and D. G. Tuck, *J. Organomet. Chem.*, 1980, **185**, 117.
65. S. C. Berk, M. C. P. Yeh, N. Jeong and P. Knochel, *Organometallics*, 1990, **9**, 3053.
66. K.-H. Thiele, E. Langguth and G. E. Müller, *Z. Anorg. Allg. Chem.*, 1980, **462**, 152.
67. F. H. van der Steen, J. Boersma, A. L. Spek and G. van Koten, *Organometallics*, 1991, **10**, 2467.
68. D. Naumann, W. Tyrra, B. Kock, W. Rudolph and B. Wilkes, *J. Fluorine Chem.*, 1994, **67**, 91.
69. S. E. Denmark, J. P. Edwards and S. R. Wilson, *J. Am. Chem. Soc.*, 1992, **114**, 2592.
70. W. Kuran and M. Czernecka, *J. Organomet. Chem.*, 1984, **263**, 1.
71. S. S. Al-Juaid *et al.*, *J. Chem. Soc., Chem. Commun.*, 1986, 908.
72. P. Górecki and W. Kuran, *J. Organomet. Chem.*, 1984, **265**, 1.
73. P. Górecki and W. Kuran, *J. Organomet. Chem.*, 1986, **312**, 1.
74. W. Kuran and E. Mazanek, *Main Group Met. Chem.*, 1989, 12241.
75. W. A. Herrmann, S. Bogdanović, J. Behm and M. Denk, *J. Organomet. Chem.*, 1992, **430**, C33.
76. T. R. Shnol, V. N. Pankratova, Yu. N. Krasnov and R. F. Galiullina, *Khim. Elementoorg. Soedin.*, 1980, **8**, 55 (*Chem. Abstr.*, **96**, 21 7969).
77. H. J. Gais, G. Bulow and G. Raabe, *J. Am. Chem. Soc.*, 1993, **115**, 7215.
78. M. R. P. van Vliet, G. van Koten, P. Buysingh, J. T. B. H. Jastrzebski and A. L. Spek, *Organometallics*, 1987, **6**, 537.
79. C. Bolm, G. Schlingloff and K. Harms, *Chem. Ber.*, 1992, **125**, 1191.
80. M. M. Olmstead, P. P. Power and S. C. Shoner, *J. Am. Chem. Soc.*, 1991, **113**, 3379.
81. M. B. Hursthouse, M. A. Malik, M. Motevalli, and P. O'Brien, *J. Chem. Soc., Chem. Commun.*, 1991, 1690.
82. I. Abrahams, M. A. Malik, M. Motevalli and P. O'Brien, *J. Chem. Soc., Dalton Trans.*, 1995, 1043.
83. R. M. Fabicon, M. Parvez and H. G. Richey, Jr., *J. Am. Chem. Soc.*, 1991, **113**, 1412.
84. G. W. Adamson, N. A. Bell and H. M. M. Shearer, *Acta Crystallogr., Sect. B*, 1982, **38**, 462.

85. K. Ruhlandt-Senge and P. P. Power, *Inorg. Chem.*, 1993, **32**, 4505.
86. S. A. Rao, T.-S. Chou, I. Schipor and P. Knochel, *Tetrahedron*, 1992, **48**, 2025.
87. D. M. Knotter *et al.*, *Inorg. Chem.*, 1991, **30**, 4361.
88. J. G. Noltes, *Recl. Trav. Chim. Pays-Bas*, 1965, **84**, 126.
89. M. B. Hursthouse, M. A. Malik, M. Motevalli and P. O'Brien, *Organometallics*, 1991, **10**, 730.
90. M. A. Malik and P. O'Brien, *Mater. Chem.*, 1991, **3**, 999.
91. M. B. Hursthouse, M. A. Malik, M. Motevalli and P. O'Brien, *J. Mater. Chem.*, 1992, **2**, 949.
92. M. A. Malik, M. Motevalli, J. R. Walsh and P. O'Brien, *Organometallics*, 1992, **11**, 3136.
93. M. Bochmann, A. P. Coleman and A. K. Powell, *Polyhedron*, 1992, **11**, 507.
94. W. Cen, K. J. Haller and T. P. Fehlner, *Inorg. Chem.*, 1993, **32**, 995.
95. W. Cen, K. J. Haller and T. P. Fehlner, *Inorg. Chem.*, 1991, **30**, 3120.
96. P. H. M. Budzelaar *et al.*, *J. Organomet. Chem.*, 1983, **243**, 137.
97. W. H. Dell and M. L. Ziegler, *Z. Naturforsch., Teil B*, 1982, **37**, 1.
98. J. T. B. H. Jastrzebski *et al.*, *J. Organomet. Chem.*, 1990, **396**, 25.
99. S. J. Dzugan and V. L. Goedken, *J. Organomet. Chem.*, 1988, **356**, 249.
100. M. J. S. Dewar and K. M. Merz, Jr., *Organometallics*, 1988, **7**, 522.
101. M. Kaupp *et al.*, *J. Am. Chem. Soc.*, 1991, **113**, 5606.
102. W. Kuhn, P. Tommack, H. Leiderer, W. Gebhardt and W. Richter, *Chemtronics*, 1989, **4**, 224.
103. C. F. Yu, F. Youngs, K. Tsukiyama, R. Bersohn and J. Preses, *J. Chem. Phys.*, 1986, **85**, 1382.
104. Y. Fujita, S. Fujii and T. Iuchi, *J. Vac. Sci. Technol. A*, 1989, **7**, 276.
105. R. L. Jackson, *J. Chem. Phys.*, 1992, **96**, 5938.
106. M. Kawasaki and N. Nishi, *Appl. Organomet. Chem.*, 1991, **5**, 247.
107. S. Ishizaka, J. Simpson and J. O. Williams, *Chemtronics*, 1986, **1**, 175.
108. J. I. Davies, M. J. Parrott and J. O. Williams, *J. Cryst. Growth*, 1986, **79**, 363.
109. W. Oppolzer, *Angew. Chem., Int. Ed. Engl.*, 1989, **28**, 38.
110. K. Soai and S. Niwa, *Chem. Rev.*, 1992, **92**, 833.
111. A. Fürstner, *Synthesis*, 1989, 571.
112. N. Oguni and T. Omi, *Tetrahedron Lett.*, 1984, **25**, 2823.
113. R. Noyori *et al.*, *J. Organomet. Chem.*, 1990, **382**, 19.
114. M. Kitamura, S. Suga, K. Kawai and R. Noyori, *J. Am. Chem. Soc.*, 1986, **108**, 6071.
115. K. Soai, S. Yokoyama and T. Hayasaka, *J. Org. Chem.*, 1991, **56**, 4264.
116. K. Soai, S. Niwa and M. Watanabe, *J. Org. Chem.*, 1988, **53**, 927.
117. K. Soai, S. Niwa and M. Watanabe, *J. Chem. Soc., Perkin Trans. 1*, 1989, 109.
118. K. Soai, M. Watanabe and A. Yamamoto, *J. Org. Chem.*, 1990, **55**, 4832.
119. S. Itsuno and J. M. J. Fréchet, *J. Org. Chem.*, 1987, **52**, 4140.
120. K. Soai and M. Watanabe, *Tetrahedron Asymm.*, 1991, **2**, 97.
121. W. Oppolzer and R. N. Radinov, *Tetrahedron Lett.*, 1988, **29**, 5645.
122. S. Niwa and K. Soai, *J. Chem. Soc., Perkin Trans. 1*, 1990, 937.
123. K. Soai and S. Niwa, *Chem. Lett.*, 1989, 481.
124. K. Soai, S. Niwa, Y. Yamada and H. Inoue, *Tetrahedron Lett.*, 1987, **28**, 4841.
125. B. Schmidt and D. Seebach, *Angew. Chem., Int. Ed. Engl.*, 1991, **30**, 99.
126. B. Schmidt and D. Seebach, *Angew. Chem., Int. Ed. Engl.*, 1991, **30**, 1321.
127. N. N. Joshi, M. Srebnik and H. C. Brown, *Tetrahedron Lett.*, 1989, **30**, 5551.
128. M. Watanabe, S. Araki, Y. Butsugan and M. Uemura, *J. Org. Chem.*, 1991, **56**, 2218.
129. M. Uemura, R. Miyake, M. Shirot and Y. Hayashi, *Tetrahedron Lett.*, 1991, **32**, 4569.
130. N. Noyori, S. Suga, K. Kawai, S. Okada and M. Kitamura, *Pure Appl. Chem.*, 1988, **60**, 1597.
131. R. Noyori and M. Kitamura, *Angew. Chem., Int. Ed. Engl.*, 1991, **30**, 49.
132. F. A. Cotton and G. Wilkinson, 'Advanced Inorganic Chemistry', 5th edn., Wiley, New York, 1988, chap. 16, p. 616.
133. W. Oppolzer and R. N. Radinov, *Tetrahedron Lett.*, 1988, **29**, 5645.
134. K. Soai, A. Ookawa, T. Kaba and K. Ogawa, *J. Am. Chem. Soc.*, 1987, **109**, 7111.
135. E. J. Corey and F. J. Hannon, *Tetrahedron Lett.*, 1987, **28**, 5233.
136. E. J. Corey and F. J. Hannon, *Tetrahedron Lett.*, 1987, **28**, 5237.
137. R. Noyori and M. Kitamura, *Angew. Chem., Int. Ed. Engl.*, 1991, **30**, 49.
138. N. Oguni, Y. Matsuda and T. Kaneko, *J. Am. Chem. Soc.*, 1988, **110**, 7877.
139. D. Guillaneux, S.-H. Zhao, O. Samuel, D. Rainford and H. B. Kagan, *J. Am. Chem. Soc.*, 1994, **116**, 9430.
140. J. W. Faller, and J. Parr, *J. Am. Chem. Soc.*, 1993, **115**, 804.
141. M. Hayashi, T. Matsuda and N. Oguni, *J. Chem. Soc., Perkin Trans. 1*, 1992, 3135.
142. M. Kitamura, S. Okada, S. Suga and R. Noyori, *J. Am. Chem. Soc.*, 1989, **111**, 4028.
143. J. Furukawa, N. Kawabata and J. Nishimura, *Tetrahedron Lett.*, 1966, 3353.
144. G. Wittig and K. Schwarzenbach, *Angew. Chem.*, 1959, **71**, 652.
145. S. Sawada, K. Takehana and Y. Inouye, *J. Org. Chem.*, 1968, **33**, 1767.
146. T. Sugimura, T. Futagawa and A. Tai, *Tetrahedron Lett.*, 1988, **29**, 5775.
147. Y. Ukaji, M. Nishimura and T. Fujisawa, *Chem. Lett.*, 1992, 61.
148. H. Takahashi, M. Yoshioka, M. Ohno and S. Kobayashi, *Tetrahedron Lett.*, 1992, **33**, 2575.
149. W. R. Vaughan, S. C. Bernstein and M. E. Lorber, *J. Org. Chem.*, 1965, **30**, 1790.
150. F. Orsini, F. Pelizzoni and G. Ricca, *Tetrahedron Lett.*, 1982, **23**, 3945.
151. J. Dekker, P. H. M. Budzelaar, J. Boersma, G. J. M. van der Kerk and A. L. Spek, *Organometallics*, 1984, **3**, 1403.
152. M. Guette, J. Capillon and J. P. Guette, *Tetrahedron*, 1992, **33**, 2895.
153. K. Soai and Y. Kawase, *Tetrahedron: Asymmetry*, 1991, **2**, 781.
154. T. J. Juhlke, R. W. Braun, T. R. Bierschenk and R. J. Lagow, *J. Am. Chem. Soc.*, 1979, **101**, 3229.
155. M. J. Almond, M. P. Beer and D. A. Rice, *Chemtronics*, 1991, **5**, 29.
156. M. J. Almond, M. P. Beer, P. Heath, C. A. Heyburn, D. A. Rice and L. A. Sheridan, *J. Organomet. Chem.*, 1994, **469**, 11.
157. D. M. Heinekey and S. R. Stobart, *Inorg. Chem.*, 1978, **17**, 1463.

158. S. Al-Hashimi and J. D. Smith, *J. Organomet. Chem.*, 1978, **153**, 253.
159. G. A. Ayako, N. H. Buttrus, C. Eaborn, P. B. Hitchcock and J. D. Smith, *J. Organomet. Chem.*, 1987, **320**, 137.
160. H. Lange and D. Naumann, *J. Fluorine Chem.*, 1984, **26**, 1.
161. D. J. Burton and D. M. Wiemers, *J. Am. Chem. Soc.*, 1985, **107**, 5014.
162. D. Naumann, K. Glinka and W. Tyrra, *Z. Anorg. Allg. Chem.*, 1991, **594**, 95.
163. K. H. Nair and D. J. Burton, *J. Fluorine Chem.*, 1992, **56**, 341.
164. J. L. Atwood, D. E. Berry, S. R. Stobart and M. J. Zaworotko, *Inorg. Chem.*, 1983, **22**, 3480.
165. O. F. Z. Khan, D. M. Frigo, P. O'Brien, A. Howes and M. B. Hursthouse, *J. Organomet. Chem.*, 1987, **334**, C27.
166. M. J. Henderson, R. I. Papasergio, C. L. Raston, A. H. White and M. F. Lappert, *J. Chem. Soc., Chem. Commun.*, 1986, 677.
167. W. J. J. Smeets, A. L. Spek, B. Fischer, G. P. M. van Mier and J. Boersma, *Acta Crystallogr., Sect. C*, 1987, **43**, 893.
168. B. Fischer, G. P. M. van Mier, J. Boersma, W. J. J. Smeets and A. L. Spek, *J. Organomet. Chem.*, 1987, **322**, C37.
169. C. C. Cummins, R. R. Schrock and W. M. Davis, *Organometallics*, 1991, **10**, 3781.
170. P. O'Brien, J. R. Walsh, A. C. Jones and S. A. Rushworth, *Polyhedron*, 1990, **9**, 1483 (*Chem. Abstr.*, **114**, 102 398).
171. M. J. Almond, M. P. Beer, K. Hagen, D. A. Rice and P. J. Wright, *J. Mater. Chem.*, 1991, **1**, 1065.
172. M. J. Almond, M. P. Beer, M. G. B. Drew and D. A. Rice, *Organometallics*, 1991, **10**, 2072.
173. M. J. Almond, M. P. Beer, M. G. B. Drew and D. A. Rice, *J. Organomet. Chem.*, 1991, **421**, 129.
174. P. R. Jacobs, E. D. Orrell, J. B. Mullin and D. J. Cole-Hamilton, *Chemtronics*, 1986, 1 (for a related Patent see *Chem. Abstr.*, **199**, 212 714).
175. N. H. Buttrus *et al.*, *J. Chem. Soc., Dalton Trans.* 1988, 381.
176. F. F. Said *et al.*, *J. Organomet. Chem.*, 1982, **224**, 121.
177. M. Weidenbruch *et al.*, *J. Organomet. Chem.*, 1989, **361**, 139.
178. I. Abrahams, M. A. Malik, M. Motevalli and P. O'Brien, *J. Organomet. Chem.*, 1994, **465**, 73.
179. B. L. Benac, A. H. Cowley, R. A. Jones, C. M. Nunn and T. C. Wright, *J. Am. Chem. Soc.*, 1989, **111**, 4986.
180. S. P. Lee and M. C. Lin, Report, Order No. AD-A237261. Avail. NTIS. From: Gov. Rep. Announce. Index (U.S.) 1991, **91**(21), Abstr. No. 158 239, 1991 (*Chem. Abstr.*, **117**, 79 760).
181. V. N. Varakin and A. P. Simonov, *Chem. Phys. Lett.*, 1992, **190**, 48.
182. V. N. Varakin and A. P. Simonov, *Laser Chem.*, 1992, **12**, 181.
183. M. A. Guerra, T. R. Bierschenk and R. J. Lagow, *J. Chem. Soc., Chem. Commun.*, 1985, 1550.
184. H. M. Marrasevit and W. I. Simpson, *J. Electrochem. Soc.*, 1973, **120**, 135.
185. G. B. Stringfellow, 'Organometallic Vapour Phase Epitaxy: Theory and Practice', Academic Press, New York, 1989.
186. P. O'Brien, *Chemtronics*, 1991, **5**, 61.
187. A. C. Jones, P. J. Wright and B. Cockayne, *J. Cryst Growth*, 1991, **107**, 297.
188. L. M. Smith and J. Thompson, *Chemtronics*, 1989, **4**, 60.
189. L. M. Smith, J. Thompson, A. C. Jones and P. R. Jacobs, *Mater. Lett.*, 1988, 722.
190. P. J. Wright, B. Cockayne, A. J. Williams, A. C. Jones and E. D. Orrell, *J. Cryst. Growth*, 1987, **84**, 552.
191. A. C. Jones, P. J. Wright and B. Cockayne, *Chemtronics*, 1988, **3**, 35.
192. B. Cockayne, P. J. Wright, A. J. Armstrong, A. C. Jones and E. D. Orrell, *J. Cryst. Growth*, 1988, **91**, 57.
193. P. J. Wright *et al.*, *J. Cryst. Growth*, 1989, **94**, 441.
194. P. J. Wright, B. Cockayne, P. J. Parbrook, A. C. Jones, P. O'Brien and J. R. Walsh, *J. Cryst. Growth*, 1990, **104**, 601.
195. H. M. Yates and J. O. Williams, *J. Cryst. Growth*, 1991, **107**, 386.
196. K. F. Jensen, A. Annapragada, K. L. Ho, J. S. Hiuh, S. Patnaik and S. Salim, *J. Phys. II*, 1991, **C2**, 243.
197. D. F. Forster *et al.*, *Adv. Mater. Optics Electron.*, 1994, **3**, 163.
198. J.-S. Huh, S. Patnaik and K. F. Jensen, *J. Electron. Mater.*, 1993, **22**, 509.
199. P. J. Wright, B. Cockayne, P. J. Parbrook, A. C. Jones, P. O'Brien and J. R. Walsh, *J. Cryst. Growth*, 1990, **104**.
200. P. J. Wright, B. Cockayne, A. C. Jones, E. D. Orrell, P. O'Brien and O. F. Z. Khan, *J. Cryst. Growth*, 1989, **94**, 97.
201. A. C. Jones, S. A. Rushworth, P. J. Wright, B. Cockayne, P. O'Brien and J. R. Walsh, *J. Cryst. Growth*, 1989, **97**, 537.
202. O. F. Z. Khan, P. O'Brien, P. A. Hamilton, J. R. Walsh and A. C. Jones, *Chemtronics*, 1989, **4**, 24 412.
203. E. I. Amma and R. E. Rundle, *J. Am. Chem. Soc.*, 1958, **80**, 4141.
204. G. E. Coates and R. A. Whitcombe, *J. Chem. Soc.*, 1956, **335**, 1204.
205. P. J. Wright, B. Cockayne, A. C. Jones and E. D. Orrell, *J. Cryst. Growth*, 1988, **91**, 63.
206. A. C. Jones, S. A. Rushworth, P. O'Brien, J. R. Walsh and C. Meaton, *J. Cryst. Growth*, 1993, **130**, 295.
207. J. E. Butler, N. Bottka, R. S. Sillmon and D. K. Gaskill, *J. Cryst. Growth*, 1986, **77**, 163.
208. R. L. Jackson, *Chem. Phys. Lett.*, 1989, **163**, 315.
209. G. E. Coates and D. Ridley, *J. Chem. Soc.*, 1965, 1870.
210. M. Motevalli, M. A. Malik and P. O'Brien, *J. Mater. Chem.*, 1995, **5**, 753.
211. A. Henglein, *Chem. Rev.*, 1989, **89**, 1861.
212. M. L. Steigerwald and L. Brus, *Acc. Chem. Res.*, 1990, **23**, 183.
213. C. B. Murray, D. J. Norris and M. G. Bawendi, *J. Am. Chem. Soc.*, 1993, **115**, 8706.
214. J. E. Bowen Katari, V. L. Colvin, and A. P. Alivisatos, *J. Phys. Chem.*, 1994, **98**, 4109.
215. C. L. Czekaj, M. S. Rau, G. L. Geoffroy, T. A. Guiton and C. G. Pantano, *Inorg. Chem.*, 1988, **27**, 3267 (*Chem. Abstr.*, **110**, 159 192a).

Author Index

This Author Index comprises an alphabetical listing of the names of the authors cited in the text and the references listed at the end of each chapter in this volume.

Each entry consists of the author's name, followed by a list of numbers, for example

Templeton, J. L., 366, 385^{233} (350, 366), 387^{370} (363)

For each name, the page numbers for the citation in the reference list are given, followed by the reference number in superscript and the page number(s) in parentheses of where that reference is cited in the text. Where a name is referred to in text only, the page number of the citation appears with no superscript number. References cited both in the text and in the tables are included.

Although much effort has gone into eliminating inaccuracies resulting from the use of different combinations of initials by the same author, the use by some journals of only one initial, and different spellings of the same name as a result of the transliteration processes, the accuracy of some entries may have been affected by these factors.

Aalten, H. L., 130^{39} (61, 66, 69, 78, 93)
Abad, J. A., 132^{172} (73, 74)
Abad, M., 54^{322} (33)
Abicht, H.-P., 51^{93} (4)
Abrahams, I., 204^{82} (185), 206^{178} (199)
Abramova, N. M., 172^{71}
Abu-Salah, O. M., 51^{115} (6), 51^{116} (5, 6), 52^{214} (18), 53^{216} (18), 53^{217} (18), 53^{218} (18), 53^{219} (18), 53^{220} (18), 53^{221} (18), 53^{222} (18), 131^{67} (61, 69, 75, 78, 109), 132^{140} (69, 109), 132^{181} (75), 132^{182} (75), 133^{238}
Adams, H.-N., 54^{333} (36)
Adamson, G. W., 204^{84} (186)
Adelhelm, M., 55^{416} (48)
Adlkofer, J., 132^{149} (71, 76), 132^{152} (71)
Afanassova, O. B., 52^{168} (12)
Agirbas, H., 173^{143} (166)
Aguirre, A., 131^{56} (61, 78, 113), 132^{127} (68, 78, 113)
Aguirre, C. J., 53^{236} (20, 22)
Ahmed, E., 54^{308} (32)
Aida, T., 49^{8} (2)
Aigbirhio, F. I., 171^{7} (136), 204^{41} (180)
Akkerman, O. S., 204^{29a} (178), 204^{29b} (178)
Albright, M. J., 133^{213} (95)
Alex, S., 172^{99} (156), 172^{100} (156)
Alexakis, A., 130^{1g} (57, 60, 65, 95, 125, 127)
Al-Hashimi, S., 206^{158} (196)
Ali, M., 172^{115} (160)
Alivisatos, A. P., 206^{214} (203)
Al-Jabar, N. A. A., 171^{19} (138), 171^{34} (143, 144)
Al-Juaid, S. S., 171^{7} (136), 204^{39} (179), 204^{40} (179), 204^{41} (180), 204^{71} (184)
Allaire, F., 172^{101} (156)
Almond, M. J., 205^{155} (196), 205^{156}

(196), 206^{171} (197, 201), 206^{172} (197), 206^{173} (197)
Al-Najjar, I. M., 51^{115} (6)
Al-Ohaly, A.-R. A., 51^{115} (6), 51^{116} (5, 6), 52^{214} (18), 53^{216} (18), 53^{217} (18), 53^{219} (18), 53^{221} (18), 132^{140} (69, 109)
Al-Showiman, S. S., 51^{115} (6)
Alt, H. G., 172^{121} (160)
Alvarez, S., 53^{257} (24)
Amma, E. I., 206^{203} (201)
Anderson, G. K., 50^{61} (3, 20, 21, 43)
Anderson, T. J., 133^{213} (95)
Andrianov, V. G., 130^{29} (61, 78, 117), 131^{68} (61, 78, 117)
Annapragada, A., 206^{196} (200, 201)
Antipin, M. Yu., 52^{190} (15), 171^{35} (144), 204^{63} (182)
Araki, S., 205^{128} (190)
Arcas, A., 51^{139} (9, 10), 51^{141} (9, 10), 54^{323} (34)
Arduengo, A. J., III, 204^{47} (180)
Armstrong, A. J., 206^{192} (200)
Arsenault, G. J., 51^{92} (4), 51^{111} (6)
Artigao, M., 54^{323} (34)
Ashby, E. C., 131^{105} (65), 133^{221} (125), 133^{222} (125), 133^{224} (125)
Atwood, J. L., 172^{112} (159), 203^{15} (177), 206^{164} (197)
Aubke, F., 55^{407} (47, 48), 172^{79} (154)
Ault, B. S., 204^{30} (178)
Aumann, R., 55^{400} (46), 55^{401} (46, 47)
Avdeef, A., 133^{217} (122, 128)
Ayako, G. A., 206^{159} (196)
Aznar, F., 171^{58} (150), 171^{59} (150)
Bach, R. D., 172^{92} (155), 172^{98} (156)
Bacher, W., 55^{416} (48)
Bäckvall, J.-E., 131^{112} (65, 75, 128)
Baenziger, N. C., 131^{96} (64, 78, 117)
Baerends, E. J., 51^{87} (4)
Bähr, B., 132^{124} (67)
Bahr, G., 130^{23} (60, 76), 130^{25} (60)
Bai, H., 204^{30} (178)

Baiker, A., 50^{22} (2)
Bakker, C. N. M., 172^{118} (160)
Balch, A. L., 50^{37} (2)
Baldi, F., 173^{137} (166)
Banasiak, U., 203^{20} (177)
Bancroft, G. M., 51^{86} (4), 55^{355} (40)
Bandaev, S. G., 172^{67}
Bandini, A. L., 51^{94} (5), 52^{165} (11), 55^{394} (45)
Banditelli, G., 52^{165} (11), 55^{393} (44, 46), 55^{394} (45), 55^{398} (46)
Banks, R. E., 132^{165} (72)
Bardají, M., 54^{294} (29)
Barluenga, J., 171^{37}, 171^{39}, 171^{47}, 171^{58} (150), 171^{59} (150)
Bartlett, P. A., 204^{55} (181)
Basil, J. D., 53^{262} (24), 53^{265} (24), 53^{279} (27, 28), 53^{280} (27, 28), 54^{301} (27)
Bassett, J.-M., 53^{256} (23)
Bassetti, M., 171^{56} (149)
Battiste, M. A., 172^{74}
Bau, R., 65, 130^{14c} (61, 65, 78, 100, 109), 131^{66} (61, 65, 78, 102, 109), 131^{77} (61, 74, 76, 78, 109, 110)
Baukova, T. V., 51^{81} (4, 11, 12), 52^{155} (9–12), 52^{156} (9, 10), 52^{161} (10, 11), 52^{163} (11, 12), 52^{168} (12), 52^{172} (13)
Baum, T. H., 50^{27} (2, 35), 54^{325} (34)
Bawendi, M. G., 206^{213} (203)
Bazhenov, D. V., 171^{43}, 172^{62} (150)
Beames, D. J., 131^{108} (65)
Beauchamp, A. L., 172^{95} (155), 172^{99} (156), 172^{100} (156), 172^{101} (156), 172^{102} (156), 172^{129} (163)
Beaulieu, P. L., 133^{223} (125)
Bechgaard, K., 50^{35} (2)
Beck, W., 55^{392} (44)
Beckhaus, R., 203^{20} (177)
Beer, M. P., 205^{155} (196), 205^{156} (196), 206^{171} (197, 201), 206^{172} (197), 206^{173} (197)
Behling, J. R., 133^{229} (127)

Behm, J., 203²¹ (177), 204⁷⁵ (184)
Belanger-Gariépy, F., 172⁹⁵ (155)
Bell, N. A., 204⁵⁷ (181), 204⁸⁴ (186)
Belli Dell'Amico, D., 55³⁸⁷ (44), 55⁴¹⁰ (48), 55⁴¹³ (48), 55⁴¹⁴ (48)
Bellon, P. L., 52²⁰³ (17), 52²⁰⁴ (17)
Benac, B. L., 206¹⁷⁹ (199)
Bennett, M. A., 51¹³¹ (8, 10), 51¹⁴² (10), 54³⁰⁷ (31)
Bergbreiter, D. E., 133²⁴⁰
BergStresser, G. L., 203²³ (177)
Berk, S. C., 204⁶⁵ (182)
Berlitz, T. F., 172¹²³ (160)
Bermúdez, M.-D., 54³¹⁰ (32), 54³¹¹ (32), 54³¹² (32, 38), 54³¹³ (32), 54³¹⁴ (32), 54³⁴² (37), 54³⁴³ (37), 54³⁴⁵ (37), 54³⁴⁶ (37), 54³⁴⁷ (37), 54³⁴⁸ (37), 54³⁴⁹ (37), 54³⁵⁰ (38), 54³⁵¹ (38), 55³⁵² (38)
Bernstein, S. C., 205¹⁴⁹ (195)
Berry, D. E., 172¹¹² (159), 203¹⁵ (177), 206¹⁶⁴ (197)
Bersohn, R., 205¹⁰³ (188)
Bertz, S. H., 60, 65, 131⁸³ (60), 131⁸⁶ (63), 131¹¹⁰ (65, 75), 131¹¹³ (65), 131¹¹⁴ (65), 131¹¹⁵ (65), 131¹¹⁶ (65), 133²⁰⁵ (78, 115), 133²²⁶ (127), 133²²⁷ (127)
Beruda, H., 50⁵⁶ (3), 52²⁰¹ (16)
Beurskens, P. T., 55⁴¹⁸ (49), 133²³⁷ (117)
Beverwijk, C. D. M., 130²¹ (60, 71–3, 76), 132¹⁶⁶ (73)
Bhargava, S. K., 51¹³¹ (8, 10), 51¹⁴² (10)
Bhat, N. G., 171¹⁰ (136)
Bible, R. H., 172⁶⁶
Bickelhaupt, F., 204²⁹ᵃ (178), 204²⁹ᵇ (178)
Bierschenk, T. R., 55³⁶² (40), 132¹⁶⁰ (72), 203⁴ (176), 205¹⁵⁴ (196), 206¹⁸³ (200)
Billups, W. E., 204⁴² (180)
Bissinger, P., 52¹⁹⁵ (15), 52¹⁹⁹ (16)
Bjørnholm, T., 50³⁵ (2)
Blaschke, G., 53²⁵⁶ (23)
Blau, S., 130¹⁸ (58)
Blenkers, J., 132¹⁷⁴ (74, 125, 127)
Block, E., 172⁸⁹ (155)
Blom, R., 204⁴⁹ (180)
Bloodworth, A. J., 147, 171⁴², 171⁴⁶, 171⁵⁰ (146), 171⁵¹ (147), 171⁵⁴ (147), 172⁶⁵, 172⁶⁸, 172⁶⁹ (151), 172⁷⁰, 172⁷²
Blumenthal, A., 50⁵⁶ (3)
Boaventura, M.-A., 172⁶³ (150)
Bocelli, G., 171⁵⁶ (149)
Boche, G., 51⁹⁵ (5)
Bochmann, M., 205⁹³ (186)
Böck, M., 55³⁹⁹ (46), 55⁴⁰⁰ (46), 55⁴⁰² (46)
Bodenbinder, M., 172⁷⁹ (154)
Boerrigter, P. M., 51⁸⁷ (4)
Boersma, J., 51⁸⁴ (4), 131⁹³ (64, 76), 131¹¹¹ (65, 66, 75, 128), 132¹⁶⁸ (73, 76, 120), 132¹⁶⁹ (73), 132¹⁷⁴ (74, 125, 127), 133¹⁸⁹ (75, 76), 172¹¹⁹ (160), 175, 176, 195, 203¹ (175, 189),

203⁹ (176), 203¹³ (177), 203¹⁴ (177), 204⁴⁵ (180), 204⁴⁸ (180), 204⁵⁰ (180), 204⁵² (180), 204⁵³ (180), 204⁵⁴ (181), 204⁵⁶ (181, 202), 204⁵⁹ (181), 204⁶⁷ (182), 205¹⁵¹ (195), 206¹⁶⁷ (197), 206¹⁶⁸ (197)
Bogdanović, S., 204⁷⁵ (184)
Böhlen, E., 131⁷⁷ (61, 74, 76, 78, 109, 110)
Bois, C., 51¹³⁶ (9, 10)
Bolm, C., 204⁷⁹ (185)
Bonati, F., 51¹⁰⁰ (5, 14), 51¹⁰⁷ (6), 51¹²⁰ (6), 51¹³⁷ (9, 11, 47), 52¹⁵¹ (10, 11, 47), 52¹⁶⁵ (11), 52¹⁶⁶ (11), 52¹⁷⁸ (13, 14), 52¹⁷⁹ (13), 52¹⁸⁴ (14), 55³⁹³ (44, 46), 55³⁹⁸ (46), 132¹⁷¹ (73), 133¹⁹³ (76)
Bond, A. M., 171¹⁶ (137)
Bondi, A., 50⁴⁹ (2)
Boone, C., 172¹⁰³ (157)
Bos, W., 55³⁷³ (43), 55⁴¹⁸ (49), 55⁴¹⁹ (49)
Bosman, W. P., 55⁴¹⁸ (49)
Bottka, N., 206²⁰⁷ (201)
Bour, J. J., 50²⁶ (2), 55³⁷⁴ (43), 55⁴¹⁸ (49), 55⁴¹⁹ (49)
Bovio, B., 51¹²⁰ (6), 51¹³⁷ (9, 11, 47), 52¹⁵¹ (10, 11, 47), 52¹⁶⁴ (11, 47), 52¹⁷⁸ (13, 14), 132¹⁷¹ (73)
Bowen Katari, J. E., 206²¹⁴ (203)
Bowmaker, G. A., 53²³⁴ (20, 22)
Boyd, P. D. W., 50⁵⁹ (3)
Brachthäuser, B., 52²⁰⁰ (16), 52²⁰¹ (16)
Brandi, A., 172⁸⁵ (155), 172⁸⁶ (155)
Braun, R. W., 205¹⁵⁴ (196)
Braunstein, P., 51¹⁰¹ (5)
Bravo, J., 172⁸⁸ (155)
Brevard, C., 133¹⁸⁵ (75, 77, 86, 121, 129)
Briant, C. E., 52²⁰⁶ (17), 52²⁰⁷ (17)
Brienne, S., 50⁵⁹ (3)
Briggs, D. A., 52¹⁵² (9, 10), 53²³⁹ (20, 30), 53²⁷⁴ (27, 28, 43)
Brinckmann, F. E., 173¹³⁷ (166)
Brinkman, G. A., 172¹¹⁸ (160)
Brito, M., 172⁸⁹ (155)
Britten, J. F., 55⁴⁰³ (47)
Brown, H. C., 171¹⁰ (136), 171⁴⁹ (145), 205¹²⁷ (190)
Brown, T., 133²²⁵ (126)
Bruce, M. I., 50⁵⁴ (3), 51¹⁰³ (6), 51¹¹³ (5, 6, 8), 51¹²⁶ (8, 12), 51¹²⁸ (5, 8), 52¹⁵⁷ (9), 52¹⁵⁸ (9, 10), 52¹⁶⁰ (9, 10), 55³⁷⁶ (43), 55³⁷⁸ (43), 133²³⁸
Brun, P., 52¹⁷³ (13)
Bruni, S., 51⁹⁴ (5)
Brus, L., 206²¹² (203)
Buchner, W., 132¹⁴⁹ (71, 76)
Buck, A. J., 55³⁸⁴ (44)
Buckingham, J., 50⁷⁰ (3)
Buckton, G., 64, 71, 131⁹⁴ (64, 71)
Budzelaar, P. H. M., 204⁴⁸ (180), 204⁵⁰ (180), 204⁵⁶ (181, 202), 204⁵⁹ (181), 205⁹⁶ (187), 205¹⁵¹ (195)
Bulow, G., 204⁷⁷ (184)
Buncel, E., 172⁹⁵ (155), 172¹⁰³ (157)
Burba, G., 130²³ (60, 76), 130²⁵ (60), 132¹²⁴ (67)

Burch, R. R., 131⁷⁰ (61, 72, 78, 107, 111, 113)
Burini, A., 51¹⁰⁰ (5, 14), 51¹⁰⁷ (6), 51¹²⁰ (6), 51¹³⁷ (9, 11, 47), 52¹⁵¹ (10, 11, 47), 52¹⁶⁴ (11, 47), 52¹⁷⁸ (13, 14), 52¹⁷⁹ (13), 52¹⁸⁴ (14), 132¹⁷¹ (73)
Burnard, R. J., 132¹⁵⁹ (72)
Burrell, A. K., 50⁵⁹ (3)
Burschka, Ch., 52¹⁵⁴ (9, 10), 131⁹⁷ (64), 132¹³⁷ (68, 115)
Burton, D. J., 130³⁰ (61, 64), 131⁹⁵ (64), 131⁹⁶ (64, 78, 117), 132¹⁴⁶ (61, 70), 206¹⁶¹ (196), 206¹⁶³ (196)
Butin, K. B., 172⁸⁰ (154)
Butler, J. E., 206²⁰⁷ (201)
Butler, W. M., 133²¹³ (95)
Butsugan, Y., 205¹²⁸ (190)
Buttrus, N. H., 206¹⁵⁹ (196), 206¹⁷⁵ (198)
Buysingh, P., 204⁷⁸ (185)
Byers, P. K., 54³³⁵ (36), 54³³⁶ (36), 54³⁴⁰ (36)

Cai, J., 171⁵⁵ (148)
Cairncross, A., 58, 130⁵ (58, 76), 132¹¹⁸ (65, 70, 76), 132¹⁴¹ (70)
Calabrese, J. C., 131⁷⁰ (61, 72, 78, 107, 111, 113)
Calderazzo, F., 55³⁸⁷ (44), 55⁴⁰⁹ (48), 55⁴¹⁰ (48), 55⁴¹³ (48), 55⁴¹⁴ (48)
Calogero, S., 51¹⁰⁷ (6), 55³⁹³ (44, 46), 55³⁹⁸ (46)
Camus, A., 130²⁴ (60), 130³¹ (61, 68, 78, 80), 130⁴⁰ (61, 88), 131¹¹⁷ (65, 110), 132¹³¹ (68, 76, 77), 132¹⁴³ (70), 133¹⁸⁷ (76), 133¹⁹⁷ (76), 133¹⁹⁸ (76, 120)
Canty, A. J., 54³³⁵ (36), 54³³⁶ (36), 54³³⁷ (36), 54³³⁸ (36), 54³³⁹ (36), 54³⁴⁰ (36), 172⁹⁶ (155), 172⁹⁷ (156)
Capillon, J., 205¹⁵² (195)
Cariati, F., 51⁹⁴ (5)
Carius, L., 71, 132¹⁵⁵ (71)
Carlson, T. F., 131⁷⁶ (61, 72, 75, 78)
Carriedo, G. A., 51¹²⁷ (5, 8), 51¹³⁶ (9, 10), 52¹⁴⁷ (10), 55³⁸⁹ (44), 131⁹⁹ (64), 133²⁰⁹ (78, 116)
Carrillo, M.-P., 54³⁴⁵ (37), 54³⁵¹ (38), 55³⁵² (38)
Carruthers, W., 171³⁸
Casey, A. T., 51¹¹⁹ (5, 6)
Castilla, M. L., 52¹⁴⁸ (10, 40, 41), 54³²⁸ (36, 40, 41)
Cattalini, L., 54³⁰⁸ (32)
Caulton, K. G., 130³³ᵃ (61, 63, 78, 110)
Cen, W., 205⁹⁴ (187), 205⁹⁵ (187)
Chackalamannil, S., 173¹⁴⁹ (168)
Champion, G. D., 50¹⁹
Chan, K. H., 172⁶⁹ (151), 172⁷²
Chan, T. C. S., 51⁸⁶ (4), 55³⁵⁵ (40)
Chang, S.-C., 204⁴² (180)
Chapdelaine, M. J., 130¹ᵉ (57, 60, 65, 95, 125, 127)
Charland, J.-P., 172¹⁰² (156)
Chastain, S. K., 55⁴²⁰ (49)
Chaudret, B., 53²⁵⁸ (24, 39)
Che, C.-M., 50¹⁴ (2), 51¹⁰⁶ (5, 6, 8, 12), 56⁴²³ (49), 56⁴²⁴ (49)

Chelius, E. C., 172[66]
Chemega, A. N., 204[63] (182)
Chen, H.-W., 53[242] (20, 22), 54[327] (35, 36)
Chenier, J. H. B., 55[382] (44), 55[383] (44), 55[406] (47)
Chernov, A. N., 171[44]
Chertskov, V. F., 171[31] (143)
Chiang, M. Y., 131[77] (61, 74, 76, 78, 109, 110)
Chicote, M.-T., 52[181] (14), 52[182] (14), 52[183] (14), 52[188] (14, 23, 24), 52[189] (14, 23, 24), 52[213] (18), 53[231] (20–3), 54[310] (32), 54[311] (32), 54[312] (32, 38), 54[313] (32), 54[314] (32), 54[323] (34), 54[324] (34), 54[326] (35, 37), 54[342] (37), 54[346] (37), 54[347] (37), 54[348] (37), 54[350] (38), 131[75] (61, 72, 78)
Chiesi-Villa, A., 51[101] (5), 52[174] (13), 52[175] (13), 130[36] (61, 78, 110), 130[37] (61, 63, 82, 84), 130[38] (61), 131[72] (61, 73, 82), 133[201] (61, 73, 78–80, 82, 84, 105, 113)
Chisholm, M. H., 53[262] (24)
Chodosh, D. F., 56[422] (49)
Choi, S. W.-K., 54[341] (37), 132[147] (70)
Chou, T.-S., 205[86] (186)
Chow, M.-S., 172[81] (154)
Chowdhury, A. K., 55[395] (45)
Chu, C.-Y., 131[84] (60)
Churchill, M. R., 133[238]
Clark, R. J. H., 54[300] (26), 54[308] (32)
Clive, D. J., 133[223] (125)
Coan, P. S., 130[33a] (61, 63, 78, 110)
Coates, G. E., 131[54] (61, 97), 186, 202, 206[204] (201), 206[209] (202)
Cockayne, B., 206[187] (200), 206[190] (200), 206[191] (200), 206[192] (200), 206[194] (200, 201), 206[199] (200), 206[200] (200), 206[201] (200), 206[205] (201)
Cole-Hamilton, D. J., 204[26] (178), 204[27] (178), 206[174] (197)
Coleman, A. P., 205[93] (186)
Colera, I., 51[133] (9)
Collum, D. B., 172[73]
Colton, R., 54[339] (36)
Colvin, V. L., 206[214] (203)
Constable, E. C., 54[315] (32, 33), 54[316] (33), 54[317] (33), 54[318] (33), 172[116] (160)
Cooksey, C. J., 171[50] (146), 172[65], 172[69] (151), 172[72]
Cooper, P. N., 171[42]
Corbeil, M.-C., 172[99] (156), 172[100] (156)
Corey, E. J., 131[108] (65), 191, 192, 205[135] (191), 205[136] (191, 192)
Corfield, P. W. R., 130[13] (61, 78, 97), 131[54] (61, 97), 131[73] (61, 78, 97, 120), 133[196] (75)
Corset, J., 171[21] (140)
Costa, G., 131[117] (65, 110), 133[197] (76), 133[198] (76, 120)
Cotton, F. A., 131[81] (60, 76–8, 115), 133[188] (75, 76), 205[132] (190)
Coucouvanis, D., 133[217] (122, 128)
Courtneidge, J. L., 171[28] (142)
Coville, N. J., 49[9] (2)

Cowley, A. H., 206[179] (199)
Cox, M. T., 171[38]
Coxon, J. M., 172[74]
Craig, P. J., 171[9] (136), 173[138] (166), 173[140] (166)
Cras, J. A., 133[237] (117)
Crespo, M., 51[92] (4)
Crespo, O., 50[36] (2)
Cross, R. J., 51[117] (5, 6), 51[118] (6)
Cummins, C. C., 206[169] (197, 199)
Cunkle, G. T., 171[18] (138)
Cupertino, D. C., 204[26] (178)
Curtis, R. J., 171[51] (147)
Czekaj, C. L., 206[215] (203)
Czernecka, M., 204[70] (184)

Dabbagh, G., 60, 65, 131[83] (60), 131[86] (63), 131[110] (65, 75), 131[113] (65), 131[114] (65), 131[116] (65), 133[226] (127)
Daniels, L. M., 172[126] (162)
Danishefsky, S., 173[149] (168)
Dantona, R., 55[387] (44)
Daotien, N., 171[12] (136, 137)
Darnall, D. W., 49[4] (1)
Dash, K. C., 50[34] (2), 53[243] (20, 22, 23)
Davidson, F., 204[47] (180)
Davidson, M. F., 51[117] (5, 6), 51[118] (6)
Davies, A. G., 171[26] (141), 171[27] (142, 167), 171[28] (142), 171[55] (148), 172[122] (160), 172[124] (161), 172[125] (161)
Davies, J. I., 205[108] (188)
Davis, L. L., 56[422] (49)
Davis, W. M., 206[169] (197, 199)
de Boer, B. G., 133[238]
de Jesús, E., 51[133] (9)
de Koning, A. J., 203[9] (176), 204[45] (180)
de la Orden, M. U., 54[309] (32)
de Sarlo, F., 172[85] (155), 172[86] (155)
de Voss, J. J., 171[52] (147)
Deacon, G. B., 171[24] (141), 171[30] (143, 144), 171[32] (143), 171[33] (143), 171[34] (143, 144), 171[36], 172[90] (155)
Dedieu, A., 51[101] (5)
Deeming, A. J., 51[109] (6)
Dekker, J., 51[84] (4), 203[13] (177), 203[14] (177), 204[51] (180), 205[151] (195)
DeKock, R. L., 51[87] (4)
Delavaux, B., 53[258] (24, 39)
Delbaere, L. T. J., 131[80] (60, 76, 78, 115)
Dell, W. H., 205[97] (187)
Demartin, F., 131[55] (61, 67, 78, 97)
DeMember, J. R., 131[71] (61, 72, 78, 100, 107)
Dempsey, D. F., 130[32] (61)
Denk, M., 204[75] (184)
Denmark, S. E., 204[69] (182)
des Tombe, F. J. A., 132[169] (73)
Deschler, U., 53[247] (21)
Desiraju, G. R., 50[39] (2, 3)
Dettorre, C. A., 53[264] (24)
Dewar, M. J. S., 205[100] (188)
Dias, H. V. R., 204[47] (180)
Diederich, F., 55[385] (44)
Dietrich, A., 52[159] (9, 10)
Diéz, J., 132[127] (68, 78, 113)

Ding, K., 172[114] (160)
Doidge-Harrison, S. M. S. V., 171[5] (136)
Donard, O. F. X., 173[138] (166)
Donovan-Mtunzi, S., 51[109] (6)
Dorigo, A. E., 133[217] (122, 128)
Döring, I., 203[7] (176), 203[10] (176)
Drew, M. G. B., 131[57] (61, 68, 78), 206[172] (197), 206[173] (197)
Drew, R. A. I., 171[52] (147)
Droege, M. W., 50[78] (4, 33, 40)
Drouin, J., 172[63] (150)
Drouin, M., 55[421] (49)
Dudis, D. S., 53[263] (24, 26, 29), 53[265] (24), 54[296] (26, 29)
Duff, D. G., 50[22] (2)
Duffy, D. N., 51[126] (8, 12)
Dufour, N., 52[193] (15)
Duisenberg, A. J. M., 204[48] (180)
Dukat, W., 72, 132[162] (72, 111)
Durana, M. E., 55[360] (40, 41)
Dyadchenko, V. P., 50[62] (3), 51[80] (4), 51[82] (4, 5, 12–14), 52[155] (9–12), 52[170] (13), 52[190] (15), 54[329] (36)
Dziwok, K., 50[52] (2)
Dzugan, S. J., 205[99] (188)

Eaborn, C., 130[14b] (61, 78, 106), 132[175] (122), 133[204] (78, 106), 171[7] (136), 171[8] (136, 137), 179, 204[36] (179), 204[40] (179), 204[41] (180), 206[159] (196)
Eaton, P. E., 171[17] (137), 171[18] (138)
Edwards, A. J., 131[100] (64, 67, 78, 100, 108)
Edwards, H. G. M., 172[76] (153), 172[77] (153)
Edwards, J. P., 204[69] (182)
Edwards, P. G., 130[14c] (61, 65, 78, 100, 109), 131[66] (61, 65, 78, 102, 109)
Edwards, P. P., 50[22] (2)
Eggleston, D. S., 56[422] (49)
Eijkelkamp, D. J. F. M., 131[112] (65, 75, 128)
Einhorn, C., 171[48]
Einhorn, J., 171[48]
Eisenstein, O., 133[202] (78, 95)
Elder, R. C., 49[4] (1)
Ellis, D. E., 50[30] (2, 3, 17), 52[211] (17)
Ellsworth, E. L., 131[106] (65, 93, 127)
Emrich, R. J., 54[298] (29, 31)
Endo, I., 55[369] (42)
Engelhardt, L. M., 54[336] (36), 204[60] (181)
Erdik, E., 130[1f] (57, 60, 65, 95, 125, 127)
Erfort, U., 172[78] (154), 173[151] (168)
Eriksson, H., 133[206] (78, 109)
Erkamp, C. J. M., 130[48] (61, 66, 78, 91)
Ernst, R. D., 203[12] (176)
Escribano, J., 54[345] (37), 54[349] (37)
Eshnazarov, Y. K., 172[67]
Esho, F. S., 131[57] (61, 68, 78)
Espinet, P., 132[183] (75), 132[184] (75)
Ettl, F., 55[385] (44)
Eujen, R., 171[29] (143)
Evans, D. G., 52[205] (17)
Evans, H. F., 131[71] (61, 72, 78, 100, 107)
Evans, W. J., 133[236] (78, 115)

Fabicon, R. M., 204^{35} (179), 204^{83} (185)

Fabrizi, L., 172^{85} (155)

Fackler, J. P., Jr., 50^{15} (2), 52^{152} (9, 10), 53^{239} (20, 30), 53^{240} (20, 29, 30), 53^{242} (20, 22), 53^{250} (22), 53^{251} (22), 53^{254} (23, 28–30), 53^{259} (24), 53^{262} (24), 53^{263} (24, 26, 29), 53^{264} (24), 53^{265} (24), 53^{267} (28–30), 53^{269} (26, 28, 29), 53^{270} (28), 53^{271} (28), 53^{272} (27, 28), 53^{273} (28), 53^{274} (27, 28, 43), 53^{275} (28, 29), 53^{276} (28), 53^{277} (28), 53^{280} (27, 28), 53^{281} (27, 28), 53^{282} (28, 29), 54^{283} (28, 29), 54^{284} (28), 54^{285} (28, 29), 54^{286} (27, 28, 30), 54^{288} (26, 29), 54^{289} (27, 29), 54^{290} (26, 29), 54^{291} (26, 29), 54^{292} (27, 29, 30), 54^{296} (26, 29), 54^{298} (29, 31), 54^{299} (29, 31), 54^{300} (26), 54^{301} (27), 54^{302} (30), 54^{303} (30), 54^{327} (35, 36), 131^{76} (61, 72, 75, 78), 133^{217} (122, 128)

Faller, J. W., 205^{140} (193)

Falvello, L. R., 54^{284} (28)

Fañanas, J., 51^{138} (9, 10, 40, 41)

Farhangi, Y., 172^{111} (159)

Fehlhammer, W. P., 55^{397} (45)

Fehlner, T. P., 205^{94} (187), 205^{95} (187)

Felici, M., 51^{100} (5, 14), 52^{184} (14)

Ferguson, R. B., 131^{80} (60, 76, 78, 115)

Fernández, E. J., 51^{104} (6, 39), 51^{145} (10), 53^{223} (18), 55^{357} (40, 41, 43), 132^{180} (74)

Fernández-Baeza, J., 53^{231} (20–3)

Fernelius, W. C., 131^{90} (63)

Fernholt, L., 203^{13} (177)

Filippelli, M., 173^{137} (166)

Finck, W., 55^{397} (45)

Fischer, B., 172^{119} (160), 204^{49} (180), 206^{167} (197), 206^{168} (197)

Fischer, E. O., 55^{399} (46), 55^{400} (46), 55^{401} (46, 47), 55^{402} (46)

Fittschen, C., 51^{138} (9, 10, 40, 41), 52^{148} (10, 40, 41), 54^{314} (32), 55^{366} (41)

Fletcher, M. T., 171^{52} (147)

Floriani, C., 51^{101} (5), 52^{174} (13), 52^{175} (13), 73, 130^{36} (61, 78, 110), 130^{37} (61, 63, 82, 84), 130^{38} (61), 131^{72} (61, 73, 82), 133^{201} (61, 73, 78–80, 82, 84, 105, 113)

Floris, B., 172^{120} (160)

Floyd, D. M., 131^{109} (65)

Folting, K., 130^{33a} (61, 63, 78, 110)

Fontaine, X. L. R., 51^{108} (6)

Font-Altaba, M., 51^{139} (9, 10), 54^{342} (37)

Forniés, J., 132^{183} (75), 132^{184} (75)

Forster, D. F., 206^{197} (200, 201)

Fraile, M. N., 51^{129} (8), 51^{130} (8, 10, 24), 52^{149} (10, 42), 54^{331} (36)

Frankland, E., 175, 203^{2} (175)

Fréchet, J. M. J., 189, 190, 205^{119} (189, 190)

Freeman, J. W., 203^{12} (176)

Freire Erdbrügger, C., 52^{188} (14, 23, 24), 53^{253} (23), 54^{331} (36)

Frenking, G., 55^{415} (48)

Frey, P. A., 50^{13} (2)

Friedrich, H. B., 51^{90} (4, 27)

Frigo, D. M., 206^{165} (197)

Frina, V., 133^{223} (125)

Fujii, S., 205^{104} (188)

Fujii, T., 133^{207} (121)

Fujisawa, T., 205^{147} (195)

Fujita, Y., 205^{104} (188)

Fukuoka, A., 51^{98} (5)

Fumanal, A. J., 52^{153} (9, 10), 132^{179} (74)

Fung, E. Y., 50^{37} (2)

Fürstner, A., 205^{111} (188)

Furukawa, J., 194, 205^{143} (194)

Futagawa, T., 205^{146} (194)

Gabor, B., 203^{7} (176)

Gais, H. J., 204^{77} (184)

Galakhov, M. V., 171^{31} (143)

Galaska, H. J., 53^{264} (24)

Galiullina, R. F., 204^{76} (184)

Gal'yanova, E. V., 171^{44}

Galyer, A. L., 203^{3} (176)

Gamasa, M. P., 130^{53} (61, 67, 78, 97), 131^{56} (61, 78, 113), 132^{127} (68, 78, 113)

Gambarotta, S., 52^{174} (13), 52^{175} (13), 130^{36} (61, 78, 110), 130^{37} (61, 63, 82, 84), 130^{38} (61), 131^{72} (61, 73, 82), 133^{201} (61, 73, 78–80, 82, 84, 105, 113)

Gamper, S., 172^{94} (155)

Garcia-Garcia, M., 54^{311} (32)

García-Granda, S., 131^{56} (61, 78, 113), 132^{127} (68, 78, 113)

Garraud, H., 173^{140} (166)

Garrell, R. L., 55^{385} (44)

Garzón, G., 53^{239} (20, 30), 53^{240} (20, 29, 30)

Gaskill, D. K., 206^{207} (201)

Gasparrini, F., 50^{11} (2)

Gasser, O., 52^{180} (14)

Gatehouse, B. M., 171^{36}, 172^{90} (155)

Gatti, L., 131^{117} (65, 110), 133^{197} (76), 133^{198} (76, 120)

Gattuso, F., 50^{20} (2)

Gebhardt, W., 205^{102} (188)

Geilich, K., 52^{162} (10, 11)

Gellert, R. W., 130^{14c} (61, 65, 78, 100, 109)

Geoffroy, G. L., 206^{215} (203)

George, C. F., 204^{31} (179)

German, L. S., 171^{31} (143)

Gernon, M., 172^{89} (155)

Gerold, A., 133^{231} (128)

Ghilardi, C. A., 65, 130^{49} (61, 65, 78, 105, 110)

Ghosh, S., 50^{32} (2)

Gibson, C. P., 131^{86} (63)

Giese, B., 168, 172^{78} (154), 173^{145} (168), 173^{146} (168), 173^{147} (168), 173^{150} (168), 173^{151} (168), 173^{152} (168), 173^{153} (168)

Gilman, H., 64, 131^{101} (64, 65, 76)

Gimeno, J., 130^{53} (61, 67, 78, 97), 131^{56} (61, 78, 113), 132^{127} (68, 78, 113)

Gimeno, M. C., 50^{36} (2), 51^{102} (6), 51^{104} (6, 39), 51^{144} (10, 22, 40, 41), 51^{145} (10), 53^{226} (19),

53^{230} (20, 22), 53^{236} (20, 22), 53^{237} (20), 53^{245} (20, 22), 53^{252} (23), 54^{294} (29), 55^{368} (42), 55^{372} (43), 132^{173} (74, 120)

Giorgini, E., 51^{120} (6), 52^{179} (13)

Giovannoli, M., 50^{11} (2)

Girolami, G. S., 130^{32} (61)

Glick, M. D., 133^{213} (95)

Glidewell, C., 173^{143} (166)

Glinka, K., 206^{162} (196)

Glockling, F., 132^{164} (72), 133^{192} (76)

Goddard, J. P., 172^{125} (161)

Goedken, V. L., 205^{99} (188)

Goel, R. G., 52^{165} (11)

Goldberg, J. E., 55^{388} (44)

Gol'ding, I. R., 132^{125} (67, 76)

Gonzales-Gomez, J. A., 173^{147} (168)

Górecki, P., 204^{72} (184), 204^{73} (184)

Görling, A., 50^{30} (2, 3, 17), 52^{211} (17)

Görlitz, F. H., 52^{159} (9, 10)

Gorrell, I. B., 204^{61} (181), 204^{62} (182)

Gorrichon-Guigon, L., 171^{21} (140)

Goti, A., 172^{86} (155)

Goubitz, K., 130^{39} (61, 66, 69, 78, 93)

Graf, W., 50^{53} (2)

Graham, G. G., 50^{19}

Grandberg, K. I., 50^{62} (3), 50^{75} (3), 50^{77} (4, 40), 51^{80} (4), 51^{81} (4, 11, 12), 51^{82} (4, 5, 12–14), 52^{155} (9–12), 52^{161} (10, 11), 52^{168} (12), 52^{170} (13), 52^{172} (13), 52^{190} (15), 54^{329} (36)

Grässle, U., 54^{332} (36), 54^{333} (36), 54^{334} (36)

Grdenić, D., 172^{104} (157), 172^{106} (158), 172^{107} (159), 172^{108} (159)

Green, M. L. H., 131^{54} (61, 97)

Greene, B., 49^{4} (1)

Gregory, C. D., 133^{194} (75, 127)

Griffiths, K. D., 51^{131} (8, 10), 51^{142} (10)

Grishin, Y. K., 171^{43}, 171^{57} (149), 172^{62} (150), 172^{84} (155, 158)

Grohmann, A., 50^{52} (2), 50^{56} (3), 52^{197} (16), 52^{198} (16, 17)

Grove, D. M., 130^{15} (58, 61, 62, 65, 66, 69, 75, 77, 78, 93, 99, 101, 128), 130^{17} (58, 61, 78, 84, 86, 105, 129), 130^{51} (61, 62, 69, 77, 78, 98), 131^{85} (62, 66, 78, 87, 93, 121, 124, 127), 131^{112} (65, 75, 128), 133^{211} (78, 91)

Grundy, K. R., 51^{128} (5, 8)

Guarna, A., 172^{85} (155), 172^{86} (155)

Guastini, C., 51^{101} (5), 52^{174} (13), 52^{175} (13), 130^{36} (61, 78, 110), 130^{37} (61, 63, 82, 84), 130^{38} (61), 131^{72} (61, 73, 82), 133^{201} (61, 73, 78–80, 82, 84, 105, 113)

Guddat, L. W., 171^{36}

Guerra, M. A., 53^{274} (27, 28, 43), 55^{362} (40), 132^{160} (72), 203^{4} (176), 206^{183} (200)

Guette, J. P., 205^{195} (195)

Guette, M., 205^{152} (195)

Guillaneux, D., 205^{139} (193)

Guiton, T. A., 206^{215} (203)

Guo, D., 173^{144} (167)

Gupta, S. K., 171^{10} (136)

Guss, J. M., 130[6b] (58, 78, 87, 117), 130[7] (58, 66, 78, 89)

Haaland, A., 51[88] (4), 133[210] (61, 82), 203[11] (176), 203[13] (177)
Habeeb, J. J., 204[64] (182)
Habtemariam, A., 171[7] (136), 204[40] (179), 204[41] (180)
Habu, H., 132[134] (68)
Hädicke, E., 52[192] (15)
Hagen, K., 206[171] (197, 201)
Hajdu, E., 172[66]
Håkansson, M., 133[202] (78, 95), 133[206] (78, 109)
Hall, K. P., 52[206] (17), 52[207] (17)
Haller, K. J., 205[94] (187), 205[95] (187)
Hallnemo, G., 133[230] (128)
Halteren, B. W. v., 172[118] (160)
Hamilton, P. A., 206[202] (200)
Hamor, T. A., 172[115] (160)
Han, Y.-H., 172[127] (162)
Hancock, R. D., 49[5] (1)
Handke, G., 131[111] (65, 66, 75, 128)
Hanna, P. J., 171[16] (137)
Hannon, F. J., 205[135] (191), 205[136] (191, 192)
Hansen, M. M., 204[55] (181)
Hansen, S. W., 132[146] (61, 70)
Hanusa, T. P., 133[236] (78, 115)
Hardcastle, K., 51[109] (6)
Hargreaves, N., 172[72]
Harlow, R. L., 204[47] (180)
Harms, K., 51[95] (5), 204[79] (185)
Harnisch, H., 173[150] (168)
Harrison, L. W., 172[64] (150)
Hartmann, C., 54[287] (27, 28, 30), 54[295] (29, 30), 54[297] (29), 54[304] (30)
Harvey, P. D., 55[421] (49)
Hashimoto, T., 132[133] (68)
Haszeldine, R. N., 132[165] (72)
Hattingh, J. T. Z., 52[167] (11)
Haubrich, A., 131[111] (65, 66, 75, 128)
Hauge, R. H., 54[321] (33), 204[42] (180)
Hawthorne, M. F., 171[6] (136, 165), 171[14] (137, 164, 165), 173[135] (164), 173[136] (165)
Hay, C. M., 55[377] (43)
Hayasaka, T., 205[115] (189, 191)
Hayashi, M., 205[141] (193)
Hayashi, Y., 205[129] (190)
He, X., 133[205] (78, 115), 133[208b] (78, 82, 84, 88, 129), 133[226] (127)
Healy, P. C., 54[337] (36)
Heath, P., 205[156] (196)
Heathcock, C. H., 204[55] (181)
Heimann, M., 132[152] (71)
Heinekey, D. M., 205[157] (196)
Heinrich, D. D., 53[273] (28), 54[291] (26, 29)
Helgesson, G., 133[212] (91, 93)
Henderson, M. J., 203[16] (177), 206[166] (197)
Henderson, R. M., 132[141] (70)
Hengefeld, A., 53[215] (18)
Hengelmolen, R., 51[87] (4)
Henglein, A., 206[211] (203)
Henkel, S., 204[32] (179)
Henkelonja, S., 204[38] (179)
Henney, R. P. G., 54[316] (33), 54[317] (33), 54[318] (33)
Hergold-Brundic, A., 172[110] (159)
Herne, T. M., 55[385] (44)

Herrmann, W. A., 203[21] (177), 204[75] (184)
Herscheid, J. D. M., 172[118] (160)
Heyburn, C. A., 205[156] (196)
Higgins, S. J., 51[108] (6)
Higuchi, R., 49[8] (2)
Hiller, W., 54[332] (36), 54[333] (36)
Hirayama, T., 132[142] (70)
Hitchcock, P. B., 130[14b] (61, 78, 106), 132[175] (122), 133[204] (78, 106), 171[7] (136), 204[40] (179), 204[41] (180), 206[159] (196)
Hiuh, J. S., 206[196] (200, 201)
Ho, K. L., 206[196] (200, 201)
Hoekstra, A., 172[118] (160)
Hoffmann, R., 53[257] (24), 133[217] (122, 128)
Hofstee, H. K., 131[93] (64, 76), 132[168] (73, 76, 120), 132[174] (74, 125, 127), 133[189] (75, 76)
Höhn, E. G., 55[416] (48)
Hollander, F. J., 133[217] (122, 128)
Holliday, A. K., 132[163] (72)
Holy, N., 53[247] (21)
Hong, X., 51[106] (5, 6, 8, 12)
Hope, H., 130[14a] (61, 78, 109), 131[64] (61, 65, 78, 106)
Hori, Y., 49[6] (2)
Horiguchi, S., 132[134] (68)
Horler, H., 173[153] (168)
Horn, E., 51[113] (5, 6, 8)
Hornbach, P., 52[186] (14)
Horton, A. D., 55[379] (43)
Hougen, J., 51[88] (4)
House, H. O., 60, 131[84] (60), 133[195] (75)
Housecroft, C. E., 50[64] (3), 50[65] (3)
Howard, J. A. K., 55[382] (44), 55[383] (44), 55[384] (44), 55[406] (47), 131[99] (64), 133[209] (78, 116)
Howell, G. N., 171[16] (137)
Howes, A., 206[165] (197)
Hu, H., 172[114] (160)
Huber, B., 52[185] (14, 15), 52[194] (15), 52[198] (16, 17), 54[295] (29, 30), 171[11] (136, 163)
Hudson, H. A., 171[16] (137)
Hueck, K., 173[152] (168)
Huffman, J. C., 130[33a] (61, 63, 78, 110)
Huh, J.-S., 206[198] (200, 201)
Huheey, J. E., 50[38] (2)
Hulce, M., 130[1e] (57, 60, 65, 95, 125, 127)
Humphrey, P. A., 52[160] (9, 10)
Humphries, R. E., 171[34] (143, 144)
Hursthouse, M. B., 172[125] (161), 203[17] (177, 197), 203[19a] (177), 203[19b] (177), 204[81] (185), 205[89] (186), 205[91] (186), 206[165] (197)
Hutchings, G. J., 49[9] (2)
Hutton, A. T., 51[110] (6)

Ibuka, T., 130[1b] (57, 60, 65, 95, 125, 127)
Ichikawa, M., 55[412] (48)
Iijima, K., 54[325] (34)
Ikariya, T., 132[132] (68, 76, 77)
Illuminati, G., 172[120] (160)
Imura, N., 171[15] (137)
Ingold, F., 51[101] (5)
Inoguchi, Y., 52[169] (12)

Inomata, K., 171[41]
Inoue, H., 205[124] (190)
Inoue, K., 55[369] (42)
Inouye, M., 51[83] (4-6), 51[99] (5)
Inouye, Y., 205[145] (194)
Isab, A. A., 50[20] (2)
Ishikawa, M., 55[359] (40, 41), 55[365] (41)
Ishizaka, S., 205[107] (188)
Ishizaki, Y., 55[369] (42)
Issleib, K., 51[93] (4)
Ito, Y., 51[83] (4-6), 51[99] (5), 132[136] (68)
Itsuno, S., 205[119] (189, 190)
Iuchi, T., 205[104] (188)
Iverfeldt, Å., 172[91] (155), 172[93] (155)
Ivkina, A. A., 172[80] (154)
Iwata, M., 51[98] (5)

Jackson, R. L., 202, 205[105] (188), 206[208] (202)
Jacob, E., 55[416] (48)
Jacobs, P. R., 204[27] (178), 206[174] (197), 206[189] (200)
Jacobsen, G. E., 204[60] (181)
Jadhav, K. P., 171[12] (136, 137)
Jagner, J., 133[202] (78, 95)
Jagner, S., 133[212] (91, 93)
Jahn, W., 50[17] (2)
Jandik, P., 53[266] (28, 29), 54[293] (28, 29)
Janssen, M. D., 131[85] (62, 66, 78, 87, 93, 121, 124, 127), 131[111] (65, 66, 75, 128)
Jarvis, J. A. J., 130[11a] (58, 60, 61, 76–8, 125), 130[11b] (58, 60, 61, 76–8, 125)
Jastrzebski, J. T. B. H., 52[176] (13), 130[10] (58, 61, 64, 75, 78, 101, 109), 130[45] (61, 66, 78, 88, 91), 130[48] (61, 66, 78, 91), 131[82] (60, 74), 131[87] (63, 75, 76, 124), 131[103] (64, 65, 69, 127), 132[122] (66, 78, 88, 91), 132[178] (74, 122, 124), 133[185] (75, 77, 86, 121, 129), 133[208a] (82, 88, 126, 129), 133[215] (109, 121), 133[234] (101, 124), 204[52] (180), 204[78] (185), 205[98] (187)
Jaw, H.-R. C., 53[260] (24)
Jawad, J. K., 55[364] (40, 41)
Jeannin, Y., 51[136] (9, 10)
Jeffery, J. C., 55[390] (44)
Jelliss, P. A., 55[390] (44)
Jemmis, E. D., 52[208] (17)
Jensen, K. F., 206[196] (200, 201), 206[198] (200, 201)
Jeong, I.-H., 132[147] (70)
Jeong, N., 204[65] (182)
Jeong, Y.-T., 132[147] (70)
Jia, G., 51[122] (5, 7), 51[123] (5, 7, 8), 51[124] (7), 51[125] (7, 8)
Jiang, Y., 53[257] (24)
Jimenez, C., 171[37], 171[39]
Jiménez, J., 52[153] (9, 10), 53[268] (26–8), 53[278] (27, 28), 54[294] (29), 54[306] (31), 55[360] (40, 41), 132[179] (74)
Jimenez, R., 54[323] (34)
Johnson, B. F. G., 55[377] (43)
Johnson, F. A., 171[20] (140)
Joly, H. A., 55[406] (47), 172[103] (157)
Jones, A. C., 200, 203[17] (177, 197),

203[18] (177), 203[19a] (177),
203[19b] (177), 204[28] (178),
206[170] (197, 200), 206[187] (200),
206[189] (200), 206[190] (200),
206[191] (200), 206[192] (200),
206[194] (200, 201), 206[199] (200),
206[200] (200), 206[201] (200),
206[202] (200), 206[205] (201),
206[206] (201)
Jones, K. L., 171[8] (136, 137)
Jones, P. G., 50[36] (2), 50[45] (2, 4, 8),
50[46] (2), 50[47] (2, 12), 50[48] (2),
51[102] (6), 51[104] (6, 39), 51[130]
(8, 10, 24), 51[132] (8), 51[138] (9,
10, 40, 41), 51[141] (9, 10), 51[143]
(10), 51[144] (10, 22, 40, 41),
51[145] (10), 52[147] (10), 52[148]
(10, 40, 41), 52[181] (14), 52[182]
(14), 52[188] (14, 23, 24), 52[191]
(15), 52[192] (15), 53[223] (18),
53[225] (19), 53[237] (20), 53[244]
(20), 53[253] (23), 53[268] (26–8),
54[294] (29), 54[305] (31), 54[306]
(31), 54[314] (32), 54[324] (34),
54[328] (36, 40, 41), 54[331] (36),
54[345] (37), 54[348] (37), 54[351]
(38), 55[352] (38), 55[353] (39),
55[354] (39), 55[357] (40, 41, 43),
55[366] (41), 55[367] (41), 55[368]
(42), 55[372] (43), 55[396] (45),
55[411] (48), 131[69] (61, 74, 77,
78, 111, 115), 131[75] (61, 72, 78),
132[173] (74, 120), 132[180] (74)
Jones, P. M., 55[405] (47)
Jones, R. A., 206[179] (199)
Jones, R. G., 131[101] (64, 65, 76)
Joshi, N. N., 205[127] (190)
Juhlke, T. J., 205[154] (196)
Jukes, A. E., 130[20] (60, 63, 75)
Jung, J.-H., 132[147] (70)

Kaba, T., 205[134] (191)
Kafafi, Z. H., 49[7] (2), 204[42] (180)
Kagan, H. B., 193, 205[139] (193)
Kai, Y., 204[25] (177)
Kaitner, B., 172[87] (155)
Kalinina, O. N., 54[329] (36)
Kalyuzhnaya, E. S., 50[75] (3)
Kamenar, B., 172[87] (155), 172[110]
(159)
Kaneko, T., 205[138] (193)
Kang, H., 172[89] (155)
Kanters, R. P. F., 52[212] (17), 55[374]
(43)
Kapteijn, G. M., 133[211] (78, 91)
Karger, G., 52[186] (14)
Karlin, K. D., 133[185] (75, 77, 86, 121,
129)
Kartashov, V. R., 171[43], 171[44],
171[45], 172[62] (150)
Kasai, N., 204[25] (177)
Kasai, P. H., 55[380] (44), 55[381] (44),
55[405] (47)
Kaska, W. C., 53[229] (19, 20, 23)
Kasuga, N., 55[359] (40, 41), 55[369] (42)
Kaupp, M., 205[101] (188)
Kawabata, N., 205[143] (194)
Kawai, K., 205[114] (189), 205[130] (190,
192)
Kawasaki, M., 205[106] (188)
Kawase, Y., 205[153] (195)
Keppler, B. K., 50[21] (2)
Khan, M. N. I., 50[15] (2), 53[259] (24),

53[269] (26, 28, 29)
Khan, O. F. Z., 206[165] (197), 206[200]
(200), 206[202] (200)
Khan, S. I., 131[66] (61, 65, 78, 102,
109)
Khanna, R. K., 173[144] (167), 173[156]
(169)
Kikuchi, K., 49[6] (2)
Kilbourn, B. T., 130[11a] (58, 60, 61,
76–8, 125)
Kim, Y.-G., 132[147] (70)
King, C., 50[15] (2), 53[259] (24)
Kingston, D., 132[164] (72), 133[192]
(76)
Kinoshita, H., 171[41]
Kitamura, M., 205[114] (189), 205[130]
(190, 192), 205[131] (190), 205[137]
(192), 205[142] (193)
Kitching, W., 171[52] (147), 173[139]
(166)
Klassen, R. B., 50[27] (2, 35)
Kläui, W., 130[18] (58)
Knachel, H. C., 53[250] (22), 53[251]
(22), 53[264] (24), 54[296] (26, 29)
Knackel, H., 54[300] (26)
Kneuper, H.-J., 51[95] (5)
Knobler, C. B., 53[220] (18), 53[221]
(18), 132[140] (69, 109), 132[181]
(75), 171[6] (136, 165), 171[14]
(137, 164, 165), 173[135] (164),
173[136] (165)
Knochel, P., 133[228] (127), 171[13] (136,
137, 148), 203[6] (176), 204[65]
(182), 205[86] (186)
Knotter, D. M., 130[15] (58, 61, 62, 65,
66, 69, 75, 77, 78, 93, 99, 101,
128), 130[48] (61, 66, 78, 91),
130[51] (61, 62, 69, 77, 78, 98),
130[52] (61, 62, 66, 69, 77, 78, 98,
124), 131[111] (65, 66, 75, 128),
132[119] (61, 65, 69, 76, 78),
205[87] (186)
Knox, K., 133[217] (122, 128)
Knunyants, I. L., 171[31] (143)
Kobayashi, S., 205[148] (195)
Kobayashi, T., 171[41]
Kochi, J. K., 171[25] (141)
Kock, B., 204[68] (182)
Kogai, B. E., 172[75]
Köhler, K., 130[18] (58)
Kokhanyuk, G. M., 51[97] (5)
Koley, A. P., 50[32] (2)
Komiya, S., 50[63] (3), 51[98] (5), 54[319]
(33), 55[356] (40, 41), 55[358] (40,
41), 55[359] (40, 41), 55[363] (40,
41), 55[365] (41), 55[369] (42),
55[371] (43)
Kompa, K. L., 50[28] (2)
König, K., 171[22] (140)
Konijn, M., 133[215] (109, 121)
Konno, H., 53[238] (20, 21), 53[249] (21)
Kopf, J., 133[203] (78, 120)
Korkodilos, D., 171[50] (146), 172[68]
Korpar-Čolig, B., 172[105] (157),
172[106] (158), 172[108] (159)
Korsunsky, V. I., 52[171] (13), 52[172]
(13)
Kostić, N. M., 172[126] (162)
Kotake, H., 171[41]
Koutsantonis, G. A., 55[376] (43)
Kozlowski, J. A., 131[107] (65, 127)
Kozlowski, J. M., 131[109] (65)
Krämer, A., 52[186] (14)

Krasnov, Yu. N., 204[76] (184)
Krause, E., 71, 132[156] (71)
Krause, N., 131[111] (65, 66, 75, 128),
133[231] (128)
Kravtsov, D. N., 52[161] (10, 11)
Krüger, C., 52[198] (16, 17), 53[233] (20,
22)
Kruger, G. J., 52[167] (11)
Kubota, M., 132[128] (68)
Kudo, H., 50[57] (3)
Kuhn, W., 205[102] (188)
Kumar, R., 132[154] (71)
Kuran, W., 204[70] (184), 204[72] (184),
204[73] (184), 204[74] (184)
Kurek, J. T., 171[49] (145)
Kuznetsova, T. S., 171[53] (147)
Kwetkat, K., 173[139] (166)
Kwong, H.-L., 50[14] (2), 56[424] (49)

Labrador, M., 55[389] (44)
Lachhein, S., 173[150] (168)
Lagow, R. J., 53[274] (27, 28, 43),
55[362] (40), 132[160] (72), 203[4]
(176), 205[154] (196), 206[183]
(200)
Laguna, A., 50[36] (2), 50[69] (3, 8, 9, 18,
19, 32, 33, 37, 39, 43), 51[102] (6),
51[104] (6, 39), 51[129] (8), 51[130]
(8, 10, 24), 51[133] (9), 51[134] (9,
10), 51[135] (9, 45), 51[138] (9, 10,
40, 41), 51[144] (10, 22, 40, 41),
51[145] (10), 52[148] (10, 40, 41),
52[149] (10, 42), 52[153] (9, 10),
52[173] (13), 53[223] (18), 53[225]
(19), 53[226] (19), 53[230] (20, 22),
53[235] (20, 21), 53[236] (20, 22),
53[237] (20), 53[245] (20, 22), 53[252]
(23), 53[253] (23), 53[268] (26–8),
53[278] (27, 28), 54[294] (29), 54[305]
(31), 54[306] (31), 54[309] (32),
54[322] (33), 54[328] (36, 40, 41),
54[330] (36, 38, 39, 41), 54[331]
(36), 55[353] (39), 55[354] (39),
55[357] (40, 41, 43), 55[360] (40,
41), 55[361] (40, 41), 55[366] (41),
55[367] (41), 55[368] (42), 55[372]
(43), 55[396] (45), 131[69] (61, 74,
77, 78, 111, 115), 132[172] (73,
74), 132[173] (74, 120), 132[179]
(74), 132[180] (74)
Laguna, M., 51[102] (6), 51[104] (6, 39),
51[129] (8), 51[130] (8, 10, 24),
51[133] (9), 51[138] (9, 10, 40, 41),
51[145] (10), 52[148] (10, 40, 41),
52[149] (10, 42), 52[153] (9, 10),
53[225] (19), 53[226] (19), 53[230]
(20, 22), 53[235] (20, 21), 53[236]
(20, 22), 53[237] (20), 53[245] (20,
22), 53[252] (23), 53[253] (23),
53[268] (26–8), 53[278] (27, 28),
54[294] (29), 54[305] (31), 54[306]
(31), 54[322] (33), 54[328] (36, 40,
41), 54[330] (36, 38, 39, 41), 54[331]
(36), 55[353] (39), 55[354] (39),
55[357] (40, 41, 43), 55[360] (40,
41), 55[366] (41), 55[367] (41),
55[368] (42), 55[372] (43), 132[173]
(74, 120), 132[179] (74)
Lagunas, M.-C., 52[181] (14), 52[182]
(14), 52[183] (14), 52[213] (18)
Lahoz, F. J., 53[278] (27, 28)
Lai, T.-F., 54[341] (37), 56[423] (49),
56[424] (49)

Laibinis, P. E., 50^{16} (2)
Lalinde, E., 132^{183} (75), 132^{184} (75)
Lambert, F., 131^{111} (65, 66, 75, 128), 131^{112} (65, 75, 128)
Lambert, J. B., 172^{66}
Lampman, G. M., 172^{70}
Lanfranchi, M., 52^{166} (11)
Lang, H., 130^{18} (58)
Lange, H., 203^5 (176), 206^{160} (196)
Langer, F., 203^6 (176)
Langguth, E., 204^{66} (182)
Langrick, C. R., 51^{108} (6)
Lappert, M. F., 130^{11a} (58, 60, 61, 76–8, 125), 130^{11b} (58, 60, 61, 76–8, 125) 203^{16} (177), 206^{166} (197)
Larock, R. C., 136, 144, 171^3 (136, 144), 171^{10} (136), 171^{60} (150), 171^{61} (150), 172^{64} (150), 172^{81} (154)
Lastra, E., 130^{53} (61, 67, 78, 97), 131^{56} (61, 78, 113)
Lau, W., 171^{25} (141)
Lauher, J. W., 50^{55} (3, 17, 43)
Laurie, S. H., 173^{140} (166)
Lautner, J., 51^{141} (9, 10), 52^{147} (10), 53^{223} (18)
Lázaro, I., 55^{366} (41), 55^{367} (41), 55^{368} (42)
Lee, S. P., 206^{180} (200)
Lee, W.-K., 54^{341} (37)
Leese, T. A., 54^{315} (32, 33), 54^{318} (33), 172^{116} (160)
Legay, F., 173^{141} (166)
Legay-Sommaire, N., 173^{141} (166)
Lehmkuhl, H., 203^7 (176), 203^{10} (176), 203^{11} (176)
Lehniger, P., 51^{93} (4)
Leiderer, H., 205^{102} (188)
Lemenovskii, D. A., 52^{168} (12)
Lenders, B., 130^{17} (58, 61, 78, 84, 86, 105, 129), 130^{18} (58)
Lenhardt, H., 173^{152} (168)
Lensch, C., 52^{191} (15)
Leoni, P., 130^{49} (61, 65, 78, 105, 110)
Leusink, A. J., 130^{6a} (58, 62, 78, 87, 117), 130^9 (58, 73–5), 130^{21} (60, 71–3, 76), 130^{43} (61, 76), 130^{46} (61, 86, 88, 120, 125), 132^{167} (73, 74, 76, 86, 125)
Leusink, J. G., 132^{120} (66, 76, 86)
Lever, J. R., 133^{219} (124)
Lewis, J. A., 55^{377} (43), 171^{52} (147)
Li, D., 50^{14} (2), 51^{106} (5, 6, 8, 12)
Liddell, M. J., 50^{54} (3), 51^{128} (5, 8)
Lin, I. J. B., 52^{177} (13, 23, 24)
Lin, J. J., 131^{105} (65), 133^{224} (125)
Lin, M. C., 206^{180} (200)
Linford, L., 52^{167} (11)
Lingnau, R., 130^{42} (61, 82)
Lipshutz, B. H., 65, 130^{1a} (57, 60, 65, 95, 125, 127), 130^2 (57), 131^{106} (65, 93, 127), 131^{107} (65, 127), 131^{109} (65)
Liu, C. W., 52^{177} (13, 23, 24)
Liu, C.-L., 171^{60} (150)
Liu, E. K. S., 203^{22} (177)
Liu, L.-K., 52^{177} (13, 23, 24)
Liz, R., 171^{58} (150), 171^{59} (150)
Lo, W.-C., 51^{106} (5, 6, 8, 12)
Lobana, T. S., 172^{113} (159)
Lock, C. J. L., 55^{403} (47)
Looney, A., 204^{61} (181), 204^{62} (182)

López-de-Luzuriaga, J. M., 51^{104} (6, 39), 51^{145} (10), 53^{236} (20, 22)
Lorber, M. E., 205^{149} (195)
Lorberth, J., 172^{123} (160)
Lorenzen, N. P., 131^{65} (60, 61, 64, 65, 69, 78, 101)
Lotz, S. D., 51^{140} (10), 203^{21} (177)
Low, P. J., 51^{103} (6)
Luche, J. L., 171^{48}
Ludwig, W., 53^{261} (24)
Luning, U., 173^{152} (168)
Lusztyk, E., 172^{124} (161)
Lusztyk, J., 172^{122} (160), 172^{124} (161)
Lynch, T. J., 133^{240}

McBride, D. W., 131^{80} (60, 76, 78, 115)
McDougall, G. J., 49^5 (1)
McGowty, D., 172^{89} (155)
McGuchan, D. C., 171^{27} (142, 167), 171^{28} (142)
McIntosh, D. F., 55^{408} (47)
Macintyre, J. E., 50^{70} (3)
McLane, R., 203^{10} (176)
McLennan, A. J., 51^{118} (6)
McLoughlin, V. C. R., 132^{145} (70)
McNeal, C. J., 52^{202} (17)
Macomber, D. W., 131^{98} (64, 77), 132^{138} (68)
McPartlin, M., 55^{379} (43)
McWhinnie, W. R., 172^{115} (160)
Madl, R., 131^{92} (63, 65)
Malik, M. A., 204^{81} (185), 204^{82} (185), 205^{89} (186), 205^{90} (186), 205^{91} (186), 205^{92} (186), 206^{178} (199), 206^{210} (202)
Manassero, M., 52^{203} (17), 52^{204} (17), 131^{55} (61, 67, 78, 97)
Mandl, J. R., 53^{256} (23)
Manoharan, P. T., 50^{32} (2)
Manojlović-Muir, L., 51^{111} (6), 51^{112} (6)
Manzano, B. R., 53^{225} (19), 55^{353} (39), 55^{354} (39)
Marais, C. F., 52^{167} (11)
Marchioro, G., 171^{18} (138)
Maresca, L., 50^{11} (2)
Margrave, J. L., 54^{321} (33), 204^{42} (180)
Markies, P. R., 204^{29a} (178), 204^{29b} (178)
Markovskii, L. N., 204^{63} (182)
Marks, M. W., 130^{14c} (61, 65, 78, 100, 109)
Marks, T. J., 133^{188} (75, 76)
Markwell, A. J., 55^{370} (43)
Maroni, P., 171^{21} (140)
Marrasevit, H. M., 206^{184} (200)
Marsich, N., 130^{24} (60), 130^{31} (61, 68, 78, 80), 130^{40} (61, 88), 131^{117} (65, 110), 132^{131} (68, 76, 77), 132^{143} (70), 133^{187} (76), 133^{197} (76), 133^{198} (76, 120)
Marsman, J. W., 130^9 (58, 73–5)
Martin, F., 173^{138} (166)
Martin, R. M., 171^{17} (137), 171^{18} (138)
Martin, S. F., 131^{63} (61, 78, 103, 128)
Martínez, F., 132^{183} (75), 132^{184} (75)
Martínez-Gallo, J. M., 171^{47}
Mason, R., 130^{6b} (58, 78, 87, 117),

130^7 (58, 66, 78, 89), 133^{218} (122)
Mason, W. R., 53^{260} (24), 55^{420} (49)
Massa, W., 172^{123} (160)
Massey, A. G., 171^{19} (138), 171^{34} (143, 144)
Massobrio, M., 133^{193} (76)
Matisons, J. G., 51^{113} (5, 6, 8)
Matković-Čalogović, D., 172^{104} (157), 172^{106} (158), 172^{107} (159), 172^{109} (159)
Matsuda, T., 205^{141} (193)
Matsuda, Y., 205^{138} (193)
Mays, M. J., 55^{379} (43)
Mazanek, E., 204^{74} (184)
Mazany, A. M., 53^{240} (20, 29, 30), 53^{254} (23, 28–30), 53^{267} (28–30), 53^{275} (28, 29), 54^{284} (28), 54^{285} (28, 29)
Meaton, C., 204^{28} (178), 206^{206} (201)
Meguro, S., 55^{371} (43)
Mehringer, G., 49^3 (1)
Mehrotra, P. K., 133^{217} (122, 128)
Melník, M., 50^{44} (2, 4, 8)
Mendia, A., 132^{180} (74)
Mennie, D., 173^{138} (166), 173^{140} (166)
Merz, K. M., Jr., 205^{100} (188)
Messmer, R. P., 55^{408} (47)
Meyer, E. M., 52^{175} (13), 130^{38} (61), 133^{201} (61, 73, 78–80, 82, 84, 105, 113)
Meyer, R., 171^{21} (140)
Meyer, W., 53^{261} (24)
Meyer-Bäse, K., 52^{188} (14, 23, 24), 54^{328} (36, 40, 41), 131^{69} (61, 74, 77, 78, 111, 115)
Michel, A., 55^{421} (49)
Miguel, D., 51^{136} (9, 10)
Mile, B., 55^{382} (44), 55^{383} (44), 55^{384} (44), 55^{406} (47)
Milewski-Mahrla, B., 53^{243} (20, 22, 23)
Miller, J. M., 171^{30} (143, 144)
Miller, W. T., 132^{159} (72), 133^{190} (75, 76)
Mills, K., 54^{340} (36)
Minchin, N. J., 54^{335} (36), 54^{337} (36), 54^{338} (36), 172^{96} (155)
Minghetti, G., 52^{165} (11), 55^{394} (45), 133^{193} (76)
Mingos, D. M. P., 50^{23} (2), 50^{24} (2), 50^{25} (2), 52^{205} (17), 52^{206} (17), 52^{207} (17), 52^{209} (17), 52^{210} (17), 52^{212} (17), 133^{217} (122, 128), 133^{218} (122)
Misiti, D., 50^{11} (2)
Miyake, R., 205^{129} (190)
Miyashita, A., 130^{33b} (61, 63, 77, 78, 110), 132^{128} (68), 132^{132} (68, 76, 77)
Miyata, K., 133^{207} (121)
Mizuno, Y., 54^{319} (33)
Mochalov, S. S., 172^{67}
Mohamadi, F., 172^{73}
Moiseev, S. K., 131^{79} (60)
Moore, C. J., 171^{52} (147)
Moore, L. S., 51^{129} (8), 51^{134} (9, 10), 54^{309} (32)
Mora, M., 51^{139} (9, 10)
Moreno, M. T., 132^{183} (75), 132^{184} (75)
Morokuma, K., 128, 133^{217} (122,

128)

Morrison, J. A., 50[79] (4, 10, 33, 40), 51[91] (4), 54[320] (33), 132[161] (72)

Moss, J. R., 51[90] (4, 27)

Motevalli, M., 203[17] (177, 197), 203[18] (177), 203[19a] (177), 203[19b] (177), 204[81] (185), 204[82] (185), 205[89] (186), 205[91] (186), 205[92] (186), 206[178] (199), 206[210] (202)

Mueting, A. M., 55[375] (43)

Muir, K. W., 51[111] (6), 51[112] (6)

Müller, B. G., 50[33] (2)

Muller, F., 130[10] (58, 61, 64, 75, 78, 101, 109)

Müller, G., 50[52] (2), 50[53] (2), 52[185] (14, 15), 52[194] (15), 52[197] (16), 53[233] (20, 22), 53[243] (20, 22, 23), 54[287] (27, 28, 30), 54[295] (29, 30), 131[61] (61, 68, 115), 132[153] (71, 78, 116), 171[11] (136, 163), 172[94] (155)

Müller, G. E., 204[66] (182)

Mullica, D. F., 55[388] (44)

Mullin, J. B., 204[26] (178), 204[27] (178), 206[174] (197)

Münninghoff, J. W., 51[84] (4), 203[14] (177)

Murakami, M., 51[83] (4–6), 51[99] (5)

Murata, A., 49[6] (2)

Murray, C. B., 206[213] (203)

Murray, H. H., III, 53[239] (20, 30), 53[240] (20, 29, 30), 53[264] (24), 53[267] (28–30), 53[274] (27, 28, 43), 53[275] (28, 29), 53[281] (27, 28), 53[282] (28, 29), 54[283] (28, 29), 54[284] (28), 54[285] (28, 29), 54[286] (27, 28, 30), 54[288] (26, 29), 54[289] (27, 29), 54[290] (26, 29), 54[292] (27, 29, 30), 54[298] (29, 31), 54[300] (26), 54[301] (27)

Musso, H., 171[22] (140)

Mutter, Z. F., 53[216] (18), 53[217] (18)

Mynott, R., 203[7] (176)

Naganuma, A., 171[15] (137)

Nagel, U., 55[392] (44)

Nair, H. K., 50[79] (4, 10, 33, 40), 132[161] (72)

Nair, K. H., 206[163] (196)

Nájera, C., 171[37], 171[39], 171[47]

Nakamura, A., 204[25] (177)

Nakatsuka, T., 132[142] (70)

Naldini, L., 52[203] (17), 131[55] (61, 67, 78, 97)

Nardin, G., 130[24] (60), 130[31] (61, 68, 78, 80), 130[35] (61, 71, 78), 130[40] (61, 88), 132[143] (70)

Nast, R., 51[114] (6), 53[215] (18)

Nasyrov, I. M., 172[67]

Natile, G., 50[11] (2)

Naumann, D., 72, 132[162] (72, 111), 203[5] (176), 204[68] (182), 206[160] (196), 206[162] (196)

Navarro, A., 51[134] (9, 10)

Needham, M., 173[138] (166)

Negishi, E., 171[12] (136, 137)

Nehl, H., 203[8] (176), 203[10] (176), 203[11] (176)

Neira, R., 54[300] (26)

Nelson, S. M., 131[57] (61, 68, 78)

Nesmeyanov, A. N., 52[168] (12),

52[190] (15), 130[29] (61, 78, 117), 131[68] (61, 78, 117), 131[79] (60)

Nesmeyanov, O. A., 172[71]

Ney, S. C., 171[36], 172[90] (155)

Ngo-Khac, T., 52[154] (9, 10), 131[97] (64), 132[137] (68, 115)

Ngoviwatchai, P., 173[154] (168)

Niccolai, N., 172[85] (155)

Nicholson, B. K., 172[117] (160)

Nicolas, G., 55[386] (44)

NiDubhghaill, O. M., 50[21] (2)

Niemann, N. C., 52[176] (13), 133[234] (101, 124)

Niiyama, H., 49[8] (2)

Nirmala, R., 50[32] (2)

Nishi, N., 205[106] (188)

Nishimura, J., 205[143] (194)

Nishimura, M., 205[147] (195)

Niwa, S., 205[110] (188), 205[116] (189), 205[117] (189), 205[122] (189), 205[123] (189), 205[124] (190)

Nkosi, B., 49[9] (2)

Nobel, D., 130[41] (61, 78, 82), 132[170] (73)

Noltes, J. G., 130[6a] (58, 62, 78, 87, 117), 130[6b] (58, 78, 87, 117), 130[7] (58, 66, 78, 89), 130[8] (58, 64), 130[9] (58, 73–5), 130[12] (61, 67, 78, 95), 130[19] (60, 65, 75, 82, 90, 97, 120, 121, 129), 130[21] (60, 71–3, 76), 130[43] (61, 76), 130[46] (61, 86, 88, 120, 125), 130[47] (61), 131[58] (61, 67, 76–8, 82, 97, 122, 124), 131[59] (61, 67, 69, 89, 97, 124), 131[60] (61, 69, 75), 131[78] (60, 67), 131[82] (60, 74), 131[87] (63, 75, 76, 124), 131[88] (63, 66), 131[89] (63), 131[103] (64, 65, 69, 127), 131[104] (64, 69, 82, 93, 124, 129), 132[120] (66, 76, 86), 132[121] (66, 76, 77, 89), 132[123] (67, 69, 78, 89, 97), 132[129] (68, 76), 132[130] (68, 77), 132[144] (70), 132[167] (73, 74, 76, 86, 125), 132[176] (74–6, 125), 132[177] (74), 132[178] (74, 122, 124), 133[191] (76), 133[199] (76), 133[200] (77, 82, 129), 133[233] (129), 186, 205[88] (186)

Norem, N. T., 54[321] (33)

Normant, J. F., 130[1g] (57, 60, 65, 95, 125, 127), 130[26] (60)

Norris, A. R., 172[95] (155)

Norris, D. J., 206[213] (203)

Noyori, N., 205[130] (190, 192)

Noyori, R., 190, 192, 205[113] (189), 205[114] (189), 205[131] (190), 205[137] (192), 205[142] (193)

Nunn, C. M., 206[179] (199)

O'Brien, P., 203[17] (177, 197), 203[18] (177), 203[19a] (177), 203[19b] (177), 204[28] (178), 204[81] (185), 204[82] (185), 205[89] (186), 205[90] (186), 205[91] (186), 205[92] (186), 206[165] (197), 206[170] (197, 200), 206[178] (199), 206[186] (200), 206[194] (200, 201), 206[199] (200), 206[200] (200), 206[201] (200), 206[202] (200), 206[206] (201), 206[210] (202)

O'Connor, M. J., 171[16] (137), 171[24] (141)

O'Donoghue, M. F., 171[30] (143, 144), 171[32] (143)

Ogawa, K., 205[134] (191)

Oguni, N., 193, 205[112] (189), 205[138] (193), 205[141] (193)

Ohnishi, R., 55[412] (48)

Ohno, M., 205[148] (195)

Ohnuma, Y., 204[25] (177)

Ohshita, J., 50[54] (3)

Okada, S., 205[130] (190, 192), 205[142] (193)

Oknińska, E., 204[43] (180), 204[44] (180)

Olbrich, F., 133[203] (78, 120)

Oliver, J. P., 133[213] (95), 172[92] (155)

Olivier, M. J., 172[129] (163)

Öller, H.-J., 171[11] (136, 163), 172[94] (155)

Olmos, E., 53[278] (27, 28)

Olmstead, M. M., 130[16] (58, 61, 64, 65, 78, 84, 101, 120, 121), 131[62] (61), 131[64] (61, 65, 78, 106), 131[102] (64, 78, 101, 113, 120, 121), 133[208b] (78, 82, 84, 88, 129), 204[80] (185, 186)

Olsen, A. W., 49[7] (2)

Olson, G. J., 173[137] (166)

Olsson, T., 133[230] (128)

Omi, T., 205[112] (189)

Omura, H., 130[5] (58, 76)

Ookawa, A., 205[134] (191)

Oppolzer, W., 189, 190, 205[109] (188), 205[121] (189), 205[133] (190)

Oram, D., 130[14a] (61, 78, 109)

Orrell, E. D., 204[26] (178), 204[27] (178), 206[174] (197), 206[190] (200), 206[192] (200), 206[200] (200), 206[205] (201)

Orsini, F., 205[150] (195)

Ortaggi, G., 172[120] (160)

Örtendahl, M., 133[202] (78, 95), 133[206] (78, 109)

Oshah, N., 173[138] (166)

Osman, A., 204[64] (182)

Ostrander, R. L., 172[127] (162)

Otte, R., 55[404] (47)

Otto, H., 52[154] (9, 10), 131[97] (64), 132[137] (68, 115)

Ozaki, S., 55[358] (40, 41), 55[359] (40, 41), 55[363] (40, 41), 55[365] (41), 55[369] (42), 55[371] (43)

Ozin, G. A., 55[408] (47)

Pajerski, A. D., 203[23] (177), 204[35] (179)

Palmieri, G., 50[11] (2)

Pankratova, V. N., 204[76] (184)

Pantano, C. G., 206[215] (203)

Paparizos, C., 53[242] (20, 22), 54[327] (35, 36)

Papasergio, R. I., 51[85] (4, 5, 12), 130[27] (61, 77, 78), 130[28] (61, 77, 78), 132[158] (72, 77), 203[16] (177), 206[166] (197)

Parbrook, P. J., 206[194] (200, 201), 206[199] (200)

Parish, R. V., 50[40] (2), 50[44] (2, 4, 8), 51[129] (8), 51[134] (9, 10), 54[309] (32)

Parkin, G., 204[61] (181), 204[62] (182)

Parr, J., 205[140] (193)

Parrott, M. J., 205[108] (188)

Parvez, M., 203[23] (177), 204[83] (185)
Pasquali, M., 130[49] (61, 65, 78, 105, 110)
Pastor, S. D., 50[12] (2)
Pasynkiewicz, S., 131[91] (63)
Patalinghug, W. C., 204[60] (181)
Pathaneni, S. S., 50[39] (2, 3)
Patnaik, S., 206[196] (200, 201), 206[198] (200, 201)
Patrick, J. M., 54[335] (36), 54[336] (36), 54[338] (36), 172[97] (156)
Paver, M. A., 131[100] (64, 67, 78, 100, 108)
Payá, J., 54[343] (37)
Payne, N. C., 51[105] (6), 51[121] (5, 6), 51[124] (7), 51[125] (7, 8)
Pearce, R., 130[11a] (58, 60, 61, 76–8, 125), 130[11b] (58, 60, 61, 76–8, 125)
Pearson, R. G., 133[194] (75, 127)
Pearson, W. B., 50[50] (2)
Peerey, L. M., 172[126] (162)
Pelizzoni, F., 205[150] (195)
Pelli, B., 55[394] (45)
Pellinghelli, M. A., 52[166] (11)
Penavic, M., 172[110] (159)
Pendlebury, R. E., 132[163] (72)
Peng, S.-M., 51[106] (5, 6, 8, 12)
Penner-Hahn, J. E., 133[228] (127)
Perevalova, E. G., 50[74] (3, 17), 50[75] (3), 50[77] (4, 40), 51[80] (4), 51[81] (4, 11, 12), 51[82] (4, 5, 12–14), 51[96] (5), 51[97] (5), 52[155] (9–12), 52[161] (10, 11), 52[168] (12), 52[170] (13), 52[190] (15), 54[329] (36)
Perreault, D., 55[421] (49)
Perry, W. D., 171[20] (140)
Persson, E. S. M., 131[112] (65, 75, 128)
Persson, I., 172[91] (155), 172[93] (155)
Perthius, J., 171[40]
Petrovskii, P. V., 172[75]
Pietroni, B. R., 51[100] (5, 14), 51[107] (6), 51[120] (6), 51[137] (9, 11, 47), 52[151] (10, 11, 47), 52[164] (11, 47), 52[178] (13, 14), 52[179] (13), 52[184] (14), 132[171] (73)
Pignolet, L. H., 55[419] (49)
Pikul, S., 131[91] (63)
Piper, T. S., 132[135] (68)
Pocev, S., 172[87] (155)
Poilblanc, R., 53[258] (24, 39)
Poisson, P., 171[40]
Poll, W., 204[37] (179)
Poplawska, J., 131[91] (63)
Popović, Z., 172[105] (157), 172[106] (158), 172[108] (159)
Porter, L. C., 53[239] (20, 30), 53[240] (20, 29, 30), 53[250] (22), 53[251] (22), 53[267] (28–30), 53[270] (28), 53[271] (28), 53[272] (27, 28), 53[274] (27, 28, 43), 53[276] (28), 53[277] (28), 54[284] (28), 54[288] (26, 29), 54[289] (27, 29), 54[290] (26, 29), 54[292] (27, 29, 30), 54[298] (29, 31)
Posner, G. H., 130[1h] (57, 60, 65, 95, 125, 127), 130[1i] (57, 60, 65, 95, 125, 127), 131[110] (65, 75)
Povey, D. C., 172[113] (159)
Powell, A. K., 205[93] (186)
Power, P. P., 130[14a] (61, 78, 109), 130[16] (58, 61, 64, 65, 78, 84, 101, 120, 121), 131[62] (61),

131[64] (61, 65, 78, 106), 131[102] (64, 78, 101, 113, 120, 121), 133[205] (78, 115), 133[208b] (78, 82, 84, 88, 129), 133[214] (100), 133[226] (127), 204[80] (185, 186), 205[85] (186)
Prasad, L. S., 50[32] (2)
Preses, J., 205[103] (188)
Pringle, P. G., 51[110] (6)
Prior, N. D., 55[377] (43)
Pritzkow, H., 52[162] (10, 11), 52[186] (14)
Puddephatt, R. J., 50[29] (2–6, 10), 50[71] (3, 6, 17, 20, 21, 31–3, 39, 43, 44, 48), 51[86] (4), 51[88] (4), 51[92] (4), 51[105] (6), 51[111] (6), 51[112] (6), 51[121] (5, 6), 51[122] (5, 7), 51[123] (5, 7, 8), 51[124] (7), 51[125] (7, 8), 55[355] (40), 55[364] (40, 41)
Puente, F., 53[236] (20, 22)
Purdy, A. P., 204[31] (179)
Pyykkö, P., 50[31] (2, 3), 50[58] (3), 50[60] (3, 17)

Qiu, S., 55[412] (48)

Raabe, G., 204[77] (184)
Rademacher, B., 204[24] (177), 204[32] (179), 204[33] (179), 204[34] (179), 204[37] (179), 204[38] (179)
Radinov, R. N., 205[121] (189), 205[133] (190)
Rahman, A. F. M. M., 172[92] (155)
Rainford, D., 205[139] (193)
Raithby, P. R., 54[316] (33), 54[317] (33), 55[377] (43), 131[100] (64, 67, 78, 100, 108)
Rajagopal, D., 171[13] (136, 137, 148)
Rajagopalan, S., 171[10] (136)
Ramachandran, R., 51[105] (6)
Randaccio, L., 130[24] (60), 130[31] (61, 68, 78, 80), 130[35] (61, 71, 78), 130[40] (61, 88)
Rao, S. A., 205[86] (186)
Rapsomanikis, S., 171[9] (136)
Rapson, W. S., 49[2] (1)
Raptis, R. G., 52[152] (9, 10), 53[239] (20, 30), 53[240] (20, 29, 30), 53[265] (24), 54[289] (27, 29), 54[290] (26, 29), 54[292] (27, 29, 30), 54[298] (29, 31)
Rassu, G., 131[55] (61, 67, 78, 97)
Raston, C. L., 51[85] (4, 5, 12), 130[27] (61, 77, 78), 130[28] (61, 77, 78), 132[158] (72, 77), 203[16] (177), 204[60] (181), 206[166] (197)
Rau, M. S., 206[215] (203)
Raubenheimer, H. G., 52[167] (11), 55[404] (47)
Rausch, M. D., 131[98] (64, 77), 132[138] (68), 172[121] (160)
Ravindranath, R., 51[121] (5, 6)
Razavi, A., 172[121] (160)
Reber, G., 54[287] (27, 28, 30)
Regitz, M., 173[146] (168)
Rei, M.-H., 171[49] (145)
Reich, R., 58, 130[4] (58)
Reid, B. D., 50[76] (4, 32)
Rennie, M. A., 131[100] (64, 67, 78, 100, 108)
Reshetova, M. D., 51[97] (5)
Retta, N., 204[36] (179)

Reutov, O. A., 172[80] (154)
Rheingold, A. L., 172[127] (162)
Rhodes, C. J., 173[143] (166)
Ricca, G., 205[150] (195)
Riccoboni, L., 133[186] (76)
Rice, D. A., 205[155] (196), 205[156] (196), 206[171] (197, 201), 206[172] (197), 206[173] (197)
Richey, H. G., Jr., 203[23] (177), 204[35] (179), 204[83] (185)
Richter, W., 130[34] (61, 71), 205[102] (188)
Rideout, J., 173[142] (166)
Ridley, D., 206[209] (202)
Riede, J., 50[56] (3), 54[295] (29, 30)
Rieke, R. D., 70, 132[148] (70)
Riera, V., 51[127] (5, 8), 51[136] (9, 10), 52[147] (10), 52[187] (14), 55[389] (44)
Rindorf, G., 50[35] (2)
Robertson, G. B., 51[131] (8, 10), 51[142] (10)
Robino, P., 55[413] (48), 55[414] (48)
Robinson, A. J., 203[18] (177)
Rodes, R., 171[58] (150)
Rodier, N., 172[63] (150)
Rodríguez, M. L., 52[147] (10)
Rogers, R. D., 53[260] (24)
Rojo, A., 52[149] (10, 42)
Rokhlin, E. M., 171[35] (144)
Roland, J. R., 132[141] (70)
Romanenko, V. D., 204[63] (182)
Roos, M., 55[404] (47)
Rösch, N., 50[30] (2, 3, 17), 52[211] (17)
Rossiter, B. E., 130[1d] (57, 60, 65, 95, 125, 127), 131[113] (65), 133[229] (127)
Rotella, F. J., 133[238]
Rotteveel, M. A., 130[45] (61, 66, 78, 88, 91)
Rozema, M. J., 171[13] (136, 137, 148)
Rozyatovskii, V. A., 172[84] (155, 158)
Rudolph, W., 204[68] (182)
Ruhlandt-Senge, K., 133[205] (78, 115), 205[85] (186)
Ruiz, A., 132[183] (75), 132[184] (75)
Ruiz, J., 52[187] (14)
Ruiz-Romero, M. E., 53[223] (18)
Rundle, R. E., 206[203] (201)
Rushworth, S. A., 204[28] (178), 206[170] (197, 200), 206[201] (200), 206[206] (201)
Russel, C. A., 131[100] (64, 67, 78, 100, 108)
Russell, G. A., 169, 173[144] (167), 173[148] (168), 173[154] (168), 173[155] (169), 173[156] (169)
Rybin, L. I., 172[82] (154)
Rypdal, K., 133[210] (61, 82)

Sadler, P. J., 50[18] (2), 50[21] (2)
Saegusa, T., 132[133] (68), 132[134] (68), 132[136] (68), 132[142] (70), 133[207] (121)
Said, F. F., 206[176] (199)
Sakhawat Hussein, M., 131[67] (61, 69, 75, 78, 109)
Salim, S., 206[196] (200, 201)
Salupo, T. A., 53[264] (24)
Samuel, O., 205[139] (193)
Sánchez, G., 51[136] (9, 10), 55[389] (44)
Sánchez-Santano, M.-J., 54[312] (32, 38), 54[343] (37), 54[346] (37),

54[347] (37), 54[348] (37), 54[350] (38)

Sandell, J., 131[64] (61, 65, 78, 106)

Sandhu, M. K., 172[113] (159)

Sanmartín, F., 51[102] (6)

Sanner, R. D., 50[78] (4, 33, 40)

Sansoni, M., 52[203] (17), 52[204] (17), 131[55] (61, 67, 78, 97)

Sappenfield, E. L., 55[388] (44)

Sarroca, C., 51[144] (10, 22, 40, 41)

Sasaoka, S., 171[41]

Satcher, J. H., Jr., 50[78] (4, 33, 40)

Saura-Llamas, I., 52[188] (14, 23, 24), 52[189] (14, 23, 24), 52[213] (18), 53[231] (20–3), 131[75] (61, 72, 78)

Savas, M. M., 53[260] (24)

Savoie, R., 172[99] (156), 172[100] (156)

Sawada, S., 205[145] (194)

Sazonenko, M. M., 51[81] (4, 11, 12)

Sazonova, V. A., 130[29] (61, 78, 117), 131[68] (61, 78, 117), 131[79] (60)

Schaal, M., 55[392] (44)

Schaap, C. A., 132[178] (74, 122, 124)

Schat, G., 204[29a] (178), 204[29b] (178)

Scheidt, W. R., 203[8] (176)

Scherbaum, F., 52[185] (14, 15), 52[194] (15), 52[197] (16), 52[198] (16, 17)

Scherer, W., 133[210] (61, 82)

Schier, A., 52[193] (15), 53[246] (21)

Schindehutte, M., 51[140] (10)

Schipor, I., 205[86] (186)

Schlebos, P. P. J., 55[374] (43)

Schlemper, E. O., 131[67] (61, 69, 75, 78, 109)

Schlingloff, G., 204[79] (185)

Schlosser, M., 130[2] (57)

Schmidbaur, H., 49[1] (1), 50[30] (2, 3, 17), 50[34] (2), 50[51] (2, 8), 50[52] (2), 50[53] (2), 50[54] (3), 50[55] (3, 17, 43), 50[56] (3), 50[72] (3), 50[73] (3, 4), 52[169] (12), 52[180] (14), 52[185] (14, 15), 52[193] (15), 52[194] (15), 52[195] (15), 52[196] (16), 52[197] (16), 52[198] (16, 17), 52[199] (16), 52[200] (16), 52[201] (16), 52[211] (17), 53[228] (19, 20, 23, 26), 53[233] (20, 22), 53[234] (20, 22), 53[243] (20, 22, 23), 53[246] (21), 53[247] (21), 53[256] (23), 53[266] (28, 29), 54[287] (27, 28, 30), 54[293] (28, 29), 54[295] (29, 30), 54[297] (29), 54[304] (30), 130[34] (61, 71), 132[149] (71, 76), 132[150] (71), 132[151] (71), 132[152] (71), 171[11] (136, 163), 172[94] (155)

Schmidt, B., 205[125] (190), 205[126] (190)

Schmitz, M., 71, 132[156] (71)

Schneller, P., 53[215] (18)

Schoone, J. C., 130[12] (61, 67, 78, 95)

Schrock, R. R., 206[169] (197, 199)

Schubert, U., 54[293] (28, 29)

Schumann, H., 52[159] (9, 10)

Schumann, U., 133[235]

Schwank, J., 50[10] (2)

Schwarz, W., 204[24] (177), 204[32] (179), 204[34] (179), 204[38] (179)

Schwarzenbach, K., 194, 205[144] (194)

Schwerdtfeger, P., 50[59] (3)

Scott, F., 55[404] (47)

Scott, J. D., 51[123] (5, 7, 8)

Sebald, A., 171[2] (135), 172[83] (154)

Sedova, N. N., 130[29] (61, 78, 117), 131[68] (61, 78, 117), 131[79] (60)

Seebach, D., 205[125] (190), 205[126] (190)

Segre, A., 55[413] (48), 55[414] (48)

Seitz, L. M., 131[92] (63, 65)

Semerano, G., 133[186] (76)

Sengupta, S., 130[1a] (57, 60, 65, 95, 125, 127)

Shabarov, Y. S., 172[67]

Shain, J., 54[284] (28)

Sharma, S., 131[106] (65, 93, 127)

Shaw, B. L., 51[108] (6), 51[110] (6)

Shaw, C. F., III, 50[20] (2)

Shearer, H. M. M., 130[13] (61, 78, 97), 131[54] (61, 97), 131[73] (61, 78, 97, 120), 133[196] (75), 204[57] (181), 204[84] (186)

Sheldrick, G. M., 51[130] (8, 10, 24), 52[191] (15), 52[192] (15), 53[225] (19), 54[314] (32), 54[324] (34), 55[353] (39), 55[354] (39), 55[357] (40, 41, 43), 55[396] (45)

Shen, L., 172[114] (160)

Shenai-Khatkhate, D. V., 204[26] (178), 204[27] (178)

Sheppard, W. A., 130[5] (58, 76), 132[118] (65, 70, 76), 132[141] (70)

Sheridan, L. A., 205[156] (196)

Shibata, S., 54[325] (34)

Shibue, A., 55[356] (40, 41), 55[358] (40, 41), 55[363] (40, 41), 55[371] (43)

Shimazu, S., 133[240]

Shin, S.-K., 132[147] (70)

Shirot, M., 205[129] (190)

Shnol, T. R., 204[76] (184)

Shoner, S. C., 204[80] (185, 186)

Shul'gin, V. F., 204[63] (182)

Shur, V. B., 172[133] (164), 173[134] (164)

Siebert, W., 52[162] (10, 11), 52[186] (14)

Sigalas, M. P., 133[202] (78, 95)

Sikirica, M., 172[104] (157), 172[105] (157), 172[106] (158), 172[107] (159), 172[108] (159)

Sillmon, R. S., 206[207] (201)

Šima, J., 51[89] (4)

Simard, M., 172[131] (163)

Simon, J., 49[3] (1)

Simonov, A. P., 206[181] (200), 206[182] (200)

Simpson, J., 205[107] (188)

Simpson, W. I., 206[184] (200)

Skelton, B. W., 51[103] (6), 52[158] (9, 10), 52[160] (9, 10), 54[335] (36), 172[96] (155), 204[60] (181)

Skorobogatova, E. V., 171[43], 171[44]

Sladkov, A. M., 132[125] (67, 76), 132[126] (67, 76)

Slee, T., 50[24] (2)

Slovokhotov, Y. L., 51[80] (4), 51[82] (4, 5, 12–14), 52[156] (9, 10), 52[161] (10, 11), 52[163] (11, 12)

Smeets, W. J. J., 130[15] (58, 61, 62, 65, 66, 69, 75, 77, 78, 93, 99, 101, 128), 130[17] (58, 61, 78, 84, 86, 105, 129), 130[50] (61, 66, 78, 95), 132[119] (61, 65, 69, 76, 78), 133[211] (78, 91), 204[29a] (178), 204[52] (180), 206[167] (197), 206[168] (197)

Smith, G. W., 172[113] (159)

Smith, J. D., 130[14b] (61, 78, 106), 132[175] (122), 133[204] (78, 106), 171[7] (136), 171[8] (136, 137), 204[36] (179), 204[40] (179), 204[41] (180), 206[158] (196), 206[159] (196)

Smith, L. M., 200, 206[188] (200), 206[189] (200)

Smith, W. E., 50[66] (3), 50[67] (3), 50[68] (3)

Smits, J. M. M., 55[418] (49)

Smyslova, E. I., 50[75] (3), 50[77] (4, 40), 51[80] (4), 51[82] (4, 5, 12–14), 52[170] (13), 52[172] (13)

Snow, M. R., 51[113] (5, 6, 8), 51[128] (5, 8)

Snyder, J. P., 133[220] (124), 133[229] (127)

Soai, K., 189, 191, 192, 205[110] (188), 205[115] (189, 191), 205[116] (189), 205[117] (189), 205[118] (189), 205[120] (189), 205[122] (189), 205[123] (189), 205[124] (190), 205[134] (191), 205[153] (195)

Sokolenko, U. A., 172[75]

Sokolov, V. I., 172[75]

Sokolova, T. N., 171[43], 172[62] (150)

Soláns, J., 51[127] (5, 8)

Soláns, X., 51[127] (5, 8), 51[139] (9, 10), 52[187] (14), 54[342] (37), 55[389] (44), 130[53] (61, 67, 78, 97)

Sone, T., 51[98] (5), 55[359] (40, 41)

Sotes, M., 132[184] (75)

Søtofte, I., 130[6b] (58, 78, 87, 117)

Sousa, L. R., 54[316] (33), 54[317] (33)

Spangler, D. P., 133[220] (124), 133[229] (127)

Spek, A. L., 51[84] (4), 130[12] (61, 67, 78, 95), 130[15] (58, 61, 62, 65, 66, 69, 75, 77, 78, 93, 99, 101, 128), 130[17] (58, 61, 78, 84, 86, 105, 129), 130[41] (61, 78, 82), 130[50] (61, 66, 78, 95), 130[51] (61, 62, 69, 77, 78, 98), 130[52] (61, 62, 66, 69, 77, 78, 98, 124), 131[58] (61, 67, 76–8, 82, 97, 122, 124), 131[85] (62, 66, 78, 87, 93, 121, 124, 127), 132[119] (61, 65, 69, 76, 78), 132[130] (68, 77), 132[170] (73), 133[211] (78, 91), 203[13] (177), 203[14] (177), 204[29a] (178), 204[29b] (178), 204[48] (180), 204[50] (180), 204[52] (180), 204[53] (180), 204[54] (181), 204[56] (181, 202), 204[59] (181), 204[67] (182), 204[78] (185), 205[151] (195), 206[167] (197), 206[168] (197)

Spencer, C. B., 204[57] (181)

Spencer, M. D., 171[46], 171[51] (147)

Speroni, F., 51[94] (5)

Spiegelmann, F., 55[386] (44)

Srebnik, M., 205[127] (190)

Stakheeva, E. N., 131[68] (61, 78, 117)

Stalteri, M. A., 55[364] (40, 41)

Stam, C. H., 52[176] (13), 130[10] (58, 61, 64, 75, 78, 101, 109), 130[39] (61, 66, 69, 78, 93), 130[45] (61, 66, 78, 88, 91), 130[48] (61, 66, 78, 91), 133[185] (75, 77, 86, 121, 129), 133[215] (109, 121), 133[234] (101, 124)

Staples, R. J., 50¹⁵ (2), 54²⁸⁹ (27, 29)
Starowieyski, K. B., 204⁴³ (180), 204⁴⁴ (180)
Steel, P. J., 172⁷⁴
Steggerda, J. J., 50²⁶ (2), 55³⁷⁴ (43), 55⁴¹⁸ (49), 55⁴¹⁹ (49), 133²³⁷ (117)
Steigelmann, O., 52¹⁹⁵ (15), 52¹⁹⁶ (16), 52¹⁹⁹ (16), 52²⁰⁰ (16), 52²⁰¹ (16)
Steigerwald, M. L., 206²¹² (203)
Stein, J., 53²⁴² (20, 22)
Stemmler, T., 133²²⁸ (127)
Sterlin, S. R., 171³¹ (143)
Sterling, J. J., 131¹¹⁰ (65, 75)
Stewart, K. R., 133²¹⁹ (124)
Still, W. C., 172⁷³
Stobart, S. R., 172¹¹² (159), 203¹⁵ (177), 205¹⁵⁷ (196), 206¹⁶⁴ (197)
Stocco, G., 50²⁰ (2)
Stojak, G. H., 173¹⁴⁰ (166)
Stone, F. G. A., 55³⁸⁸ (44), 55³⁹⁰ (44), 55³⁹¹ (44), 131⁹⁹ (64), 133²⁰⁹ (78, 116)
Strähle, J., 54³³² (36), 54³³³ (36), 54³³⁴ (36), 55³⁸⁷ (44), 130⁴² (61, 82)
Stretton, G. N., 171²⁴ (141), 171³⁰ (143, 144), 171³³ (143)
Stringfellow, G. B., 206¹⁸⁵ (200)
Strologo, S., 130³⁶ (61, 78, 110)
Strother, S., 171¹⁶ (137)
Struchkov, Yu. T., 51⁸⁰ (4), 51⁸² (4, 5, 12–14), 52¹⁵⁶ (9, 10), 52¹⁶¹ (10, 11), 52¹⁶³ (11, 12), 52¹⁹⁰ (15), 130²⁹ (61, 78, 117), 131⁶⁸ (61, 78, 117), 171³⁵ (144), 204⁶³ (182)
Stumpf, K., 52¹⁶² (10, 11)
Suga, S., 205¹¹⁴ (189), 205¹³⁰ (190, 192), 205¹⁴² (193)
Sugimura, T., 205¹⁴⁶ (194)
Suginome, M., 51⁸³ (4–6), 51⁹⁹ (5)
Sullivan, A. C., 130¹⁴ᵇ (61, 78, 106), 132¹⁷⁵ (122), 133²⁰⁴ (78, 106)
Sun, K. K., 133¹⁹⁰ (75, 76)
Sundararajan, G., 131¹¹³ (65)
Sutcliffe, R., 55³⁸² (44)
Suzuki, S., 49⁶ (2)
Swepston, P. N., 203¹² (176)
Swingle, N. M., 130¹ᵈ (57, 60, 65, 95, 125, 127), 131¹¹³ (65)
Sýkora, J., 51⁸⁹ (4)
Symons, M. C. R., 173¹⁴² (166)
Szafranski, C. A., 55³⁸⁵ (44)

Tai, A., 205¹⁴⁶ (194)
Takahashi, H., 205¹⁴⁸ (195)
Takats, J., 131⁸¹ (60, 76–8, 115)
Takehana, K., 205¹⁴⁵ (194)
Tallant, N. A., 147, 171⁵¹ (147), 171⁵⁴ (147)
Tariverdian, P. A., 131⁷¹ (61, 72, 78, 100, 107)
Tartón, M. T., 54³⁰⁵ (31)
Tashtoush, H., 173¹⁵⁵ (169)
Tauler, E., 52¹⁸⁷ (14)
Tavakkoli, K., 171⁸ (136, 137)
Taylor, D. R., 132¹⁶⁵ (72)
Taylor, R. J. K., 160, 171²³ (141, 160), 130¹ᶜ (57, 60, 65, 95, 125, 127)

Taylor, S. E., 172⁹⁵ (155)
ten Hoedt, R. W. M., 130¹² (61, 67, 78, 95), 130⁴⁷ (61), 131⁵⁸ (61, 67, 76–8, 82, 97, 122, 124), 131⁵⁹ (61, 67, 69, 89, 97, 124), 131⁶⁰ (61, 69, 75), 131⁷⁸ (60, 67), 133¹⁹¹ (76)
Theobald, F., 172⁶³ (150)
Thiel, W. R., 133²¹⁰ (61, 82)
Thiele, K.-H., 201, 203²⁰ (177), 204⁶⁶ (182)
Thomas, I. M., 54³³⁹ (36)
Thomas, K. M., 130⁷ (58, 66, 78, 89)
Thompson, J., 200, 206¹⁸⁸ (200), 206¹⁸⁹ (200)
Thompson, K. L., 171⁴⁹ (145)
Thöne, C., 51¹⁴³ (10)
Thorup, N., 50³⁵ (2)
Tiekink, E. R. T., 51¹²⁸ (5, 8), 55³⁷⁶ (43)
Tipsword, G. H., 133²²⁰ (124)
Tiripicchio, A., 52¹⁶⁶ (11)
Titcombe, L., 54³⁴⁰ (36)
Tobe, M. L., 54³⁰⁸ (32)
Tocher, D. A., 53²⁸² (28, 29), 54²⁸³ (28, 29), 172¹¹⁶ (160)
Tocher, J. H., 54³⁰⁰ (26)
Togni, A., 50¹² (2)
Tomás, M., 132¹⁸³ (75)
Tomietto, M., 55⁴⁰⁶ (47)
Tomita, S., 132¹³⁶ (68)
Tommack, P., 205¹⁰² (188)
Torocheshnikov, V. N., 172⁸⁴ (155, 158)
Torregiani, E., 51¹⁰⁷ (6)
Traldi, P., 55³⁹⁴ (45)
Treurnicht, I., 50²⁹ (2–6, 10), 51¹⁰⁵ (6), 51¹¹¹ (6), 51¹¹² (6), 51¹²¹ (5, 6)
Trovato, M. P., 171⁵⁶ (149)
Trzcinska-Bancroft, B., 53²⁶⁹ (26, 28, 29), 53²⁸¹ (27, 28), 54²⁹⁹ (29, 31)
Tse, J. S., 51⁸⁶ (4)
Tsuda, T., 132¹³³ (68), 132¹³⁴ (68), 132¹⁴² (70), 133²⁰⁷ (121)
Tsukiyama, K., 205¹⁰³ (188)
Tsutsui, M., 133²⁰⁰ (77, 82, 129)
Tuck, D. G., 132¹⁵⁴ (71), 204⁶⁴ (182)
Tucker, C. E., 171¹³ (136, 137, 148)
Turpin, J., 53²³¹ (20–3)
Tyrra, W., 204⁶⁸ (182), 206¹⁶² (196)

Uang, B.-J., 173¹⁴⁹ (168)
Uemura, M., 205¹²⁸ (190), 205¹²⁹ (190)
Ugo, R., 133²⁰⁰ (77, 82, 129)
Ukaji, Y., 205¹⁴⁷ (195)
Ukhin, L. Yu., 132¹²⁶ (67, 76)
Ulibarri, T. A., 133²³⁶ (78, 115)
Ullenius, C., 133²³⁰ (128)
Umen, M. J., 131⁸⁴ (60)
Urano, T., 171¹⁵ (137)
Usón, A., 53²³⁰ (20, 22), 53²³⁵ (20, 21), 53²⁴⁵ (20, 22), 53²⁵² (23), 53²⁵³ (23), 131⁶⁹ (61, 74, 77, 78, 111, 115), 132¹⁷³ (74, 120)
Usón, R., 50⁶⁹ (3, 8, 9, 18, 19, 32, 33, 37, 39, 43), 51¹²⁹ (8), 51¹³⁰ (8, 10, 24), 51¹³³ (9), 51¹³⁴ (9, 10), 51¹³⁵ (9, 45), 51¹⁴⁶ (10, 18), 52¹⁴⁸ (10, 40, 41), 52¹⁵⁰ (10, 41), 52¹⁷³ (13), 53²²³ (18),

53²²⁴ (19), 53²²⁵ (19), 53²²⁶ (19), 53²²⁷ (19), 53²³⁰ (20, 22), 53²³⁵ (20, 21), 53²⁴¹ (20, 22), 53²⁴⁵ (20, 22), 53²⁵² (23), 53²⁵³ (23), 53²⁵⁵ (23, 40, 41), 53²⁶⁸ (26–8), 54³⁰⁵ (31), 54³⁰⁶ (31), 54³⁰⁹ (32), 54³²² (33), 54³²⁴ (34), 54³²⁶ (35, 37), 54³²⁸ (36, 40, 41), 54³³¹ (36), 55³⁵³ (39), 55³⁵⁴ (39), 55³⁵⁷ (40, 41, 43), 55³⁶⁰ (40, 41), 55³⁶¹ (40, 41), 55³⁶⁶ (41), 55³⁶⁷ (41), 55³⁹⁶ (45), 131⁶⁹ (61, 74, 77, 78, 111, 115), 132¹⁵⁷ (72), 132¹⁷² (73, 74), 132¹⁷³ (74, 120), 132¹⁸⁰ (74)
Ustynyuk, Y. A., 172⁸⁴ (155, 158)

Valle, G., 55³⁹³ (44, 46), 55³⁹⁸ (46)
van Dam, H., 133²³⁹ (120)
van den Berg, E., 55³⁷⁴ (43)
van der Kerk, G. J. M., 130²¹ (60, 71–3, 76), 131⁹³ (64, 76), 132¹⁶⁶ (73), 132¹⁶⁸ (73, 76, 120), 132¹⁶⁹ (73), 132¹⁷⁴ (74, 125, 127), 133¹⁸⁹ (75, 76), 203⁹ (176), 204⁴⁵ (180), 204⁴⁸ (180), 204⁵⁰ (180), 204⁵⁶ (181, 202), 204⁵⁹ (181), 205¹⁵¹ (195)
van der Sluis, P., 130¹⁷ (58, 61, 78, 84, 86, 105, 129)
van der Steen, F. H., 204⁵³ (180), 204⁵⁴ (181), 204⁶⁷ (182)
van der Velden, J. W. A., 50²⁶ (2)
van Klaveren, M., 131¹¹¹ (65, 66, 75, 128), 131¹¹² (65, 75, 128)
van Koten, G., 52¹⁷⁶ (13), 62, 130⁶ᵃ (58, 62, 78, 87, 117), 130⁶ᵇ (58, 78, 87, 117), 130⁷ (58, 66, 78, 89), 130⁸ (58, 64), 130⁹ (58, 73–5), 130¹⁰ (58, 61, 64, 75, 78, 101, 109), 130¹² (61, 67, 78, 95), 130¹⁵ (58, 61, 62, 65, 66, 69, 75, 77, 78, 93, 99, 101, 128), 130¹⁷ (58, 61, 78, 84, 86, 105, 129), 130¹⁹ (60, 65, 75, 82, 90, 97, 120, 121, 129), 130³⁹ (61, 66, 69, 78, 93), 130⁴¹ (61, 78, 82), 130⁴³ (61, 76), 130⁴⁵ (61, 66, 78, 88, 91), 130⁴⁶ (61, 86, 88, 120, 125), 130⁴⁷ (61), 130⁴⁸ (61, 66, 78, 91), 130⁵¹ (61, 62, 69, 77, 78, 98), 130⁵² (61, 62, 66, 69, 77, 78, 98, 124), 131⁵⁸ (61, 67, 76–8, 82, 97, 122, 124), 131⁵⁹ (61, 67, 69, 89, 97, 124), 131⁶⁰ (61, 69, 75), 131⁷⁸ (60, 67), 131⁸² (60, 74), 131⁸⁵ (62, 66, 78, 87, 93, 121, 124, 127), 131⁸⁷ (63, 75, 76, 124), 131⁸⁸ (63, 66), 131⁸⁹ (63), 131¹⁰³ (64, 65, 69, 127), 131¹⁰⁴ (64, 69, 82, 93, 124, 129), 131¹¹¹ (65, 66, 75, 128), 131¹¹² (65, 75, 128), 132¹¹⁹ (61, 65, 69, 76, 78), 132¹²⁰ (66, 76, 86), 132¹²¹ (66, 76, 77, 89), 132¹²² (66, 78, 88, 91), 132¹²³ (67, 69, 78, 89, 97), 132¹²⁹ (68, 76), 132¹³⁰ (68, 77), 132¹³⁹ (69, 77, 82, 86, 121, 128), 132¹⁴⁴ (70), 132¹⁶⁷ (73, 74, 86, 125), 132¹⁷⁰ (73), 132¹⁷⁶ (74–6, 125), 132¹⁷⁷ (74), 132¹⁷⁸

(74, 122, 124), 133[185] (75, 77, 86, 121, 129), 133[191] (76), 133[199] (76), 133[200] (77, 82, 129), 133[208a] (82, 88, 126, 129), 133[211] (78, 91), 133[215] (109, 121), 133[232] (128), 133[233] (129), 133[234] (101, 124), 133[239] (120), 172[119] (160), 204[52] (180), 204[53] (180), 204[54] (181), 204[67] (182), 204[78] (185)

van Mier, G. P. M., 172[119] (160), 206[167] (197), 206[168] (197)

van Rijn, P. E., 203[9] (176), 204[45] (180)

van Rooyen, P. H., 51[140] (10), 52[167] (11)

van Vliet, M. R. P., 204[58] (181), 204[78] (185)

Varakin, V. N., 206[181] (200), 206[182] (200)

Vardhan, H. B., 172[92] (155), 172[98] (156)

Vaugeois, J., 172[131] (163)

Vaughan, W. R., 205[149] (195)

Vecchio, A. M., 51[119] (5, 6)

Veldkamp, A., 55[415] (48)

Verne, H. P., 133[210] (61, 82)

Veya, P., 51[101] (5)

Vicente, J., 51[139] (9, 10), 51[141] (9, 10), 52[181] (14), 52[182] (14), 52[183] (14), 52[188] (14, 23, 24), 52[189] (14, 23, 24), 52[213] (18), 53[231] (20–3), 53[232] (20–3), 54[310] (32), 54[311] (32), 54[312] (32, 38), 54[313] (32), 54[314] (32), 54[323] (34), 54[324] (34), 54[326] (35, 37), 54[342] (37), 54[343] (37), 54[344] (37), 54[345] (37), 54[346] (37), 54[347] (37), 54[348] (37), 54[349] (37), 54[350] (38), 54[351] (38), 55[352] (38), 131[74] (61, 72, 78, 115), 131[75] (61, 72, 78)

Villacampa, M. D., 51[135] (9, 45), 55[361] (40, 41), 55[396] (45)

Villacorta, G. M., 131[110] (65, 75)

Visser, G. W. M., 172[118] (160)

Vittal, J. J., 51[122] (5, 7), 51[123] (5, 7, 8), 51[124] (7), 51[125] (7, 8)

Volden, H. V., 51[88] (4)

Volgin, Yu. V., 130[29] (61, 78, 117)

Volkonskii, A. Yu., 171[35] (144)

Waas, J., 203[6] (176)

Wade, K., 131[54] (61, 97)

Wagner, F. E., 51[107] (6), 54[304] (30), 55[393] (44, 46)

Wagner, R., 133[231] (128)

Wald, K., 50[55] (3, 17, 43)

Walker, N. P. C., 172[125] (161)

Wallace, F. A., 131[71] (61, 72, 78, 100, 107)

Walsh, J. R., 203[17] (177, 197), 203[18] (177), 203[19a] (177), 203[19b] (177), 204[28] (178), 205[92] (186), 206[170] (197, 200), 206[194] (200, 201), 206[199] (200), 206[201] (200), 206[202] (200), 206[206] (201)

Walton, J. K., 52[158] (9, 10)

Wang, C., 172[79] (154)

Wang, J.-C., 53[259] (24)

Wang, S., 54[302] (30), 54[303] (30), 131[76] (61, 72, 75, 78)

Wang, X., 172[114] (160)

Wang, Z., 55[403] (47)

Wanklyn, J. A., 71, 132[155] (71)

Wardell, J. L., 171[1] (135, 136, 144, 151, 153, 162), 171[4] (136), 171[5] (136)

Watanabe, K., 133[207] (121)

Watanabe, M., 205[116] (189), 205[117] (189), 205[118] (189), 205[120] (189), 205[128] (190)

Watkins, J. J., 133[221] (125), 133[222] (125), 133[224] (125)

Watkins, J. W., II, 49[4] (1)

Watson, M. J., 50[23] (2)

Webb, G., 132[165] (72)

Webb, R. L., 56[422] (49)

Wehman, E., 130[44] (61, 78, 88, 120), 130[45] (61, 66, 78, 88, 91), 130[48] (61, 66, 78, 91), 132[122] (66, 78, 88, 91)

Wehman-Ooyevaar, I. C. M., 133[211] (78, 91)

Wehmeyer, R. M., 70, 132[148] (70)

Weidenbruch, M., 206[177] (199)

Weidlein, J., 204[32] (179)

Weigand, W., 55[392] (44)

Weijers, F., 132[169] (73)

Weiss, E., 64, 131[65] (60, 61, 64, 65, 69, 78, 101), 133[203] (78, 120), 133[235]

Weiss, H., 55[387] (44)

Weiss, W., 171[22] (140)

Weissig, V., 203[20] (177)

Welch, A. J., 50[76] (4, 32), 132[183] (75), 132[184] (75)

Wells, A. F., 133[216] (122)

Wen, Y.-S., 52[177] (13, 23, 24)

Werner, H., 52[154] (9, 10), 131[97] (64), 132[137] (68, 115)

Westerhausen, M., 204[24] (177), 204[32] (179), 204[33] (179), 204[34] (179), 204[37] (179), 204[38] (179)

Whangbo, M.-H., 133[219] (124)

Wheeler, A. C., 52[207] (17)

Whetten, R. L., 55[385] (44)

Whitcombe, R. A., 206[204] (201)

White, A. H., 51[85] (4, 5, 12), 51[103] (6), 52[157] (9), 52[158] (9, 10), 52[160] (9, 10), 54[335] (36), 54[336] (36), 54[337] (36), 54[338] (36), 130[27] (61, 77, 78), 130[28] (61, 77, 78), 132[158] (72, 77), 172[96] (155), 172[97] (156), 203[16] (177), 204[60] (181), 206[166] (197)

Whitesides, G. M., 50[16] (2)

Whitmire, K. H., 54[321] (33)

Whitten, C. E., 131[110] (65, 75)

Whittington, B. I., 172[74]

Whitton, A. J., 172[117] (160)

Wickramasinghe, W. A., 51[142] (10)

Wiemers, D. M., 130[30] (61, 64), 131[95] (64), 206[161] (196)

Wijnhoven, J., 55[374] (43)

Wilhelm, R. S., 131[107] (65, 127), 131[109] (65)

Wilkes, B., 204[68] (182)

Wilkins, C. L., 55[395] (45)

Wilkins, J. M., 131[84] (60)

Wilkinson, D. L., 171[11] (136, 163)

Wilkinson, G., 132[135] (68), 203[3] (176), 205[132] (190)

Willert-Porada, M. A., 131[96] (64, 78, 117)

Williams, A. J., 206[190] (200)

Williams, J. O., 205[107] (188), 205[108] (188), 206[195] (200, 201)

Williams, M. J., 171[38]

Williams, M. L., 52[160] (9, 10)

Willis, A. C., 51[142] (10)

Willner, H., 55[407] (47, 48), 55[417] (48), 172[79] (154)

Wilson, D. R., 203[12] (176)

Wilson, S. R., 204[69] (182)

Winter, C. H., 172[127] (162)

Wipf, P., 130[3] (57)

Wittig, G., 194, 205[144] (194)

Wong, W.-T., 55[377] (43), 56[423] (49), 56[424] (49)

Woods, L. A., 131[101] (64, 65, 76)

Wordel, R., 55[393] (44, 46)

Wrackmeyer, B., 172[83] (154)

Wright, D. S., 131[100] (64, 67, 78, 100, 108)

Wright, P. J., 206[171] (197, 201), 206[187] (200), 206[190] (200), 206[191] (200), 206[192] (200), 206[193] (200, 201), 206[194] (200, 201), 206[199] (200), 206[200] (200), 206[201] (200), 206[205] (201)

Wright, T. C., 206[179] (199)

Wu, Y., 172[114] (160)

Wuest, J. D., 172[128] (163), 172[129] (163), 172[130] (163), 172[131] (163), 172[132] (163)

Wyatt, P. B., 203[18] (177)

Xu, X., 131[64] (61, 65, 78, 106)

Yam, V. W.-W., 50[14] (2), 54[341] (37)

Yamada, Y., 205[124] (190)

Yamamoto, A., 130[33b] (61, 63, 77, 78, 110), 132[128] (68), 132[132] (68, 76, 77), 205[118] (189)

Yamamoto, H., 133[207] (121)

Yamamoto, T., 132[128] (68), 132[132] (68, 76, 77)

Yamamoto, Y., 53[238] (20, 21), 53[248] (21), 53[249] (21), 130[1b] (57, 60, 65, 95, 125, 127), 130[1f] (57, 60, 65, 95, 125, 127), 132[150] (71), 132[151] (71), 204[46] (180)

Yang, H., 50[13] (2)

Yang, X., 171[6] (136, 165), 171[14] (137, 164, 165), 173[135] (164), 173[136] (165)

Yang, Z.-Y., 130[30] (61, 64)

Yasuda, H., 204[25] (177)

Yasuoka, N., 204[25] (177)

Yates, H. M., 206[195] (200, 201)

Yazawa, T., 133[207] (121)

Yazdi, S. N., 171[28] (142)

Yeh, M. C. P., 204[65] (182)

Yip, H.-K., 56[423] (49)

Yokoyama, S., 205[115] (189, 191)

Yoshioka, M., 205[148] (195)

Youngs, F., 205[103] (188)

Yu, C. F., 205[103] (188)

Yuan, H. S. H., 131[66] (61, 65, 78, 102, 109)

Yus, M., 171[37], 171[39], 171[47]

Zacharie, B., 172[128] (163), 172[129] (163), 172[130] (163), 172[132] (163)

Zangrando, E., 130[35] (61, 71, 78)

Zaworotko, M. J., 172[112] (159),
    206[164] (197)
Zaworotzo, M. J., 203[15] (177)
Zefirov, N. S., 171[43], 171[44], 172[62]
    (150)
Zeller, E., 50[54] (3)
Zhao, S.-H., 205[139] (193)
Zhao, Y., 50[31] (2, 3), 50[60] (3, 17)
Zheng, Z., 171[6] (136, 165), 171[14]
    (137, 164, 165), 173[136] (165)
Zhenyang, L., 50[24] (2)
Zhu, L., 172[126] (162)
Ziegler, J. B., 50[19]
Ziegler, M. L., 205[97] (187)
Zimmer-Gasser, B., 53[256] (23)
Zoroddu, M. A., 131[55] (61, 67, 78,
    97)
Zotova, S. V., 172[71]
Zubieta, J., 172[89] (155)
Zubieta, J. J., 133[185] (75, 77, 86, 121,
    129)
Zybill, C. E., 53[233] (20, 22), 131[61]
    (61, 68, 115), 132[153] (71, 78,
    116)

Subject Index

JOHN NEWTON

David John (Services), Slough, UK

This Subject Index contains individual entries to the text pages of Volume 3. The index covers general types of organometallic compound, specific organometallic compounds, general and specific organic compounds where their synthesis or use involves organometallic compounds, types of reaction (insertion, oxidative addition, etc.), spectroscopic techniques (NMR, IR, etc.), and topics involving organometallic compounds.

Because authors may have approached similar topics from different viewpoints, index entries to those topics may not always appear under the same headings. Both synonyms and alternatives should therefore be considered to obtain all the entries on a particular topic. Commonly used synonyms include alkyne/acetylene, compound/complex, preparation/synthesis, etc. Entries where the oxidative state of a metal has been specified occur after all the entries for the unspecified oxidation state, and the same or similar compounds may occur under both types of heading. Thus $Cr(C_6H_6)_2$ occurs under Chromium, bis(η-benzene) and again under Chromium(0), bis(η-benzene). Similar ligands may also occur in different entries. Thus a carbene–metal complex may occur under Carbene complexes, Carbene ligands, or Carbenes, as well as under the specific metal. Individual organometallic compounds may also be listed in the Cumulative Formula Index in Volume 14.

Acetic acid
  reaction with mercury(II)
    nitrate, 158
Acetone
  mercuriation, 139
Alanine
  reaction with methylmercury
    cations, 156
Aldehydes
  reaction with organozinc
    compounds, 190–2
Alkenes
  solvomercuriation, 144–8
Alkylaminomercuriation
  alkenes, 145
Alkylmercuriation
  alkynes, 137
Alkynes
  alkylmercuriation, 137
  solvomercuriation, 148–50
Aminomercuriation
  alkenes, 145
  alkynes, 150
Arylcopper magnesium arenethiolates
  structures, 104
Asymmetric synthesis
  organozinc compounds, 189,
    190
Auracyclopentadienes
  synthesis, 37
Aurates, trichloro(phenyl)-
  synthesis, 32
Aurates(I), bis(alkynyl)-, 19
Aurates(I), dialkyl-
  synthesis, 17
Aurates(I), dialkynyl-
  synthesis, 18

Aurates(I), polyhalogenophenyl-,
    18
Aurates(III), diaryl
  synthesis, 33
Aurates(III), tetraalkyl-
  synthesis, 42
Aurates(III), tetraaryl-
  synthesis, 42
Azidomercuriation
  alkenes, 145

Biphenylene
  mercuriation, 142
Bonds
  mercury–carbon
    formation, 136–54
    formation, transmetallation,
      136–54

Cadmium, bis(trifluorosilyl)-
  stability, 176
Cadmium, dimethyl-
  photochemistry, 200–3
Cadmium, trifluoromethyl-
  organocopper compound
    preparation from, 64
Carbenes
  insertion
    organomercury compounds
      synthesis, 153, 154
Carbonyl compounds
  α-mercuriated
    structures, 157–9
Chiral amplification
  organozinc compounds, 192,
    193
Copper(II) acetylacetonate

organocopper compound
    preparation from, 63
Copper, alkynyl-
  synthesis
    metallation, 67
Copper(I) bromide
  organocopper compound
    preparation from, 63
Copper, bromo(dimethyl sulfide)-
  organocopper compound
    preparation from, 60
Copper(I) carboxylate
  decarboxylation, 70
Copper complexes
  cyclopentadienyl
    organocopper compound
      preparation from, 64
Copper compounds
  alkyl
    synthesis, 71
  perfluoroalkyl
    synthesis, 70
  vinyl
    synthesis, 70
Copper(II) halides
  organocopper compound
    preparation from, 63
Copper, mesityl-
  organocuprate preparation
    from, 69
Copper, perfluoro-*t*-butyl-
  synthesis, 70
Copper, phenyl-
  insolubility, 120
Copper *t*-butoxide
  organocuprate preparation
    from, 68
Copper triflate

organocopper compound
preparation from, 62
Copper, (trifluoromethyl)-
organocopper compound
preparation from, 64
Crystal growth
organocadmium compounds,
201
Cyclopentadienylmercury
compounds
structures, 160–2
Cyclopropanes
solvomercuriation, 151, 152

Diketones
mercuriation, 140
Diorganocadmium compounds,
196, 197
Diorganogold(III) halides
synthesis, 33, 34
Diorganogold(III) pseudohalides
synthesis, 33, 34
Diorganozinc compounds, 176,
177
Diorganylgold(III) complexes
group 15 donor ligands, 36–9
group 16 donor ligands, 34–6

Enantiomer recognition
organozinc compounds, 192,
193

Fluorodemercuriation
aromatic mercury compounds,
160

Gold
applications, 2
reclamation, 1
Gold clusters
heterometallic, 43
homometallic, 43
Gold complexes, 1–49
alkenes
synthesis, 44
alkyne
synthesis, 44
carbene
reactions, 44–7
synthesis, 44–7
carbonyl
properties, 47–9
synthesis, 47–9
isocyanide
properties, 49
synthesis, 49
polynuclear, 6
trifluoromethyl-
chemical vapor deposition,
33
Gold(I) complexes
acetylides
polymeric, 6
alkyl phosphine
synthesis, 3
alkynyl
applications, 6
dinuclear, 13
aryl, 8
synthesis, 19
cyclopentadienyl phosphine, 9
fluorenyltriphenylphosphine, 9
halogenoalkyl

synthesis, 4
imidazolyl, 11
iminoalkyl, 11
phenylacetylides, 6
two gold–carbon bonds, 17–19
ylides, 19–31
dinuclear, 23–31
mononuclear, 20–3
polynuclear, 23–31
Gold(II) complexes
ylides, 19–31
dinuclear, 23–31
mononuclear, 20–3
polynuclear, 23–31
Gold(III) complexes
alkyl, 31
carbenes, 43
diaryl
synthesis, 37
monoaryl
synthesis, 32
one gold–carbon bond, 31–3
two gold–carbon bonds, 33–9
three gold–carbon bonds,
39–42
four gold–carbon bonds, 42,
43
ylides, 19–31, 43
dinuclear, 23–31
mononuclear, 20–3
polynuclear, 23–31
Gold compounds
ferrocenyl, 11
Gold(III) nitrate, dimethyl-
synthesis, 36

Hydridodemercuriation, 165
Hydrodemercuriation
hydroperoxides, 146
Hydromercuriation
alkynes, 137
Hydroperoxymercuriation
alkenes, 145

Ketones
mercuriation, 140

Lead, tetraalkyl-
reaction with silver nitrate, 71

Mercurial compounds, phenyl-
synthesis, 138, 140
Mercurials, aryl-
radical reactions, 167
Mercury(II) carboxylates
decarboxylation, 142
Mercury chloride, methyl-
reduction, 166
Mercury(II) compounds
aryl
structures, 159, 160
methyl-
interaction with biologically
important molecules,
156
Lewis acids, 155
structures, 155–7
Mercury(II) fluorosulfonate
reaction with carbon
monoxide, 154
Mercury–hydrogen exchange
aliphatic, 139, 140
aromatic, 141, 142

Methane, polyaurio-
synthesis, 16
Methane, tetraaurio-
hypercoordinate, 15
Methane, tetrakis[(tricyclohexylphosphine)gold(I)]-
synthesis, 16
Methane, triaurio-
synthesis, 15

Nanocrystallites
semiconductors
synthesis, 203
Nitratomercuriation
alkenes, 145

Organoargentates
anionic mononuclear
heteroleptic
solid-state structures, 105–8
anionic mononuclear homo-
leptic
solid-state structures, 105–8
aryl
bonding, 129
stereochemistry, 129
solid-state
structures, 100–9
synthesis, 73
Organocadmium compounds,
175–203
adducts, 197
alkyl
adducts, 177, 178
in electronics industry, 200–3
fluoroalkyls, 196
physical studies, 200
theoretical studies, 200
zinc–heteroatom bonds, 198,
199
Organocopper compounds,
57–129
alkenyl
solid-state structures, 95
alkyl
solid-state structures, 77–80
alkynyl
bonding, 124
heteroleptic solid-state
structures, 96–100
homoleptic solid-state
structures, 96–100
amine stabilized cationic
solid-state structures,
110–15
amine stabilized neutral
solid-state structures,
110–15
aryl
functionalized heteroatom-
containing substituents,
solid-state structures,
86–8
noncoordinating
substituents, solid-state
structures, 82–6
solid-state structures, 82–95
bonding, 75–129
cyclopentadienyl
structures, 115
ferrocenyl
structures, 117
heteroleptic

solid-state structures, 89–95
metal salt complexes, 66
phosphine stabilized cationic
  solid-state structures,
    110–15
phosphine stabilized neutral
  solid-state structures,
    110–15
phosphorylides
  solid-state structures, 115
polymeric
  structures, 120, 121
solid-state
  structures, x-ray analysis,
    77–117
solution
  structures, 125
stereochemistry, 121–9
structure, 75–129
synthesis, 60–75
  electrochemical, 71
  transmetallation, 60–4
thermal stability, 75–7
Organocuprates
  anionic mononuclear
    heteroleptic
    solid-state structures, 105–8
  anionic mononuclear homo-
    leptic
    solid-state structures, 105–8
  anionic polynuclear hetero-
    leptic
    solid-state structures, 108,
      109
  anionic polynuclear homoleptic
    solid-state structures, 108,
      109
  aryl
    bonding, 129
    stereochemistry, 129
  conjugate addition to
    unsaturated carbonyl
    compounds, 127
  neutral heteroleptic
    solid-state structures, 101–5
  neutral homoleptic
    solid-state structures, 101–5
  organocopper compound
    preparation, 64
  solid-state
    structures, 100–9
  synthesis
    interaggregate exchange, 69
    metallation, 67–9
    self-assembly, 69
Organogold complexes
  σ-bonded
    reactions, 3–17
    structure, 3–17
    synthesis, 3–17
Organolithium reagents
  organocopper compound
    preparation from, 63
Organomagnesium halides
  organocopper compound
    preparation from, 63
Organomagnesium reagents
  organocuprate preparation
    from, 65
Organomercury compounds,
    135–69
  polydentate Lewis acids,
    162–5

properties, 155–69
radical reactions, 166
structures, 155–69
Organomercury salts
  radical anions, 167
Organosilver compounds, 57–129
  alkenyl
    synthesis, 71, 72
  alkyl
    solid-state structures, 77–80
    synthesis, 71
  alkynyl
    heteroleptic solid-state
      structures, 96–100
    homoleptic solid-state
      structures, 96–100
    synthesis, 71, 75
  amine stabilized cationic
    solid-state structures,
      110–15
  amine stabilized neutral
    solid-state structures,
      110–15
  aryl
    functionalized heteroatom-
      containing substituents,
      solid-state structures,
      86–8
    noncoordinating
      substituents, solid-state
      structures, 82–6
    solid-state structures, 82–95
    synthesis, 71, 73
  bonding, 75–129
  imidazoyl
    synthesis, 73
  phosphine stabilized cationic
    solid-state structures,
      110–15
  phosphine stabilized neutral
    solid-state structures,
      110–15
  phosphorylides
    solid-state structures, 115
  polymeric
    structures, 120, 121
  solid-state
    structures, x-ray analysis,
      77–117
  solution
    structures, 125
  stereochemistry, 121–9
  structure, 75–129
  synthesis, 60–75
  thermal stability, 75–7
  trifluoromethyl
    synthesis, 72
Organosilver lithium compounds
  diaryl
    synthesis, 74
Organothallium compounds
  organocopper compound
    preparation from, 64
Organozinc compounds, 175–203
  adducts
    in electronics industry, 200,
      201
  alkyl
    adducts, 177, 178
    alkyl–zinc–oxygen bonds, 184,
      185
    alkyl–zinc–selenium groups,
      186

alkyl–zinc–sulfur groups, 186
amides, 181, 182
bulky silyl ligands, 179, 180
carbamates
  synthesis, 185
carbenes, 180
carbynes, 180
diselenocarbamates, 186
dithiocarbamates, 186
in electronics industry, 200–3
heterometallic, 187, 188
in organic synthesis, 188–95
photodissociation, 188
physical studies, 188
theoretical studies, 188
unsaturated ligands, 180
zinc–halide bonds, 182
Oroganomercury hydrides
  structures, 165, 166
Oxymercuriation
  alkenes, 145

Peroxymercuriation
  intramolecular
    hydroperoxides, 147

Q-particles
  semiconductors
    synthesis, 203
Quantum dots
  semiconductors
    synthesis, 203

Ruthenocenes
  mercuriation, 141
  permercuriation, 162

Semiconductors
  nanocrystallites
    synthesis, 203
  organocadmium compounds,
    200
Silver borofluoride
  transmetallation, 72
Silver nitrate
  reaction with tetraalkyllead, 71
  transmetallation, 74
Silver, perfluoroisopropyl-
  synthesis, 72
Silver, phenyl-
  structure, 120
Silver, styrenyl-
  synthesis, 72
Simmons–Smith reaction
  organozinc compounds, 194,
    195
Solvomercuriation, 144–52
Sulfinatomercuriation
  alkenes, 145

Transmetallation
  organocopper compound
    preparation, 64–7
1,3,5-Triazine, 1,3,5-trimethyl-
  hexahydro-
  adduct with dimethylzinc, 177

Xanthine
  reaction with methylmercury
    cations, 156

Ylides
  diaurated, 14

Zinc, bis[3-dimethyl-
    (amino)propyl]-
  structure, 176
Zinc, bis(2,4-pentadienyl)-
  tetramethylethylenediamine
    adduct, 177
Zinc, bis(trifluoromethyl)-
    bis(pyridine)-
  structure, 177
Zinc, bis[tris(trimethylsilyl)-
    methyl]-
  preparation, 179
Zinc, dialkynyl-
  preparation, 180
Zinc, diaryl-
  organocopper compound
    preparation from, 64
Zinc, diethyl-
  organosilver compound
    synthesis from, 71
Zinc, dimethyl-
  adduct with 18-crown-6, 177
  adduct with sparteine, 177
  preparation, 176
Zinc, diphenyl-
  adduct with crown ethers, 178
  reaction with Lewis bases, 178
Zinc enolates
  preparation, 180, 181
Zincates
  preparation, 179
Zincocene
  structure, 180